Ecotoxicology of Nanoparticles in Aquatic Systems

Editors

Julián Blasco
Department of Ecology and Coastal Management
Institute of Marine Sciences of Andalusia (CSIC)
Spain

Ilaria Corsi
Department of Physical, Earth and Environmental Sciences
University of Siena
Italy

CRC Press
Taylor & Francis Group
Boca Raton London New York

CRC Press is an imprint of the
Taylor & Francis Group, an **informa** business
A SCIENCE PUBLISHERS BOOK

Cover credit: Cover illustrations reproduced by kind courtesy of Anita Jemec Kokalj (Slovenia) and Andrews Krupinski Emerenciano (Brazil).

CRC Press
Taylor & Francis Group
6000 Broken Sound Parkway NW, Suite 300
Boca Raton, FL 33487-2742

Printed on acid-free paper
Version Date: 20190506

International Standard Book Number-13: 978-1-138-06726-4 (Hardback)

Library of Congress Cataloging-in-Publication Data

Names: Blasco, Juliâan, editor. | Corsi, Ilaria, 1971- editor.
Title: Ecotoxicology of nanoparticles in aquatic systems / editors, Juliâan Blasco (Department of Ecology and Coastal Management, Institute of Marine Sciences of Andalusia (CSIC), Spain), Ilaria Corsi (Department of Physical, Earth and Environmental Sciences, University of Siena, Italy).
Description: Boca Raton, FL : CRC Press, 2019. | "A science publishers book." | Includes bibliographical references and index.
Identifiers: LCCN 2019020004 | ISBN 9781138067264 (hardback)
Subjects: LCSH: Nanoparticles--Environmental aspects. | Water--Pollution. | Marine pollution. | Environmental toxicology. | Aquatic ecology.
Classification: LCC TD196.N36 E36 2019 | DDC 577.6/27--dc23
LC record available at https://lccn.loc.gov/2019020004

Visit the Taylor & Francis Web site at
http://www.taylorandfrancis.com

and the CRC Press Web site at
http://www.crcpress.com

Preface

Nanoscale materials and particles are used extensively in a wide range of emerging technologies and commercial applications including biomedicine, pharmaceuticals and personal care products, renewable energy and electronic devices. These materials possess distinctive physicochemical properties as compared to their larger-scale counterparts, including greater strength, lighter weight, increased control within the visible spectrum and greater chemical reactivity. As the production, use and disposal of nanomaterials continue to grow, their presence in the environment is expected to increase. It is therefore necessary to understand the impact of nanomaterials on the aquatic environment, which is a common receptor of environmental pollutants. Recent environmental fate models emphasize that nanowastes will enter aquatic environments, thus potentially affecting natural ecosystems.

Many books have been published which address the effects of nanomaterials; however, none have focused on aquatic ecosystems and their resident biota. This book will serve as a valuable reference for undergraduate and graduate students, and for postgraduate and senior researchers who are investigating this emerging topic.

The eleven chapters included herein provide an overview of the ecotoxicity of nanoparticles in aquatic systems, and will thus be useful for attaining current knowledge on the fate, behaviour and ultimate toxicity of nanoparticles in aquatic ecosystems. The state of the art of innovative technologies relevant to nanoparticle analysis is also addressed in considerable details.

In Chapter 1 by Sendra, Moreno and Blasco, the toxicity of metal and metal oxide engineered nanoparticles (Me-ENPs and MeO-ENPs) to phytoplankton microalgae is reported. Perturbations of phytoplankton communities due to the presence of toxic compounds can affect both the structure and function of entire ecosystems. Chapter 1 summarizes current knowledge on the intrinsic (e.g., particle size, specific surface area, shape, zeta potential, water solubility, photocatalytic activity, crystallization, purity, redox potential, coating composition) and extrinsic factors (chemical transformations, agglomeration/aggregation, redox stage, eco-corona formation) that affect the toxicity of Me-ENPs and MeO-ENPs. Finally, discussion related to improvement of ENP toxicity tests, climate change effects and trophic transfer of ENPs are presented.

In Chapter 2 by Canesi, Auguste and Bebianno, data on sub-lethal effects and mechanisms of action of different types of NPs on aquatic invertebrates is reviewed. For each main class of NPs (zero valent metal NPs, nano-oxides, carbon based NPs, quantum dots), *in vitro* and *in vivo* data are summarized, including those on embryo/larval development. Data on NP particle size, shape, coating and behaviour in media are presented in relation to the biological effects observed in different experimental models.

Chapter 3 by Katsumiti and Cajaraville reviews recent data on the use of *in vitro* toxicity testing with cells of aquatic vertebrate (fish) and invertebrate (bivalve mollusk) species, focusing on cells from marine mussels *Mytilus galloprovincialis*, a worldwide sentinel species of pollution and potential target for nanoparticle toxicity. The use of screening tests has allowed for classification of NPs' relative toxicity and the identification of key properties that influence their toxicity, which are primarily driven by the chemical composition and behaviour of NPs in exposure media. Other factors such as size, mode of synthesis, shape and presence of additives also influence NP toxicity. Overall, *in vitro* results generally agree with effects found *in vivo*, suggesting that *in vitro* tests can be used as a rapid, reproducible and sensitive tool for environmental risk assessment of nanomaterials in aquatic environments.

Chapter 4 by Schmidt, Kühnel and Jemec-Kokalj offers an overview of toxicity tests and bioassays and their current status of development and adoption, specifically for NM toxicity testing involving organisms commonly used in environmental research. The core of the chapter is a detailed presentation of toxicity tests embracing unicellular organisms (microbes, algae, protozoa), invertebrates (crustaceans, mollusks, insects, nematodes), vertebrates (amphibians, fish), vertebrate cells (fish cells or cell line) and higher aquatic plants. For each of the test organisms, an overview of the scope, test parameters, common species used, advantages and limitations for NMs testing, as well as challenges and needs for further optimization and developments are presented. Unique challenges in toxicity testing of NMs such as properties, behaviour, transformations in liquid media, consequences of attachment, bioaccumulation and kinetics in organisms and interferences with toxicity test components or with detection methods are addressed. Finally, the current state of the art regarding standardization of these tests for NMs testing is presented.

In Chapter 5 by Lee and Browning, recent studies showing species-specific susceptibility by fish to toxicity during developmental periods is reported. Also addressed are how concentration, size, surface modification and type of NMs can affect teratogenicity and mortality rate.

In Chapter 6 by Handy and Al-Bairuty, the classical approach of target organ pathology is proposed to identify the types of pathology observed from NM exposures in fish *in vivo*. Substantial 'cross-talk' is known to occur between the organ systems of fish, which enables physiological integration for optimal organism function. Most pathologies have been documented for conventional chemicals, and the etiology is often different for nano forms. Many data gaps exist, with information limited to only a few species of fish and mostly pristine NMs.

Chapter 7 by Jimeno-Romero, Kohl, Marigómez and Soto addresses how high-resolution visualization of NPs in target tissue and cell compartments of aquatic organisms combined with *in situ* quantification of NPs by digital microscopy can enhance our understanding of NP-cell interactions. During their journey from the environment to the cell, NPs experience numerous physical and chemical modifications, which affect their speciation, bioavailability, fate and possible toxic potential to aquatic organisms. A thorough review of microscopy techniques for characterization of environmental NPs and, most importantly, for determining how they pass across biological barriers, how they end up being internalized by cells and where and how can they cause toxic effects is presented.

In Chapter 8 by Poynton, 'omic technologies are reviewed for their application to advancing our understanding of the exposure and effects of NMs on aquatic organisms. 'Omic technologies have recently been applied to identify exposure pathways that may distinguish the toxicity of NMs from bulk materials and ions. The primary aim of the chapter is to provide an overview of omics in NM ecotoxicology by highlighting major findings and potential applications facilitated by genomics. Emerging areas of research and future directions for the field are introduced.

In Chapter 9 by Little and Fernandes, a review of nanomaterial fate, behaviour and effects on aquatic sedimentary systems is provided alongside methods for analysis of NMs in sediments and the extent and mode through which they exert toxicity towards key benthic species. Where conditions favour deposition, aquatic sediments receive an influx of NMs. Despite being identified as likely recipients of nano-wastes, studies of the ecotoxicological effects on benthic organisms are severely lacking in comparison with studies of aquatic exposures using pelagic species. Given the uncertainty of NM toxicity and the ecological importance of sediments, understanding potential interactions between the two is of significant interest.

In Chapter 10 by Corsi and Grassi, nanotechnological solutions for remediation of water pollution, also known as nanoremediation, are presented. Nanoremediation, using the appropriate strategies, offers far-reaching benefits, allowing us to reduce human and environmental risks and to satisfy regulatory requirements, boost circular economies and support a fully effective expansion of the sector. In this chapter, the following aspects are discussed: (i) ecosafety of ENMs using an eco-design approach as a priority feature; (ii) ecotoxicity testing to be more environmentally consistent by including realistic exposure scenarios; and (iii) research and innovation for greener and ecosafe ENMs with the aim of developing more sustainable and ecofriendly nanoremediation solutions.

In Chapter 11 by Picó and Andreu, techniques for detecting ENP/NM/MNs in water are provided. Detection is a demanding task not only because of the extremely small size of the particles and their potential sequestration and agglomeration, but also because of their unique physical and chemical characteristics. The aim of this chapter is to critically review the state-of-art ENP/NM/MN detection in the environment, highlighting the advances of the last five years (up to 2018). The available analytical techniques used for detection and characterization of nanoparticles in water including particle size analysis, particle fraction concentration counts, surface area analysis, morphology and particle chemical composition analysis will be critically analyzed to describe their advantages and pitfalls. Aspects of sample preparation, imaging techniques (electron microscopy, scanning electron microscopy, X-ray microscopy), separation methods (flow field fractionation, liquid chromatography, hydrodynamic chromatography) and detection/characterization techniques (e.g., single particle inductively coupled plasma, mass spectrometry) are discussed in depth. Chapter 11 also introduces the application of these techniques to enhance our knowledge of the fate, behaviour, disposition and toxicity of ENMs in the environment.

The editors have learned a great deal from the chapters provided, and anticipate that readers will likewise gain substantial value from the data, trends and technologies provided herein. This book should provide important tools for researchers to advance their knowledge base regarding the ecotoxicological effects of nanomaterials on aquatic systems.

Acknowledgements

The editors would like to acknowledge all of the chapter authors who contributed to the realization of the book with their excellent and hard work. It would have been impossible to write this book without their contributions. If this book will have an impact on promoting the growth of the field, then our project will have been successful.

We are extremely grateful to the external reviewers selected by the editors for providing invaluable remarks to improve our original version of the chapters submitted by the authors.

We thank the CRC Press for agreeing to undertake our book project and the editorial staff for guiding us through the publishing process.

Finally, we thank our families for their support and understanding of the time we spent to go ahead with this project.

Contents

1

Toxicity of Metal and Metal Oxide Engineered Nanoparticles to Phytoplankton

Marta Sendra, Ignacio Moreno and *Julián Blasco**

Introduction

Microalgae (unicellular or filamentous microscopic primary producers) is a miscellaneous ecological group which include cyanobacteria and prochlorophyceae, green algae, euglenophyceae, dinoflagellates, some species of red algae, and a wide diversity of chromists (such as diatoms, coccolitophorides, eustigmatophiceae, and diverse brown-golden algae), among other groups. It means that organisms called "microalgae" could belong to at least three natural kingdoms. Those organisms conform the basis of the aquatic trophic nets and support at least a half of the planetary primary production (Falkowski 1980). Recent studies reveal that those data could underestimate the real importance of marine phytoplankton, as techniques such as flow cytometry point out that the abundance of micro and nanoplankton could be higher than expected in the past (González-García et al. 2018).

In any case, perturbations in phytoplankton communities due to toxic processes caused by any xenobiotic could propagate to the rest of ecosystems. Not only is the abundance of microalgal biomass important, as taxonomic diversity of the microalgal assemblage has also been demonstrated to be a key issue in the maintenance of certain grazers (Moreno-Garrido et al. 1999).

Nowadays, the occurrence of the legacy and emergent pollutants in the aquatic ecosystems can represent a real risk for aquatic organisms depending on their concentrations, environmental conditions, and species sensitivity. Among emergent pollutants, nanomaterials and specifically engineered nanoparticles (ENPs) have aroused a great attention due to their wide use and applications.

Nanotechnology is one of the more important emerging technologies identified in the European Union (EU) 2020 Strategy (EU Commission 2013). It has demonstrated to promote innovation, development, and economic growth (Hristozov et al. 2016). The goal of nanotechnology research funding is to improve industrial applications and commercial products through new physicochemical

Institute of Marine Sciences of Andalusia (CSIC). Campus Rio San Pedro, 11510 Puerto Real (Cádiz), Spain.
* Corresponding author: julian.blasco@csic.es

properties of engineered nanomaterials (ENMs) that may influence their kinetics, bioavailability, toxicity, and fate. Among the applications of ENMs found, some are: catalysis, lubricants, paints, cosmetic products, sensor device, bio-remediation, antibacterial and antifungal agents, drugs delivery, medical diagnosis, aquaculture, water treatments, air disinfection, food packing, clothes, toys, goods sports, plastic and archaeological stones (Moreno-Garrido et al. 2015). The fascinating physico-chemical properties (such as size, surface area, photocatalytic and redox capacity among others) (Hendren et al. 2011, Piccinno et al. 2012) need to be investigated during different steps: the release in the environment, transformation in different media, and exposure to organisms and hazard assessment in order to increase the level of certainty in the risk assessment results (Stone et al. 2014). The requirements for an environmental risk assessment of ENMs require the knowledge of their environmental concentrations. Despite significant advances in analytical methods, it is a complicated task, because the methods are not sensitive enough for current environmentally realistic concentration, and they cannot distinguish natural NMs from engineered ones. The improvement of recent studies in evolving from static to dynamic mathematical models has allowed researchers to know the Predicted Environmental Concentration (PEC) through materials flow and environment (Sun et al. 2016).

ENMs are releasing during their life cycle, from (1) synthesis and production, (2) handling, (3) transport, (4) incorporation to application and products, (5) use of products or applications that incorporate them (such as textile, sunscreen, sport goods...), to (6) disposal of products or applications which incorporate them (incineration, landfill, sewage treatments) (Caballero-Guzman and Nowack 2016).

Intrinsic Characteristic of ENPs that Influence Microalgae Toxicity

The novel and intrinsic physicochemical properties of ENPs with respect to conventional materials (bulk) are of great importance and should be considered in toxicological test with microalgae. They are: (1) particle nominal size, (2) specific surface area, (3) shape, (4) zeta potential, (5) water solubility, (6) photocatalytic activity, (7) crystallization, (8) purity of samples, (9) redox potential, and (10) composition coating, among the most studied characteristics.

Nominal Size of ENPs

Nominal size of ENPs is the most remarkable characteristic which make ENPs a perfect candidate in new applications and commercial products. However, differences in toxicity taking into account EC_{50} of growth inhibition for different NPs size will be closely related with the composition of ENPs.

With respect to titanium dioxide NP (TiO_2 NPs), one of most used chemicals in cosmetic products, with low toxicity to microalgae when they are not exposed to UV-A/B and C, it is possible to observe how smaller size of ENPs did not show important differences in toxicity to both freshwater and marine microalgae (Table 1) (Hund-Rinke and Simon 2006, Hartmann and Baun 2010, Ji et al. 2011, Xia et al. 2015, Sendra et al. 2017f). However, in other studies, small differences are found between NPs form and bulk (Aruoja et al. 2009, Clément et al. 2013, Sendra et al. 2017f).

In the case of silver (Ag) NPs, some authors have demonstrated that particles size is important in their toxicity, finding higher toxicity in small Ag NPs in different freshwater microalgae (Angel et al. 2013, Ivask et al. 2014, Sendra et al. 2017d); however, the same trend was not observed in marine microalgae where small Ag NPs were not related with smaller values of EC_{50} (Sendra et al. 2017d). Contrary to TiO_2 NPs, Ag NPs are reactive, showing a rate of dissolution inversely proportional to NPs size. From this dissolution of Ag NPs are released ions (Ag^{+}) to the culture media, making them mainly responsible for Ag NPs toxicity to microalgae (Navarro et al. 2008b).

In relation to cerium oxide (CeO_2) NPs, most of the authors found a relation between toxicity and particle size for different microalgae species (Hoecke et al. 2009, Rogers et al. 2010, Rodea-Palomares et al. 2011). Sendra et al. (2017) found no relation between toxicity and CeO_2 NPs size. In this study,

Table 1. EC_{50} (growth inhibition) in freshwater and marine microalgae exposed TiO_2, Ag and CeO_2 NPs of different size.

Media	NPs	Size (nm)	Species	EC_{50} ($mg \cdot L^{-1}$)	Light	Test time (h)	References
Freshwater	TiO_2	25–70 bulk	*Pseudokirchneriella subcapitata*	5.83 35.9	White and UV-A	72	(Aruoja et al. 2009)
		5–10 50 600 ± 200		241 71.1 145	white	72	(Hartmann and Baun 2010)
		25 100	*Chlorella* sp.	120 and non-toxic to 50 and 600 nm particles	white	72	(Ji et al. 2011)
		38 ± 12 423 ± 154	*Desmodesmus subspicatus*	> 50 > 50	white	72	(Hund-Rinke and Simon 2006)
		38 ± 12 423 ± 154	*Chlamydomonas reinhardtii*	551.7 423.70	white	72	(Sendra et al. 2017b)
		21 60 400		2.30 1.35	UV-A	72	(Sendra et al. 2017b)
Marine		38 ± 12 423 ± 154	*Nitzschiac losterium*	88.7 118.80 19.05	white	96	(Xia et al. 2015)
		38 ± 12 423 ± 154	*Phaeodactylum tricornutum*	132 185	white	72 h	(Sendra et al. 2017b)
		15 25 32 44000		1.98 6.58	UV-A	72 h	(Sendra et al. 2017b)
		4.5 16.7 46.7		10.91 11.30 14.30 35.51	white	72 h	(Clément et al. 2013)

Table 1 contd. ...

...Table 1 contd.

Media	NPs	Size (nm)	Species	EC_{50} ($mg \cdot L^{-1}$)	Light	Test time (h)	References
Freshwater	Ag	10 20 40 60 80	*Chlamydomonas reinhardtii*	< 10 $\mu g \cdot L^{-1}$ < 10 $\mu g \cdot L^{-1}$ > 300 $\mu g \cdot L^{-1}$	white	72	(Sendra et al. 2017d)
		14 15 200–3500	*Pseudokirchneriella subcapitata*	0.18 0.52 0.82 0.94 1.94	white	72	(Ivask et al. 2014)
	Ag	4.5 16.7 46.7		3 19.5 966	white	72	(Angel et al. 2013)
Marine		9.4 26 38	*Phaeodactylum tricornutum*	> 300 162.5 > 300	white	72 h	(Sendra et al. 2017d)
Freshwater	CeO_2	14 20 29 Bulk	*Chlamydomonas reinhardtii*	> 200 88.8 > 200	white	72 h	(Sendra et al. 2017f)
		10 25 50 60 5000	*Pseudokirchneriella subcapitata*	10.2 11.7 19.1 n.d.	white	72 h	(Hoecke et al. 2009)
		10–20 nm Bulk		29.6 9.7 4.4 16.4 56.7	white	72 h	(Rodea-Palomares et al. 2011)
		9.4 26 38		10.3 66	white	72 h	(Rogers et al. 2010)
Marine	CeO_2	9.4 26 38	*Phaeodactylum tricornutum*	> 200 2.21 51.47	white	72 h	(Sendra et al. 2017f)
			Nannochloris atomus	> 200 > 200 > 200	white	72 h	(Sendra et al. 2017f)

the NPs with the middle size showed to be to more toxic that the other two tested microalgae in the different environment. This NPs differed in one and important intrinsic property, that is, it showed positive zeta potential in the culture media (Sendra et al. 2017f). In another study, with freshwater microalgae *Pseudokirchneriella subcapitata* exposed to different CeO_2 NPs with different nominal size, zeta potential, shape, and Ce^{3+}/Ce^{4+} ratio in NPs surface, the main drivers of toxicity was the percentage of Ce^{3+} in NPs surface (Pulido-Reyes et al. 2015).

Sometimes in studies which compare NPs with different nominal size given by suppliers or even measured by TEM, the smallest NPs have not always had correlation with the mean size and peak of hydrodynamic radii measured by DLS in the culture media. Therefore, agglomerate size for NPs in a certain size range could not be related with a nominal size. To assess the importance of NPs size in toxicity test to microalgae is necessary to know size distribution of NPs and the percentage of NPs agglomerates lower than 20 nm. The importance of knowing the volume of NPs < 20 nm is related to the fact that the microalgae cell wall has a pore size between 5–20 nm (Moore 2006, Navarro et al. 2008a).

Specific Surface Area of NPs

Specific surface area (S_{BET}) is a very important intrinsic characteristic of ENPs and it is inversely proportional to NPs size. ENPs with higher S_{BET} will be more reactive with their surrounding environment. For instance, ENPs with high values of S_{BET} will have higher dissolution rate, photocatalytic and redox activity. Furthermore, higher values of S_{BET} provokes higher contact between ENPs and organisms and molecules. Most of the articles published which assess the toxicity of intrinsic characteristic of ENPs are not focused on S_{BET} (Baer 2011).

Shape of ENPs

Another contributing intrinsic characteristic of ENPs toxicity is their shape (Nangia and Sureshkumar 2012, Mortazavi et al. 2017).

Information about toxicity of different shape of NPs in microalgae is scarce. Most of the studies are developed in macrophages. A recent study demonstrated that rod shape CeO_2 NPs was more toxic than cubic/octahedral CeO_2 NPs (Forest et al. 2017). Nangia et al. (2012) observed translocation rate constants of functionalized cone, cube, rod, rice, pyramid, and sphere shaped NPs through lipid membranes. The results indicate that depending on the NP's shape and surface charge, the translocation rates can span 60 orders of magnitude. Unlike isotropic NPs, positively charged, faceted, rice-shaped NPs undergo electrostatics-driven reorientation in the vicinity of the membrane to maximise their contact area and translocate instantaneously, disrupting lipid self-assembly and thereby causing significant membrane damage (Nangia and Sureshkuma 2012).

Yamamoto et al. (2004) studied the mechanical cytotoxicity of different shape (dendritic, spindle, and spherical) and non-soluble NPs in two cell lines (L929 and J774A.1). The authors concluded that NPs with more sharp edges provoked higher cytotoxicity (dendritic) and NPs cytotoxicity may be related to the phagocytic process of particles (Yamamoto et al. 2004). Controversial results were found for some authors: they demonstrated that rounded NPs could be internalized 500% more than rod-shape NPs due to greater membrane wrapping time required for the elongated particles in mammalian cells (Chithrani et al. 2006, Chithrani and Chan 2007).

With respect to limited literature about the importance of NPs' shape in microalgae toxicity, Peng et al. (2011) demonstrated the importance of the nanostructure in three marine microalgae species: *Thalassiosira pseudonana*, *Chaetoceros gracilis* and *Phaeodactylum tricornutum*. The authors demonstrated that toxicity was not only related to the Zn^{+2} released by ZnO NPs' surface with higher S_{BET}, these authors evidenced the highest toxicity to one dimensional structure NMs (Peng et al. 2011).

On the other hand, Pulido-Reyes et al. (2015) assessed the toxicity of CeO_2 NPs with different physicochemical characteristics in *P. subcapitata*. The characteristics evaluated were nominal size, zeta potential, shape (rod, cubic and spheric NPs) and percentage of Ce^{3+} in CeO_2 NPs' surface. The authors concluded that shape per se did not affect the toxicity to microalgae neither in size nor in zeta potential; the only factor to show negative effects was the percentage of Ce^{3+} (Pulido-Reyes et al. 2015). The toxicity associated with the shape of NPs depend on the composition of particles used and its stability in aqueous media which allows to NPs to conserve the shape and not change its conformational structure through agglomeration/aggregation process. This aspect should be taken into account through TEM images over time exposure, to study this intrinsic characteristic of NPs. On the other hand, the cell line in the case of animals and species in the case of microalgae is key. The same NPs can provoke different effects depending on mechanism of toxicity. Thus, in the case of the cell wall and lipid membrane, the ultrastructure, composition and receptors are fundamentals in relation to the endocytosis process.

Zeta Potential

Zeta potential is a measure of the magnitude of the electrostatic or charge repulsion/attraction between particles, and is one of the fundamental parameters that affects the colloid's stability. That is to say, colloids have stability and instability zones. Thus, NPs with zeta potential higher to 30 mV and lower to –30 mV are in a stable zone maintaining its electrostatic repulsion. On the other hand, NPs with a zeta potential from –30 to 30 mV are in a zone of instability.

Charge of NPs is going to determine the heteroagglomeration NPs-cell; this physical interaction between NPs and cellular membranes is mainly governed by their surface charges (El Badawy et al. 2011). Positive NPs will provoke higher heteroagglomeration and therefore, the scientific community has defined it as a "proton sponge". The hypothesis suggests that positively charged NPs bind with high affinity to lipid bilayers on the cell membrane in favour of cellular uptake via endocytosis (Nel et al. 2009, Van Lehn and Alexander-Katz 2011, Lin and Alexander-Katz 2013) and thus, generating cell toxicity (Bexiga et al. 2011, Salvati et al. 2011, Wang et al. 2013).

Microalgae usually have negative surface charge as a result of the presence of carboxylic, phosphoryl, amine, and hydroxyl groups on their cell surface (Hadjoudja et al. 2010). Due to the negative charge of the main groups on microalgae surface, attractive forces will be higher with positively charged NPs than with negative ones. These interaction NPs-cell can trigger membrane cell damage leading NPs from extracellular environment to cytosol and therefore, giving rise to cellular stress and toxicity. In addition, the negative charge on the cell surface is vital for microalgae growth, and especially for preventing the natural aggregation of microalgae cells in suspension (Grima et al. 2003). The concentration and reactivity of these negative charges could alter the growth phase and metabolic condition of microalgae, as one of the functions of a negative surface charge is known to be the adsorption of essential elements (Uduman et al. 2010).

Some studies have cleared up the toxicity of positively charged NPs; in the study of Badawayet et al. (2010), different types of AgNPs were selected, and it was observed that the NPs with most negative zeta potential and citrate coating were the least toxic, whereas the positively charged with BPEI coating were the most toxic to *Bacillus* sp. (El Badawy et al. 2011). In the work of Hauck et al. (2008), the highest uptake, bioaccumulation and toxicity analysed gene expression were Au NPs charged positively in comparison to Au NPs charged negatively (Hauck et al. 2008).

With respect to toxicity of different surface charged NPs in microalgae, the number of studies have been lower for phytoplankton species than from higher organisms. Sendra et al. (2017) demonstrated how CeO_2 NPs with similar size and different zeta potential provoked different effects (growth, cell viability, percentage of ROS and autofluorescence) in freshwater microalgae *C. reinhardtii* and marine microalgae *P. tricornutum*. For the cited microalgae, there were a difference of one and two order of magnitude for EC_{50} of growth inhibition for *C. reinhardtii* and *P. tricornutum* (Sendra et al. 2017f). In addition, to the highest toxicity for positive NPs, in the same study, it was observed how positive NPs generated higher changes in cell complexity. Cell complexity measured by flow cytometry is directly related to internalization of NPs; therefore, apart from damages associated with the physical interaction between the NPs-Cell and membrane cell, the internalization of NPs inside the cell is another cause of CeO_2 NPs toxicity (Sendra et al. 2017f). Although, some authors have demonstrated that the charge of NPs is one of the cornerstones of the intrinsic properties of NPs without taking into account the composition of NPs, controversial results were found with respect to the study of CeO_2 NPs in microalgae. Pullido-Reyes et al. (2015) reported that zeta potential did not affect the toxicity of CeO_2 NPs in the freshwater microalgae *P. subcapitata* but other characteristics such as Ce^{+3}/Ce^{+4} played a more important role in toxicity (Pulido-Reyes et al. 2015).

Nowadays, due to the abundance of micro and nanoplastics in aquatic ecosystems, the intrinsic properties of these materials have been studied by the scientific community. One of the topics more studied about nanoplastics is their surface (coating and charge of the particle). In respect to charge of nanoplastics, some works have demonstrated different effects depending on positive and negative charge. Thus, positive polystyrene nanoplastics provoked higher effects in the marine microalgae *Dunaliella tertiolecta* (Bergami et al. 2017). Similar to the previous study, a growth inhibition in freshwater microalga *P. subcapitata* was reported by Casado et al. (2013) after exposure to 55 and 110 nM positive surface-charged PEI-modified PS NPs, with EC_{50} of 0.58 and 0.54 $\mu g \cdot mL^{-1}$. Nolte et al. (2017) showed that adsorption of neutral and positively charged plastic NPs onto the cell wall of *P. subcapitata* was stronger than that of negatively charged particles. Bhattacharya et al. (2010) observed how positive plastic NPs provoked higher electrostatic interaction, hydrogen bonding, and hydrophobic interaction between *Chlorella* and *Scenedesmus* and the plastic NPs. Positive NPs hindered algal photosynthetic activities and promoted ROS production (Bhattacharya et al. 2010).

Dissolution Rate of NPs

The main characteristic of NPs related to acute toxicity in different organisms is the composition of NPs. Usually, the unwanted effects of toxic NPs are related to dissolution of NPs which release ions to the environment from NPs' surface (Fabrega et al. 2009). Most of the studies about NPs' toxicity in microalgae are related with its rate dissolution so Ag, CuO, and ZnO are the NPs more toxic in the literature (Brunner et al. 2006, Xia et al. 2008). The reactivity of small NPs with high S_{BET} is potentially much greater than that of larger particles, so ions released to environment are greater; this has been demonstrated in different studies for different NPs and different culture media (Auffan et al. 2009, Ma et al. 2011, Angel et al. 2013, Sendra et al. 2017d). Aruoja et al. (2008) studied the toxicity in *P. subcapitata* of non-dissolved NPs as TiO_2 NPs and dissolved NPs such as CuO and ZnO NPs. Toxicity of ZnO was due to Zn^{2+} ions and the same amount of Zn^{2+} was found for ZnO NPs and ZnO bulk; therefore, both showed the same microalgae toxicity. The toxicity of CuO was different for CuO NPs and bulk as the percentage of Cu^{+2} ions dissolved from particles was larger in CuO NPs than in the CuO bulk. The mechanism of TiO_2 NPs toxicity was different with respect CuO NPs and ZnO NPs in dissolving capacity (Aruoja et al. 2009).

Intrinsic characteristics such as zeta potential and coating of NPs are related to the agglomeration stage and ultimately, with its reactivity surface. Angel et al. (2013) showed that AgNPs (Ag–Cit) had a faster rate of dissolution than PVP-coated AgNPs (Ag–PVP) in freshwater and seawater culture media.

Di-He et al. (2012) showed that Ag NPs toxicity in marine microalgae *Chattonella marina* was due to Ag^+. When cysteine (ligand of Ag^+) was added in both treatments, toxicity in *C. marina* was removed (He et al. 2012). These authors found higher toxicity in *C. marina* for Ag^+ than Ag NPs. However, Navarro et al. (2008) used the same amount of Ag^+ released from Ag NPs to study the effect of NPs, and they reported that the toxicity of Ag NPs was not only associated to Ag NPs dissolution but also an effect of Ag NPs *per se* (Navarro et al. 2008b).

In others studies, the authors concluded that the toxicity of ZnO NPs was solely due to Zn^{2+} in marine microalgae *Thalassiosira pseudomona* (Miao et al. 2010). In the study of Miao et al. (2009), the marine diatom *Thalassiosira weissflogii* showed to suffer less unwanted effects when AgNPs where suspended with natural organic carbon (NOC). NOC had dual effects on nanoparticle toxicity by controlling their solubilization and dispersion in aquatic environments.

Although it was previously reported that differences in dissolution are found due to size of particles, these differences were more pronounced in relation to the culture media (this aspect will be explained in the Section: Chemical Transformation). In this section, the secondary characteristics of NPs-leading by extrinsic factors such as physicochemical properties of culture media and environmental variables—will be considered in relation to the dissolution rate.

Despite the number of publications in the last years in relation to toxicity of metal and metal oxide NPs (which can dissolve in aquatic environment such as Ag, ZnO and CuO), there are still many discrepancies among studies and toxicity in freshwater and marine microalgae.

Photocatalytic Activity of NPs

TiO_2 NPs are widely used in different application and commercial products from cosmetic, paints, coating, plastics, sunscreens, foods, water remediation and pharmaceuticals (Vance et al. 2015). TiO_2 NPs under ultraviolet radiation (UVR) are photocatalytic; TiO_2 is able to generate ROS in aquatic environments producing adverse effects (such as cell damages, unregulated cell signalling, change in cell motility, apoptosis, DNA damages) in organisms (Sayes et al. 2006, Montiel-Dávalos et al. 2012, Shi et al. 2013, Fu et al. 2014).

Although the photocatalytic capacity of TiO_2 NPs is known, the role of UVR is not being considered in many studies which test the toxicity of TiO_2 NPs in microalgae; therefore, the toxicity of photo-reactive NPs is being underestimated (Sendra et al. 2017b).

In order to discuss this point, the results of some studies carried out to determine the toxicity of TiO_2 NPs in microalgae, coupled with UVR are discussed. In the study of Lee et al. (2013), growth inhibition of *P. subcapitata* were conducted under visible light, as well as UVA and UVB pre-irradiation conditions for two photocatalytic NPs: ZnO and TiO_2. The growth of *P. subcapitata* was inhibited under visible light, UVA and UVB irradiation conditions, with no significant differences among the light conditions for both NPs tested (Lee and An 2013). Another study from Hund-Rinke et al. (2006) did not find difference in toxicity either when the microalgae *Desmodesmus subspicatus* was exposed to TiO_2 NPs (Hund-Rinke and Simon 2006).

On the other hand, in the study of Roy et al. (2016), different results were found. These authors tested the toxicity of TiO_2 NPs in two freshwater microalgae *Chlorella* and *Scenedesmus* under dark, visible light and UVA. Values of EC_{50} showed differences among regime lights (EC_{50}: 5.95, 2.16 and 1.56 $mg \cdot L^{-1}$ for *Chlorella* and 7.63, 4.13 and 2.75 $mg \cdot L^{-1}$ for *Scenedesmus,* respectively). With respect to EC_{50} values for growth inhibition, differences among treatments were reported. The higher toxicity was observed under UVA exposure conditions (Roy et al. 2016).

Dalai et al. (2013), also found differences in toxicity of *S. obliquus* when it was exposed to TiO_2 NPs in dark and UV light conditions. *S. obliquus* showed more sensitivity when TiO_2 NPs were irradiated by UV, the percentage of ROS and LDH increased and percentage of viable cells decreased (Dalai et al. 2013).

Sendra et al. (2017) assessed the toxicity of TiO_2 NPs under two regime lights (visible light and UVA) for two microalgae from different environments, freshwater *C. reinhardtii* and marine *P. tricornutum.* The authors also evidenced the higher toxicity of TiO_2 NPs in both microalgae when the toxicity test was developed under UVA; the differences in EC_{50} was two order of magnitude lower in tests developed under UVA as compared to tests under visible light (Sendra et al. 2017b).

Crystallization of NPs

Crystallization can determine some specific characteristics of NPs. In the case of TiO_2 NPs, the crystallization affects the photocatalytic capacity as the most significant characteristic of these NPs. Commercial TiO_2 are found in two form of crystallization: one of them anatase (the crystallization is more photocatalytic), and the other one rutile (TiO_2 in photostable form); however, the mix of 80%

anatase and 20% rutile is the most photocatalytic form used in most of the applications (Uchino et al. 2002, Sayes et al. 2006).

Jin et al. (2011) demonstrated that ROS generation induced by anatase can be used as a paradigm to assess TiO_2 NP toxicity. Besides, TiO_2 NPs in the rutile phase do not exhibit the surface properties that allow spontaneous ROS generation. It could lead to cellular toxicity if the level of ROS production overwhelms the antioxidant defence of the cell or induces the mitochondrial apoptotic mechanisms (Jin et al. 2011).

Iswarya et al. (2015) assessed the toxicity of anatase and rutile TiO_2 NPs in *Chlorella* sp. under UV-A radiation. In this work, it was the first time that the effects produced by different anatase/rutile proportions were analysed in *Chlorella* sp. Depending on the proportion of anatase/rutile, the effect can be additive or antagonist. When the percentage of anatase is higher than rutile, the effect is additive, else the effect is antagonist. In this study, it is demonstrated that rutile is more stable than anatase and anatase/rutile. Through TEM images it was observed that toxicity mechanisms for anatase and rutile were different. In the case of anatase, the cell damage was in the nucleous and cell membrane; however, in the case of rutile, there was chloroplast and internal organelle damage. Cell growth inhibition for 1 $mg \cdot L^{-1}$ of anatase was 38.59 ± 1.28 $mg \cdot L^{-1}$ and for rutile 29.71 ± 5.9 $mg \cdot L^{-1}$; however, in the case of chlorophyll, the yield inhibition was 31.64 ± 6.22 and 54.26 ± 5.43 $mg \cdot L^{-1}$ for anatase and rutile, respectively (Iswarya et al. 2015).

The different effects of anatase and rutile under UVR have been reported in the scientific literature and they are associated with photocatalytic and photostable properties, respectively. However, some authors have studied the effects of both crystallization phases without excitation with UVR. In the study of Ji et al. (2011), compare the EC_{50} of anatase and rutile in *Chlorella* sp. For anatase, the EC_{50} was 120 $mg \cdot L^{-1}$, while in the case of rutile, there was no inhibition in population growth. TEM images showed that anatase entrapped more microalgae than rutile, so the toxicity can be from physical damage on cell walls to limitation of light by shading effects (Ji et al. 2011).

Ma et al. (2015) tested the heteroagglomeration between anatase and rutile NPs with cells of *Chlorella pyrenoidosa.* The results showed by authors were that anatase had higher heteroagglomeration (NPs-Cells) with respect to rutile (Ma et al. 2015). The higher heteroagglomeration of anatase as compared to rutile could explain the higher cytotoxicity of this form of crystallization on cells (Braydich-Stolle et al. 2009, Ji et al. 2011). Troppová et al. (2017) studied the toxicity of different crystallization of TiO_2 NPs and ZnO NPs synthetised by unconventional processing (pressurized hot water) against standard calcination procedure. And they observed that unconventional synthesis of NPs provoked an increase in the rate of nanosized crystalline aggregation and dissolution of Ti^{4+} from TiO_2 and Zn^{2+} from ZnO affecting the toxicity of studied organisms (Troppová et al. 2017).

Porosity of NPs

Taking into account that porosity is an intrinsic property of NPs, particles with the same diameter can differ in their surface area, therefore affecting their reactivity, behaviour in the environment and effects on organisms.

Because of the aforementioned, porosity characterization is considered to provide valuable information, but is not recommended as essential due to constraints associated with complexity, cost and availability (Oberdörster et al. 2005).

Purity of Samples

Purity of samples is another intrinsic property of NPs considered in the toxicity test. This information is given by suppliers of NPs and the purity of samples almost always is higher than 99%. Scarce data about purity of NPs are available in toxicity tests in the literature (Kumar et al. 2014).

Redox Potential of NPs

ENPs such as CeO_2, Fe_2O_3 and Fe^0, unlike their micron-sized equivalents, generate ROS by chemical redox reactions with biomolecules in the surrounding medium (Luna-Velasco et al. 2011). Certain photoactive ENMs such as the well-known TiO_2, but also CeO_2, SiO_2, Al_2O_3, ZnO and Fe_2O_3, can also act as photosensitizers by absorbing photoenergy from light and transferring it to molecular oxygen present inside cells which can directly yield ROS such as •OH, H_2O_2 and O_2 (Unfried et al. 2007, Auffan et al. 2009, Li et al. 2012).

In the case of CeO_2, NPs have been considered as excellent antioxidants and have become a focus of numerous studies. The presence of CeO_2 NPs can protect cells by competitively reacting with the •OH hydroxyl radical. This radical scavenging activity is due to high amount of Ce^{+3} on NPs surface respect to bulk form (Celardo et al. 2011).

Recently, many scientific groups have developed ecotoxicological tests in aquatic organisms involving CeO_2 NPs. The results from these ecotoxicological tests are controversial, with the effects of CeO_2 NPs generally being either toxic (Hoecke et al. 2009, Rogers et al. 2010, Rodea-Palomares et al. 2011, Zhang et al. 2011, Rodea-Palomares et al. 2012, Leung et al. 2015, Deng et al. 2017a, Sendra et al. 2017f) and/or protective (Xia et al. 2008, Deng et al. 2017a, Sendra et al. 2017f).

In case of microalgae, such role of CeO_2 NPs (toxic or protective) has not been fully elucidated. Apart from one single study which reports no effects of CeO_2 NPs on algae (Velzeboer et al. 2008), other studies reported different effects on inhibition growth of algae in similar concentration ranges, from 4.4 to 29.6 $mg \cdot L^{-1}$ (Hoecke et al. 2009, Manier et al. 2011, Rodea-Palomares et al. 2011, Manier et al. 2013). Concentration lower than 5 $mg \cdot L^{-1}$ of CeO_2 NPs allowed an increase of *P. tricornutum* growth population higher than controls (Deng et al. 2017a).

Sendra et al. (2017) tested three particles of CeO_2 NPs from different size and charge in three microalgae from different environments. CeO_2 NPs in this study showed dual results (toxic and protective) to the selected microalgae, and these results were related to the changes in the cell complexity of the microalgae. Greater changes in the cell complexity with respect to the controls provoked the inhibition of cell growth and increase of intracellular ROS production; however, low and moderate changes in cell complexity provoked a stimulation of growth of the microalgae population. These changes in cell complexity and therefore, toxic and protective effects were related to the charge of CeO_2 NPs, ultrastructure and composition of cell wall of microalgae tested (Sendra et al. 2017f).

Coating of NPs

NPs coating has become an important issue in ecotoxicological tests on aquatic organisms. This fact is related to the following hypothesis; (i) coating on NPs can prevent a fast dissolution, with a release of ions that is much more dilatory (ii) on the other side, coating inhibits the attractive forces among NPs, so agglomeration process are more scarce, and therefore NPs are more stable, bioavailable and toxic to organisms (Levard et al. 2012).

Angel et al. (2013) tested the toxicity of Ag NPs, citrate-coated Ag NPs and PVP-coated Ag NPs in the freshwater microalgae *P. subcapitata* and marine diatom *P. tricornutum*. Both coatings mitigated the effects of Ag NPs on microalgae. Furthermore, both coated-Ag NPs did not show the same efficiency in inhibiting toxicity. The citrate-coated AgNPs were more toxic than the PVP-coated AgNPs and the presence of humic acid or chloride helped to mitigate the toxicity (Angel et al. 2013).

Sometimes it is important to assess the role and the toxicity of the coating of NPs. In the study of Schiavo et al. (2017) the toxicity of PVP/PI coated Ag NPs, Ag NPs and only PVP/PI in three marine microalgae *I. galbana*, *T. suecica* and *P. tricornutum* was evaluated. These authors demonstrated that PVP/PI coated Ag NPs showed the same toxicity as PVP/PI alone and this toxicity was higher than Ag NPs; therefore, the toxicity of these experiments was due to PVP/PI (Schiavo et al. 2017).

The dispersion of NPs plays an important role in toxicity bioassay, because it will determine the bioavailability and toxicity of NPs. Many organic stabilizers have been used recently. Andreani et al. (2017) used PVA (for Ag NPs) and DMSO (for Cu NPs) as stabilisers to test the toxicity in the microalgae *Raphidocelis subcapitata.* When the Ag NPs were stabilized with PVA, the microalgae showed less toxicity as compared to Ag NPs alone (EC_{50} 0.25 and 6.9 mg·L^{-1} for Ag NPs and PVA-Ag NPs, respectively). However, Cu NPs stabilized with DMSO showed higher toxicity than non-stabilized NPs. The presence of DMSO around the particles may also have contributed to reducing the toxicity of the nano-Cu to *R. subcapitata* by decreasing the oxidation of oleic acid. In fact, the concentration of copper persisting in the suspension after deposition was low (Andreani et al. 2017).

The toxicity of metal NPs is a complex combination of average size, chemical composition, solubilisation or persistence in suspension of the metal forms, interaction with test medium components and sensitivity of test species and cell lines. The combination of all of these factors makes the toxicity of metal NPs unpredictable and points for the need of an extensive evaluation of each new formulation (Andreani et al. 2017).

Transformation of Nanoparticles by Extrinsic Environmental Factors

Due to small size of NPs (< 100 nm) and high specific surface area, the NPs are very reactive with their surrounding environment. Physicochemical properties of culture media (pH, ionic strength, temperature, UVR, organic matter and protein among others) can lead to chemical, physical and biological transformation. Therefore, a secondary characterization of NPs is needed (hydrodynamic radii, agglomeration stage, homoagglomeration, heteroagglomeration) because they can influence and determine the bioavailability, transport, persistence, fate and toxic effects of NPs (Handy et al. 2008, Misra et al. 2012, Amde et al. 2017). These transformations largely depend on colloid forces such as: hydrodynamic interactions, electrodynamic interaction, electrostatic interaction, solvent interactions, steric interactions and polymer bridging interactions (Lynch and Dawson 2008, Nel et al. 2009).

Chemical Transformation

Dissolution of NPs is an important property that influences their mode of action (Misra et al. 2012). In some cases, dissolution of NPs can lead to delivery of highly toxic ions, for example, when NPs are composed of elements which, in solution, are known to be toxic (Zn^{2+}, Cu^{2+}, Cd^{2+} and Ag^{+}) (Brunner et al. 2006, Xia et al. 2008). Intrinsic or primary characteristics of NPs such as size, surface area, surface morphology, porosity, crystallinity and coating influence the dissolution rate (Borm et al. 2006, Misra et al. 2012). Moreover, extrinsic characteristics of aquatic environments such as presence of organic and/or inorganic components in the exposure media which can catalyse the solubility of NPs need to be considered (Stark 2011, Gondikas et al. 2012).

The ionic strength can catalyse the dissolution of some NPs. In marine water, the dissolution of Ag NPs is about 20 times higher than in freshwater media (Oukarroum et al. 2012, Sendra et al. 2017d). The explanation for the higher dissolution of AgNPs in marine water than freshwater was the higher concentration of NaCl (420 mM) in marine water as compared to freshwater (Kent and Vikesland 2012, Sendra et al. 2017d). The chemical species formed in freshwater from Ag^{+} released from Ag NPs were Ag^{+} (26.7%) and $AgCl^{-}$(aq) (64.3%). In contrast, the species formed in marine water were $AgCl_2^{-}$ (53.7%) and $AgCl_3^{-2}$ (45.2%). In marine culture media, the chemical species present are not as bioavailable to marine microalgae as in freshwater culture media (Sendra et al. 2017d). The chemical species formed in both culture media could be translated to a higher toxic effect of AgNPs on freshwater species. In the freshwater microalgae, *C. reinhardtii*, the toxicity was size dependent of NPs as smaller NPs showed higher dissolution (EC_{50}: < 10, < 10 and > 300 μg·L^{-1} from smaller to larger NPs). On the other hand, although in marine water the dissolution of Ag NPs was size dependent, the toxicity did not show relation with this factor being related to agglomerate size

(EC_{50} of smallest agglomerate: 162.5 $\mu g \cdot L^{-1}$ respect to largest agglomerates > 300 $\mu g \cdot L^{-1}$ (Sendra et al. 2017d).

Physical Transformation

Agglomeration/Aggregation

Due to colloid forces among NPs, the instability of NPs in aqueous natural matrices (seawater, lagoon, river and groundwater) results in the homoagglomeration process (NPs-NPs). The homoagglomeration processes of NPs is dominated by ionic strength (IS), pH and occurrence of natural organic matter (NOM). From some NPs such as TiO_2 and CeO_2, the agglomeration stage was studied over 48 h in artificial fresh and marine culture media. Both NPs (TiO_2 and CeO_2) showed similar hydrodynamic radii when were measured by DLS and static light scattering. However, these differences in agglomeration stage were relevant when these NPs were suspended in ultrapure water (Sendra et al. 2017b, Sendra et al. 2017f); therefore the authors concluded that differences between 0 to 50 mM of IS triggered high differences in agglomeration stage; however differences between 50 and 700 mM of IS did not show differences in size agglomerates. This range of IS among 0–50 and 50–500 mM was also studied to examine the homoagglomeration and heteroagglomeration processes by Ma et al. (2015); TiO_2 NPs showed the same conclusion elucidated by Sendra et al. (2017b).

The meanings of aggregate and agglomerate have sometimes been interchanged in the literature (Nichols et al. 2002). To avoid this confusion, British Standards Institution (BSI) defined these two terms (BSI 2007). Aggregate is a "particle comprising strongly bonded or fused particles where the resulting external surface area may be significantly smaller than the sum of calculated surface areas of the individual components". The BSI notes that "the forces holding an aggregate together are strong forces, for example, covalent bonds, or those resulting from sintering or complex physical entanglement". The BSI defines agglomerate as a "collection of loosely bound particles or aggregates or mixtures of the two where the resulting external surface area is similar to the sum of the surface areas of the individual components" and notes that "the forces holding an agglomerate together are weak forces, for example van der Waals forces, as well as simple physical entanglement". Due to concentration of salts in saltwater as compared to freshwater and the decrease of electronegative double layer between NPs-NPs in seawater, the aggregation process will be higher in the marine environment than in the freshwater environment where agglomerate process governed the interaction between particles.

Some authors use natural organic matter to stabilize NPs (Yang et al. 2009, Quik et al. 2010, Keller et al. 2013). In the study by Cerrillo et al. (2016), Suwanee River natural organic matter (SR-NOM) was employed to maintain stable TiO_2 NPs and CeO_2 NPs over 72 h in freshwater microalgae culture media; conversely, NPs were unstable in culture media without SR-NOM through settling process. The toxicity of selected NPs were tested in freshwater microalgae *P. subcapitata;* the authors demonstrated that 8 $mg \cdot L^{-1}$ of SR-NOM act as "camouflage" in NPs toxicity. The EC_{50} was reduced from 1.24 and 0.27 $mg \cdot L^{-1}$ to 31.9 for CeO_2 and even increase growth rate for TiO_2 NPs. The "camouflage" can be due for complexation, adsorption, electrostatic forces (zeta potential more electronegative) and oxidation/reduction (Cerrillo et al. 2016). When experiments were developed with synthetic organic matter, no clear influence of the stability in ecotoxicology was observed (Schwabe et al. 2013, Zhu et al. 2014, Grillo et al. 2015).

Redox Stage

Studies have shown that a decreasing in NPs size provokes higher formation of oxygen vacancies (i.e., CeO_2 NPs) (Deshpande et al. 2005). The redox properties and the formation of oxygen vacancies play an important role for the application of NPs and is of interest due to the effects of this kind of NP on aquatic organisms. Sendra et al. (2017d) studied the scavenging of H_2O_2 by different CeO_2

NPs in freshwater and marine water observed that in marine water the scavenging of H_2O_2 by CeO_2 was higher than in freshwater (Sendra et al. 2017f).

Biological Transformation

Biological transformation such as biodegradation (bacterias, lignin peroxidase, fungae bacteria), bioxidation, bioreduction, coating degradation and phosphate transformation can produce new NMs with a different toxic profile from pristine NMs. The knowledge of transformation of NMs once in contact with its surrounding environments is of critical importance for assessing the biological impacts of NPs and the kinetics of such impacts (Amde et al. 2017, Pulido-Reyes et al. 2017).

Eco-coronas

Another of ENPs' transformation processes in the environment is the formation of **eco-coronas**. Eco-corona is the hard and/or soft corona around ENPs. These eco-coronas can to be form by extracellular proteins, such as the digestive proteins excreted by algae and bacteria, hemicellulose, cellulose, extracellular polymeric substances, tannic acid and humid acid among others (Lynch et al. 2014).

The role of the microorganisms in eco-corona formation is really important. The large amount of exopolymeric substances (EPS) with high molecular weight excreted by microorganisms can stabilise NPs. These transformations can change the charge of NPs, their fate and toxicity (Monopoli et al. 2012, Morelli et al. 2018). In a recent work, it was demonstrated how biogenic organic matter released by marine microalgae *Dunaliella tertiolecta* favoured the NPs suspension stability in the first hours of the experiment, preventing the TiO_2 NPs from settling to the bottom of the test vessels (Morelli et al. 2018).

An important consideration for environmental studies is that NPs are not just one class of potential pollutant. Given that NPs contain a wide range of different materials with different physical, chemical, and toxicological properties, they should not be considered a single homogeneous group (Quigg et al. 2013).

Mechanisms of NPs Toxicity in Microalgae

Direct Mechanisms of NPs Toxicity in Microalgae

Adsorption of NPs on Microalgae

Usually the culture media used for microalgae toxicity test (artificial fresh and marine water) (Guillard and Ryther 1962, ASTM 1975) have got a high amount of salts; therefore, zeta potential values of NPs suspended in culture media are very different to those of NPs suspended in ultrapure water, so NPs can be found in the instability zone (Jiang et al. 2009, Sendra et al. 2017b, Sendra et al. 2017d, Sendra et al. 2017f). The chemical species found in culture media are adsorbed by NPs; it decreases the electronegative double layer between the NPs-cell (heteroagglomeration process) (Ma et al. 2015, Sendra et al. 2017e).

Microalgae entrapped by NPs have been observed in many studies with different NPs composition. Different authors have checked heteroagglomeration of TiO_2 NPs in freshwater microalgae such as *C. reinhardtii*, *P. subcapitata*, *S. oblicuous* (Wang et al. 2008, Aruoja et al. 2009, Chen et al. 2012, Dalai et al. 2013). Heteroagglomeration was also found between carbon nanotubes (CNT) and *C. vulgaris* and *P. subcapitata* (Schwab et al. 2011) and SiO_2 NPs and *S. obliquus* (Wei et al. 2010).

Heteroagglomeration between microalgae and NPs in saltwater have been studied recently. Some authors have demonstrated heteroagglomeration between TiO_2 NPs and marine microalgae *P. tricornutum* (Deng et al. 2017b, Minetto et al. 2017), *Nitzschia closterium* (Xia et al. 2015), *Karenia brevis* and *Skeletonema costatum* (Li et al. 2015). In the case of CeO_2 NPs, it entrapped microalgae with *P. tricornutum* has been reported (Deng et al. 2017b, Minetto et al. 2017).

Others cases of heteroagglomeration were observed between *Dunaliella salina* and PbS NPs (Zamani et al. 2014) and *P. tricornutum* and *D. tertiolecta* and CdSe/ZnS QDs (Morelli et al. 2013).

In the work of Sendra et al. (2017c), the heteroagglomeration process in one freshwater microalgae *C. vulgaris* and two marine microalgae *Isochrysis galbana* (motil microalgae) and *P. tricornutum* through co-settling experiments between TiO_2 NPs and cells were studied. The results showed the greater capacity of TiO_2 NPs to catch microalgae than TiO_2's bulk form and the stronger heteroagglomeration between TiO_2 NPs with marine microalgae than TiO_2 NPs with freshwater microalgae. Stronger heteroagglomeration of marine species can be due to higher IS of the culture media, the salts present in the culture media could be compressing the electro-double layer of TiO_2 NPs and cells, and would impede the electrostatic repulsion force (Sendra et al. 2017e).

DLVO theory has been employed to explain the heteroagglomeration process between NPs-cells; however, this theory does not take account other complex processes which are developed in aqueous suspensions such as NPs-Cells and NPs-NPs together. Furthermore, EPS are outside the scope of DLVO theory. The EPS can act as a "glue" between NPs-Cells or, in contrast, release NPs into environment through a detoxification mechanism (Ma et al. 2015, Sendra et al. 2017e).

On the other hand, not only are the intrinsic characteristics of NPs and/extrinsic characteristics of culture media related to the heteroagglomeration process but are also characteristic of microalgae species such as (i) size of microalgae, smaller microalgae have more S_{BET} than larger microalgae and increase the contact with the external environment for a greater lapse of time; (ii) composition and ultrastructure of cell wall NPs can be accumulated in the ridges on the cell wall of microalgae; (iii) motility of microalgae. Sendra et al. (2017c) and Li et al. (2017) observed that non-swimming microalgae showed higher heteroagglomeration than motile microalgae. Although non-motile microalgae were more sensitive to heteroagglomeration, in *Tetraselmis suecica,* tested NPs were found in the flagella, which restricted the motility of microalgae (Li et al. 2017).

In contrast, motile microalgae-over the initial hours of the test-increase the possibility of interaction with NPs suspended in culture media (Schiavo et al. 2017, Sendra et al. 2017e, Zhao et al. 2017). However, homoagglomeration and the settling process happen after first hours of experiments. This fact can provokes than benthic microalgae will be more exposed to NPs agglomerates (Moreno-Garrido et al. 2015).

Direct Effects of NPs in Microalgae that are Related to NPs' Adsorption onto the Cell Surface

Increase in Cellular Weight due to the Adsorption of NPs

An increase in cellular weight of microalgae due to heteroagglomeration (NPs-Cells process) could remove microalgae from the photic zone, affecting photosynthesis. In the experiments of Ma et al. (2015) and Sendra et al. (2017c), the settling process of microalgae were affected when NPs were present in culture suspensions. Huang et al. (2005) demonstrated that *P. subcapitata* cells adsorbed onto their surface TiO_2 NPs and folded 2.3 times their own weight.

Shading Effects

NPs can absorb available light leading to a negative impact on photosynthesis and therefore, on the growth of the population. Some authors have confirmed that the heteroagglomeration can affects photosynthesis (Aruoja et al. 2009, Hartmann and Baun 2010, Gong et al. 2011, Ji et al. 2011, Schwab et al. 2011, Chen et al. 2012, Xia et al. 2015, Huang et al. 2016, Li et al. 2017).

Nutrient Limitation

Nutrient limitations and interruption of energy transductions (related to ATP synthesis) has also been suggested as a possible effect of the mechanism of adsorption of NPs in cell wall (Aruoja et al. 2009, Yeung et al. 2009, Hartmann and Baun 2010, Ji et al. 2011, Schwab et al. 2011).

Cell wall and Cell Membrane Damage

Due to adsorption of NPs onto the cell surface, physical damage in cell wall and membrane cell have been reported by many authors. TiO_2 NPs have been revealed to cause cell wall damage in *P. tricornutum* (Wang et al. 2016b), *Chlorella pyrenoidosa* (Zhang et al. 2017). Cell wall damage usually is confirmed by lactate dehydrogenase assay, viability tests and visually by SEM analysis.

Some authors observed loss of membrane integrity and disruption of membrane potential by Ag NPs, which may lead to the inactivation of associated enzymes and the destabilization of the lipid bilayer (El Badawy et al. 2011, Tsiola et al. 2017). Furthermore, the formation of pits on cell surfaces may lead to structural alterations and intrusion of NPs in the cytoplasm, with subsequent responses on the cellular level, such as generation of ROS, disruption of DNA replication, and ATP synthesis (Durán et al. 2016).

As the loss of cell membrane integrity by NPs is well known, a recent work used Ag NPs as an application in cell wall disruption to release carbohydrate and lipid from *C. vulgaris* for biofuel production (Razack et al. 2016).

In the studies of Braydich-Stolle et al. (2009) and Hirakawa et al. (2004), it was observed that anatase TiO_2 NPs provoked cell necrosis and membrane leakage in microalgae cells.

Plasmolysis

Morphological alteration into cytosol is produced by NPs and the plasma membrane is detached from the cell wall. In some studies, the cytosol is leaked due to the disruption of plasma membrane and the plasma membranes as well as the cell wall are degraded. In some works, it has been observed—through TEM images—that there is a separation between the lipid membrane cell and membrane cell wall; this separation is irreversible and it is a permanent plasmolysis. These characteristics of membrane cells and cell wall triggered by NPs were found by Gong et al. (2011) and Oukarroum et al. (2017) in *C. vulgaris* exposed to NiO NPs and by Zhao et al. (2016) in *C. pyrenoidosa* exposed to CuO NPs and by Chen et al. (2012) in *C. reinhardtii* exposed to TiO_2 NPs.

Increase of EPS Production

As it was previously reported, EPS can provide different functions in toxicity with regard to NPs: (i) stabilize NPs dispersions, (ii) induce their aggregation and thus, either exacerbate or reduce the direct toxicity of NPs (Wilkinson and Reinhardt 2005), (iii) provide abundant binding ligands for the toxicants (e.g., trace metal ions) released from NPs. Among the EPS functions in microalgae organisms are: water retention, sorption of organic compounds and inorganic ions, adhesion, cohesion of biofilm and cell aggregation, nutrient source, enzyme activity, enzyme generation, antivirus, antioxidant, antitumor and anti-inflammatory, surfactant, emulsifier and export of cell components (Xiao and Zheng 2016).

The role of EPS in toxicological tests with NPs, is not restricted to their contribution of detoxification mechanisms in microalgae (Miao et al. 2009), but also the effect that they provoke when are removed from culture media. Thus, it has been reported that mortality of *Cylindrotheca closterium* and *Cylindrotheca fusiformis* increased a 15 and 30%, respectively, when EPS were removed from the culture media (Pletikapić et al. 2012).

In Table 2 are shown a selection of experiments carried out to analyse the role of EPS in the toxicity of NPs to microalgae.

Uptake of NPs into Microalgae

In microalgae, plasma membrane is usually surrounded by cell wall, which represent the first barrier. Therefore, NPs uptake will require the passage through the cell wall and then internalization. Microalgae walls are made of glycoproteins and polysaccharides (cellulose, hemicellulose and pectin) such as carrageenan and agar. Diatoms also present in the cell wall (frustules) which are composed

Table 2. Summary of experiments carried out to analyse the role of EPS in microalgae-NPs toxicity tests.

Environment	Species	NPs	References
Freshwater	*Scenedesmus obliquus*	TiO_2	(Dalai et al. 2013)
	Chlamydomonas reinhardtii	TiO_2	(Sendra et al. 2017b)
	Chlamydomonas reinhardtii	Ag	(Taylor et al. 2016a)
	Chlorella pyrenoidosa	CuO	(Zhao et al. 2016)
Marine	*Phaeodactylum tricornutum*	TiO_2	(Sendra et al. 2017b)
	Phaeodactylum tricornutum	TiO_2	(Deng et al. 2017b)
	Thalassiosira weissflogii	Ag	(Miao et al. 2009)
	Cylindrotheca fusiformis	Ag	(Pletikapić et al. 2012)
	Cylindrotheca closterium	Ag	(Pletikapić et al. 2012)
	Phaeodactylum tricornutum	CeO_2	(Deng et al. 2017b)
	Amphora sp.	CdSe	(Zhang et al. 2012)
	Dunaliella tertiolecta	CdSe	(Zhang et al. 2012)
	Phaeocystis globosa	CdSe	(Zhang et al. 2012)
	Thalassiosira pseudonana	CdSe	(Zhang et al. 2012)
	Thalassiosira pseudonana	CdSe	(Zhang et al. 2013)
	Amphora sp.	Polystyrene	(Chen et al. 2011)
	Ankistrodesmus angustus	Polystyrene	(Chen et al. 2011)
	Phaeodactylum tricornutum	Polystyrene	(Chen et al. 2011)

of biogenic silica (Popper et al. 2011). According to the size of pores across the cell walls 5–20 nm for green microalgae and 3 to 50 nm in diatoms (Vrieling et al. 1999), only the smaller NPs could reach the cell membrane and to be internalized by diatoms (Moore 2006, Navarro et al. 2008a). The highest concentration in internal cell compartments of some non-soluble NPs such as TiO_2 NPs has been observed after 24 h of exposure. The mechanisms of uptake NPs into cytosol are related to membrane cell damage and its disruption (Behra et al. 2013, Dalai et al. 2013, Sendra et al. 2017b).

In general, the main mechanisms of NPs uptake into cells are: passive diffusion, transport through ion channels, carrier-mediated transport, phagocytosis micropinocytosis and caveolar/endocytic routes (Verma et al. 2008). Once inside the cells, NPs can be incorporated within the functional machinery of the cells and cause deleterious effects (Moore 2006). On the other side, dissolved metal from NPs such as Ag^+ from metallic Ag NPs, can be transported by mechanisms such as pointed out in the free-ion activity model (FIAM) and the biotic ligand model (BLM) (Di Toro et al. 2001, Campbell et al. 2002, Slaveykova and Wilkinson 2003, Campbell and Fortin 2013).

In addition to its intrinsic cell wall characteristics, the permeability of the wall can change during cell division, with the newly synthesized cell wall being more permeable to NPs (Wessels 1994).

For all the aforementioned, the external structure, such as the presence or absence of a cell wall, could determine the sensitivity of microalgae to NPs (Oukarroum et al. 2012). Some works have studied the importance of cell wall as physical barrier in internalization and toxicity of NPs in microalgae. The works of Piccapietra et al. (2012) and Röhder et al. (2014) assessed the toxicity of Ag and CeO_2 NPs in a wild type and a cell wall free mutant of the green microalga *C. reinhardtii*. The cell wall of *C. reinhardtii* seems to act as protective barrier limiting the internalization of AgNPs when the cells of the same species—with and without a cell wall—were compared (Piccapietra et al. 2012). In contrast, sensitivity to CeO_2 NPs and $Ce(NO_3)_3$ between a wild type and a cell wall free mutant of *C. reinhardtii* was similar, indicating that the cell wall did not prevent the toxic effects (Röhder et al. 2014).

Others studies have focused on the toxicity of NPs in wild microalgae—with and without cell wall-from natural aquatic environments. Most of the studies have used as target species the microalgae *Dunaliella* sp., due to absence of the cell wall. The marine microalgae *D. tertiolecta* has been shown to be more sensitive to different NPs (zinc oxide, single-walled carbon nanotubes, silicon dioxide and carbon black) than other marine microalgae as *I. galbana* and *Tetraselmis suecica* (Miglietta et al. 2011). A higher sensitivity was also observed for *D. tertiolecta* exposed to Ag NPs as compared to freshwater microalgae *Chlorella vulgaris* (Oukarroum et al. 2012). This last study showed that sensitivity of both species—due to the presence or absence of the cell wall—and the characteristics of culture media provoked significant differences in the behaviour between NPs and the microalgae.

In a recent work carried out to study the sensitivity to Ag and CeO_2 NPs and dissolved forms through different endpoints in *Dunaliella salina* and *Chlorella autothrophica* (Sendra et al. 2017a), it was observed that the lack of a cell wall in *D. salina* did not obligatorily suppose higher vulnerability to the internalization of contaminants, probably because other defence mechanisms such as the production of EPS (Mishra et al. 2011) and of β-carotene (Cowan et al. 1992) might prevent or counteract a possible higher internalization. Each species might have its own mechanism to prevent or reduce the toxic effects caused by the exposure to contaminants, and those mechanisms may be linked to: (i) chemical processes at the level of the cell membrane/wall that entraps the compound and prevents the absorption; (ii) physiological processes that internally bind and reduce the bioavailability of the compounds; and (iii) processes of elimination of the compounds to the external environment (Visviki and Rachlin 1994, Nassiri et al. 1997, Wu and Wang 2011). The different results among authors is because of the complexity of the NPs' behaviour in the culture suspensions, interaction between NPs-cells due to characteristics of culture media and species tested and sensitivity and detoxification mechanisms by microalgae.

Techniques to Detect NPs Internalization in Microalgae

Currently, in most of the work developed in the literature, NPs are detected in the intracellular microalgae compartment by different techniques:

i) TEM images. The most direct method to detect single or agglomerates of NPs. Moreover, this techniques identify the organelles which internalize NPs. Metal and oxide metal composition are detected by EDX.
ii) Fluorescent microscope. Some NPs in the market are conjugated with fluorescent molecules and it can be detected in the cytosol of the cells.
iii) Analytical methods. Internalization of NPs can be detected through a protocol of washes that provide the proper way to identify NPs in the supernatant, NPs adsorbed onto cell surface and, the last one, that is, internalized NPs.
iv) Flow cytometry technique. Once the cell population has been submitted to the washes protocol, the cells in suspensions are analysed in a flow cytometer. The use of detectors FSC and SCC related with cell volume and cell granularity, respectively in order to assess NPs internalization.

In Table 3, a selection of studies about NPs internalization in microalgae using different methodologies is reported.

Effects of NPs internalized

If NPs are able to be internalized by microalgae, they can generate structural and functional damage in different organelles; furthermore, if NPs reach the nucleus, as in some studies has been observed by TEM, they can produce DNA damage by breaking the H bonding (Klaine et al. 2008).

The NPs which are able to dissolve in culture media have taken part in a new paradigm in the studies which assess the internalization of NPs. The recent concept about green synthesis of NPs by microalgae through metal dissolved form queries about the origin of NPs found in the cytosol of microalgae. Therefore, the paradigm is that the particulate metal found inside the microalgae is really due to the internalization of the NP itself should be reviewed, because some questions remain open:

Table 3. Summary of publications about methodologies employed to study internalization of NPs in microalgae.

Species	NPs	Method	References
Chlamydomonas reinhardtii	TiO_2	Analytical method	(Sendra et al. 2017b)
Phaeodactylum tricornutum	TiO_2	Analytical method	(Sendra et al. 2017b)
Anabaena variabilis	TiO_2	TEM	(Cherchi et al. 2011)
Scenedesmus obliquus	TiO_2	TEM and analytical	(Dalai et al. 2013)
Chlorella vulgaris	TiO_2	TEM	(Xia et al. 2018)
Chlorella sp.	TiO_2	TEM	(Iswarya et al. 2015)
Chlamydomonas reinhardtii	TiO_2	Flowcytometry	(Gunawan et al. 2013)
Nitzschia closterium	TiO_2	Flow cytometry and analytical methods	(Xia et al. 2015)
Chlamydomonas reinhardtii	Ag	Flow cytometry	(Sendra et al. 2017d)
Phaeodactylum tricornutum	Ag	Flow cytometry	(Sendra et al. 2017d)
Chlorella autotrophica	Ag	Flow cytometry	(Sendra et al. 2017d)
Dunaliella salina	Ag	Flow cytometry	(Sendra et al. 2017d)
Chlorella pyrenoidosa	Ag	TEM	(Zhou et al. 2016)
Chlamydomonas reinhardtii	Ag	High-resolution secondary ion mass spectrometry (NanoSIMS) images and TEM	(Wang et al. 2016a)
Raphidocelis subcapitata	Ag	multimodal imaging approach incorporating dark-field light microscopy, high-resolution electron microscopy, and nanoscale secondary ion mass spectrometry (NanoSIMS)	(Sekine et al. 2017)
Chlorella autotrophica	CeO_2	Flow cytometry	(Sendra et al. 2017a)
Dunaliella salina	CeO_2	Flow cytometry	(Sendra et al. 2017a)
Chlamydomonas reinhardtii	CeO_2	Flow cytometry	(Sendra et al. 2017f)
Phaeodactylum tricornutum	CeO_2	Flow cytometry	(Sendra et al. 2017f)
Nannochloris atomus	CeO_2	Flow cytometry	(Sendra et al. 2017f)
Pseudokirchneriella subcapitata	CeO_2	Flowcytometry	(Manier et al. 2013)
Chlamydomonas reinhardtii	CeO_2	TEM; vacuole and NPs around the nucleus	(Taylor et al. 2016b)
Microcystis aeruginosa	CuO	HRTEM	(Wang et al. 2011)
Chlorella pyrenoidosa	CuO	TEM	(Zhao et al. 2016)
Euglena gracilis	ZnO	TEM	(Brayner et al. 2010)
Chlamydomonas reinhardtii	ZnO	Flow cytometry	(Gunawan et al. 2013)
Anabaena flos-aqua	CdS	TEM	(Brayner et al. 2011)
Euglena gracilis	CdS	TEM	(Brayner et al. 2011)
Chlorella sp.	Iron oxide	TEM and EDX	(Toh et al. 2014)

Are internal nanoparticles originated in detoxification processes? or even how can we differentiate both origin -internalization of formation? What is percentage of each processes?

Many studies have demonstrated how microalgae can accumulate metals in internal granules as a mechanism of detoxification (Maeda and Sakaguchi 1990). Freshwater and marine microalgae have been used for Ag NPs production such as *C. reinhardtii*, *Scenedesmus* sp., *Chaetoceros calcitrans*, *Chlorella salina*, *I. galbana* and *Tetraselmis suecica* (Merin et al. 2010, Barwal et al. 2011, Jena et al. 2014). With respect to Au NPs synthesis, *Spirulina*, *Leptolyngbyav alderianum*, *Tetraselmis suecica, Klebsormidium flaccidum* and *Chlorella vulgaris* have been employed (Shakibaie et al.

2010, Kalabegishvili et al. 2011, Luangpipat et al. 2011, Dahoumane et al. 2012, Kalabegishvili et al. 2012, Roychoudhury and Pal 2014). In conclusion, due to the ability of some microalgae to produce internal NPs from dissolved metals, the origin of the intracellular NPs observed in certain studies should be carefully investigated.

Nanoparticles and Oxidative Stress

One of the most important mechanism of NPs toxicity is the increase in the oxidative stress by stimulating the respiratory burst in phagocytic cells with increased oxygen consumption, resulting in production of free radicals such as O^-, H_2O_2, and NO (Vallyathan et al. 1992, Blackford Jr et al. 1994, Vallyathan and Shi 1997, Manke et al. 2013). The generation of free radicals is produced by reactivity of NPs' surface by the following mechanisms (Singh 2015):

1) Membrane cell damage by NPs allows the influx of free radicals from the extracellular fluid into the cytoplasm.
2) Mitochondrial damage allows the release of free radicals (Xia et al. 2006).
3) Chloroplast damage allow the release of free radicals.
4) Reaction of the oxidant with the NP surface generates free radicals that remain bound to the surface.
5) NPs, such as CeO_2, possess structural defects with altered electronic properties that can interact with molecular O_2 to generate or scavenge ROS via Fenton-type reactions.
6) Free radicals bound to the NP surface are released in aqueous suspensions, generating H_2O_2, OH^- and O_2.
7) Metal and metal oxide NPs promote the activation of an intercellular radicals generating system, such as the MAPK and NF-kB pathways.

The oxidative stress has been pointed out as the main mechanism of NP toxicity (Nel et al. 2006):

1) NPs generate ROS and RNS that increase oxidative stress and ensuing protein, DNA, and membrane injury.
2) Oxidative stress induces phase II enzymes, activation of antioxidative enzymes inflammation, mitochondrial damage and activation of the reticule-endothelial system (Hazani et al. 2013, Lindgren 2014).
3) Mitochondrial damage may include damage of the inner membrane, pore opening, energy failure, cytotoxicity, and apoptosis.
4) Oxidative stress increases expression of proinflammatory cytokines.
5) NPs may activate the reticule endothelial system that increases oxidative stress.

Effect of NPs Dependent on Oxidative Stress

Changes in the Ultrastructure of Microalgae

NPs in microalgae can provoke changes at surface and intracellular levels in the cells. With respect to surface levels, changes in cell wall and cell membrane can be observed. Li et al. (2017), observed via TEM images, how nickel oxide (NiO) NPs in *C. vulgaris*, at 31 mg·L^{-1} severe shrink in the cell shape, the disruption in plasma membrane leading to leakage of cytosol. When *Anabaena variabilis* was exposed to 1 mg·L^{-1} of TiO_2 NPs, membrane disruption in vegetative heterocyst cells and an increase on membrane roughness was observed (Cherchi et al. 2011).

The percentage of ROS can increase in the cell wall and surrounding environments caused by photocatalytic NPs such as TiO_2 NPs exposed to ultraviolet radiation (UVR). This fact can trigger damage in an indirect manner in cell wall and organic compounds (Miller et al. 2012).

The phototoxicity of TiO_2 NPs have been demonstrated in different culture media from freshwater to saltwater through extracellular ROS production that can injure the cell wall of microalgae (Dalai et al. 2013, Sendra et al. 2017b). In contrast to photocatalytic NPs, other NPs which form heteroagglomerates NPs-cells demonstrated a direct relation between percentage of ROS and membrane cell damage.

In relation to changes at intracellular levels, some studies found swelling up of vacuoles and pigment bodies and injury of thylakoid grana (Li et al. 2017). Gong et al. (2011) found that thylakoids showed disordered grana lamella and the appearance of membrane limited crystalline inclusions which indicated that the photosynthesis could be affected. Furthermore, TEM images have shown improper nucleus, phosphate granule depositions, degradation of starch–pyrenoid complex and dentate membrane and degradation of chloroplast (Dalai et al. 2013). Therefore, disruption of the chloroplasts and mitochondria may generate intermediate signals involved in programmed cell death and induces the apoptosis of cells (Apel and Hirt 2004, Van Breusegem and Dat 2006).

Decrease in Cell Viability

The membrane cell damage by oxidative stress of NPs is translated into a loss of cell viability leading to death of cells by apoptosis (Hirakawa et al. 2004, Braydich-Stolle et al. 2009). These authors demonstrated that oxidative stress caused by ROS generation of anatase phase TiO_2 NPs triggers pathway signalling through caspase activation (Hirakawa et al. 2004, Braydich-Stolle et al. 2009). The most common endpoint of an ecotoxicological test that assess the cell viability in certain way is the growth of the microalgae population.

DNA Damage

Several previous reports suggested that ROS generation induces damage to DNA (Dick et al. 2003, Olmedo et al. 2005). However, the damage is not only to the DNA, but also to the proteins, lipids and other metabolites in the cells (Hirakawa et al. 2004, Tucci et al. 2013) during the TiO_2 photocatalysis. All these may lead to the release of proapoptotic factors and cause programmed cell death (Jin et al. 2011).

Taylor et al. (2016b) demonstrated that CeO_2 NPs in *C. reinhardtii* provoked down-regulation in genes related to photosynthesis and carbon fixation associated with effects on energy metabolism. On the other hand, upregulation of genes related to oxidative stress such as sod1, gpx, cat and ptox2 were found when *C. reinhardtii* was exposed to TiO_2 NPs (Wang et al. 2008).

Deleterious Effects in Photosynthetic Apparatus

One of the more sensitive toxicological responses to increase of ROS by NPs is the unwanted effects in the photosynthesis and this have been reported in the genes affected by NPs related to photosynthesis and carbon fixation (Taylor et al. 2016a). The effective quantum yield of photosystem II (PSII) under stress conditions is affected when microalgae are exposed to NPs. This is consequence that the energy that arrives through antenna systems is not passed to the photochemical reaction (da Costa et al. 2016). Therefore, it is a proper measured to assess the toxicity of NPs in photosynthetic microorganisms (Miller et al. 2017).

In several studies (Sendra et al. 2017a, Sendra et al. 2017b, Sendra et al. 2017d), the toxicity of TiO_2, Ag and CeO_2 NPs was assessed in *C. reinhardtii*, *P. tricornutum*, *C. autotrophica* and *D. salina*. Ag NPs was the NPs which showed higher negative effects in this response while TiO_2 and CeO_2 NPs had an effect at high concentration of NPs. The results about TiO_2 NPs and CeO_2 NPs are in agreement with others works, where authors found inhibition of effective quantum yield of PSII at higher concentration that they are found in the environment (Chen et al. 2012, Gao et al. 2013, Dao and Beardall 2016, Hu et al. 2016, Deng et al. 2017a).

With respect to toxicity of Ag NPs, Navarro et al. (2008b) observed in *C. reinhardtii* a decrease in curves of photosynthetic yield at short time of exposure (1–2 h) although this was time-dependent during the first 2 h. Oukarroum et al. (2012) reported that Ag NPs seemed to induce structural damage to the photosystem II in *C. vulgaris* and *D. tertiolecta*.

Autofluorescence of chlorophyll *a* is often affected by NPs' toxicity. An increase in autofluorescence could be explained by a blockage of the electron transport chain at the PSII level (Cid et al. 1995), indicating an inhibitory effect located on the oxidant side, probably due to inactivation of PSII reaction centres (Samson and Popovic 1988). Sendra et al. (2017f) observed that the more sensitive species such as marine diatom *P. tricornutum's* exposure to CeO_2 NPs had higher autofluorescence with respect to the controls than the other species tested. In the work of Deng et al. (2017a), an increase in chlorophyll contents in *P. tricornutum* when it was exposed to TiO_2 NPs was also found.

On the other hand, the loss of chlorophyll is related to degradation or downregulation in pigments' synthesis due to damage in chloroplast and an inhibition in the electron transport chain in the donor centre. A decrease in chlorophyll "a" was found in many studies of microalgae exposed to NPs (Mayfield et al. 1995, Miao et al. 2009, Liu et al. 2012, Oukarroum et al. 2012, Rodea-Palomares et al. 2012). Furthermore, due to failures of photosynthetic apparatus (Sendra et al. 2017a), a chlorotic population of *D. salina* was found when they are exposed to AgNPs. The chlorosis state consists of a low level of residual photosynthesis, in which both photosystems (PI and PII) gradually lose their activity, and chlorophyll and phycobiliproteins are degraded (Sauer et al. 2001).

Additional to the effects due to ROS that were previously mentioned, in microalgae exposed to NPs disturbances in the cellular phosphate management, and inhibition of DNA synthesis, collapse of proton pump, denaturation of ribosomes, degradation of lipopolysaccharide molecules and inactivation of proteins and enzymes by bonding on active sites have also been found (Klaine et al. 2008).

Indirect Mechanisms of NPs Toxicity in Microalgae

Release of Ions from NPs

Concerning the toxicity of NPs, the direct effect of NPs and indirect effects mediated by ions released from their surface need to be differentiated (Wijnhoven et al. 2009). Many studies have demonstrated that one of the main mechanisms of toxicity of metal and oxide metal NPs (such as Ag, CuO and ZnO) is related to the dissolution of NPs though the release of ions into the environment (Hiriart-Baer et al. 2006, Navarro et al. 2008b, Fabrega et al. 2011). The dissolution rate of NPs is related to the composition of the culture media and physicochemical characteristics of NPs previously explained in the Section: Chemical Transformation. Therefore, NPs with higher surface area are more reactive with the environment and the dissolution rate will be higher than bulk form; furthermore, composition of culture media will determine the chemical species found, its availability and toxicity.

In relation to toxicity due to NPs' dissolution, Notter et al. (2014) carried out a meta-analysis about a comparison between toxicity of NPs and dissolved metals. These authors established different toxicity ratios (TR); TR equal to 1 means that particles are as toxic as dissolved metal ions. TR lower than 1 means that nano form is less toxic than the dissolved metal based on total concentration. TR higher than 1 means toxicity of NPs is greater than the toxicity of the dissolved ion alone. In this work, 93.8% of scientific articles about Ag toxicity, Ag NPs was less toxic than dissolved metals, a 100% for Cu and 81% for Zn (Notter et al. 2014). Some of the effects originated by ions from NPs surface are disruption of cell membrane, ROS generation and interaction with proteins and DNA (Durán et al. 2016).

Photocatalytic and Redox Activity of NPs

Some commercial NPs, such as TiO_2 NPs are used to increase the whiteness and opacity of many consumer products; TiO_2 NPs are photocatalytic with exposure to UV radiation (wavelengths < 400 nm). Due to the photocatalytic properties of TiO_2 NPs, a coating is required to suppress their photocatalytic capacity. Therefore, under solar radiation, TiO_2 NPs are able to generate ROS in surrounding environments and these oxygen free radicals interact with organic functional groups of the cell wall and lipid membrane. The toxicity of this photocatalytic NPs depends on UV radiation (Watlington 2005, Hund-Rinke and Simon 2006, Dalai et al. 2013, Lee and An 2013, Li et al. 2014, Wallis et al. 2014, Coll et al. 2015, Sendra et al. 2017b). In a recent work, it was estimated that intracellular ROS was a small fraction of total ROS, the highest amount being extracellular (Morelli et al. 2018).

In relation to redox capacity, CeO_2 NPs can store oxygen and scavenging ROS; therefore, recently, CeO_2 NPs have been used as antioxidant therapies (Celardo et al. 2011). Sendra et al. (2017f) assessed the capacity of different CeO_2 NPs for scavenging H_2O_2 in freshwater and marine culture media; in both culture media, CeO_2 NPs demonstrated their ability to eliminate H_2O_2. Due to this ability, recent studies concerning to CeO_2 NPs and microalgae have found a protective effect of CeO_2 NPs in microalgae, with decrease in ROS and stimulated population growth (Xia et al. 2008, Deng et al. 2017b, Sendra et al. 2017f).

Interaction NPs-pollutants

NPs themselves may serve as pollutant carriers (indirect effect). In this manner, NPs may enhance or reduce the bioavailability of other toxic substances to algae (Navarro et al. 2008a, Ferry et al. 2009, Hou et al. 2013).

Supporting such concerns, some authors have showed a "Trojan horse effect" between classical or emergent pollutants with NPs and microalgae. Yang et al. (2012) demonstrated interactions between legacy pollutants (Cd^{2+}), TiO_2 NPs and the *C. reinhardtii*. They found that TiO_2 NPs could alleviate the effects of Cd^{2+} on the green alga. Cd^{2+} adsorption by TiO_2-ENPs decreased its ambient free ion concentration and its intracellular accumulation in the cells as well as its toxicity. Other studies have also shown that TiO_2 NPs can remove heavy metals (Pb, Cd, Cu, Zn, and Ni) from water (Debnath and Ghosh 2011).

The recent work of Zhang et al. (2017) evidenced different toxic interaction between TiO_2 NPs and different organochloride compounds in the freshwater microalgae *Chlorella pyrenoidosa*. Synergic effects were found between Atrazine-TiO_2 NPs, antagonist effects were found in Hexachlorobenzene-TiO_2 NPs and 3,3',4,4'-tetrachlorobiphenyl-TiO_2 NPs. Additive effects were reported between pentachlorobenzene and TiO_2 NPs. The TiO_2 NPs showed the same physicochemical characteristics with and without organochloride in the suspensions. With respect to accumulation of OC (organochloride), these pollutants increased their bioaccumulation in the microalgae when TiO_2 NPs were added to the culture media (Zhang et al. 2017).

In the study of Tang et al. (2013), two trends were observed when the interaction between TiO_2 NPs and Zn^{+2} with the freshwater microalgae *Anabaena* sp. was studied. These trends were concentration-dependent. The occurrence of TiO_2 NPs at low concentrations (< 1.0 mg·L^{-1}) significantly enhanced the toxicity of Zn^{2+} and consequently, reduced the EC_{50} value. On the other hand, the toxicity of the Zn^{2+} and TiO_2 NPs system decreased with increasing TiO_2 NPs concentration because of the substantial adsorption of Zn^{2+} by NPs surface. The toxicity curve of the Zn^{2+}/TiO_2 system as a function of incremental nano-TiO_2 concentrations was parabolic. The toxicity significantly increased at the initial stage, reached its maximum, and then decreased with increasing TiO_2 NPs concentration (Tang et al. 2013).

Adsorption of pollutants by the NPs' surface has been explained previously; however, absorption by NPs can also be the nutrients found in the environment leading to nutrient limitations. In the study of Hoecke et al. (2009), it was reported that adsorption of phosphate was about a 50% in 32 mg·L^{-1} of CeO_2 NPs.

Recommendation of NPs Toxicity Test in Microalgae

Due to special characteristics of ENMs, modifications of OECD test guidelines for NPs ecotoxicity tests have been proposed based on laboratory experience with Ag and TiO_2 NPs. Among the proposed changes, the following recommendations have been suggested: (i) shaking over all experiment time and (ii) to avoid the use EDTA in the culture medium because it is chelated with metal (Hund-Rinke et al. 2016). Some results indicate that exposure time and test conditions can have a significant role in toxicity response of marine algae to TiO_2 NPs. As recently underlined, there is an urgent need to properly set experimental conditions in running bioassays for ENMs risk assessment, due to peculiar properties of the materials and their interactions with exposure media (Petersen et al. 2015). Indeed, the use of standardized test conditions should be further adapted to better mimic realistic environmental exposure scenarios in which marine species will come in contact with ENMs in their natural environment. One of the mimic realistic environmental exposure scenarios is the UV radiation in ecotoxicity experiments with photoactive ENMs (Lee et al. 2009, Miller et al. 2012, Sendra et al. 2017b, Sendra et al. 2017c). Solar radiation, an important component of most ecosystems, can also play a role as an ecological stressor. Exposure to specific wavelength of sunlight, such as ultraviolet radiation, can result in increased oxidative stress and damage to biological macromolecules allowing easier uptake NPs and higher toxicity (Babu et al. 2003, Casati and Walbot 2004, Häder and Sinha 2005, Obermüller et al. 2005, Olson et al. 2006, Roberts et al. 2017). The sunlight can also interact with certain xenobiotic compounds in a phenomenon known as photo-induced, photo-enhanced, photo-activated, or photo-toxicity and to have an additional effect in complex system with NPs and organic compounds (Arfsten et al. 1996, Cho et al. 2003, Diamond 2003, Diamond et al. 2006, Weinstein and Diamond 2006).

One of the first works that used UV light in microalgae test exposure to photocatalytic material was Miller et al. (2012). Miller et al. (2012) exposed four species of marine phytoplankton to TiO_2 NPs and UV light intensity equivalent to average oceanic surface irradiance. The authors reported ROS production of 10–20 fold increase over background hydroxyl radical generation in seawater. The authors also reported reduced growth of phytoplankton under TiO_2 NPs exposure.

The combination of UV radiation and CuO NPs resulted in synergistic effects of these two stressors for all biological endpoints studied growth, chlorophyll and membrane cell damage in freshwater microalgae *C. reinhartii* (Cheloni et al. 2016).

In a recent work, the 'camouflages' of natural organic matter (NOM) in the adverse effects of NPs such as TiO_2 and CeO_2 NPs on green microalgae was studied (Cerrillo et al. 2016). These authors proposed the use of NOM in the standardization of nanoecotoxicology test.

Kennedy et al. (2017) have proposed a flow chart to realize ecotoxicological tests according to the case of NPs and culture media. In Fig. 1, the flow chart of the toxicity test with aquatic species is shown.

Climate Change and Toxicity of NPs in Microalgae

Recently, Xia et al. (2018) have studied the importance of ocean acidification in the toxic effects of TiO_2 NPs on marine microalgae *Chlorella vulgaris*. Concerns about the environmental effects and toxic mechanisms of NPs on marine ecosystems have increased over last decade. Meanwhile, ocean acidification has become a global environmental problem, seawater pH has fallen 0.1 units since pre-industrial times and it is predicted to fall 0.3–0.5 units by the year 2100 and 0.5–0.7 units by the year 2300 (Caldeira and Wickett 2005). However, the combined effects of NPs and ocean acidification on marine organisms are still not well understood. In the study of Xia et al. (2018), the effects of different pH values of 7.77 and 7.47 related to ocean acidification on the behaviour, bioavailability and toxicity of TiO_2 NPs to the marine microalga *Chlorella vulgaris* was investigated. The results showed that ocean acidification enhanced the growth inhibition of algal cells caused by TiO_2 NPs with synergistic interactive effects of pH and TiO_2 NPs on oxidative stress, thus indicating that lower pH significantly increased the oxidative damage of TiO_2 NPs. The reason to explain the higher amount of oxidative damage of TiO_2 associated with ocean acidification was attributed to the slighter aggregation and more suspended NPs in acidified seawater. Overall, these findings provide useful information on marine environmental risk assessments of NPs under near future ocean acidification conditions.

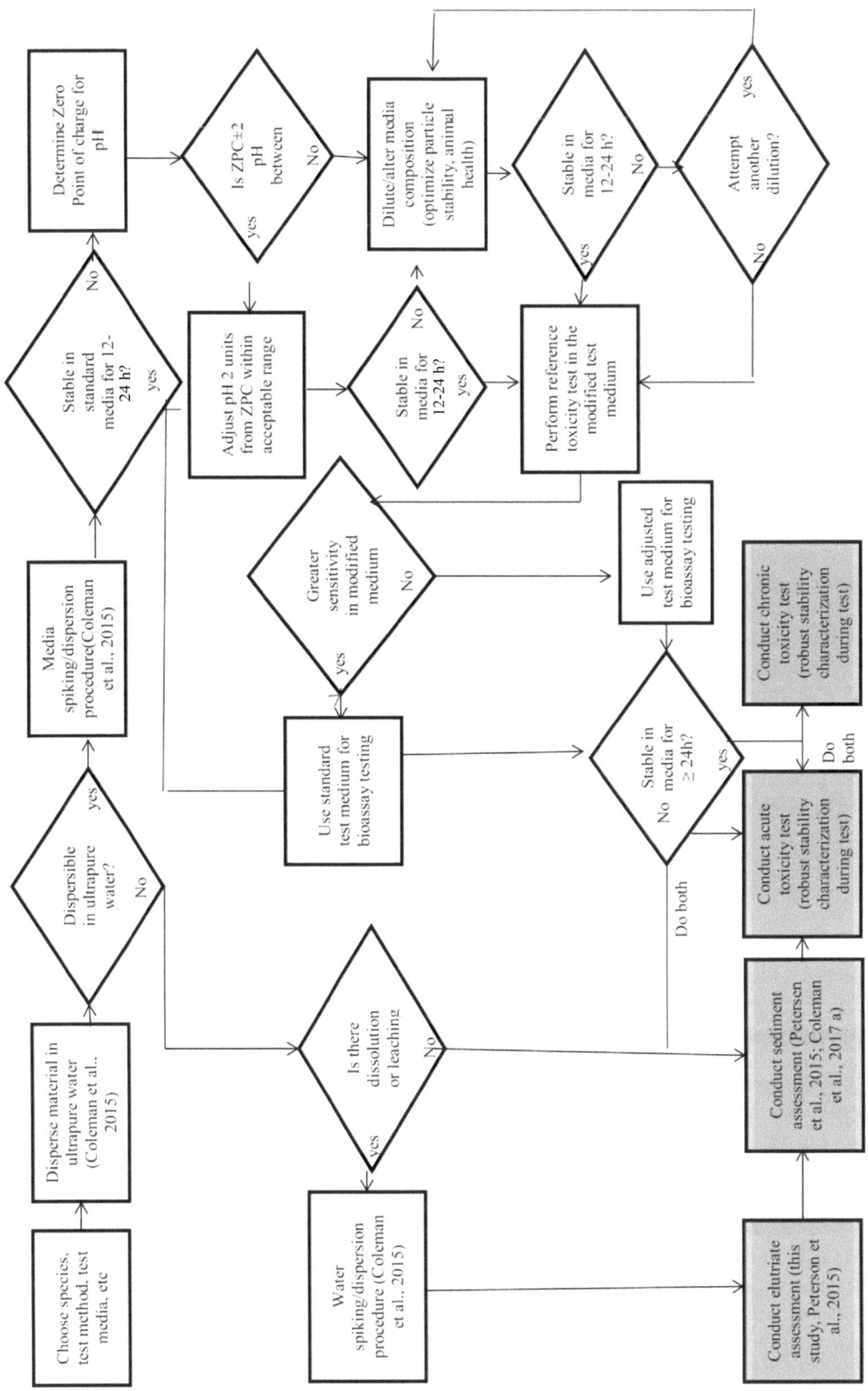

Figure 1. Hazard NPs evaluation in aquatic species (adapted from Petersen et al. 2015, OECD 2012, Coleman et al. 2015, Coleman et al. 2017a,b).

Further studies with changes in the temperature, salinity and pH could be addressed to know the behaviour, bioavailability and toxicity of emergent pollutants under climate change scenarios.

Trophic Transfer

This term is employed to define the transfer of a substance up a food chain from one trophic level to a higher one (e.g., to a predator from its prey). This process is well documented for metals (Luoma and Rainbow 2008). In contrast, the biomagnification process—which involves an increasing of metal concentration at higher trophic levels—is not so general and it has been verified only for some cases (e.g., methylmercury). In order to check this process, the biomagnification factor (BMF) is employed, analyzing the outcomes of metal transfer between prey and predator. Three situations can be reported: (a) predator concentration > prey concentration, (b) no differences between predator and prey concentrations and (c) prey concentration > predator concentration. Only in the first case can biomagnification be spoken of.

The assimilation of metals from diet by consumer organisms is one of the main routes for metal bioaccumulation. The information about metal and metal oxide engineered nanoparticles (Me-ENPs) is limited, although some review has been carried out on this topic (Stine Rosendal et al. 2016), these authors pointed out the four key factors of trophic transfer of Me-ENPs: (a) environmental transformation, (b) uptake and accumulation in the prey, (c) internal fate and localization in the prey and (d) digestive physiology of the predator. In this chapter's subsection, trophic transfer is focused on phytoplankton species, because they represent a key compartment in the aquatic ecosystems supporting the food web. In fact, the key factor of trophic transfer of Me-ENPs (as it is mentioned previously) are affected by the characteristics, cellular and physiological process of phytoplankton.

In the Table 4 are summarized a significant number of studies carried out to assess the trophic transfer between phytoplankton species and their predators (e.g., crustaceans and bivalves) for Me-ENPs, from freshwater and seawater ecosystems. Other studies with complex systems as mesocosms are included.

The trophic transfer has been verified for a wide number of Me-ENPs and food web, including them as prey fresh and seawater microalgae species (see references in Table 4). The bioaccumulation of AuNPs in phytoplankton species depends on the physical structure and surface availability for NPs interaction and the trophic transfer to a consumer (*Daphnia magna*) has been observed (Lee et al. 2015). AuNPs trophic transfer from biofilm to *Gammarus fosarum* has been assessed and the results obtained by metal analysis and electron microscopy confirmed this transfer showing relation with concentration in the contaminated biofilm (Baudrimont et al. 2018). The ingestion and digestion of AuNPs have been showed in the mollusk *Corbicula fluminea*, nanoparticles were located in branchial and digestive epithelial cells (Renault et al. 2008). Cd/Se/Zn QDs are accumulated in the copepod *Ceriodaphnia dubia* from *Pseudokirchneriella subcapitata*, acts as a protector for the nanoparticle coating, reducing nanoparticle toxicity and allowing an increasing of bioaccumulation and provoking higher toxicity as consequence of the breakdown of the QDs in the copepod (Bouldin et al. 2008). The exposure of *Ceriodaphnia dubia* to *Chlorella elipsoides* fed with Al_2O_3 NPs showed impairment of natural feeding and effect in the flow of energy (Pakrashi et al. 2014). Co_3O_4 and Mn_2O_3 NPs, although less studied that other types of NPs, show high toxic potential. The post-exposure feeding with *Daphnia magna* exposed NPs showed that metal body burden remained high and can provoke accumulation of metals in the food chain (Heinlaan et al. 2017). TiO_2 NPs have shown biomagnification (BMF > 1) in two levels of the food web (*Scenedesmus obliquus* and *D. magna*) through dietary exposure (Chen et al. 2015). TiO_2 NPs in two crystalline phases (rutile and anatase) was assessed in relation to the potential trophic transfer from *Chlorella* sp. to *Ceriodaphnia dubia*; individual NPs showed a significant biomagnification factor (BMF) under two conditions UV-A and visible, whereas the binary mixture of both kind of NPs showed a decrement which has been associated to the antagonistic effect of the mixture reducing Ti accumulation (Iswarya et al. 2018). The application of the biodynamic approach to study *D. magna* Ag assimilation from dietary exposure (AgNPs) or

Table 4. Selected references of trophic transfer of Me ENPs between algae and higher trophic levels.

ENP Type	Trophic Levels	Media	Exposure Time	References
Cd/Se/ZnSQDs	*Pseudokirchneriella subcapitata-Ceriodaphnia dubia*	FW	48–96 h	Boudin et al. 2008
Au (amine coated)	*Scenedesmus subspicatus-Corbicula fluminea*	FW	24 h–7 d	Renault et al. 2008
Au rods	*Marine mesocosm (food web)*	Estuarine	12 d	Ferry et al. 2009
ZnO	*Thalassiosira weissflogii-Acartia tonsa*	SW	7 d	Jarvis et al. 2013
CeO_2 (rods)	*Isochrysis galbana-Mytilus galloprovincialis*	SW	37 d	Conway et al. 2014
SnO_2, CeO_2, Fe_3O_4, SiO_2	*Cricosphaera elongata-Paracentrotus lividus*	SW	48 h–15 d	Gambardella et al. 2014
Au (citrate or PEG coating)	*Dunaliella salina-Mytilus galloprovincialis*	SW	24 h	Larguinho et al. 2014
Al_2O_3	*Chlorella elipsoides-Ceriodaphnia dubia*	FW	48–72 h	Pakrashi et al. 2014
TiO_2	*Scenedesmus obliquus-Daphnia magna*	FW	72 h–35 d	Chen et al. 2015
Ag (PVP, PEG, citrate coatings)	*Chlorella vulgaris-Daphnia magna*	FW	4–24 h	Kalman et al. 2015
Au	*Chlamydomonas reinhardtii* or *Euglena gracilis-Daphnia magna*	FW	24–48 h	Lee et al. 2015
CuO	*Chlorella vulgaris-Daphnia magna*	FW	14 d	Wu et al. 2017
nZVI	*Scenedesmus* sp.-*Ceriodaphnia dubia*	FW	24–72 h	Bhuvaneshwari et al. 2017
CoO_3, Mn_2O_3	*Chlamydomonas reinhardtii-Daphnia magna*	FW	48 h	Heinlan et al. 2017
CeO_2	*Mesocosm*	FW		Zhao et al. 2017
Ag	*R. subcapitata-Daphnia magna*	FW	48 h	Ribeiro et al. 2017
TiO_2	*Chlorella* sp.-*Ceriodaphnia dubia*	FW	48 h	Iswarya et al. 2018
Au	*Periphytic biofilm-Gammarus fossarum*	FW	7 d	Baudrimont et al. 2018
TiO_2 P25	*Dunaliella salina-Artemia salina*	SW	48 h	Bhuvaneshwari et al. 2018

aqueous source ($AgNO_3$) showed that food is the main route of accumulation and no biomagnification was observed (Kalman et al. 2015). Ribeiro et al. (2017) reported the accumulation of Ag from $AgNO_3$ and AgNPs by diverse routes: water, diet and simultaneous water and dietary exposure. The combined route provoked higher Ag concentrations in *D. magna* in comparison to $AgNO_3$ and the uptake from water explained the most of the increase of Ag bioaccumulation from NPs. *D. magna* exposure to CuO NPs from waterborne or by indirect way (feeding) showed differences in the toxicity being higher for direct exposure and the mechanisms involved in the detoxification are related to delivery conditions (Wu et al. 2017). The transfer of nZVI from chemical and biological origin between algae and daphnid did not show biomagnification (Bhuvaneshwari et al. 2017). In the complex freshwater system (water, sediment, water lettuce, water silk, Asian clams, snails, water fleas, Japanese medaka, and Yamato shrimp), CeO_2 NPs showed a preferential accumulation in sediment and negative correlation between trophic position and Ce accumulation and no biomagnification (Zhao et al. 2017).

The number of studies about the trophic transfer for seawater aquatic species are lower than freshwater. For AuNPs in a food wed with two levels, algae (*Dunaliella salina*) and mollusk (*Mytilus galloprovincialis*), gills and digestive gland showed variable Au content depending on the NP coating (Larguinho et al. 2014). P25 TiO_2 NPs are employed in many applications and the trophic transfer from microalgae to crustacean *Artemia salina* was analyzed by Bhuvaneshwari et al. (2018), who reported that waterborne exposure was the main route in comparison to dietary exposure and no biomagnification and trophic transfer from microalgae to *Artemia*. CeO_2 NPs are sorbed on phytoplankton in a short time and the exposure of *M. galloprovincialis* from direct or sorbed did not show significant differences in tissues or pseudofeces Ce concentration (Conway et al. 2014). For ZnO NPs, trophic transfer has been demonstrated in a marine plankton community (*Thalassiosira weissflogii-Acartia tonsa*) and the effects have been detected at realistic concentrations which can affect to ecosystem productivity (Jarvis et al. 2013). MeO-NPs (SiO_2, SnO_2, CeO_2, Fe_3O_4) trophic transfer and effect was analyzed in the food web marine microalgae—larvae sea urchin showing metal bioavailability and alteration in the development (Gambardella et al. 2014). In an estuarine mesocosm containing sea water, sediment, sea grass, microbes, biofilms, snails, clams, shrimp and fish, Au nanorods were introduced and they passed from the water column to the marine food web (Ferry et al. 2009).

Surface affinity has been pointed out as a strong predictor of uptake in a simple food web showing a good relationship between internalized NPs in *D. magna* and contaminated algae (*Chlorella vulgaris*) and it can be useful as screening tool (Geitner et al. 2016). The relation between trophic levels, community composition, ecosystem and evolutionary processes and the transfer across ecosystem boundaries should be considered to improve the knowledge about the impact of nanoparticles (Bundschuh et al. 2016).

Concluding Remarks

Microalgae are a key indicator in fresh and seawater ecosystems to know the impact of Me-NPs and MeO-NPs. The occurrence of these nanomaterials in aquatic ecosystems has increased as resulting from their wide use. Although, the knowledge about their effect on phytoplankton is restricted to specific types of NPs and lab experimental conditions. However, the intrinsic and extrinsic factors which can affect the toxicity and the mechanism involved in the toxic response and protection mechanisms have been elucidated partially. The standardization of toxicity tests is a key aspect that it should be taken in account because it can help to increase the reproducibility and to decrease the uncertainty of risk assessment of these nanomaterials and to predict their effects. Nevertheless, complex systems as mesocosms should be employed when the effect on the whole ecosystem want to be assessed because of their higher environmental relevance, including the analysis of trophic transfer of NPs along the food web. Finally, aspects such as global change can significantly affect both behaviour and toxicity of nanomaterials to phytoplankton species. Thus, changes in temperature, pH or carbonic system species can provoke the release of metal from NPs and modify the their behaviour in the environment affecting the structure or physiological process of phytoplankton species.

Acknowledgements

This work has been funded by the European Regional Funds (ERF) and by the regional government project of Andalusia PE2011-RNM-7812 and the Spanish Government Plan Nacional I+D+I CTM2016-75908-R.

References

Amde, M., J.-f. Liu, Z.-Q. Tan and D. Bekana. 2017. Transformation and bioavailability of metal oxide nanoparticles in aquatic and terrestrial environments. A review. Envrion. Pollut. 230: 250–267.
Andreani, T., V. Nogueira, V.V. Pinto, M.J. Ferreira, M.G. Rasteiro, A.M. Silva et al. 2017. Influence of the stabilizers on the toxicity of metallic nanomaterials in aquatic organisms and human cell lines. Sci. Total Environ. 607: 1264–1277.
Angel, B.M., G.E. Batley, C.V. Jarolimek and N.J. Rogers. 2013. The impact of size on the fate and toxicity of nanoparticulate silver in aquatic systems. Chemosphere 93: 359–365.
Apel, K. and H. Hirt. 2004. Reactive oxygen species: metabolism, oxidative stress, and signal transduction. Ann. Rev. Plant Biol. 55: 373–399.
Arfsten, D.P., D.J. Schaeffer and D.C. Mulveny. 1996. The effects of near ultraviolet radiation on the toxic effects of polycyclic aromatic hydrocarbons in animals and plants: a review. Ecotox. Environ. Safe. 33: 1–24.
Aruoja, V., H.-C. Dubourguier, K. Kasemets and A. Kahru. 2009. Toxicity of nanoparticles of CuO, ZnO and TiO_2 to microalgae Pseudokirchneriella subcapitata. Sci. Total Environ. 407: 1461–1468.
ASTM, 1975. Standard specification for substitute ocean water. Designation D 1141–75.
Auffan, M., J. Rose, J.-Y. Bottero, G.V. Lowry, J.-P. Jolivet and M.R. Wiesner. 2009. Towards a definition of inorganic nanoparticles from an environmental, health and safety perspective. Nat. Nanotechnol. 4: 634–641.
Babu, T.S., T.A. Akhtar, M.A. Lampi, S. Tripuranthakam, D.S. Dixon and B.M. Greenberg. 2003. Similar stress responses are elicited by copper and ultraviolet radiation in the aquatic plant Lemna gibba: Implication of reactive oxygen species as common signals. Plant Cell Physiol. 44: 1320–1329.
Baer, R. 2011. Surface characterization of nanoparticles: Critical needs and significant challenges. Surf. Anal. 17: 163–169.
Barwal, I., P. Ranjan, S. Kateriya and S.C. Yadav. 2011. Cellular oxido-reductive proteins of Chlamydomonas reinhardtii control the biosynthesis of silver nanoparticles. J. Nanobiotechnol. 9: 56.
Baudrimont, M., J. Andrei, S. Mornet, P. Gonzalez, N. Mesmer-Dudons, P.Y. Gourves et al. 2018. Trophic transfer and effects of gold nanoparticles (AuNPs) in Gammarus fossarum from contaminated periphytic biofilm. Environ. Sci. Pollut. Res. 25: 11181–1191.
Behra, R., L. Sigg, M.J.D. Clift, F. Herzog, M. Minghetti, B. Johnston et al. 2013. Bioavailability of silver nanoparticles and ions: from a chemical and biochemical perspective. J. Roy. Soc. Interface 10.
Bergami, E., S. Pugnalini, M. Vannuccini, L. Manfra, C. Faleri, F. Savorelli et al. 2017. Long-term toxicity of surface-charged polystyrene nanoplastics to marine planktonic species Dunaliella tertiolecta and Artemia franciscana. Aquat. Toxicol. 189: 159–169.
Bexiga, M.G., J.A. Varela, F. Wang, F. Fenaroli, A. Salvati, A. Lynch et al. 2011. Cationic nanoparticles induce caspase 3-, 7- and 9-mediated cytotoxicity in a human astrocytoma cell line. Nanotoxicology 5: 557–567.
Bhattacharya, P., S. Lin, J.P. Turner and P.C. Ke. 2010. Physical adsorption of charged plastic nanoparticles affects algal photosynthesis. J. Physi. Chem. C 114: 16556–16561.
Bhuvaneshwari, M., D. Kumar, R. Roy, S. Chakraborty, A. Parashar, A. Mukherjee et al. 2017. Toxicity, accumulation, and trophic transfer of chemically and biologically synthesized nano zero valent iron in a two species freshwater food chain. Aquat. Toxicol. 183: 63–75.
Bhuvaneshwari, M., V. Thiagarajan, P. Nemade, N. Chandrasekaran and A. Mukherjee. 2018. Toxicity and trophic transfer of P25 TiO_2 NPs from Dunaliella salina to Artemia salina: Effect of dietary and waterborne exposure. Environ. Res. 160: 39–46.
Blackford Jr, J.A., J.M. Antonini, V. Castranova and R.D. Dey. 1994. Intratracheal instillation of silica up-regulates inducible nitric oxide synthase gene expression and increases nitric oxide production in alveolar macrophages and neutrophils. Am. J. Respir Cell Mol. Biol. 11: 426–431.
Borm, P., F.C. Klaessig, T.D. Landry, B. Moudgil, J.r. Pauluhn, K. Thomas et al. 2006. Research strategies for safety evaluation of nanomaterials, part V: role of dissolution in biological fate and effects of nanoscale particles. Toxicol. Sci. 90: 23–32.
Bouldin, J.L., T.M. Ingle, A. Sengupta, R. Alexander, R.E. Hannigan and R.A. Buchanan. 2008. Aqueous toxicity and food chain transfer of Quantum Dots™ in freshwater algae and Ceriodaphnia dubia. Environ. Toxicol. Chem. 27: 1958–1963.
Braydich-Stolle, L.K., N.M. Schaeublin, R.C. Murdock, J. Jiang, P. Biswas, P., J.J. Schlager et al. 2009. Crystal structure mediates mode of cell death in TiO_2 nanotoxicity. J. Nanopart. Res. 11: 1361–1374.
Brayner, R., S.A. Dahoumane, C. Yéprémian, C. Djediat, M. Meyer, A. Couté et al. 2010. ZnO nanoparticles: Synthesis, characterization, and ecotoxicological studies. Langmuir 26: 6522–6528.
Brayner, R., S.A. Dahoumane, J.N.-L. Nguyen, C. Yéprémian, C. Djediat, A. Couté et al. 2011. Ecotoxicological studies of CdS nanoparticles on photosynthetic microorganisms. J. Nanosci. and Nanotechnol. 11: 1852–1858.
Brunner, T.J., P. Wick, P. Manser, P. Spohn, R.N. Grass, L.K. Limbach et al. 2006. *In vitro* cytotoxicity of oxide nanoparticles: comparison to asbestos, silica, and the effect of particle solubility. Environ. Sci. Technol. 40: 4374–4381.
Bundschuh, M., F. Seitz, R.R. Rosenfeldt and R. Schulz. 2016. Effects of nanoparticles in fresh waters: Risks, mechanisms and interactions. Freshw. Biol. 61: 2185–2196.
Caballero-Guzman, A. and B. Nowack. 2016. A critical review of engineered nanomaterial release data: Are current data useful for material flow modeling? Environ. Pollut. 213: 502–517.
Caldeira, K. and M.E. Wickett. 2005. Ocean model predictions of chemistry changes from carbon dioxide emissions to the atmosphere and ocean. J. Geophys. Res. 110.
Campbell, P.G., O. Errécalde, C. Fortin, V.P. Hiriart-Baer and B. Vigneault. 2002. Metal bioavailability to phytoplankton—applicability of the biotic ligand model. Comp. Biochem. Physiol. C 133: 189–206.

Campbell, P.G. and C. Fortin. 2013. Biotic ligand model. Encyclopedia of Aquatic Ecotoxicology. Springer, pp. 237–246.

Casado, M.P., A. Macken and H.J. Byrne. 2013. Ecotoxicological assessment of silica and polystyrene nanoparticles assessed by a multitrophic test battery. Environ. Int. 51: 97–105.

Casati, P. and V. Walbot. 2004. Crosslinking of ribosomal proteins to RNA in maize ribosomes by UV-B and its effects on translation. Plant Physiol. 136: 3319–3332.

Celardo, I., J.Z. Pedersen, E. Traversa and L. Ghibelli. 2011. Pharmacological potential of cerium oxide nanoparticles. Nanoscale 3: 1411–1420.

Cerrillo, C., G. Barandika, A. Igartua, O. Areitioaurtena and G. Mendoza. 2016. Towards the standardization of nanoecotoxicity testing: Natural organic matter 'camouflages' the adverse effects of TiO_2 and CeO_2 nanoparticles on green microalgae. Sci. Total Environ. 543: 95–104.

Cid, A., C. Herrero, E. Torres and J. Abalde. 1995. Copper toxicity on the marine microalga Phaeodactylum tricornutum: effects on photosynthesis and related parameters. Aquat. Toxicol. 31: 165–174.

Clément, L., C. Hurel and N. Marmier. 2013. Toxicity of TiO_2 nanoparticles to cladocerans, algae, rotifers and plants–effects of size and crystalline structure. Chemosphere 90: 1083–1090.

Coleman, J.G., Kennedy, A.J. and Harmon, A.R. 2015. Environmental Consequences of Nanotechnologies: Nanoparticles dispersion in aqueous media: SOP-T-1. ERDC/EL SR-15-2-Vicksburg MS: U.S. Army Engineer Research and Development Center.

Coleman, J.G., A.J. Kennedy, J.K. Stanley and L. Rabalais. 2017a. Environmental Consequences of Nanotechnologies: Sediment spiking methodologies for nanomaterial powders and aqueous suspensions for aquatic toxicity bioassay: SOP-T-3. ERDC/EL SR-17-X. Vicksburg MS: U.S. Army Engineer Research and Development Center.

Coleman, J.G., A. Butler, A. Harmon and A. Kennedy. 2017ab. Environmental Consequences of Nanotechnologies: Soil spiking methodologies for nanomaterial: SOP-T-4. ERDC/EL SR-17-X. Vicksburg MS: U.S. Army Engineer Research and Development Center.

Coll, C., D. Notter, F. Gottschalk, T. Sun, C. Som and B. Nowack. 2015. Probabilistic environmental risk assessment of five nanomaterials (nano-TiO_2, nano-Ag, nano-ZnO, CNT, and fullerenes). Nanotoxicology 1–9.

Commission, E. 2013. Europe 2020: A European strategy for smart, sustainable and inclusive growth, Brussels.

Cowan, A., P. Rose and L. Horne. 1992. Dunaliella salina: a model system for studying the response of plant cells to stress. J. Exp. Bot. 43: 1535–1547.

Cheloni, G., E. Marti and V.I. Slaveykova. 2016. Interactive effects of copper oxide nanoparticles and light to green alga Chlamydomonas reinhardtii. Aquat. Toxicol. 170: 120–128.

Chen, C.-S., J.M. Anaya, S. Zhang, J. Spurgin, C.-Y. Chuang, C. Xu et al. 2011. Effects of engineered nanoparticles on the assembly of exopolymeric substances from phytoplankton. PLoS One 6: e21865.

Chen, L., L. Zhou, Y. Liu, S. Deng, H. Wu and G. Wang. 2012. Toxicological effects of nanometer titanium dioxide (nano-TiO_2) on Chlamydomonas reinhardtii. Ecotox. Environ. Saf. 84: 155–162.

Chen, J., H. Li, X. Han and X. Wei. 2015. Transmission and Accumulation of Nano-TiO_2; in a 2-Step Food Chain (Scenedesmus obliquus to Daphnia magna). Bull. Environ. Contam. Toxicol. 95: 145–149.

Cherchi, C., T. Chernenko, M. Diem and A.Z. Gu. 2011. Impact of nano titanium dioxide exposure on cellular structure of Anabaena variabilis and evidence of internalization. Environ. Toxicol. Chem. 30: 861–869.

Chithrani, B.D., A.A. Ghazani and W.C. Chan. 2006. Determining the size and shape dependence of gold nanoparticle uptake into mammalian cells. Nano Letters 6: 662–668.

Chithrani, B.D. and W.C. Chan. 2007. Elucidating the mechanism of cellular uptake and removal of protein-coated gold nanoparticles of different sizes and shapes. Nano Lett. 7: 1542–1550.

Cho, E.-a., A.J. Bailer and J.T. Oris. 2003. Effect of methyl tert-butyl ether on the bioconcentration and photoinduced toxicity of fluoranthene in fathead minnow larvae (Pimephales promelas). Environ. Sci. Technol. 37: 1306–1310.

Conway, J.R., S.K. Hanna, H.S. Lenihan and A.A. Keller. 2014. Effects and implications of trophic transfer and accumulation of CeO_2 nanoparticles in a marine mussel. Environ. Sci. Technol. 48: 1517–1524.

da Costa, C.H., F. Perreault, A. Oukarroum, S.P. Melegari, R. Popovic and W.G. Matias. 2016. Effect of chromium oxide (III) nanoparticles on the production of reactive oxygen species and photosystem II activity in the green alga Chlamydomonas reinhardtii. Sci. Total Environ. 565: 951–960.

Dahoumane, S.A., C. Djediat, C. Yéprémian, A. Couté, F. Fiévet, T. Coradin et al. 2012. Species selection for the design of gold nanobioreactor by photosynthetic organisms. J. Nanopart. Res. 14: 883.

Dalai, S., S. Pakrashi, M. Joyce Nirmala, A. Chaudhri, N. Chandrasekaran, A.B. Mandal et al. 2013. Cytotoxicity of TiO_2 nanoparticles and their detoxification in a freshwater system. Aquat. Toxicol. 138-139: 1–11.

Dao, L.H. and J. Beardall. 2016. Effects of lead on growth, photosynthetic characteristics and production of reactive oxygen species of two freshwater green algae. Chemosphere 147: 420–429.

Debnath, S. and U.C. Ghosh. 2011. Equilibrium modeling of single and binary adsorption of Cd (II) and Cu (II) onto agglomerated nano structured titanium (IV) oxide. Desalination 273: 330–342.

Deng, X.-Y., J. Cheng, X.-L. Hu, L. Wang, D. Li and K. Gao. 2017a. Biological effects of TiO_2 and CeO_2 nanoparticles on the growth, photosynthetic activity, and cellular components of a marine diatom Phaeodactylum tricornutum. Sci. Total Environ. 575: 87–96.

Deshpande, S., S. Patil, S.V. Kuchibhatla and S. Seal. 2005. Size dependency variation in lattice parameter and valency states in nanocrystalline cerium oxide. Appl. Phy. Lett. 87: 133113–133113.

Di Toro, D.M., H.E. Allen, H.L. Bergman, J.S. Meyer, P.R. Paquin and R.C. Santore. 2001. Biotic ligand model of the acute toxicity of metals. 1. Technical basis. Environ. Toxicol. Chem. 20: 2383–2396.

Diamond, S.A. 2003. Photoactivated toxicity in aquatic environments. UV Effects in Aquatic Organisms and Ecosystems. New York: Johm Wiley & Sons, Inc 219–250.

Diamond, S.A., D.R. Mount, V.R. Mattson, L.J. Heinis, T.L. Highland, A.D. Adams et al. 2006. Photoactivated polycyclic aromatic hydrocarbon toxicity in medaka (Oryzias latipes) embryos: relevance to environmental risk in contaminated sites. Environ. Toxicol. Chem. 25: 3015–3023.

Dick, C.A., D.M. Brown, K. Donaldson and V. Stone. 2003. The role of free radicals in the toxic and inflammatory effects of four different ultrafine particle types. Inhal. Toxicol. 15: 39–52.

Durán, N., M. Durán, M.B. de Jesus, A.B. Seabra, W.J. Fávaro and G. Nakazato. 2016. Silver nanoparticles: A new view on mechanistic aspects on antimicrobial activity. Nanomedicine: NBM 12: 789–799.

El Badawy, A.M., R.G. Silva, B. Morris, K.G. Scheckel, M.T. Suidan and T.M. Tolaymat. 2011. Surface charge-dependent toxicity of silver nanoparticles. Environ. Sci. Technol. 45: 283–287.

Fabrega, J., S.R. Fawcett, J.C. Renshaw and J.R. Lead. 2009. Silver nanoparticle impact on bacterial growth: effect of pH, concentration, and organic matter. Environ. Sci. Technol. 43: 7285–7290.

Fabrega, J., S.N. Luoma, C.R. Tyler, T.S. Galloway and J.R. Lead. 2011. Silver nanoparticles: behaviour and effects in the aquatic environment. Environ. Int. 37: 517–531.

Falkowski, P.G. 1980. Primary productivity in the sea. Environmental Science Research, 19. Falkowski Ed. Plenum press, New York and London.

Ferry, J.L., P. Craig, C. Hexel, P. Sisco, R. Frey, P.L. Pennington et al. 2009. Transfer of gold nanoparticles from the water column to the estuarine food web. Nat. Nanotechnol. 4: 441.

Forest, V., L. Leclerc, J.-F. Hochepied, A. Trouvé, G. Sarry and J. Pourchez. 2017. Impact of cerium oxide nanoparticles shape on their *in vitro* cellular toxicity. Toxicol. *In Vitro* 38: 136–141.

Fu, P.P., Q. Xia, H.-M. Hwang, P.C. Ray and H. Yu. 2014. Mechanisms of nanotoxicity: Generation of reactive oxygen species. J. Food Drug Anal. 22: 64–75.

Gambardella, C., L. Gallus, A.M. Gatti, M. Faimali, S. Carbone, L.V. Antisari et al. 2014. Toxicity and transfer of metal oxide nanoparticles from microalgae to sea urchin larvae. Chem. Ecol. 30(4).

Gao, J., G. Xu, H. Qian, P. Liu, P. Zhao and Y. Hu. 2013. Effects of nano-TiO_2 on photosynthetic characteristics of Ulmus elongata seedlings. Environ. Pollut. 176: 63–70.

Geitner, N.K., S.M. Marinakos, C. Guo, M. O´Brian and M.K. Wiesner. 2016. Nanoparticle surface affinity as predictor of trophic transfer. Environ. Sci. Technol. 50: 6663–6669.

Gondikas, A.P., A. Morris, B.C. Reinsch, S.M. Marinakos, G.V. Lowry and H. Hsu-Kim. 2012. Cysteine-induced modifications of zero-valent silver nanomaterials: implications for particle surface chemistry, aggregation, dissolution, and silver speciation. Environ. Sci. Technol. 46: 7037–7045.

Gong, N., K. Shao, W. Feng, Z. Lin, C. Liang and Y. Sun. 2011. Biotoxicity of nickel oxide nanoparticles and bio-remediation by microalgae Chlorella vulgaris. Chemosphere 83: 510–516.

González-García, C., J. Forja, M.C. González-Cabrera, M.P. Jiménez and L.M. Lubián. 2018. Annual variations of total and fractioned chlorophyll and phytoplankton groups in the Gulf of Cádiz. Sci. Total Environ. 613-614: 1551–1565.

Grillo, R., A.H. Rosa and L.F. Fraceto. 2015. Engineered nanoparticles and organic matter: a review of the state-of-the-art. Chemosphere 119: 608–619.

Grima, E.M., E.-H. Belarbi, F.A. Fernández, A.R. Medina and Y. Chisti. 2003. Recovery of microalgal biomass and metabolites: Process options and economics. Biotech. Adv. 20: 491–515.

Guillard, R.R.L. and J.H. Ryther. 1962. Studies of marine planktonic diatoms: I. Cyclotella nana Hustedt, and Detonula confervacea (cleve) Gran. Can. J. Microbiol. 8: 229–239.

Gunawan, C., A. Sirimanoonphan, W.Y. Teoh, C.P. Marquis and R. Amal. 2013. Submicron and nano formulations of titanium dioxide and zinc oxide stimulate unique cellular toxicological responses in the green microalga Chlamydomonas reinhardtii. J. Hazard. Mat. 260: 984–992.

Häder, D.-P. and R.P. Sinha. 2005. Solar ultraviolet radiation-induced DNA damage in aquatic organisms: potential environmental impact. Muta. Res 571: 221–233.

Hadjoudja, S., V. Deluchat and M. Baudu. 2010. Cell surface characterisation of Microcystis aeruginosa and Chlorella vulgaris. J. Colloid Interface Sci. 342: 293–299.

Handy, R., R. Owen and E. Valsami-Jones. 2008. The ecotoxicology of nanoparticles and nanomaterials: current status, knowledge gaps, challenges, and future needs. Ecotoxicol. 17: 315–325.

Hartmann, N.B. and Baun, A. 2010. The nano cocktail: ecotoxicological effects of engineered nanoparticles in chemical mixtures. Integr. Environ. Assess. Manag. 6: 311.

Hauck, T.S., A.A. Ghazani and W.C. Chan. 2008. Assessing the effect of surface chemistry on gold nanorod uptake, toxicity, and gene expression in mammalian cells. Small 4: 153–159.

Hazani, A.A., M.M. Ibrahim, A.I. Shehata, G.A. El-Gaaly, M. Daoud, M. Fouad et al. 2013. Ecotoxicity of Ag-nanoparticles on two microalgae, Chlorella vulgaris and Dunaliella tertiolecta. Arch. Biol. Sci. 65: 1447–1457.

He, D., J.J. Dorantes-Aranda and T.D. Waite. 2012. Silver nanoparticle—algae interactions: oxidative dissolution, reactive oxygen species generation and synergistic toxic effects. Environ. Sci. Technol. 46: 8731–8738.

Hendren, C.O., X. Mesnard, J. Dröge and M.R. Wiesner. 2011. Estimating production data for five engineered nanomaterials as a basis for exposure assessment. Environ. Sci. Technol. 45(9): 4190–4190.

Hirakawa, K., M. Mori, M. Yoshida, S. Oikawa and S. Kawanishi. 2004. Photo-irradiated titanium dioxide catalyzes site specific DNA damage via generation of hydrogen peroxide. Free Radical Res. 38: 439–447.
Hiriart-Baer, V.P., C. Fortin, D.-Y. Lee and P.G. Campbell. 2006. Toxicity of silver to two freshwater algae, Chlamydomonas reinhardtii and Pseudokirchneriella subcapitata, grown under continuous culture conditions: influence of thiosulphate. Aquat. Toxicol. 78: 136–148.
Hoecke, K.V., J.T. Quik, J. Mankiewicz-Boczek, K.A.D. Schamphelaere, A. Elsaesser, P.V.d. Meeren et al. 2009. Fate and effects of CeO_2 nanoparticles in aquatic ecotoxicity tests. Environ. Sci. Technol. 43: 4537–4546.
Hou, W.-C., P. Westerhoff and J.D. Posner. 2013. Biological accumulation of engineered nanomaterials: a review of current knowledge. Environ. Sci.: Process. Impacts 15: 103–122.
Hristozov, D., A. Zabeo, K. Alstrup Jensen, S. Gottardo, P. Isigonis, L. Maccalman et al. 2016. Demonstration of a modelling-based multi-criteria decision analysis procedure for prioritisation of occupational risks from manufactured nanomaterials. Nanotoxicology 10: 1215–1228.
Hu, H., Y. Deng, Y. Fan, P. Zhang, H. Sun, Z. Gan et al. 2016. Effects of artificial sweeteners on metal bioconcentration and toxicity on a green algae Scenedesmus obliquus. Chemosphere 150: 285–293.
Huang, C.-P. 2005. Short-Term Chronic Toxicity of Photocatalytic Nanoparticles to Bacteria, Algae, and Zooplankton. Nanotechnology and the Environment: Applications and Implications Progress Review Workshop III, p. 89.
Hund-Rinke, K. and M. Simon. 2006. Ecotoxic effect of photocatalytic active nanoparticles (TiO_2) on algae and Daphnids (8 pp). Environ. Sci. Pollut. Res. 13: 225–232.
Huang, J., J. Cheng and J. Yi. 2016. Impact of silver nanoparticles on marine diatom Skeletonema costatum. J. Appl. Toxicol. 36: 1343–1354.
Hund-Rinke, K., A. Baun, D. Cupi, T.F. Fernandes, R. Handy, J.H. Kinross et al. 2016. Regulatory ecotoxicity testing of nanomaterials–proposed modifications of OECD test guidelines based on laboratory experience with silver and titanium dioxide nanoparticles. Nanotoxicology 10: 1442–1447.
British Standard Institution. 2007. Natural stone test methods: determination of real density and apparent density, and of total and open porosity. British Standards Institution.
Iswarya, V., M. Bhuvaneshwari, S.A. Alex, S. Iyer, G. Chaudhuri, P.T. Chandrasekaran et al. 2015. Combined toxicity of two crystalline phases (anatase and rutile) of titania nanoparticles towards freshwater microalgae: *Chlorella* sp. Aquat. Toxicol. 161: 154–169.
Iswarya, V., M. Bhuvaneshwari, N. Chandrasekaran and A. Mukherjee. 2018. Trophic transfer potential of two different crystalline phases of TiO_2 NPs from *Chlorella* sp. to Ceriodaphnia dubia. Aquat. Toxicol. 197: 89–97.
Ivask, A., I. Kurvet, K. Kasemets, I. Blinova, V. Aruoja, S. Suppi et al. 2014. Size-dependent toxicity of silver nanoparticles to bacteria, yeast, algae, crustaceans and mammalian cells *in vitro*. PloS One 9: e102108.
Heinlaan, M., M. Muna, K. Juganson, O. Oriekhova, S. Stoll, A. Kahru et al. 2017. Exposure to sublethal concentrations of Co_3O_4 and Mn_2O_3 nanoparticles induced elevated metal body burden in Daphnia magna. Aquatic Toxicology 189: 123–133.
Jarvis, T.A., R.J. Miller, H.S. Lenihan and G.K. Bielmyer. 2013. Toxicity of ZnO nanoparticles to the copepod Acartia tonsa, exposed through a phytoplankton diet. Environ. Toxicol. Chem. 32: 1264–1269.
Jena, J., N. Pradhan, R.R. Nayak, B.P. Dash, L.B. Sukla, P.K. Panda et al. 2014. *Microalga Scenedesmus* sp.: a potential low-cost green machine for silver nanoparticle synthesis. J. Microbiol. Biotechnol. 24: 522–533.
Ji, J., Z. Long and D. Lin. 2011. Toxicity of oxide nanoparticles to the green algae *Chlorella* sp. Chem. Eng. J. 170: 525–530.
Jiang, J., G. Oberdörster and P. Biswas. 2009. Characterization of size, surface charge, and agglomeration state of nanoparticle dispersions for toxicological studies. J. Nanoparticle Res. 11: 77–89.
Jin, C., Y. Tang, F.G. Yang, X.L. Li, S. Xu, X.Y. Fan et al. 2011. Cellular toxicity of TiO_2 nanoparticles in anatase and rutile crystal phase. Biol. Trace Elem. Res. 141: 3–15.
Kalabegishvili, T., E. Kirkesali, I. Murusidze, G. Tsertsvadze, M. Frontasyeva, I. Zinicovscaia et al. 2011. Characterization of microbial synthesis of silver and gold nanoparticles with electron microscopy techniques. J. Adv. Microsc. Rese. 6: 313–317.
Kalabegishvili, T., A. Faanhof, E. Kirkesali, M. Frontasyeva, S. Pavlov and I. Zinicovscaia. 2012. Synthesis of gold nanoparticles by blue-green algae Spirulina platensis. Proceedings of the international conference nanomaterials: Applications and properties. Sumy State University Publishing, pp. 02NNBM09-02NNBM09.
Kalman, J., K.B. Paul, F.R. Khan, V. Stone and T. Fernandes. 2015. Characterisation of bioaccumulation dynamics of three differently coated silver nanoparticles and aqueous silver in a simple freshwater food chain. Environ. Chem. 12: 662–672.
Keller, A.A., S. McFerran, A. Lazareva and S. Suh. 2013. Global life cycle releases of engineered nanomaterials. J. Nanopart. Res. 15: 1692.
Kennedy, A.J., J.G. Coleman, S.A. Diamond, N.L. Melby, A.J. Bednar, A. Harmon et al. 2017. Assessing nanomaterial exposures in aquatic ecotoxicological testing: Framework and case studies based on dispersion and dissolution. Nanotoxicology 11: 546–557.
Kent, R.D. and P.J. Vikesland. 2012. Controlled evaluation of silver nanoparticle dissolution using atomic force microscopy. Environ. Sci. & Technol. 46: 6977–6984.
Klaine, S.J., P.J. Alvarez, G.E. Batley, T.F. Fernandes, R.D. Handy et al. 2008. Nanomaterials in the environment: behavior, fate, bioavailability, and effects. Environ. Toxicol. Chem. 27: 1825–1851.
Kumar, A., P. Kumar, A. Anandan, T.F. Fernandes, G.A. Ayoko and G. Bisko. 2014. Engineered Nanomaterials: Knowledge Gaps in Fate, Exposure, Toxicity, and Future Directions J. Nanomat.2014: 1–16.

Larguinho, M., D. Correia, M.S. Diniz and P.V. Baptista. 2014. Evidence of one-way flow bioaccumulation of gold nanoparticles across two trophic levels. J. Nanopart. Res. 16: 2549.
Lee, S.-W., S.-M. Kim and J. Choi. 2009. Genotoxicity and ecotoxicity assays using the freshwater crustacean Daphnia magna and the larva of the aquatic midge Chironomus riparius to screen the ecological risks of nanoparticle exposure. Environ. Toxicol. Pharm. 28: 86–91.
Lee, W.-M. and Y.-J. An. 2013. Effects of zinc oxide and titanium dioxide nanoparticles on green algae under visible, UVA, and UVB irradiations: No evidence of enhanced algal toxicity under UV pre-irradiation. Chemosphere 91: 536–544.
Lee, W.M., S.J. Yoon, Y.J. Shin and Y.J. An. 2015. Trophic transfer of gold nanoparticles from Euglena gracilis or Chlamydomonas reinhardtii to Daphnia magna. Environ. Pollut. 201: 10–16.
Leung, Y.H., M.M. Yung, A.M. Ng, A.P. Ma, S.W. Wong, C.M. Chan et al. 2015. Toxicity of CeO_2 nanoparticles—the effect of nanoparticle properties. J. Photochem. and Photobiol. B 145: 48–59.
Levard, C., E.M. Hotze, G.V. Lowry and G.E. Brown Jr. 2012. Environmental transformations of silver nanoparticles: Impact on stability and toxicity. Environ. Sci. Technol. 46: 6900–6914.
Li, F., Z. Liang, X. Zheng, W. Zhao, M. Wu and Z. Wang. 2015. Toxicity of nano-TiO_2 on algae and the site of reactive oxygen species production. Aquat. Toxicol. 158: 1–13.
Li, J., S. Schiavo, G. Rametta, M.L. Miglietta, V. La Ferrara, C. Wu et al. 2017. Comparative toxicity of nano ZnO and bulk ZnO towards marine algae Tetraselmis suecica and Phaeodactylum tricornutum. Environ. Sci. Pollut. Res. 24: 6543–6553
Li, K., Y. Chen, W. Zhang, Z. Pu, L. Jiang and Y. Chen. 2012. Surface interactions affect the toxicity of engineered metal oxide nanoparticles toward Paramecium. Chem. Res. Toxicol. 25: 1675–1681.
Li, S., L.K. Wallis, H. Ma and S.A. Diamond. 2014. Phototoxicity of TiO_2 nanoparticles to a freshwater benthic amphipod: Are benthic systems at risk? Sci. Total Environ. 466-467: 800–808.
Lin, J. and A. Alexander-Katz. 2013. Cell membranes open "doors" for cationic nanoparticles/biomolecules: Insights into uptake kinetics. ACS Nano 7: 10799–10808.
Lindgren, A.L. 2014. The effects of silver nitrate and silver nanoparticles on Chlamydomonas reinhardtii: A proteomic approach. Degree Project, Department of Biology and Environmental Sciences, University of Gothenburg, Germany.
Liu, W., Y. Ming, Z. Huang and P. Li. 2012. Impacts of florfenicol on marine diatom Skeletonema costatum through photosynthesis inhibition and oxidative damages. Plant Physiol. Biochem. 60: 165–170.
Luangpipat, T., I.R. Beattie, Y. Chisti and R.G. Haverkamp. 2011. Gold nanoparticles produced in a microalga. J. Nanopart. Res. 13: 6439–6445.
Luna-Velasco, A., J.A. Field, A. Cobo-Curiel and R. Sierra-Alvarez. 2011. Inorganic nanoparticles enhance the production of reactive oxygen species (ROS) during the autoxidation of l-3, 4-dihydroxyphenylalanine (l-dopa). Chemosphere 85: 19–25.
Luoma, S.N. and P.S. Rainbow. 2008. Metal Contamination in Aquatic Environments, Science and Lateral Mangement. Cambridge University Press, Cambridge.
Lynch, I. and K.A. Dawson. 2008. Protein-nanoparticle interactions. Nano Today 3: 40–47.
Lynch, I., K.A. Dawson, J.R. Lead and E. Valsami-Jones. 2014. Macromolecular coronas and their importance in nanotoxicology and nanoecotoxicology. Frontiers of Nanoscience. Elsevier, pp. 127–156.
Ma, R., C. Levard, S.M. Marinakos, Y. Cheng, J. Liu, F.M. Michel et al. 2011. Size-controlled dissolution of organic-coated silver nanoparticles. Environ. Sci. Technol. 46: 752–759.
Ma, S., K. Zhou, K. Yang and D. Lin. 2015. Heteroagglomeration of oxide nanoparticles with algal cells: Effects of particle type, ionic strength and pH. Environ. Sci. Technol. 49: 932–939.
Maeda, S. and T. Sakaguchi. 1990. Accumulation and detoxification of toxic metal elements by algae. Introduction to Applied Phycology 109–136.
Mayfield, S.P., C.B. Yohn, A. Cohen and A. Danon. 1995. Regulation of chloroplast gene expression. Annual Review of Plant Physiol. Plant Mol. Biol. 46: 147–166.
Manier, N., A. Bado-Nilles, P. Delalain, O. Aguerre-Chariol and P. Pandard. 2013. Ecotoxicity of non-aged and aged CeO_2 nanomaterials towards freshwater microalgae. Environ. Pollut. 180: 63–70.
Manier, N., M. Garaud, P. Delalain, O. Aguerre-Chariol and P. Pandard. 2011. Behaviour of ceria nanoparticles in standardized test media–influence on the results of ecotoxicological tests. Journal of Physics: Conference Series. IOP Publishing, p. 012058.
Manke, A., L. Wang and Y. Rojanasakul. 2013. Mechanisms of nanoparticle-induced oxidative stress and toxicity. BioMed Res. Int. Article ID 942916, 15 pages http://dx.doi.org/10.1155/2013/942916.
Merin, D.D., S. Prakash and B.V. Bhimba. 2010. Antibacterial screening of silver nanoparticles synthesized by marine micro algae. Asian Pac. J. Trop. Med. 3: 797–799.
Miao, A.-J., K.A. Schwehr, C. Xu, S.-J. Zhang, Z. Luo, A. Quigg et al. 2009. The algal toxicity of silver engineered nanoparticles and detoxification by exopolymeric substances. Environ. Pollut. 157: 3034–3041.
Miao, A.J., X.Y. Zhang, Z. Luo, C.S. Chen, W.C. Chin, P.H. Santschi et al. 2010. Zinc oxide–engineered nanoparticles: Dissolution and toxicity to marine phytoplankton. Environ. Toxicol. Chem. 29: 2814–2822.
Miglietta, M., G. Rametta, G. Di Francia, S. Manzo, A. Rocco, R. Carotenuto et al. 2011. Characterization of nanoparticles in seawater for toxicity assessment towards aquatic organisms. Sensors and microsystems. Springer, pp. 425–429.
Miller, R.J., S. Bennett, A.A. Keller, S. Pease and H.S. Lenihan. 2012. TiO_2 nanoparticles are phototoxic to marine phytoplankton. PloS One 7: e30321.
Miller, R.J., E.B. Muller, B. Cole, T. Martin, R. Nisbet, G.K. Bielmyer-Fraser et al. 2017. Photosynthetic efficiency predicts toxic effects of metal nanomaterials in phytoplankton. Aquatic Toxicol. 183: 85–93.

Minetto, D., G. Libralato, A. Marcomini and A.V. Ghirardini. 2017. Potential effects of TiO_2 nanoparticles and $TiCl_4$ in saltwater to Phaeodactylum tricornutum and Artemia franciscana. Sci. Total Environ. 579: 1379–1386.

Mishra, A., K. Kavita and B. Jha. 2011. Characterization of extracellular polymeric substances produced by micro-algae Dunaliella salina. Carbohydr. Polym. 83: 852–857.

Misra, S.K., A. Dybowska, D. Berhanu, S.N. Luoma and E. Valsami-Jones. 2012. The complexity of nanoparticle dissolution and its importance in nanotoxicological studies. Sci. Total Environ. 438: 225–232.

Monopoli, M.P., C. Åberg, A. Salvati and K.A. Dawson. 2012. Biomolecular coronas provide the biological identity of nanosized materials. Nat. Nanotechnol. 7: 779–786.

Montiel-Dávalos, A., J.L. Ventura-Gallegos, E. Alfaro-Moreno, E. Soria-Castro, E. García-Latorre et al. 2012. TiO_2 Nanoparticles Induce Dysfunction and Activation of Human Endothelial Cells. Chem. Res. Toxicol. 25: 920–930.

Moore, M.N. 2006. Do nanoparticles present ecotoxicological risks for the health of the aquatic environment? Environ. Int. 32: 967–976.

Morelli, E., E. Salvadori, R. Bizzarri, P. Cioni and E. Gabellieri. 2013. Interaction of CdSe/ZnS quantum dots with the marine diatom Phaeodactylum tricornutum and the green alga Dunaliella tertiolecta: a biophysical approach. Biophys.Chem. 182: 4–10.

Morelli, E., E. Gabellieri, A. Bonomini, D. Tognotti, G. Grassi and I. Corsi. 2018. TiO_2 nanoparticles in seawater: Aggregation and interactions with the green alga Dunaliella tertiolecta. Ecotoxicol. Environ. Saf. 148: 184–193.

Moreno-Garrido, I., L.M. Lubián and A.M.V.M. Soares. 1999. Growth differences in cultured populations of *Brachionusplicatilis* Müller caused by heavy metal stress as function of microalgal diet. Bull. Environ. Contam. Toxicol. 63: 392–398.

Moreno-Garrido, I., S. Pérez and J. Blasco. 2015. Toxicity of silver and gold nanoparticles on marine microalgae. Mar. Environ. Res. 111: 60–73.

Mortazavi, S.M., M. Khatami, I. Sharifi, H. Heli, K. Kaykavousi, M.H.S. Poor et al. 2017. Bacterial biosynthesis of gold nanoparticles using *Salmonella enterica* subsp. enterica serovar Typhi isolated from blood and stool specimens of patients. J. of Clust Science 28: 2997–3007.

Nangia, S. and R. Sureshkumar. 2012. Effects of nanoparticle charge and shape anisotropy on translocation through cell membranes. Langmuir 28: 17666–17671.

Nassiri, Y., J. Mansot, J. Wéry, T. Ginsburger-Vogel and J. Amiard. 1997. Ultrastructural and electron energy loss spectroscopy studies of sequestration mechanisms of Cd and Cu in the marine diatom Skeletonema costatum. Arch. Environ. Contam. Toxicol. 33: 147–155.

Navarro, E., A. Baun, R. Behra, N.B. Hartmann, J. Filser, A.-J. Miao et al. 2008a. Environmental behavior and ecotoxicity of engineered nanoparticles to algae, plants, and fungi. Ecotoxicol. 17: 372–386.

Navarro, E., F. Piccapietra, B. Wagner, F. Marconi, R. Kaegi, N. Odzak et al. 2008b. Toxicity of silver nanoparticles to Chlamydomonas reinhardtii. Environ. Sci. Technol. 42: 8959–8964.

Nel, A., T. Xia, L. Mädler and N. Li. 2006. Toxic potential of materials at the nanolevel. Science 311: 622–627.

Nel, A.E., L. Mädler, D. Velegol, T. Xia, E.M. Hoek, P. Somasundaran et al. 2009. Understanding biophysicochemical interactions at the nano–bio interface. Nat. Mater. 8: 543–557.

Nichols, W.T., G. Malyavanatham, D.E. Henneke, D.T. O'brien, M.F. Becker and J.W. Keto. 2002. Bimodal nanoparticle size distributions produced by laser ablation of microparticles in aerosols. J. Nanopart. Res. 4: 423–432.

Nolte, T.M., N.B. Hartmann, J.M. Kleijn, J. Garnæs, D. van de Meent and A.J. Hendriks. 2017. The toxicity of plastic nanoparticles to green algae as influenced by surface modification, medium hardness and cellular adsorption. Aquat. Toxicol. 183: 11–20.

Notter, D.A., D.M. Mitrano and B. Nowack. 2014. Are nanosized or dissolved metals more toxic in the environment? A meta-analysis. Environ. Toxicol. Chem. 33: 2733–2739.

Oberdörster, G., E. Oberdörster and J. Oberdörster. 2005. Nanotoxicology: An emerging discipline evolving from studies of ultrafine particles. Environ. Health Perspect. 823–839.

Obermüller, B., U. Karsten and D. Abele. 2005. Response of oxidative stress parameters and sunscreening compounds in Arctic amphipods during experimental exposure to maximal natural UVB radiation. J. Exp. Mar. Biol. Ecol. 323: 100–117.

Olmedo, D.G., D.R. Tasat, M.B. Guglielmotti and R.L. Cabrini. 2005. Effect of titanium dioxide on the oxidative metabolism of alveolar macrophages: an experimental study in rats. J. Biomed. Mater. Res. A 73: 142–149.

Olson, M.H., M.R. Colip, J.S. Gerlach and D.L. Mitchell. 2006. Quantifying ultraviolet radiation mortality risk in bluegill larvae: effects of nest location. Ecological Appl. 16: 328–338.

Organization of Economic Cooperation and Development (OECD). 2012. TestNo. 211: Daphnia magna reproduction test. Paris. France.

Oukarroum, A., S. Bras, F. Perreault and R. Popovic. 2012. Inhibitory effects of silver nanoparticles in two green algae, Chlorella vulgaris and Dunaliella tertiolecta. Ecotox. Environ. Saf. 78: 80–85.

Oukarroum, A., W. Zaidi, M. Samadani and D. Dewez. 2017. Toxicity of nickel oxide nanoparticles on a freshwater green algal strain of chlorella vulgaris. BioMed Res. Int. Article ID 9528180.

Pakrashi, S., S. Dalai, N. Chandrasekaran and A. Mukherjee. 2014. Trophic transfer potential of aluminium oxide nanoparticles using representative primary producer (*Chlorella ellipsoides*) and a primary consumer (*Ceriodaphnia dubia*). Aquat. Toxicol. 152: 74–81.

Peng, X., S. Palma, N.S. Fisher and S.S. Wong. 2011. Effect of morphology of ZnO nanostructures on their toxicity to marine algae. Aquat. Toxicol.102: 186–196.

Petersen, E.J., S.A. Diamond, A.J. Kennedy, G.G. Goss, K. Ho, J. Lead et al. 2015. Adapting OECD aquatic toxicity tests for use with manufactured nanomaterials: Key issues and consensus recommendations. Environmental Science & Technol. 49: 9532–9547.

Piccapietra, F., C.G. Allué, L. Sigg and R. Behra. 2012. Intracellular silver accumulation in Chlamydomonas reinhardtii upon exposure to carbonate coated silver nanoparticles and silver nitrate. Environ. Sci. Technol. 46: 7390–7397.

Piccinno, F., F. Gottschalk, S. Seeger and B. Nowack. 2012. Industrial production quantities and uses of ten engineered nanomaterials in Europe and the world. J. Nanopart. Res. 14: 1–11.

Pletikapić, G., V. Žutić, I. Vinković Vrček and V. Svetličić. 2012. Atomic force microscopy characterization of silver nanoparticles interactions with marine diatom cells and extracellular polymeric substance. J. Mol. Recognit. 25: 309–317.

Popper, Z.A., G. Michel, C. Hervé, D.S. Domozych, W.G. Willats, M.G. Tuohy et al. 2011. Evolution and diversity of plant cell walls: from algae to flowering plants. Ann. Review Plant Biol. 62: 567–590.

Pulido-Reyes, G., I. Rodea-Palomares, S. Das, T.S. Sakthivel, F. Leganes, R. Rosal et al. 2015. Untangling the biological effects of cerium oxide nanoparticles: The role of surface valence states. Sci. Rep. 5.

Pulido-Reyes, G., F. Leganes, F. Fernández-Piñas and R. Rosal. 2017. Bio-nano interface and environment: A critical review. Environ. Toxicol. Chem. 36: 3181–3193.

Quigg, A., W.-C. Chin, C.-S. Chen, S. Zhang, Y. Jiang, A.-J. Miao et al. 2013. Direct and indirect toxic effects of engineered nanoparticles on algae: Role of natural Organic Matter. ACS Sust. Chem. Eng. 1: 686–702.

Quik, J.T., I. Lynch, K. Van Hoecke, C.J. Miermans, K.A. De Schamphelaere, C.R. Janssen et al. 2010. Effect of natural organic matter on cerium dioxide nanoparticles settling in model fresh water. Chemosphere 81: 711–715.

Razack, S.A., S. Duraiarasan and V. Mani. 2016. Biosynthesis of silver nanoparticle and its application in cell wall disruption to release carbohydrate and lipid from *C. vulgaris* for biofuel production. Biotechnol. Rep. 11: 70–76.

Renault, S., M. Baudrimont, N.M. Dudons, P. Gonzalez, S. Mornet and A. Brisson. 2008. Impacts of gold nanoparticle exposure on two freshwater species: a phytoplanktonic alga (*Scenedesmus subspicatus*) and a benthic bivalve (*Corbicula fluminea*). Gold Bull. 41: 116–126.

Ribeiro, F., C.A.M. Gestel, M.D. Pavlaki, S. Azevedo, A.M.V.M. Soares and S. Loureiro. 2017. Bioaccumulation of silver in Daphnia magna: Waterborne and dietary exposure to nanoparticles and dissolved silver. Sci. Total Environ. 574: 1633–1639.

Roberts, A.P., M.M. Alloy and J.T. Oris. 2017. Review of the photo-induced toxicity of environmental contaminants. Comp. Biochem. Physiol. 191: 160–167.

Rodea-Palomares, I., K. Boltes, F. Fernández-Piñas, F. Leganés, E. García-Calvo, J. Santiago et al. 2011. Physicochemical characterization and ecotoxicological assessment of CeO_2 nanoparticles using two aquatic microorganisms. Toxicol. Sci. 119: 135–145.

Rodea-Palomares, I., S. Gonzalo, J. Santiago-Morales, F. Leganés, E. García-Calvo, R. Rosal et al. 2012. An insight into the mechanisms of nanoceria toxicity in aquatic photosynthetic organisms. Aquat. Toxicol. 122: 133–143.

Rogers, N.J., N.M. Franklin, S.C. Apte, G.E. Batley, B.M. Angel, J.R. Lead et al. 2010. Physico-chemical behaviour and algal toxicity of nanoparticulate CeO_2 in freshwater. Environ. Chem. 7: 50–60.

Röhder, L.A., T. Brandt, L. Sigg and R. Behra. 2014. Influence of agglomeration of cerium oxide nanoparticles and speciation of cerium(III) on short term effects to the green algae *Chlamydomonas reinhardtii*. Aquat. Toxicol. 152: 121–130.

Roy, R., A. Parashar, M. Bhuvaneshwari, N. Chandrasekaran and A. Mukherjee. 2016. Differential effects of P25 TiO_2 nanoparticles on freshwater green microalgae: *Chlorella* and *Scenedesmus* species. Aquat. Toxicol. 176: 161–171.

Roychoudhury, P. and R. Pal. 2014. Spirogyra submaxima–a green alga for nanogold production. J. Algal Biomass Utln. 5: 15–19.

Salvati, A., C. Åberg, T. dos Santos, J. Varela, P. Pinto, P. Lynch et al. 2011. Experimental and theoretical comparison of intracellular import of polymeric nanoparticles and small molecules: Toward models of uptake kinetics. Nanomedicine: NBM 7: 818–826.

Samson, G. and R. Popovic. 1988. Use of algal fluorescence for determination of phytotoxicity of heavy metals and pesticides as environmental pollutants. Ecotoxicol. Environ. Saf. 16: 272–278.

Sauer, J., U. Schreiber, R. Schmid, U. Völker and K. Forchhammer. 2001. Nitrogen starvation-induced chlorosis insynechococcus pcc 7942. low-level photosynthesis as a mechanism of long-term survival. Plant Physiol. 126: 233–243.

Sayes, C.M., R. Wahi, P.A. Kurian, Y. Liu, J.L. West, K.D. Ausman et al. 2006. Correlating nanoscale titania structure with toxicity: a cytotoxicity and inflammatory response study with human dermal fibroblasts and human lung epithelial cells. Toxicol. Sci. 92: 174–185.

Schiavo, S., N. Duroudier, E. Bilbao, M. Mikolaczyk, J. Schäfer, M. Cajaraville et al. 2017. Effects of PVP/PEI coated and uncoated silver NPs and PVP/PEI coating agent on three species of marine microalgae. Sci. Total Environ. 577: 45–53.

Schwab, F., T.D. Bucheli, L.P. Lukhele, A. Magrez, B. Nowack, L. Sigg et al. 2011. Are carbon nanotube effects on green algae caused by shading and agglomeration? Environ. Sci. Technol. 45: 6136–6144.

Schwabe, F., R. Schulin, L.K. Limbach, W. Stark, D. Bürge and B. Nowack. 2013. Influence of two types of organic matter on interaction of CeO_2 nanoparticles with plants in hydroponic culture. Chemosphere 91: 512–520.

Sekine, R., K.L. Moore, M. Matzke, P. Vallotton, H. Jiang, G.M. Hughes et al. 2017. Complementary imaging of silver nanoparticle interactions with green algae: dark-field microscopy, electron microscopy, and nanoscale secondary ion mass spectrometry. ACS Nano 11: 10894–10902.

Sendra, M., J. Blasco and C.V. Araújo. 2017a. Is the cell wall of marine phytoplankton a protective barrier or a nanoparticle interaction site? Toxicological responses of *Chlorella autotrophica* and *Dunaliella salina* to Ag and CeO_2 nanoparticles. Ecol. Indic. 95: 1053–1067.

Sendra, M., I. Moreno-Garrido, M.P. Yeste, J.M. Gatica and J. Blasco. 2017b. Toxicity of TiO2, in nanoparticle or bulk form to freshwater and marine microalgae under visible light and UV-A radiation. Environ. Pollut. 227: 39–48.

Sendra, M., D. Sánchez-Quiles, J. Blasco, I. Moreno-Garrido, L.M. Lubián, S. Pérez-García et al. 2017c. Effects of TiO2 nanoparticles and sunscreens on coastal marine microalgae: Ultraviolet radiation is key variable for toxicity assessment. Environ. Int. 98: 62–68.

Sendra, M., M.P. Yeste, J.M. Gatica, I. Moreno-Garrido and J. Blasco. 2017d. Direct and indirect effects of silver nanoparticles on freshwater and marine microalgae (*Chlamydomonas reinhardtii* and *Phaeodactylum tricornutum*). Chemosphere 179: 279–289.

Sendra, M., M.P. Yeste, J.M. Gatica, I. Moreno-Garrido and J. Blasco. 2017e. Homoagglomeration and heteroagglomeration of TiO_2, in nanoparticle and bulk form, onto freshwater and marine microalgae. Sci. Total Environ. 592: 403–411.

Sendra, M., P.M. Yeste, I. Moreno-Garrido, J.M. Gatica and J. Blasco. 2017f. CeO_2 NPs, toxic or protective to phytoplankton? Charge of nanoparticles and cell wall as factors which cause changes in cell complexity. Sci. Total Environ. 590-591: 304–315.

Shakibaie, M., H. Forootanfar, K. Mollazadeh-Moghaddam, Z. Bagherzadeh, N. Nafissi-Varcheh, A.R. Shahverdi et al. 2010. Green synthesis of gold nanoparticles by the marine microalga *Tetraselmis suecica*. Biotechnol. App. Biochem. 57: 71–75.

Shi, H., R. Magaye, V. Castranova and J. Zhao. 2013. Titanium dioxide nanoparticles: A review of current toxicological data. Part. Fibre Toxicol. 10: 1–33.

Singh, A.K. 2015. Engineered nanoparticles: Structure, properties and mechanisms of toxicity. Academic Press.

Slaveykova, V.I. and K.J. Wilkinson. 2003. Effect of pH on Pb biouptake by the freshwater alga *Chlorella kesslerii*. Environ. Chem. Lett. 1: 185–189.

Stark, W.J. 2011. Nanoparticles in biological systems. Angew. Chem Int. Ed. 50: 1242–1258.

Stine Rosendal, T., S. Henriette, W.-N. Margrethe and R.K. Farhan. 2016. Trophic transfer of metal-based nanoparticles in aquatic environments: A review and recommendations for future research focus. Environ. Sci.-Nano 3: 966–981.

Stone, V., S. Pozzi-Mucelli, L. Tran, K. Aschberger, S. Sabella, U. Vogel et al. 2014. ITS-NANO-Prioritising nanosafety research to develop a stakeholder driven intelligent testing strategy. Part. Fibre Toxicol. 11: 9.

Sun, T.Y., N.A. Bornhöft, K. Hungerbühler and B. Nowack. 2016. Dynamic probabilistic modeling of environmental emissions of engineered nanomaterials. Environ. Sci. Technol. 50: 4701–4711.

Tang, Y., S. Li, J. Qiao, H. Wang and L. Li. 2013. Synergistic effects of nano-sized titanium dioxide and zinc on the photosynthetic capacity and survival of *Anabaena* sp. Int.l J. Mol. Sci. 14: 14395–14407.

Taylor, C., M. Matzke, A. Kroll, D.S. Read, C. Svendsen and A. Crossley. 2016a. Toxic interactions of different silver forms with freshwater green algae and cyanobacteria and their effects on mechanistic endpoints and the production of extracellular polymeric substances. Environ. Sci.-Nano 3: 396–408.

Taylor, N.S., R. Merrifield, T.D. Williams, J.K. Chipman, J.R. Lead and M.R. Viant. 2016b. Molecular toxicity of cerium oxide nanoparticles to the freshwater alga *Chlamydomonas reinhardtii* is associated with supra-environmental exposure concentrations. Nanotoxicology 10: 32–41.

Toh, P.Y., B.W. Ng, C.H. Chong, A.L. Ahmad, J.-W. Yang, C.J.C. Derek et al. 2014. Magnetophoretic separation of microalgae: the role of nanoparticles and polymer binder in harvesting biofuel. RSC Adv. 4: 4114–4121.

Troppová, I., L. Matějová, H. Sezimová, Z. Matěj, P. Peikertová and J. Lang. 2017. Nanostructured TiO_2 and ZnO prepared by using pressurized hot water and their eco-toxicological evaluation. J. Nanopart. Res. 19: 198.

Tsiola, A., P. Pitta, A.J. Callol, M. Kagiorgi, I. Kalantzi, K. Mylona et al. 2017. The impact of silver nanoparticles on marine plankton dynamics: Dependence on coating, size and concentration. Sci. Total Environ. 601: 1838–1848.

Tucci, P., G. Porta, M. Agostini, D. Dinsdale, I. Iavicoli, K. Cain et al. 2013. Metabolic effects of TiO_2 nanoparticles, a common component of sunscreens and cosmetics, on human keratinocytes. Cell Death & Dis. 4: e549.

Uchino, T., H. Tokunaga, M. Ando and H. Utsumi. 2002. Quantitative determination of OH radical generation and its cytotoxicity induced by TiO(2)-UVA treatment. Toxicol. *in vitro* 16: 629–635.

Uduman, N., Y. Qi, M.K. Danquah, G.M. Forde and A. Hoadley. 2010. Dewatering of microalgal cultures: a major bottleneck to algae-based fuels. J. Ren. Sustain. Ener. 2: 012701.

Unfried, K., C. Albrecht, L.-O. Klotz, A. Von Mikecz, S. Grether-Beck and R.P. Schins. 2007. Cellular responses to nanoparticles: Target structures and mechanisms. Nanotoxicology 1: 52–71.

Vallyathan, V., V. Castranova, N.S. Dalal and K. Van Dyke. 1992. Prevention of the acute cytotoxicity associated with silica containing minerals. Google Patents.

Vallyathan, V. and X. Shi. 1997. The role of oxygen free radicals in occupational and environmental lung diseases. Environ. Health Perspect. 105: 165.

Van Breusegem, F. and J.F. Dat. 2006. Reactive oxygen species in plant cell death. Plant Physiol. 141: 384–390.

Van Lehn, R.C. and A. Alexander-Katz. 2011. Penetration of lipid bilayers by nanoparticles with environmentally-responsive surfaces: Simulations and theory. Soft Matter 7: 11392–11404.

Vance, M.E., T. Kuiken, E.P. Vejerano, S.P. McGinnis, M.F. Hochella Jr, D. Rejeski et al. 2015. Nanotechnology in the real world: Redeveloping the nanomaterial consumer products inventory. Beilstein J. Nanotechnol. 6: 1769–1780.

Velzeboer, I., A.J. Hendriks, A.M. Ragas and D. Van de Meent. 2008. Nanomaterials in the environment aquatic ecotoxicity tests of some nanomaterials. Environ. Toxicol. Chem. 27: 1942–1947.

Verma, A., O. Uzun, Y. Hu, Y. Hu, H.-S. Han, N. Watson et al. 2008. Surface-structure-regulated cell-membrane penetration by monolayer-protected nanoparticles. Nat. Materi. 7: 588–595.

Visviki, I. and J. Rachlin. 1994. Acute and chronic exposure of *Dunaliella salina* and *Chlamydomonas bullosa* to copper and cadmium: effects on ultrastructure. Arch. Environ. Contam. Toxicol. 26: 154–162.

Vrieling, E.G., T.P. Beelen, R.A. van Santen and W.W. Gieskes. 1999. Diatom silicon biomineralization as an inspirational source of new approaches to silica production. J. Biotechnol. 70: 39–51.

Wallis, L.K., S.A. Diamond, H. Ma, D.J. Hoff, S.R. Al-Abed and S. Li. 2014. Chronic TiO_2 nanoparticle exposure to a benthic organism, Hyalella azteca: impact of solar UV radiation and material surface coatings on toxicity. Sci. Total Environ. 499: 356–362.

Wang, F., L. Yu, M.P. Monopoli, P. Sandin, E. Mahon, A. Salvati et al. 2013. The biomolecular corona is retained during nanoparticle uptake and protects the cells from the damage induced by cationic nanoparticles until degraded in the lysosomes. Nanomedicine: NBM 9: 1159–1168.

Wang, J., X. Zhang, Y. Chen, M. Sommerfeld and Q. Hu. 2008. Toxicity assessment of manufactured nanomaterials using the unicellular green alga *Chlamydomonas reinhardtii*. Chemosphere 73: 1121–1128.

Wang, S., J. Lv, J. Ma and S. Zhang. 2016a. Cellular internalization and intracellular biotransformation of silver nanoparticles in *Chlamydomonas reinhardtii*. Nanotoxicology 10: 1129–1135.

Wang, Y., X. Zhu, Y. Lao, X. Lv, Y. Tao, W. Huang et al. 2016b. TiO_2 nanoparticles in the marine environment: Physical effects responsible for the toxicity on algae *Phaeodactylum tricornutum*. Sci. Total Environ. 565: 818–826.

Wang, Z., J. Li, J. Zhao and B. Xing. 2011. Toxicity and internalization of CuO nanoparticles to prokaryotic alga *Microcystis aeruginosa* as affected by dissolved organic matter. Environ. Sci. Technol. 45: 6032–6040.

Watlington, K. 2005. Emerging nanotechnologies for site remediation and wastewater treatment. Environmental Protection Agency.

Wei, C., Y. Zhang, J. Guo, B. Han, X. Yang and J. Yuan. 2010. Effects of silica nanoparticles on growth and photosynthetic pigment contents of Scenedesmus obliquus. J. Environ. Sci. 22: 155–160.

Weinstein, J.E. and S.A. Diamond. 2006. Relating daily solar ultraviolet radiation dose in salt marsh-associated estuarine systems to laboratory assessments of photoactivated polycyclic aromatic hydrocarbon toxicity. Environ. Toxicol. Chem. 25: 2860–2868.

Wessels, J. 1994. Developmental regulation of fungal cell wall formation. Ann. Rev. Phytopath. 32: 413–437.

Wijnhoven, S.W., W.J. Peijnenburg, C.A. Herberts, W.I. Hagens, A.G. Oomen and E.H. Heugens et al. 2009. Nano-silver—a review of available data and knowledge gaps in human and environmental risk assessment. Nanotoxicology 3: 109–138.

Wilkinson, K. and A. Reinhardt. 2005. Contrasting roles of natural organic matter on colloidal stabilization and flocculation in freshwaters. CRC Press: Boca Raton, pp. 143–170.

Wu, Y. and W.-X. Wang. 2011. Accumulation, subcellular distribution and toxicity of inorganic mercury and methylmercury in marine phytoplankton. Environ. Pollut. 159: 3097–3105.

Wu, F., A. Bortvedt, B.J. Harper, L.E. Crandon and S.L. Harper. 2017. Uptake and toxicity of CuO nanoparticles to Daphnia magna varies between indirect dietary and direct waterborne exposures. Aquat. Toxicol. 190: 78–86.

Xia, B., B. Chen, X. Sun, K. Qu, F. Ma and M. Du. 2015. Interaction of TiO_2 nanoparticles with the marine microalga *Nitzschia closterium*: Growth inhibition, oxidative stress and internalization. Sci. Total Environ. 508: 525–533.

Xia, B., Q. Sui, X. Sun, Q. Han, B. Chen, L. Zhu et al. 2018. Ocean acidification increases the toxic effects of TiO_2 nanoparticles on the marine microalga Chlorella vulgaris. J. Hazardous Mat. 346: 1–9.

Xia, T., M. Kovochich and A. Nel. 2006. The role of reactive oxygen species and oxidative stress in mediating particulate matter injury. Clin. Occup. Environ. Med. 5: 817–836.

Xia, T., M. Kovochich, M. Liong, L. Mädler, B. Gilbert, H. Shi et al. 2008. Comparison of the mechanism of toxicity of zinc oxide and cerium oxide nanoparticles based on dissolution and oxidative stress properties. ACS Nano 2: 2121–2134.

Xiao, R. and Y. Zheng. 2016. Overview of microalgal extracellular polymeric substances (EPS) and their applications. Biotechnol. Adv. 34: 1225–1244.

Yamamoto, A., R. Honma, M. Sumita and T. Hanawa. 2004. Cytotoxicity evaluation of ceramic particles of different sizes and shapes. J. Biomedical Mater. Res. A 68: 244–256.

Yang, K., D. Lin and B. Xing. 2009. Interactions of humic acid with nanosized inorganic oxides. Langmuir 25: 3571–3576.

Yang, W.-W., Y. Li, A.-J. Miao and L.-Y. Yang. 2012. Cd^{2+} toxicity as affected by bare TiO_2 nanoparticles and their bulk counterpart. Ecotoxicol. Environ. Saf. 85: 44–51.

Yeung, K.L., W.K. Leung, N. Yao and S. Cao. 2009. Reactivity and antimicrobial properties of nanostructured titanium dioxide. Catal. Today 143: 218–224.

Zamani, H., A. Moradshahi, H.D. Jahromi and M.H. Sheikhi. 2014. Influence of PbS nanoparticle polymer coating on their aggregation behavior and toxicity to the green algae *Dunaliella salina*. Aquat. Toxicol. 154: 176–183.

Zhang, H., X. He, Z. Zhang, P. Zhang, Y. Li, Y. Ma et al. 2011. Nano-CeO_2 exhibits adverse effects at environmental relevant concentrations. Environ. Sci. Technol. 45: 3725–3730.

Zhang, S., Y. Jiang, C.-S. Chen, J. Spurgin, K.A. Schwehr, A. Quigg et al. 2012. Aggregation, dissolution, and stability of quantum dots in marine environments: Importance of extracellular polymeric substances. Environ. Sci. Technol. 46: 8764–8772.

Zhang, S., Y. Jiang, C.-S. Chen, D. Creeley, K.A. Schwehr, A. Quigg et al. 2013. Ameliorating effects of extracellular polymeric substances excreted by Thalassiosira pseudonana on algal toxicity of CdSe quantum dots. Aquat. Toxicol. 126: 214–223.

Zhang, S., R. Deng, D. Lin and F. Wu. 2017. Distinct toxic interactions of TiO_2 nanoparticles with four coexisting organochlorine contaminants on algae. Nanotoxicology 11: 1115–1126.

Zhao, J., X. Cao, X. Liu, Z. Wang, C. Zhang, J.C. White et al. 2016. Interactions of CuO nanoparticles with the algae Chlorella pyrenoidosa: adhesion, uptake, and toxicity. Nanotoxicology 10: 1297–1305.

Zhao, X., M. Yu, D. Xu, A. Liu, X. Hou, F. Hao et al. 2017. Distribution, bioaccumulation, trophic transfer, and influences of CeO_2 Nanoparticles in a constructed aquatic food web. Environ. Sci. Technol. 51: 5205–5214.

Zhou, K., Y. Hu, L. Zhang, K. Yang and D. Lin. 2016. The role of exopolymeric substances in the bioaccumulation and toxicity of Ag nanoparticles to algae. Sci. Rep. 6: 32998.

Zhu, M., H. Wang, A.A. Keller, T. Wang and F. Li. 2014. The effect of humic acid on the aggregation of titanium dioxide nanoparticles under different pH and ionic strengths. Sci. Total Environ. 487: 375–380.

Conflict of interest: The authors declare no conflict of interest.

2

Sublethal Effects of Nanoparticles on Aquatic Invertebrates, from Molecular to Organism Level

Laura Canesi,[1,*] *Manon Auguste*[1] and *Maria Joao Bebianno*[2]

Introduction

The increasing production and usage in various fields of different types of nanoparticles (NPs), estimated to grow to over half a million tons by 2020, would lead to their release in substantial amounts in different types of ecosystems, with possible impacts on environmental and human health (Corsi et al. 2014). NPs are expected to enter soil and waterways and eventually reach the marine environment. Invertebrates, which represent more than 90 percent of animal species, widespread in terrestrial, freshwater and marine ecosystems, are emerging both as suitable target organisms and as models for evaluating the environmental impact of NPs (Baun et al. 2008, Canesi et al. 2012, Canesi and Prochàzovà 2013, Hayashi and Engelmann 2013, Rocha et al. 2015a, Canesi and Corsi 2016). In particular, aquatic invertebrates are at potential risk of exposure, as indicated by the increasing amount of data obtained in different model species traditionally utilized in ecotoxicity tests. However, evidence is accumulating on the complex interactions occurring between NPs and environmental media, that affect their behaviour (aggregation/agglomeration, redox reactions, dissolution, exchange of surface moieties, and reactions with biomacromolecules) and consequent fate/bioavailability and impact on aquatic organisms. This in particular applies to those types of NPs produced in higher amounts, such as metal-oxides and carbon-based (Matranga and Corsi 2012, Gottschalk et al. 2013, Baker et al. 2014, Corsi et al. 2014). These data underlined the importance of establishing the criteria for reliability and relevance evaluation of quali-quantitative ecotoxicity data for nanomaterials (Petersen et al. 2015, Selck et al. 2016, Hund-Rinke et al. 2016, Hartmann et al. 2017). However, although data from ecotoxicological assays are essential for regulatory purposes, increasing information is

[1] DISTAV, University of Genoa, Corso Europa 26, 16126 Genoa, Italy.
[2] CIMA, University of Algarve, Campus de Gambelas, 8000-397 Faro, Portugal.
Emails: Manon.Auguste@edu.unige.it; mbebian@ualg.pt
* Corresponding author: laura.canesi@unige it

now available on the sub-lethal effects and modes of action of different NPs measured at different levels of biological organization (from molecular to whole organism level), obtained from *in vitro* and *in vivo* studies in selected model species. This approach provides crucial information on NP uptake and fate within the organism, the identification of target molecules and biological functions that can be affected by different NP types and the implications of these effects on the health of individuals, in terms of metabolism and reproductive outcome. These data can greatly enhance our understanding of the possible impact of NP exposure on different species in different environments. In this chapter, available data on the sub-lethal effects and modes of action of the main classes of NPs obtained from *in vitro* and *in vivo* studies in aquatic invertebrates are summarised. Data are also reported from studies involving exposure during early life stages, which can be highly sensitive to environmental perturbations, that would greatly help in the identification of those NPs that represent a major threat to aquatic species.

However, for a correct evaluation of the biological impact of NPs in different freshwater and marine invertebrates, information on behaviour of different types of NPs with different physico-chemical properties in different exposure media is needed. This will affect particle uptake and accumulation, and consequent interactions with target cells and tissues, resulting in different modes of action and biological outcomes (Canesi and Corsi 2016). Although NP behaviour in environmental and experimental media is outside the scope of this chapter, some basic information should be considered. For example, the effects of certain types of NPs (like zero valent metal NPs and metal based nano-oxides), may be partly due to the release of free metal ions both in the external medium and within the cell/organism. Other types of NPs (non metal based nano-oxides and carbon based NPs) generally show a limited solubility in aqueous solutions, leading to the formation of agglomerates of different sizes and with different stability in freshwater and seawater media, which will affect particle bioavailability and effects. These latter may also be influenced by the presence of organic and particulate matter or micro-organisms (bacteria or microalgae) which invertebrates feed on.

In this light, data on characterization of NP suspensions in exposure media (such as aggregation/agglomeration state, changes in surface charge, release of ions) has become an essential piece of information when discussing the biological effects in different experimental settings. Although these were not generally collected in the first papers on the effects of NPs on aquatic invertebrates, the results on NP behaviour in exposure media, when available, will be mentioned throughout this chapter in relation with the biological effects and mode of actions observed in different experimental models.

Zero Valent Metal NPs

The most common zero-valent NPs are metal-based NPs such as gold and silver NPs, that are produced for different applications. Gold nanoparticles (AuNPs), due to their peculiar properties (bulk gold is chemically inert and electron dense and therefore considered to be "safe"), offer a large range of biomedical applications such as bio-detection, bio-imaging or targeted drug delivery (Tiwari et al. 2011). Despite the almost non existence of natural AuNPs, manufactured AuNPs, without proper treatment after use, will likely end up in the water compartment. Predicted environmental concentrations (PEC) of gold in UK surface water are 470 pg/L (Mahapatra et al. 2015).

The biocidal action of silver has been known for several decades and therefore it has been used in a variety of consumer products and in food industry for its antibacterial activity. However, in its nanosized forms (AgNPs), it offers new applications in microelectronics and medical imaging areas, due to its high electrical and thermal conductivity (McGillicuddy et al. 2017). PEC values of AgNPs in European surface waters are between 0.59 and 2.16 ng/L (Gottschalk et al. 2013). With the increasing production and use, in the longer term, the aquatic environment will face higher concentration of zero-valent NPs that will potentially affect the health of different organisms.

In Vitro Effects of AuNPs and AgNPs

In light of its non-environmental relevance, little information is available on the effects of AuNP on aquatic organisms, in particular using *in vitro* testing methods. However, few studies have shown detrimental effects on cells of marine bivalve cells.

In the hemocytes of marine mussel *Mytilus galloprovincialis* exposed to three types of AuNPs of different sizes (5, 15, 40 nm) for 24 hr, a slight decrease in viability was observed at concentrations > 50 mg/L. Smaller size AuNPs showed stronger effects, although in general, the AuNPs were not highly toxic (Katsumiti et al. 2015a). Moreover, the citrate coating appeared to play a role in the toxicity observed. Explant of gills from the clam *Ruditapes philippinarum* exposed to 750 µg/L AuNP (24 nm) for 1 and 6 hr, showed a time dependent accumulation within gill cells. Secretory granules containing AuNPs were observed in the apex of ciliary cells or outside the gill epithelium attached to microvilli and cilia. Epithelial cells also internalized AuNPs, as seen attached to the outer membrane of the mitochondria and to the nuclear envelope. In response to AuNP treatment, gill cells showed alterations that could possibly affect the physiological function of the tissue, such as high vacuolation, thickened basal lamina and a loss of cilia and microvilli in the apex (García-Negrete et al. 2015).

Stronger effects were observed on the same cell types exposed to maltose coated AgNPs of different sizes (20–100 nm) for 24 hr, with decreased cell viability from 10 mg/L for both hemocytes and gill cells. Increased ROS production, catalase activities and DNA damage were also observed, with different time dependent effects in each cell type, at concentrations between 0.62 and 1.5 mg/L of AgNPs. In comparison, ionic silver (Ag^+) was effective at much lower concentrations (0.03–0.06 mg/L), suggesting little contribution of released ionic silver from the NP form to the biological effects observed (Katsumiti et al. 2015b).

Effects of AuNPs and AgNPs on Embryo Development

Sea urchin embryos exposed to AgNPs showed different sensitivity according to the species and to the developmental stage when the exposure is initiated. Embryos of *Arbacia lixula, Paracentrotus lividus* and *Sphaerechinus granularis* were exposed to 60 nm AgNPs at different times during post-fertilization (pf), depending on the time course of development for each species. *A. lixula* was the most susceptible species to AgNPs together with *S. granularis*, since exposure to 50 and 100 µg/L caused almost total developmental arrest at any time pf. In contrast, *P. lividus* displayed a lower sensitivity, with few retarded embryos at all concentrations (1–100 µg/L) and developmental stages tested (Burić et al. 2015). *Strongylocentrotus droebachiensis* at the echinopluteus and juvenile (1 month old pluteus) stages exposed to poly(allylamine)-coated silver nanoparticles (PAAm-AgNPs; ~ 15 nm) for 24 hr at 100 µg/L, showed evidence of uptake and interferences in larval development, with developmental arrest only at the highest concentration (Magesky et al. 2016). Other types of AgNPs (5–35 nm) affected *P. lividus* embryos (2 hr pf) after 24 and 48 hr exposure. Higher concentrations (3 mg/L) provoked developmental arrest, while lower concentration (0.3 mg/L) produced strong skeletal alterations (very short arms and small calcite spicules) (Šiller et al. 2013).

Only few data are available on zero valent metal NPs in bivalve species. Newly fertilized oyster *Crassostrea virginica* embryos exposed to AgNP (15 nm) for 48 hr showed decrease in normal development only at the highest concentration tested (1.6 µg/L) (Ringwood et al. 2010). In *M. galloprovincialis,* exposure to stabilized zero-valent nanoiron (nZVI) (~ 50 nm), at concentrations from 0.1 to 10 mg/L, resulted in a decrease in normal D-shaped embryos. In the same conditions, sperm exposure induced mortality of spermatozoa, DNA damage and decrease in fertilization success (Kádár et al. 2011).

With regards to uptake and modes of action within the embryo, information is available only on AgNPs. In *S. droebachiensis*, the digestive tract was the main route for uptake (Magesky et al. 2016). Sea urchin (1 month old) treated with AgNPs (14 nm) showed engulfment of large AgNP agglomerates (> 200 nm) by circulating coelomocytes, forming phagocytic vesicles that later fused with large and dense lysosomes to form phagolysosomes. When AgNPs were found inside the cytoplasm, endoplasmic reticulum and mitochondria seemed to be preferentially targeted (Magesky et al. 2017). Moreover, smaller agglomerates (around 100 nm) were found in endo-lysosomal compartments in peritoneocytes (cells from the coelomic epithelium), which suggest an intracellular digestion pathway and a secretory pathway. These cells produced secretory vesicles ready for being released containing AgNP agglomerates, which can represent the natural secretory pathway of undigested material by the cells (Magesky et al. 2017). This is an example of the possible defense response to AgNP exposure. Young sea urchins (3 months old) treated with AgNP (100 μg/L) for 48 hr showed increased expression of Hsp70 stress protein, indicating the establishment of cellular protective mechanisms (Magesky et al. 2017). Another example of induction of the detoxification process in response to AgNP exposure was found in bivalves. Oyster embryos exposed to 0.16 μg/L AgNP (48 hr) showed a large increase in metallothionein (MT) expression. However, some doubts persist whether MT induction was initiated by AgNPs or by the Ag^+ released from NP dissolution (Ringwood et al. 2010).

Taken together, available data indicate that zero valent metal NPs can affect developmental processes in marine invertebrate at μg/L concentrations, to a different extent depending on the type of particle, time of exposure, developmental stage and species. Although the effects could be partly ascribed to the release of metal ions, the results show that protective and detoxification responses can be activated in embryos like in adults.

In Vivo Effects of AuNPs and AgNPs

Filter-feeding and suspension-feeding organisms are significant targets for NP exposure, thus being appropriate species to evaluate nano-ecotoxicology in the aquatic environment. Although most of the information is available from marine organisms and among those from *in vivo* experiments (about 80%), NPs are accumulated by different sources (water, sediments and diet) as individual particles, homo or hetero-aggregates/agglomerates and the accumulation is dependent on the nano-specific composition, behaviour and fate as well as environmental characteristics (Rocha et al. 2015a). Several NPs, such as AgNPs and BSA-Au NPs, are preferentially accumulated in the digestive system (Ward and Kach 2009, Hull et al. 2011, Zuykov et al. 2011a, Al-Sid-Cheikh et al. 2013). The scallop *Chlamys islandica* exposed to Ag NPs (10–80 nm; 110–151 ng/L; 12 hr) accumulate more AgNPs in the digestive system (digestive gland, intestine, crystalline style and anus) than in other tissues (gills, mantle, gonads, kidney and muscle) (Al-Sid-Cheikh et al. 2013). Moreover, the exposure of the mussel *Mytilus edulis* to radiolabelled Ag (< 40 nm, 110 mAg, 0.7 μg/L, 3 h 30 min–72 hr) showed a higher concentration of AgNPs in the digestive gland, that was further transported to the hemocytes that play an important role in the translocation of AgNPs to other tissues (Zuykov et al. 2011b). In mussels *Mytilus galloprovincialis,* AgNPs (42 ± 10 nm, 10 μg Ag/L; 15 days), accumulation begins in the gills and after a certain time, is eliminated. Ag levels were similar to those exposed to the same concentration in the ionic Ag form indicating that Ag ions are released from NPs (Gomes et al. 2012). In the digestive gland of *M. galloprovincialis*, the accumulation of AgNPs increases with the time of exposure and like for *M. edulis* and *C. islandica,* the accumulation was higher (5-fold) than in the gills indicating that, the digestive gland plays a key role in AgNPs accumulation and detoxification (Gomes et al. 2012). Furthermore, mussels were more efficient in eliminating AgNPs from this tissue, probably due to a higher propensity of these particles to dissolve with time. The elimination rate of AgNPs (40–50 nm; 10 μg/L; 14 hr), in the suspension feeder *Scrobicularia plana*, was higher after waterborne exposure when compared to the exposure through the diet (Buffet et al. 2013). In the ragworm *Nereis diversicolor*, AgNPs (30 ± 5 nm, 250 ng/g, 10 days) were internalised by endocytosis into the gut epithelia, mainly in the form of aggregates (a large number of endosomes and vesicles were detected near the cellular membrane) predominantly associated with inorganic granules, organelles and heat-denatured proteins suggesting that in this species, Ag is predominantly accumulated in the nano form (García-Alonso et al. 2011).

NP characteristics, namely size and surface area, are a key factor in the induction of oxidative stress. Although preferentially accumulated in the digestive system, AgNPs induce less antioxidant responses in this tissue compared to the gills, suggesting a lower effect of ROS induced by AgNPs in this tissue. AgNPs (42 ± 10 nm, 10 µg Ag/L; 15 days) induce oxidative stress in *M. galloprovincialis* tissues that is tissue and time dependent. In the gills, AgNPs enhance the production of $^1O_2$ and H_2O_2 that increase the activity of SOD, CAT and GPx to counteract the effect of ROS and reduce the possibility of oxidative damage (Gomes et al. 2014b). Contrarily in the digestive gland, although enzymatic activities were enhanced in the beginning of exposure to AgNPs, changes in SOD and GPx activities were transitory and CAT activity inhibited suggesting a lesser damaging effect of ROS in the digestive system (Gomes et al. 2014b). On the other hand, the role of metallothioneins (MTs) proposed for Ag homeostasis and detoxification (Amiard et al. 2006, Lansgton et al. 1998) is tissue dependent in *M. galloprovincialis* exposed to AgNPs. In *M. galloprovincialis* gills, AgNPs induce an increase of MT levels indicating that Ag ions released from AgNPs were bound to MTs to regulate Ag bioavailability, detoxify Ag ions or scavenge ROS generated by AgNPs (Gomes et al. 2014b). The same was detected in the oyster *C. virginica* exposed to AgNPs (15 ± 6 nm, 0.0016 µg/L, 48 hr) along with adverse effects in lysosomal activity (Ringwood et al. 2010). Conversely, in the digestive gland after an initial induction of MT, other mechanisms of detoxification occur, indicating that the mechanism of storage of AgNPs could be non-toxic or that insoluble silver-sulphide that precipitate make silver ions unavailable for binding to MT and inhibit the potential of deleterious effects of these NPs. Given the capacity of AgNPs to generate ROS (Lapresta-Férnandez et al. 2012) and cause membrane damage by increasing lipid peroxidation (LPO), it was not surprising that AgNPs also cause changes in LPO levels in *M. galloprovincialis* gills indicating that the antioxidant defence system (antioxidant enzymes plus MT) were unable to detoxify ROS production and prevent oxidative damage. However, no changes in LPO were detected in the digestive gland so the presence of stable Ag-sulphide complexes might be responsible for the absence of oxidative damage in this tissue (Gomes et al. 2014b). Exposure to AgNPs also causes DNA damage in mussel haemocytes that increases with time of exposure (Gomes et al. 2013a) indicating that the genotoxic potential of AgNPs is a result of direct interaction with DNA or of indirect interaction (oxidative stress and ROS) that result in DNA strand breaks, among other DNA lesions. Genotoxicity of AgNPs seems to be mediated by oxidative stress in *Macoma balthica* after exposure to sediment spiked Ag NPs (20, 80 nm) (Dai et al. 2013). If not taken up by the cells, Ag NPs could mediate toxicity by attacking the cellular membrane surface and release Ag^+, compromising cell integrity and permeability or originate extracellular ROS and oxidative stress by surface processes due to NP effect (Navarro et al. 2008). A Trojan horse-type mechanism that enables the transport of metal ions into cells (Limbach et al. 2007) or surface oxidation of AgNPs after contact with proteins in the cytoplasm that liberates Ag^+ ions, can also amplify AgNPs toxicity but this needs to be confirmed in this species. Although behavioural changes and neurotoxicity effects induced by NPs are still controversial, it is interesting to note that in *S. plana* exposed to AgNPs (40–50 nm; 10 µg/L; 14 hr) no neurotoxic effects were observed (Buffet et al. 2013).

A proteomic approach used to assess the effects of AgNPs (42 ± 10 nm, 10 µg/L, 15 d) in *M. galloprovincialis* gills and digestive gland identified, beside oxidative stress, different tissue-specific protein pathways that reflect differences in uptake, tissue-specific functions, redox requirements and modes of action. These include cytoskeleton and cell structure (catchin protein, myosin heavy chain) disturbances, stress response (heat-shock protein 70), oxidative stress (GST), transcription regulation (nuclear receptor subfamily 1G), adhesion and mobility (precollagenP), and energy metabolism (ATP synthase F0 subunit 6 and NADH dehydrogenase subunit 2). One of the explanations for the alteration of proteins associated with the structure and function of the cytoskeleton (down-regulation of catchin in gills and upregulation of myosin in digestive gland) is oxidative stress leading to cytoskeleton disorganization. The down-regulation of HSP70 in *M. galloprovincialis* gills reinforce the role of these proteins in cell protection and repair from damage induced by oxidative stress also associated with apoptosis. GST plays a role in the antioxidant defence system of bivalves including ROS metabolism, so the up-regulation of this enzyme in the gills accounts for a cellular compensation mechanism to protect cells against ROS-induced damage (Gomes et al. 2014b). Furthermore, the up-regulation of the nuclear receptor subfamily 1G in the same tissue indicates that AgNPs have the capacity to interfere with signal transduction in DNA-related functions and induce genotoxicity, confirming the genotoxic potential previously detected by the Comet assay (Gomes et al. 2013a). The up-regulation of precollagen-P involved in the detoxification of metals in mussel tissues is related with the Ag dissolution and elimination in *M. galloprovincialis* gills as a response of Ag^{+} ions released from the particles, highlighting the importance of particle solubility in NP behaviour and toxicity (Gomes et al. 2013b, Bebianno et al. 2017). AgNPs also affect ATP synthase F0 subunit 6 (up-regulated) and NADH dehydrogenase subunit 2 (up-regulated) in the gills and digestive gland, altering the normal mitochondrial electron transport system and consequently, the production of ATP and ROS. Apart from affecting proteins related to cytoskeleton (paramyosin), exposure to Ag NPs also affects proteins involved in stress response (major vault protein in the gills and ras partial in the digestive gland). The induction of ras partial is associated with the uptake of AgNPs and with a consequent cytoskeleton disruption. These results confirm that AgNPs not only release Ag^{+} but also induce oxidative stress signal transduction pathways mediated by the mitochondria and nucleus that can lead to apoptosis (Gomes et al. 2014b, Bebianno et al. 2017).

As for AgNPs, size is also an important feature in the effect of AuNPs. Smaller AuNPs (5.3 nm; 750 µg/L, 24 hr) induce in *M. edulis* higher oxidative stress than larger AuNPs (13 nm; 750 µg/L, 24 hr) suggesting that size is a key factor in biological responses of this type of NPs (Tedesco et al. 2010). Size-dependent effects were also detected in the burrowing capacity of *S. plana* where a higher burrowing impairment occurred for larger Au NPs (40 nm) when compared to smaller ones (5–15 nm; 100 µg/L; 7 d) (Pan et al. 2012). These NPs also induce different gene expression and protein changes in the tissues of aquatic invertebrates. As in the case of AgNPs, AuNPs (1.6×10^{3}–1.6×10^{5} Au NPs/cell, 10.0 ± 0.5 nm, 7 d) accumulated in branchial and digestive epithelial cells of the freshwater clam *C. fluminea*, also induce MT and oxidative stress in the gills and visceral mass by altering the gene expression of CAT, SOD, GST and cytochrome *C* oxidase subunit-1 (Renault et al. 2008). AuNP-citrate (13 nm; 750 µg/L; 24 hr) also induce higher ubiquitination and CAT activity in the digestive gland of *M. edulis* and higher carbonylation in the gills compared to the digestive gland (Tedesco et al. 2010). The decrease in size of AuNPs (from 13 to 5.3 ± 1 nm) induce a decrease in the ratio reduced/oxidized glutathione along with a reduction in protein thiol oxidation and an increase LPO, along with a decrease in membrane lysosomal stability in the hemocytes (Tedesco et al. 2010). Unlike for AgNPs, no genotoxic effects were observed in *M. balthica* after exposure to sediment spiked AuNPs (200 µg/g d. w. sed.; 35 d) (Dai et al. 2013). Beside oxidative stress, Au NPs also increase AChE activity in the clam *S. plana* (5–40 nm; 100 µg/L; 16 d) (Pan et al. 2012), which was associated with a phenomenon of overcompensation.

Nano-oxides

A number of metal and non metal containing nano-oxides (ZnO, TiO_2, CeO_2, CuO, Fe_2O_3, SiO_2, CrO_2, MoO_3, Bi_2O_3, $BaTiO_3$, $LiCoO_2$, InSnO) are produced for a variety of applications, from commercial products to environmental and biomedical applications. Their entry in the environment can be intentional or not (e.g., as nano-waste); however, released in different compartments, they can undergo several types of transformations and affect the surrounding life (reviewed by Amde et al. 2017).

In Vitro Effects of Nano-Oxides

In vitro data are only available for marine bivalves, the mussel *M. galloprovincialis* and the clam *R. philippinarum*, using haemocytes and gills excisions. The first report is on *in vitro* exposure of *Mytilus* hemocytes to nano-Fe_2O_3 (50 nm) (30 min, 1 µg/mL), showing a decrease in lysosomal membrane stability (LMS) (Kádár et al. 2010). The effects of short term exposure to different types of nano-oxides (n-TiO_2, n-SiO_2; n-ZnO; n-CeO_2) were thoroughly investigated in haemocytes at concentrations of 1, 5 and 10 μg/mL. Significant effects on LMS, phagocytic activity, ROS and nitric oxide production, and lysozyme release depend on the nano-oxide type, time of exposure and endpoint measured; only n-ZnO induced pre-apoptotic processes (reviewed in Canesi and Procházková 2013, Canesi and Corsi 2016). In the same concentration range, n-TiO_2 and n-ZnO, at different times of exposure, interfere with the phagocytic activity of hemocytes from *R. philippinarum* and *M. galloprovincialis* (Marisa et al. 2015, Santillán-Urquiza et al. 2017).

Other studies investigated the effects of particle size on mussel haemocytes. Exposure to ZnO nanorods (< 130 and < 280 nm) indicated a stronger effect on cell viability of smaller particles. In both haemocytes and gill cells, viability decreased from 10 μg/mL; however, a similar effect was observed after exposure to ionic Zn, suggesting that the effects were mainly due to the release of Zn^{2+} (Katsumiti et al. 2015a). Other studies investigated the effect of both size and coating (n-ZnO, Fe_2O_3 coated ZnO; Santillán-Urquiza et al. 2017) on hemocytes and gill cells at different times of exposure. Overall, only high concentrations (50 to 100 μg/mL) affected cell viability, independent of the time of exposure, particle size and coating. However, the effect of certain types of coating was demonstrated for n-ZnO; the stabilizer Ecodis P90 alone showed some toxicity, whereas inulin coating can change the availability of NP to the cells, favouring recognition by haemocytes and resulting in enhanced uptake (Katsumiti et al. 2015a, Santillán-Urquiza et al. 2017).

Cellular uptake was observed in both haemocytes and excised gills. After exposure to nano-Fe_2O_3 (50 nm, 12 hr), electron-dense deposits throughout the cytoplasm within epithelial gill cells were observed; haemocytes exhibited large lysosomes with an electron-dense content and some electron-dense deposits within the cytoplasm. Nano-Fe_2O_3 agglomerates could be internalized by pinocytosis. However, parallel experiments carried out with hydrated $FeCl_3$ salt showed little difference in the responses (in terms of LMS, lipid peroxidation and cholinesterase activity) between exposure to the ionic and the nano Fe forms (Kádár et al. 2010). Rapid internalization in haemocyte endosomes was also observed for both n-ZnO and n-TiO_2 (Canesi and Procházková 2013, Canesi and Corsi 2016, Santillán-Urquiza et al. 2017). In contrast, internalization of n-SiO_2 was not observed up to 24 hr of exposure (Katsumiti et al. 2015a). Granulocytes and hyalinocytes of the clam *R. philippinarum* internalized n-TiO_2 (21 nm) after 60 min within vacuoles and cytoplasm, and n-TiO_2 were also found in contact with the cell surface membrane (Marisa et al. 2015).

The mechanisms of cellular stress response were investigated in mussel haemocytes and excised gills in response to n-TiO_2. In haemocytes, activation of stress signalling was observed (as increased levels of phosphorylated p38 MAP Kinase); in gills, the efflux activity of ABC transporter increased (reviewed in Canesi and Procházková 2013, Canesi and Corsi 2016).

Overall, the results obtained *in vitro* underlined that in the haemocytes of marine bivalves, the endo-lysosomal system and related phagocytic activity represent significant targets for nano-oxide exposure, independent of the type of the particles. Moreover, data obtained with n-TiO_2 as a model nano-oxide, indicate activation of a stress response in both haemocytes and gills. These observations represent the basis for further *in vivo* studies on the immunomodulatory effects and on activation of defence mechanisms in response to nano-oxides.

Effects of Nano-Oxides on Embryo/Larval Development

A number of studies are available showing dose-dependent delay and/or arrest of development in different species. Marine macroinvertebrates (sea urchins and mussels) were the most common model species utilised.

The effects of different types of nano-oxides (n-TiO_2; n-ZnO; n-CuO) were investigated in sea urchins. Sperm exposure did not affect fertilization success but subsequent embryo development. In different exposure conditions, skeletal anomalies were observed at different developmental stages (Maisano et al. 2015, Torres Duarte et al. 2016, Manzo et al. 2013, Gambardella et al. 2013). In contrast, in the mussel *M. galloprovincialis* no effect on embryo development were observed at concentrations < 1 mg/L of n-TiO_2 (Libralato et al. 2013, Balbi et al. 2014).

With regards to planktonic species, decreased hatching rate was observed in *A. salina* exposed to n-CuO (~ 114 nm) at concentrations from 50 to 100 mg/L, suggesting that n-CuO can disturb the activity of proteolytic enzymes present in *Artemia* cysts that in normal condition lease the chorion layer and enable hatching (Madhav et al. 2017). A dose-dependent reduction in swimming speed was observed after 24 hr exposure to Fe_3O_4-NPs and α-Fe_2O_3-NPs. Moreover, small NP agglomerates were seen attached onto the gill and body surface, also creating holes in the body surface (Zhu et al. 2017c, Wang et al. 2017). As in sea urchin plutei, that displayed abnormal morphology of the skeleton arms, impairment of swimming ability can affect survival as food searching is impacted (Torres-Duarte et al. 2016, Maisano et al. 2015). In *Tigriopus fulvus* nauplii (< 24 hr-old) CuO NPs also induced molt failure at sub-lethal concentrations (3–25 mg/L) (Rotini et al. 2018).

Despite the number of studies on embryo-larval toxicity of different types of nano-oxides in different model species, little information is available on the mechanisms underlying the developmental effects observed. Interference with neurotransmission system has been reported. Abnormalities were observed in sea urchin *A. lixula* embryos together with changes in acetylcholinesterase (AChE) activity as a marker of neurotoxicity. Inhibition of AChE by CuO NPs (7–20 μg/L) was reported, suggesting an effect on the deposition of the spicules during morphogenesis. A drop in serotonin was also observed, which can be highly negative for embryos development in reason of its role in regulation of morphology and growth of serotonergic neurons (Maisano et al. 2015). A dose-dependent decrease in choline and N-acetyl serotonin was also observed in response to CuO NPs (Capello et al. 2017). Similarly, in *P. lividus* n-TiO_2 decreased AChE activity (≤ 1 mg/L) (Gambardella et al. 2013). Most of the studies concluded that there is a link between the morphological alterations observed and changes in cholinesterase activities (ChEs), due to the role of these enzymes during neuronal differentiation in sea urchin embryo development (e.g., in neurotransmission regulation and during cell proliferation and migration).

Nonetheless, in *Artemia* nauplii exposed to n-CeO_2 (15 nm) for 48 hr, AChE activity increased in a dose dependent manner (10–75 mg/L (Suganthāraj David et al. 2017). This opposite pattern could be due to the high concentration of n-CeO_2 used in comparison to other studies or may possibly represent a species-specific response.

Changes in oxidative stress related biomarkers induced by nano-oxide exposure were reported for several larvae species and NP types. Changes in antioxidant enzyme activities, ROS production, and lipid peroxidation were observed in sea urchins (Torres Duarte et al. 2016, Wu et al. 2015) and planktonic crustaceans (Madhav et al. 2017, Suganthāraj David et al. 2017, Zhu et al. 2017c). However, oxidative stress conditions were induced only by high concentrations (mg/L).

Finally, nano-oxides have been shown to affect membrane transporters in the sea urchin embryo. MDR (Multidrug resistance) transporters of *L. pictus* (cleavage stage) decreased in activity after 90 min CuONP exposure (25 μg/ml) and an accumulation of Cu^+ was observed in embryos. Complementary experiments using endocytosis inhibitors confirmed that endocytosis is an important route for nano-copper uptake by micromeres but indicated the presence of other mechanisms (Torres Duarte et al. 2017). Similarly, in *L. pictus*, exposure to CuO NPs and ZnO NPs (0.5–10 μg/ml) resulted in inhibition of the ABC efflux transporter (Wu et al. 2015).

These studies underline the potential harmful effects of nano-oxides on early developmental stages of aquatic species. The effects can be due not only to NP uptake but also by contact or attachment of NP agglomerates to larvae, thus affecting normal development and behaviour.

In Vivo Effects of Nano-Oxides

CuO NPs are one of the most commonly used NPs today and the exposure is an important approach to assess the ecotoxicological responses in aquatic species. The digestive gland of *M. galloprovincialis* is the preferential site for CuO NP accumulation (2-fold higher than the gills), reflecting the key role of this tissue in the NP bioaccumulation and detoxification. However, *M. galloprovincialis* gills are more susceptible to oxidative stress and MT induction by CuO NPs (< 50 nm; 10 µg/L; 7–15 d) than the digestive gland (Gomes et al. 2012). In the polychaete *H. diversicolor,* CuO NPs (40–500 nm, 10 µg/L, 16 d) induced an increase in antioxidant defences, namely CAT and GST, but no signs of neurotoxicity (AChE) and damage were detected (Buffet et al. 2011). In the clam *S. plana* exposed to the same CuO NPs, oxidative stress also occurred, confirming the oxidative potential of CuO NPs towards aquatic organisms (Buffet et al. 2011). Changes in carbonyls and protein thiols were also reported in *M. edulis* in response to CuO NPs (50 nm, 400–1000 mg/L, 1 hr), with a decrease in reduced protein thiols and an increase in protein carbonyls in the gills (Hu et al. 2014). Six proteins were identified: alpha- and beta-tubulin, actin, tropomyosin, triosephosphate isomerase and Cu-Zn superoxide dismutase, indicative of significant protein oxidation of cytoskeleton and key enzymes in response to CuO NPs.

The association between NPs exposure, behavioural changes and neurotoxicity effects is controversial. Regarding the putative neurotoxicity effects, only Gomes et al. (2013a) detected an inhibition of AChE in *M. galloprovincialis* gills exposed to CuO NPs (31 nm; 10 µg/L; 15 d). As in the case of other metal NPs, genotoxicity in the haemocytes of *M. galloprovincialis* exposed to CuO NPs (31 nm; 10 µg/L) are mediated by oxidative stress. Similarly, genotoxicity of CuO NPs (29.5 nm) was also observed in *S. plana* exposed to the same CuO NP concentration (10 µg/L). In *S. plana*, exposed to larger CuO NPs (40–500 nm, 10 µg/L, 16 d), Cu accumulated, affecting the burrowing and feeding behaviour as well as causing an increase of SOD, CAT and GST enzymatic activities, along with MT-like protein induction, suggesting a specific NP effect (Buffet et al. 2011).

CuO NPs (31 ± 10 nm, 10 µg/L, 15 d) exposure in *M. galloprovincialis* tissues (gills and digestive gland), induced down- or up-regulation in proteins related to cytoskeleton and cell structure, transcription regulation, energy metabolism, oxidative stress, stress response, apoptosis and proteolysis, indicating that CuO NPs play a putative role in cellular toxicity and consequent cell death. Apart from the traditional molecular targets of CuO NPs exposure in *M. galloprovincialis* tissues (e.g., ATP synthase, GST and HSPs), the expression pattern of proteins involved in apoptosis (caspase 3/7-1), proteolysis (cathepsin L) and stress response (heat shock cognate 71) revealed that CuO NPs of this size induce an apoptotic pathway that is caspase dependent. Moreover, caspase 3/7-1, cathepsin L, Zn-finger protein and precollagen-D are potential novel targets to assess the effects of these NPs (Gomes et al. 2014a).

The *in vivo* effects of n-TiO_2 have been thoroughly investigated in *M. galloprovincialis* at different concentrations (1–100 mg/L) and times of exposure (1 and 4 days). Biomarker responses at the molecular, cellular and tissue level were investigated in haemocytes, gills and digestive gland, together with titanium accumulation in the tissues. Data allowed to formulate a hypothesis on the possible pathways of uptake of n-TiO_2 agglomerates formed in seawater by the gills, that are partly accumulated to the digestive gland and partly rejected as pseudofaeces. NPs can be then potentially translocated from the digestive system to the haemolymph, and to circulating haemocytes, where n-TiO_2 induced changes in functional parameters and gene expression, leading to immunomodulation. In addition, changes in oxidative stress and lysosomal biomarkers were observed at tissue level; these effects were also observed with n-SiO_2 and n-CeO_2 (reviewed in Canesi et al. 2012, Canesi and Procházková 2013, Canesi and Corsi 2016). However, studies with n-TiO_2 showed that only a minor fraction (< 10%) of the nominally added concentrations were available to the mussels in the suspended form, due to both particle sedimentation and agglomeration/adsorption processes occurring in the presence of organic particles produced by the animals during exposure (mucus, faecal pellets, gametes). This would explain the low, although tissue-specific, concentrations of Ti measured in the gills and digestive gland. Subsequent studies carried out in other marine bivalves, freshwater gastropods, and sea urchins, confirmed that this type of nano-oxide, in the concentration range of mg/L, substantially induced similar responses with immunomodulation, activation of stress signalling pathways, changes in oxidative-stress related parameters and DNA damage (Ali et al. 2015, Doyle et al. 2015, Pinsino et al. 2015, Girardello et al. 2016, Shi et al. 2017). The time course and extent of these responses were dependent on the species and exposure conditions.

Overall, when evaluating the effects of nano-oxides observed in aquatic invertebrates in different experimental models and conditions, the behaviour of nano-oxides, characterised by a low solubility in aqueous media, must be taken into account. In particular, in studies with marine invertebrate cells, the high ionic strength of exposure medium would further increase particle instability leading to agglomeration; this in turn will depend on the type of particle, primary size, concentration and time of exposure. Similarly, many of the effects observed for metal based nano-oxides could be ascribed to the release of soluble metal ions. The exposure medium is also expected to affect the influence of particle coating that will depend on the type of particle, but also on its stability in high ionic strength media.

Semiconductor Materials ("Quantum Dots")

Quantum dots (QDs) are semiconductor metalloid-crystal structures with unique optical and electronic properties and a wide range of applications (treatment and diagnosis of cancer, targeted drug delivery). Used in imaging analysis since 1988, sixteen years after being discovered (Ekimov and Onushchenko 1982), QDs can have a variety of metal ions in their chemical composition including metal complexes of group II-IV (CdSe, CdTe, CdSeS, ZnS, ZnSe and PbSe) or of group III-V series (InP, InAs, GaAs and GaN). The most common QDs are Cd-based QDs and CdSe and CdTe are the most commonly used in biological and medical applications (Smith et al. 2008). The production and use of QDs considerably increased in the last years which inevitably result in their presence in the aquatic environment where, depending on the ionic strength, their behaviour will change and toxicity occur when interacting with aquatic invertebrates because these NPs have in their composition, Cd atoms, a well know carcinogenic compound whose introduction in the aquatic environment is forbidden in many countries.

The main processes that occur in the aquatic environment are due to the presence of homo- or hetero-agglomeration/aggregation of QDs that are dependent on the ionic strength of the aquatic environment. In saline waters, QDs tend to aggregate and settle in the sediments more easily than in fresh water (Rocha et al. 2015a, 2017). Therefore, the impact of QDs in the aquatic environment is a cause of concern due to their size and reactivity. The abiotic (pH, visible and ultraviolet electromagnetic radiation) and biotic (stabilization with organic and suspended particulate matter, interaction with macromolecules and/or organisms and biological transformation) characteristics of the aquatic environment may induce physico-chemical transformations (oxidation, sulfidation, settling, bioturbation, resuspension, photo-activation ROS generation) of QDs. These include changes in surface coating (including degradation and charge changes), aggregation/agglomeration and advection and/ or release of Cd. Most of these properties are size, surface charge, area and coating dependent and are possible sources for toxicity of QDs. Thus the knowledge of the behaviour, transformation and fate of QDs in the aquatic environment is needed to clarify the impact of these NPs. Details of these processes are described in a recent review by Rocha et al. (2017).

They are accumulated by aquatic invertebrates either in the water column or in the sediments,mainly by the direct passage across the gills and through the absorption by the digestive gland (Canesi et al. 2012). When accumulated in aquatic invertebrates due to changes in pH inside their digestive system and lysosomes, transformations occur that can result in QDs dissolution along with NPs effects.

In Vitro Effects of Quantum Dots

Although *in vitro* studies with mammalian cells indicate that QDs generate ROS, data about *in vitro* bioassays using aquatic invertebrates is scarce. Bruneau et al. (2013) compared the immunotoxicity *in vitro* of a mixture of CdS/Cd-TeQDs (1–10 nm; 0.86 mg/L) and identified a reduction in the phagocytic activity in the haemocytes of *M. edulis*. In the haemocytes of *M. galloprovincialis* exposed *in vitro* to a wide range of CdS QDs (5 nm in size caped with GSH) concentrations (10−4 to 102 mgCd/L; 24

hr), the QDs' accumulation inside the vesicles of the endocytic-lysosomal system increase with the time. The LOEC calculated for the haemocytes exposed to CdS/CdTe QDs (0.86 mg/L) were similar to those exposed to CdS QDs capped with GSH (1 mg Cd/L) (Bruneau et al. 2013, Katsumiti et al. 2014). Because gill cells are the first tissue to be in contact with many contaminants, in particular QDs, gill epithelial cells are an interesting model to assess QDs' toxicity *in vitro*. Besides cytotoxicity in the haemocytes, gill cells are more sensitive than haemocytes when exposed to CdS QDs capped with GSH (1 mg Cd/L). These QDs were genotoxic to both cell types with haemocytes more sensitive than gill cells. Genotoxic effects of QDs were also observed in *M. edulis* haemocytes after *in vitro* exposure to CdS QDs (4 nm; 10 mg/L; 4 hr) covered with a capping agent MPEG-SH (thiol-terminated methyl polyethylene glycol), used to avoid agglomeration (Munari et al. 2014). Regarding the generation of ROS, production occurred earlier in the haemocytes compared with the gills cells and its pathway seems to be related not only by the release of Cd ion from CdS QDs (up to 10%) but also due to the cell uptake of CdS QDs through endocytic pathways and subsequent interactions with different cell organelles (Katsumiti et al. 2014).

Effects of Quantum Dots on Embryo Development

Little information is available regarding the effects of QDs in embryos. In *C. elegans* exposed to CdSe/ZnS mercaptosuccinic acid (MSA)-capped quantum dots (QDs-MSA) (1 μM), growth from larvae to adults was unaffected. However, adults produced more eggs but laid them prematurely. These eggs have a phenotype defect that result in a high percentage of mortality and reduced lifespan which suggest that QDs-MSA might have a direct effect in the disruption of motor neurons during the reproduction process which could have negative effect in the maintenance of the population (Hsu et al. 2012).

In Vivo Effects of Quantum Dots

Although the information of QDs accumulation and ecotoxicity in aquatic invertebrates is still limited, accumulation and ecotoxicity data is available for micro-crustaceans, Annelida, platyhelminthes nematodes, bivalves and gastropods. For further information, see a review of Rocha et al. (2017).

In micro-crustaceans, QDs are directly taken up from the water by the digestive system where they remain confined to the gut (Jackson et al. 2012) or are translocated to other cells (Feswick et al. 2013). The accumulation of CdSe/ZnS (10–20 nm; 24 hr) in *D. magna* and in *C. dubia* depends on the capping layer and its toxicity on the functionalization and environmental conditions. An increase in UV light intensity has a synergistic effect in the toxicity of *D. magna* exposed to CdSe/ZnSe QDs coated with tri-n-octylphosphine oxide/gum arabic (GA) or mercaptopropionic acid QD (^{MPA}QD), and gum arabic/tri-*n*-octylphosphine oxide QD ($^{GA/TOPO}$QD) (Lee et al. 2015).

Sediment dwelling species also accumulate QDs due to the aggregation/agglomeration in the water column. In *H. diversicolor*, QDaccumulation depends on the route of exposure (higher accumulation of CdS QDs in water exposed compared to diet (5–10 nm; 10 μg/L; 14 d). Oxidative stress, genotoxic effects and behavioural impairment are induced by QDs in this species. Energy consumption was impaired after seven days of waterborne or dietary exposure that was dependent on the exposure source (Buffet et al. 2014). In contrast, in the nematode *C. elegans*, although the body surface is in direct contact with the water, no CdTe/ZnS QDs (3.6–4 nm; 0.1–1 μg/L; 3.5 d) internalization by dermal route was observed. QDs were distributed into the digestive lumen, but also attached to the pharyngeal inter surface and present in the lysosomes (Qu et al. 2011). From the digestive system, QDs were transferred to the gonads and motor neurons (indicating potential reproductive and neurotoxic effects) (Qu et al. 2011, Zhao et al. 2015). These effects were not related to the release of Cd ions from QD dissolution and when the QD shell is composed of ZnS, toxicity is reduced (Zhao et al. 2015, Zhou et al. 2015). The main elimination route of QDs in this species seems to be the anal and vulvar pathway. In fact, Zhou et al. (2015) indicate that autophagy is a defensive strategy to clear and recycle QD-damaged organelles in *C. elegans* exposed to CdTe and CdTe/CdS/ZnS QDs (6.1–6.7 nm; 1×10^{-5} mol/L; 3 hr–15 d).

Bivalves have a great capacity to accumulate QDs in their tissues from the water phase where, as was previously mentioned, aggregation/agglomeration are dependent on the ionic strength and are related to the bioavailability and uptake route of QDs (Rocha et al. 2015a). In the freshwater mussel *Elliptio complanata*, CdTe QDs (1.6–8 mg/L; 24 hr) were present in the digestive gland and oxidative damage occurred in the gills. Immunosuppressive and inflammatory effects were also detected in the peripheral haemolymph associated with oxidative stress and genotoxicity in the gills and digestive gland (Gagné et al. 2008). When *E. complanata* was exposed to the same range of CdTe concentrations, MT levels increased in the digestive gland and decreased in the gills suggesting that in this tissue, MT was oxidized as a result of oxidative stress (Peyrot et al. 2009). Oxidative stress and behaviour impairment was also detected in the marine clam *S. plana* exposed to CdS QDs (5–6 nm; 10 µg Cd/L; 14 d) through water or diet. ROS production and behaviour impairment was dependent on the source of QDs exposure. ROS in cells alter the processes of energy production leading to behaviour disturbances that could lead to changes in its biology (growth, reproduction). Furthermore, changes in the antioxidant defence system of *S. plana* were not enough to prevent apoptosis (Buffet et al. 2014). In seawater, where QDs seems to be less bioavailable compared to freshwater, aggregated/agglomerated QDs is mainly taken up in mussel gills by endocytosis and/or phagocytosis in the digestive gland. QDs are transferred from the digestive gland to the haemolymph and distributed among mussel tissues. The uptake, toxicokinetics and metabolism of *M. galloprovincialis* exposed to CdTe QDs (6 nm; 10 µg/L) have different tissue responses. The gills have an important role in the QDs uptake, the digestive gland a major role in QDs metabolism while the haemolymph transport the QDs by circulating haemocytes (Rocha et al. 2015b, 2016). QDs are linked to the biologically active metal form that is potentially toxic for mussel tissues due to the capacity to generate ROS and induce oxidative stress. These toxic effects are tissue and time dependent and involve, besides ROS production, changes in antioxidant enzymes activity and protein expression, immune response and genotoxicity, as well as cell-type replacement in digestive tubules (Rocha et al. 2016). The genotoxic effect induced by QDs seems more important than for other metal based NPs such as Ag and Cu (Gomes et al. 2012, 2013a,b, 2014) and similar to other Cd-based NPs of similar size and concentration (CdS QDs 4 nm; 10 mg/L; 4 hr; Munari et al. 2014), (5 nm; 0.001–100 mg/L; 24 hr; Katsumiti et al. 2014), CdTe QDs (1.6–8 mg/L; 24 hr; Gagné et al. 2008). Tissue specific *mt* transcription patterns were observed in QDs exposed gills which were more sensitive than the digestive gland that was associated with the biological active metal form (Rocha et al. 2017). QDs also interact with DNA or nuclear proteins where intracellular release of Cd^{2+} ions from the QDs core occur and/or DNA strand-breaks are produced through an indirect mechanism of ROS production and oxidative stress. However, this hypothesis needs to be confirmed. No elimination of QDs was observed in *M. galloprovincialis* tissues (after 30 days of depuration), indicating that QDs have a high retention rate in mussel tissues which is a cause of concern when these NPs are present in the marine environment (Rocha et al. 2016).

Although different QDs accumulation strategies exist in different aquatic invertebrate species, there are some common features in toxicokinetics and elimination, oxidative stress and genotoxicity related with the release of Cd ions from the QDs but also to the nanosize specific properties. Therefore, Rocha et al. (2017) proposed a mode of action for QDs based on the results for the marine mussel *M. galloprovincialis* that can be applied to freshwater species. The uptake of QDs (monomeric and aggregated/agglomerated forms) depends on the size, tissue and exposure time and the ecotoxicology effects involve inter- and extra-cellular ROS production that induces oxidative stress that leads to an inflammatory process, alters the immune system and induces DNA damage (Rocha et al. 2015a,b, 2017). An interesting feature is that the alterations observed in the antioxidant defence system of fresh and marine mussels exposed to QDs are higher when compared to mussels exposed to the equivalent Cd concentration in the dissolved form indicating, that the oxidative stress induced by QDs accumulation is not only due to the release of Cd ions from the QD but also associated to the nano-specific effect (Gagné et al. 2008, Buffet et al. 2014, Rocha et al. 2015a,b, 2016, 2017). Results suggest that bivalves could be recommended as good indicators to assess the effects of Cd-based QDs in the aquatic environment.

The information presently available on QDs bioavailability, accumulation and effects in aquatic invertebrates indicate that they are dependent on their nano-specific properties (size, shape, surface charge, functionalization and coating) and on changes in environmental conditions. Besides the nano-specific effects, the release of Cd ions from QDs also occur and for this reason, there is an urgent need to develop Cd-free QDs or new "core-shell-conjugate" QD structures to avoid the release of Cd ions to the aquatic environment.

Carbon-based NPs

Carbon is one of the most abundant element on Earth, and it can also be naturally found under nanosized forms with different shapes and sizes generated under several conditions. Among manufactured carbon based NPs (CNPs), the most used are graphene oxide (GO), C60 fullerene, carbon nanotubes CNTs (multiwall (MWCNT) or single (SWCNTs), Nano carbon black (NCB), and nanodiamonds (NDs). Due to their multiple properties, they promised highly interesting applications, from nanomedicine as drug carriers to electrical applications. CNTs, due to their sorption capacity of organic components, are also very attractive for their potential applications for pollution remediation in the ocean, in particular for sediment remediation (Gui et al. 2010). Many antifouling paintings contain carbon-based NPs in their mixture, and little information is available on their leaching into the water compartment (De Volder et al. 2013). For the main classes of CNPs, Predicted Environmental Concentrations (PECs) are low, with 0.004 ng/L for CNT and 0.017 ng/L for fullerene in European surface waters. However, annual deposition in sediments is expected to be higher, with 241 ng/kg/yr for CNT and 17.1 ng/kg/yr for fullerene (Gottschalk et al. 2013). Therefore, sediment dwelling and demersal species are susceptible to be more impacted.

In experimental settings, exposure concentrations identified from literature are from 0.001 to 100 mg/L, with most studies carried out at concentrations much higher (10^6) than PEC values and no data are available at levels comparable with PECs. CNPs also comprise nanoplastics, whose contribution to the impact of plastic pollution on environmental and human health represents an emerging concern (Eriksen et al. 2014). Most of our knowledge is on the uptake and biological impact of microplastics (< 5 mm), whereas smaller nanosized carbon polymeric particles (nanoplastics) are by far less studied, despite the fact that they can originate from the weathering and/or fragmentation from larger debris, mainly driven by physical-chemical but also biological factors in the receiving environment (Lambert and Wagner 2016). Although, due to mesh netting sizes, nanoplastics are not detected with commonly used methods for plastic pollution surveys, they are likely to be pervasive in the aquatic environment (Hartmann et al. 2016). To date, experimental data are available only on the effects and mechanisms of action of commercial nanoplastics (mainly polystyrene). The results show that model nanoplastics can cause severe damages in aquatic invertebrates both *in vitro* and *in vivo*.

In Vitro Effects of CNPs

In vitro data are at present available only for the haemocytes of the marine mussel *M. galloprovincialis*. Increased ROS production was observed after short term exposure (30 min) to C60 at lower concentrations (1 and 5 μg/ml) but not at the higher concentration tested (10 μg/ml). In addition, both C60 fullerene and NCB increased extracellular lysozyme and nitric oxide (NO) release, indicating inflammatory processes (Canesi et al. 2012, Canesi and Prochàzovà 2013). At longer times of exposure (24 hr), increased ROS production was observed at concentration of 25 and 15 μg/ml of GO and reduced PVP coated GO (rGO-PVP), respectively, but not at concentrations < 1.5 and 2.5 μg/ml for both GO types (Katsumiti et al. 2017).

Uptake by mussel haemocytes was observed for both GO and NCB. After 1 h exposure to NCB, hemocytes showed a concentration-dependent uptake (Canesi et al. 2012). Katsumiti et al. (2017) suggested that endocytosis and/or phagocytosis are involved in the uptake of GO and rGO-PVP; free GO was also found in the cytoplasm, that could be eventually released by breaking the membranes involved in endo-lysosomal pathways. However, it seems that this behaviour is characteristic of GO but not of CNPs in general, because no lysosomal damage was observed after exposure to C60 nor NCB (Canesi et al. 2012). This is likely due to the shape of GO as nanoplatelets. At both short-term (< 1 h) and longer-term (up to 24 h) exposure, mitochondria seem to be targeted by several types of CNPs. NCB (10 μg/ml, 45 min exposure) induced a decrease in mitochondrial mass/number and membrane potential (Canesi et al. 2008). The mitochondrial damage also observed after exposure to GO could be induced by the increased ROS production (Katsumiti et al. 2017).

The *in vitro* effects of amino modified nano polystyrene (PS-NH_2) (50 nm) have been recently investigated (Canesi et al. 2017, and refs quoted therein). Lysosomes were shown as the main subcellular target: a dose dependent decrease in lysosomal membrane stability was recorded (1; 5; 50 μg/ml), followed by lysozyme release. Extracellular ROS and NO production were also observed. At the highest concentration tested, PS-NH_2 also induced pre-apoptotic processes. Moreover, the role of soluble haemolymph proteins in the interaction with NPs and cells was investigated. In the presence of haemolymph serum (HS), the biological fluid of mussel, PS-NH_2 showed stronger effects on haemocytes functional parameters (LMS, phagocytosis activity and ROS production), compared to the artificial sea water medium. Also, the effect on phosphorylation state of the stress activated p38 MAPK and of PKC was different depending on the medium (Canesi et al. 2017). Studies with other types of NPs revealed specific interactions between soluble protein components in the hemolymph and NPs with different surface chemistry. These interactions would affect particle recognition by hemocytes resulting in distinct effects. The serum protein involved in mediating specific interactions between PS-NH_2 and haemocytes was identified as an immune related, cation binding protein (Canesi et al. 2017). These results, that represent the first data on the formation of a NP-protein corona in biological fluids of aquatic invertebrates, underline the importance of the medium used for studying the effects and mechanisms of action of NPs in *in vitro* testing with invertebrate cell models.

Effects of CNPs on Embryo and Larval Development

Information on the effects of CNPs on embryo/larval development of aquatic invertebrates is still limited to few NP types. Accordingly, most papers mainly report data on CNP uptake and accumulation, and on ecotoxicological endpoints evaluated by standard toxicity tests utilising high exposure concentrations (in the range of mg/L). However, also from these studies, information on sub-lethal responses and on the mechanisms of action of CNPs are starting to emerge.

Exposure of *A. salina* cysts to GO (from 25 to 600 mg/L for 24 hr) resulted in a decreasing hatching rate. GO was observed within the gut of young larvae and in yolk. Moreover, the swimming speed of the larvae was inhibited and the body length was decreased in a dose-dependent manner. For *A. salina* larval, mobility is a key function that enables them to find food. At higher concentrations (from 200 mg/L), lipid peroxidation, ROS production and SOD activities were significantly increased (Zhu et al. 2017a). Similar responses (inhibition of swimming behaviour and decreased hatching rate) were observed with oxidized MWCNTs (25 to 600 mg/L for 72 hr) (Zhu et al. 2017b). In this study, biomarkers of oxidative stress were also evaluated: increased ROS production and antioxidant enzymes activities (CAT, SOD, GPx) were recorded.

In the sea urchin embryo, exposure to functionalized-SWCNTs (f-SWCNTs) for 96 hr (2 to 10 mg/L) resulted in particle uptake. However, no abnormalities in the early developmental stages were observed. This absence of embryotoxicity was apparently due to the sea urchin capacity to quickly depurate, as observed during a recovery period (24 hr with food addition) (Magesky and Pelletier 2015).

Uptake of nanoplastics was observed in several species at their early life stage. In *A. franciscana* nauplii (Instar I nauplius stage < 24 hr old) exposed to cationic (PS-NH_2) and anionic (PS-COOH) polystyrene NPs (48 hr; concentration 5–100 μg/ml), particle agglomerates were observed inside the gut lumen. Moreover, PS-NH_2 were stuck on the external surface of sensorial appendages of larvae, thus indicating a possible hampering effect on natation and food uptake. Within the same experiment, increase in molts after exposure to PS-NH_2 was observed (Bergami et al. 2017). Moreover, *A. franciscana* nauplii exposed for 48 hr to PS-NH_2 (1 μg/ml) showed up regulation of *clap* and *cstb* genes, that are involved in growth, molting and tissue remodelling. Activation of this mechanism could play a role in limiting multiple moulting that could be very costly for the organism in terms of energy (Bergami et al. 2017). Embryos of *D. magna* exposed to polystyrene NPs (25 nm) at 5 mg/L showed NP accumulation within fat droplets of neonates. At this stage, embryos do not feed, which suggests absorption through the body epithelium (Brun et al. 2017). Uptake of polystyrene (16 and 870 nm) was observed in oyster larvae aged of 3, 10, and 24 days pf. Younger larvae favoured smaller size NPs and older ones preferred to feed on the high nm and μm size. Moreover, PS-NH_2 were present and retained in higher amounts in the intestine compared to PS-COOH (Cole and Galloway 2015).

Studies on sea urchin embryos (*P. lividus*) exposed to the same two types of nanopolystyrene (aminated and carboxylated) have shown that PS-NH_2 is more distributed on the outer surface membrane of the morula, while PS-COOH on the external surface at the morula and blastula stage. PS-COOH was present in the gut after 24 hr exposure to 25 μg/ml. The *Cas8* gene was up regulated by 24 hr PS-NH_2 exposure at low concentration (3 μg/ml) but not by PS-COOH. This gene is involved in apoptotic processes, activated possibly in consequence of cell membrane damage induced by PS-NH_2 exposure (Della Torre et al. 2014). Other work on sea urchin embryos has shown that exposure to PS-NH_2 could activate an alternative signalling pathway as a defense mechanism for survival. Abnormalities like poorly developed skeleton rods or underdeveloped arms were observed after exposure to 4 μg/ml at 24 hr hpf, and embryos with a normal structure showed a delay in development. PS-NH_2 exposure at low concentrations (3 and 4 μg/ml) induced upregulation of gene expression in particular at the pluteus stage (at 48 hr), of P-p38 MAPK, Hsp60 and Hsp70, Pl-Hsp60 and Pl-Univin (Pinsino et al. 2017).

In *M. galloprovincialis*, exposure to PS-NH_2 (50 nm) (from 0.001 to 20 mg/L) showed a dose-dependent decrease in normal larval development at 48 hr hpf with a total developmental arrest at higher concentrations. At 0.150 mg/L, PS-NH_2 was shown to affect the biomineralization process, as shown by changes in expression of related genes (Chitin synthase and extrapallial protein). The molecular data were supported by morphological observations of the growing shell observed by Scanning Electron Microscopy, showing thin, semi-transparent valves with irregular shell surfaces. PS-NH_2 also affected transcription of genes involved in multi xenobiotic resistance (MXR) and immune response (Balbi et al. 2017).

In Vivo Effects of Carbon Based NPs

In crustaceans, uptake of CNPs was observed in the digestive tract. The copepod *Paracyclopina nana* exposed for 96 hr to MWCNTs (2.5 to 20 mg/L) was able to ingest and excrete CNTs through the gut (Kim et al. 2016). Similarly, the copepod *Tigriopus japonicus* and the rotifer *Brachionus koreanus* exposed to MWCNTs (12 nm) for 96 hr and 24 hr, respectively, showed evidence of ingestion of MWCNTs throughout the gut tract at higher doses (100 mg/L) but not at lower doses (Lee et al. 2016a,b).

Bivalves have been shown to take up different types of CNPs. Uptake of nanodiamonds (NDs) by clams was observed, with degeneration of digestive gland cells and increasing vacuolation (Cid et al. 2015). In *Mytilus*, C60 fullerene was accumulated in a higher quantity in the digestive gland than in the gills (Di et al. 2017). Within *Mytilus* digestive gland, the lysosomal system may represent a significant target for NCB and C60 (Canesi et al. 2012). LMS of digestive gland cells showed a decreasing trend at increasing concentrations of both NCB and C60 after 24 hr exposure. Accumulation of lysosomal lipofuscin was also observed at higher concentrations (1 and 5 mg/L). However, NCB seems to have higher adverse effects than C60, since in addition a significant increase in lysosomal accumulation of neutral lipids was observed, indicating also a dysregulation of lipid metabolism in the tissue (Canesi et al. 2012).

In the clam *R. philippinarum*, an alteration of energy reserves (as protein and glycogen content) was observed after chronic exposure to MWCNTs (0.1–1 mg/L, 28 days). The authors stated that to fuel defence mechanisms was costly and cells used glycogen and proteins for this purpose (De Marchi et al. 2017b). These data indicate that CNPs are ingested by different aquatic invertebrates, apparently independent of the feeding mechanisms, and may therefore affect the digestive processes, with consequent effects on metabolism.

With regard to the mode of action, exposure to CNPs has been mainly shown to affect pro-antioxidant processes, depending on the species studied. In general, antioxidant enzymes activities increased after exposure, although differences observed depend on the species, type of CNP and exposure conditions. In the rotifer *B. koreanus* exposed to 100 mg/L MWCNT (12 nm) for 24 hr, ROS generation increased (Lee et al. 2016a); in contrast, in the copepod *T. japonicus* exposed to the same MWCNT type and concentration, a decrease in ROS was observed (Lee et al. 2016b). CAT activity increased in both species, whereas an opposite trend was shown for GST (increased for the rotifer but decreased for the copepod). An increase in LPO was reported after exposure to MWCNTs (28 days) of the clam *R. philippinarum* or after nanodiamond (ND) exposure (14 days) of *Corbicula fluminea* even at low concentrations (0.01 mg/L) (DeMarchi et al. 2017b, Cid et al. 2015). Increases in both CAT and GST activities were also recorded in *C. fluminea* exposed to NDs. In *M. galloprovincialis* exposed to NCB for 24 hr, CAT and GST activities in the digestive gland increased at higher concentrations (1 and 5 mg/L) (Canesi et al. 2012). Similarly, a general increasing trend was observed for SOD and GPx in *R. philippinarum* exposed to MWCNTs. Oxidative stress conditions result in genotoxicity: exposure of *M. galloprovincialis* to C60 fullerene (680 ± 19 nm) for three days at 1 mg/L induced DNA damage in haemocytes accompanied by an induction in the expression of p53. However, after three days of recovery, the level of DNA damage decreased, with a persistent level of p53 expression, suggesting the activation of DNA repair (Di et al. 2016). In the rotifer *B. koreanus* and copepod *P. nana*, MWCNT induced adverse effects through an ERK-dependent MAPK signalling pathway (Lee et al. 2016b, Kim et al. 2016).

In the polychaete *Diopatra neopolitana*, chronic exposure (28 days) to GO flakes (200–400 nm; from 0.01 to 1 mg/L) resulted in increased LPO and higher antioxidant enzyme activities; however, such a defence mechanism was apparently not sufficient to prevent further damage, as indicated by the reduction of the regeneration capacity of segments, in particular at higher doses (De Marchi et al. 2017a). In the clam *R. philippinarum*, chronic MWCNT exposure (28 days) induced neurotoxic effects with an inhibition of cholinesterase activity at all the concentrations tested (from 0.01 to 1 mg/L), possibly due to high affinity of carbon nanotubes for AChE (De Marchi et al. 2017b). Alterations of neurotransmission may result in behavioural changes: the polychaete *Tubifex tubifex*, exposed to GO (~ 116 nm) for five days showed an increased burrowing time compared to non exposed animals, making them more susceptible to predation (Zhang et al. 2017).

With regards to nanoplastics, most studies have reported that different types of nanopolystyrene were ingested by aquatic invertebrates such as crustaceans and rotifers. Nanoplastics were found in the gut at several concentrations and times of exposure. In the copepod *T. japonicas* exposed to two types of nanopolystyrene of different sizes (50 and 500 nm) for 96 hr, agglomerates were present in the gut. For the freshwater cladoceran, *D. magna*, exposed to nanopolystyrene (100 nm) for 24 hr at concentration 1 mg/L, the presence of food during the experiment time helped to reduce the amount of NPs retained within the gut (Rist et al. 2017). The authors suggested that NPs in the presence of food can form bigger agglomerates and are selectively taken up by the organisms. In the rotifer *B. plicatilis* (> 24 hr old), two types of surface modified polystyrene, PS-COOH (40 nm) and PS-NH_2 (50 nm), induced different effects. After 48 hr exposure to 5 μg/ml, PS-COOH were retained in the gut, even after a recovery period, suggesting a longer time needed for egestion. PS-NH_2 caused mortality in the rotifer, which was not observed for PS-COOH (Manfra et al. 2017). The aggregation state of PS-COOH as microscale agglomerates compared with the smaller size agglomerates formed by PS-NH_2 in seawater could explain the lower toxicity observed for PS-COOH, which progressed along the digestive tract until the end, whereas smaller agglomerates were able to enter the cells and tissues, causing stronger deleterious effects. This hypothesis was also put forward for nanopolystyrenes of different sizes, where smaller particles (50 and 500 nm) were suggested to have a longer retention time in organism (here the rotifer *B. koreanus*) and thus, a longer time to exert negative effects (Jeong et al. 2016).

Proteins released by *D. magna* neonates create an eco-corona around COOH- or NH_2-polystyrene NPs that caused destabilization of the NP dispersions over the subsequent 6 hr (Nasser and Lynch 2016). Secreted proteins identified by mass spectrometry included Type VI secretion system, a stress response protein, and QseC sensor protein used in cell-to-cell signalling. Independent of the protein composition, the amount of secreted protein increased over time, and this progressively increased particle instability and agglomeration. Interestingly, the eco-corona coated NPs resulted in a lower EC_{50} than equivalent uncoated NPs, and were less effectively removed from the gut. The authors hypothesized that larger agglomerates are probably more attractive as a food source, leading to higher accumulation and increasing toxicity (Nasser and Lynch 2016).

In the rotifer *B. koreanus* exposed 12 days nanopolystyrene of different sizes (50 and 500 nm), an increase in antioxidant enzyme activities (e.g., SOD, GR, GPx) was observed, likely due to the increase of ROS production. Moreover, phosphorylation of C-Jun Nterminal kinase (p-JNK) and p38 was increased after exposure to smaller NPs size polystyrene compared to the bigger size. The stress activated MAPK response was suggested to be correlated to the increase in intra cellular ROS level observed (Jeong et al. 2016).

Overall, *in vivo* studies show that CNPs are ingested by several aquatic invertebrates and can causes damage to digestive systems, in particular to the gut epithelium. This may consequently affect feeding processes and ultimately lead to death. However, studies where exposure was followed by a recovery period showed that some organisms were able to depurate or activate efficient defence mechanisms towards chronic and low doses of CNPs. Both *in vivo* and *in vitro* data indicate that oxidative stress related parameters are significantly affected by exposure to CNPs. These effects were accompanied by membrane damage (plasma, lysosome, mitochondria) in almost all studies, using several types of invertebrate models. Preliminary data also suggest that the biological effects observed may be also related to the shape of the CNPs, in particular nanotubes or nano platelets. With regards to nanoplastic, the only available data are on different types of polystyrene, in its anionic (PS-COOH) and cationic (PS-NH_2) forms. In all studies, PS-NH_2 has been found to be more toxic than the anionic form, as in mammalian systems. However, this difference in toxicity could be due not only to the particle surface charge *per se*, but to difference in the size of agglomerates formed in exposure media. This observation is directly linked with the general conclusion on size-related toxicity of NPs, where the smaller NPs have shown to be more hazardous than their bigger size counterparts. Surface charge may also play a role in particle recognition and in mediating the effect at the cellular level. Finally, the results obtained in model marine invertebrate species point out how certain types of nanoplastics can affect embryo development acting on a number of target genes involved in crucial physiological functions.

Conclusions and Perspectives

In the last decade, knowledge on the sublethal effects of different classes of NPs in aquatic invertebrates has considerably increased. Data obtained from *in vitro* testing are helpful to understand in a simpler manner the interactions of NPs with cells and to draw the main mechanisms of action. Invertebrate hemocytes and gill cells have proven as suitable models for investigating the effects and mechanisms of action of NPs. In Fig. 1, available *in vitro* data are summarized for different classes on NPs. As in nanotoxicology, also in nanoecotoxicology, among urgent research needs and priorities it is of critical importance to incorporate nanosafety into the development of novel nanotechnologies and products

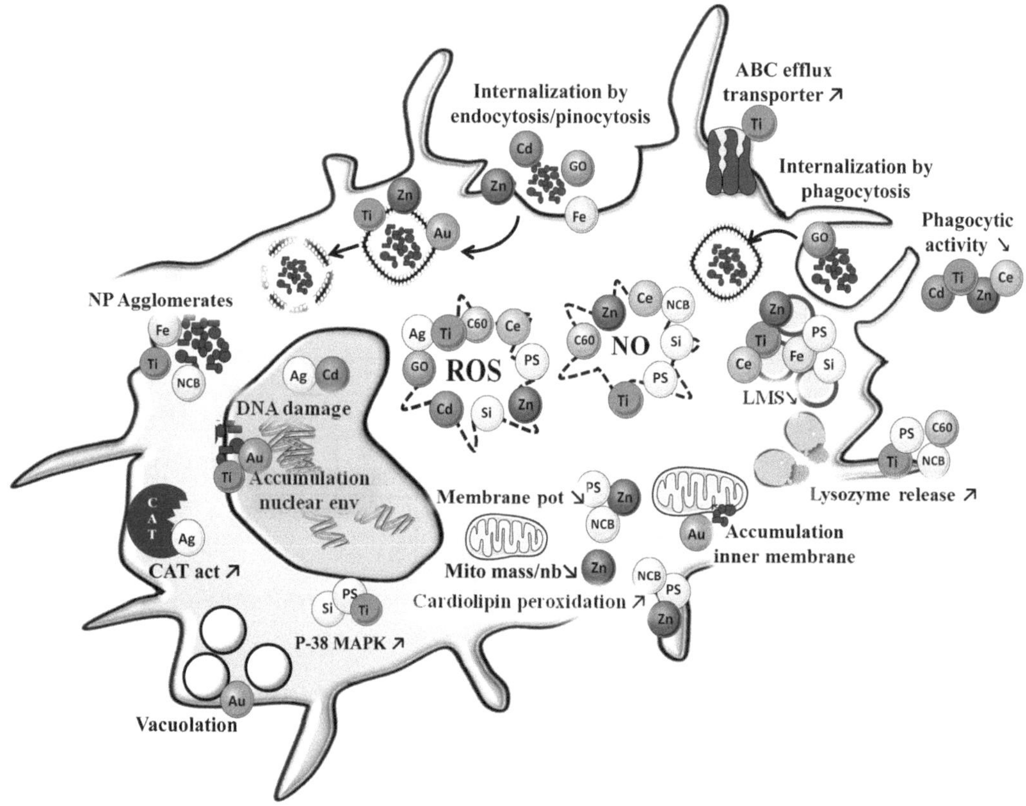

Figure 1. Schematic diagram of main *in vitro* effects and mechanisms of action of NPs at cellular level in bivalves (hemocytes or gill cells).

Abbreviated NPs names: AgNP, AuNP, nZVI, nZnO, nTiO, nCeO, nSiO, nCuO, nFeO, CNT (MWCNT and SWCNT), GO, C60, PS (PS-NH_2 and PS-COOH). For other abbreviations, refer to the text.

(safety by design). In this light, the utilisation of cell models of invertebrate groups representative of different environmental compartments can represent a valuable tool for the *in vitro* screening of the potential toxicity of different NPs. Unfortunately, *in vitro* studies on aquatic invertebrates are limited by the absence of commercially available immortalised cell lines. Moreover, information is emerging on the interactions of NPs occurring at the molecular level in both the external and internal environment (i.e., exposure medium and biological fluids within the organisms), leading to the formation of NP-protein coronas, that will in turn affect NP uptake and toxicity. In this light, for a correct evaluation of the *in vitro* effects of NPs in invertebrate cells, information is needed on particle behaviour in the biological fluids of different species (Canesi and Corsi 2016, Canesi et al. 2017).

With regards to data obtained in *in vivo* exposure experiments, carried out both in developing stages and in adults, similar outcomes were found for different kinds of NPs (zero valent, nano oxides, quantum dots and carbon based), namely uptake and/or accumulation within tissues, ROS production, oxidative stress, membrane and DNA damage. Particle composition, size and distribution, functional groups or coatings, solubility, aggregation/agglomeration are key elements for determining their ecotoxicity in both freshwater and marine species. Studies involving exposure to concentrations closer to PECs and longterm exposure experiments will contribute in the ecotoxicity assessment of NPs. Moreover, data obtained on edible invertebrates would be of additional interest in the light of the economic importance of aquaculture species as well as of their potential role as a source of contamination to humans.

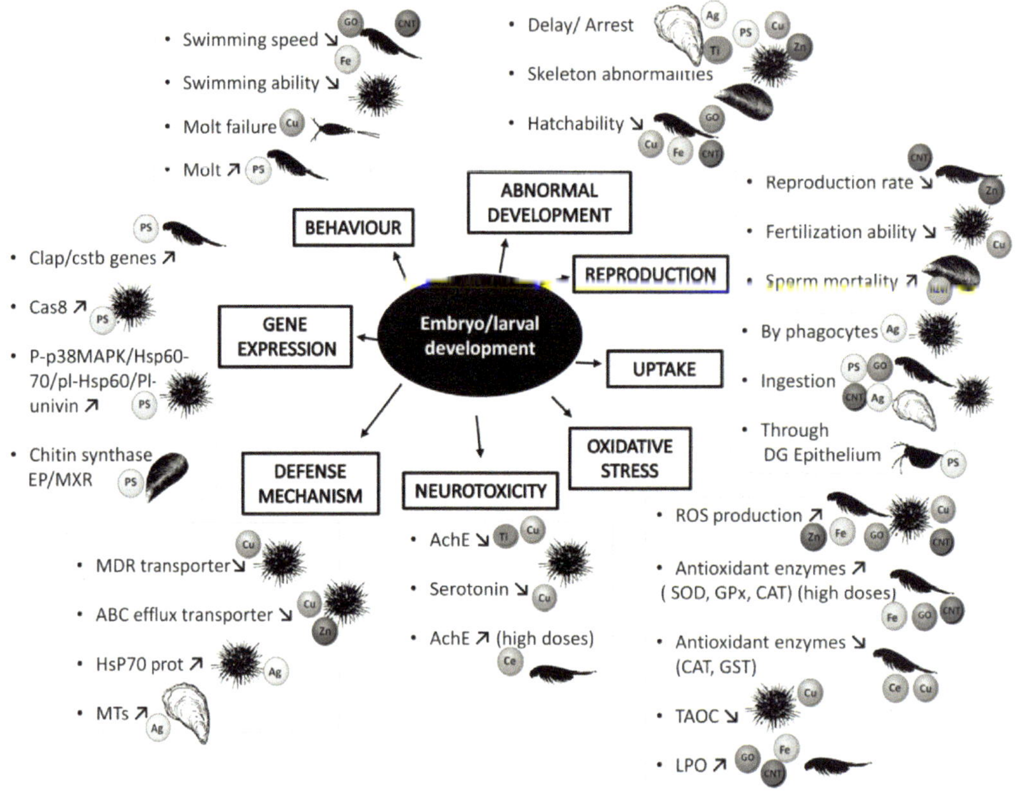

Figure 2. Schematic diagram of main effects of NPs observed on embryos and larval development.

Legend: *Crassostrea* sp.; *D. magna*; *Artemia* sp.; *Tigriopus fulvus*; *M. galloprovincialis*; *Sea urchin* (*S. granularis*, *A. lixula*, *L. pictus*, *P. lividus*, *S. droebachisiensis*).

Abbreviated NPs names: AgNP, AuNP, nZVI, nZnO, nTiO, nCeO, nSiO, nCuO, nFeO, CNT (MWCNT and SWCNT), GO, C60, PS ($PS-NH_2$ and PS-COOH). For other abbreviations, refer to the text.

From a strictly ecological perspective, the growing body of evidence on the developmental effects of NPs in aquatic invertebrates can represent an additional cause for concern. NPs can affect developmental processes to a different extent, depending on the type of particle, concentration, time of exposure, developmental stage and species (Fig. 2). The effects of NPs on early stages would be reflected at higher level of biological organisation, thus affecting population sustainability. These studies will contribute in bridging the gap between mechanistic molecular understanding of the effects of NPs and the larger scale ecosystem responses.

References

Ali, D., H. Ali, S. Alarifi, S. Kumar, M. Serajuddin, A.P. Mashih et al. 2015. Impairment of DNA in a Freshwater Gastropod (*Lymnea luteola* L.) after exposure to Titanium dioxide nanoparticles. Arch Environ. Contam. Toxicol. 68: 543–552.

Al-Sid-Cheikh, M., C. Rouleau and E. Pelletier. 2013. Tissue distribution and kinetics of dissolved and nanoparticulate silver in Iceland scallop (*Chlamys islandica*). Mar. Environ. Res. 86: 21–28.

Amde, M., J. Liu, Z.Q. Tan and D. Bekana. 2017. Transformation and bioavailability of metal oxide nanoparticles in aquatic and terrestrial environments. A review. Environ. Pollut. 230: 250–267.

Amiard, J.–C., C. Amiard–Triquet, S. Barka, J. Pellerin and P.S. Rainbow. 2006. Metallothioneins in aquatic invertebrates: Their role in metal detoxification and their use as biomarkers. Aquat. Toxicol. 76: 160–202.

Baker, T.J., C.R. Tyler and T.S. Galloway. 2014. Impacts of metal and metal oxide nanoparticles on marine organisms. Environ. Pollut. 186: 257–271.

Balbi, T., A. Smerilli, R. Fabbri, C. Ciacci, M. Montagna, E. Grasselli et al. 2014. Co-exposure to n-TiO_2 and Cd^{2+} results in interactive effects on biomarker responses but not in increased toxicity in the marine bivalve *M. galloprovincialis*. Sci. Total Environ. 493: 355–364.

Balbi, T., G. Camisassi, M. Montagna, R. Fabbri, S. Franzellitti, C. Carbone et al. 2017. Impact of cationic polystyrene nanoparticles (PS-NH_2) on early embryo development of *Mytilus galloprovincialis*: Effects on shell formation. Chemosphere 186: 1–9.

Baun, A., N.B. Hartmann, K. Grieger and K.O. Kusk. 2008. Ecotoxicity of engineered nanoparticles to aquatic invertebrates: A brief review and recommendations for future toxicity testing. Ecotoxicology 17: 387–395.

Bebianno, M.J., T. Gomes and D. Sheehan. 2017. Protein expression profiles in marine organisms exposed to nanoparticles. pp. 173–189. *In*: T.G. Barrera and J.L. Gomez-Ariza (eds.). Environmental Problems in Marine Biology: Methodological Aspects and Applications. CRC Press.

Bergami, E., S. Pugnalini, M.L. Vannuccini, L. Manfra, C. Faleri, F. Savorelli et al. 2017. Long-term toxicity of surface-charged polystyrene nanoplastics to marine planktonic species *Dunaliella tertiolecta* and *Artemia franciscana*. Aquat. Toxicol: 189: 159–169.

Brun, N.R., M.M.T. Beenakker, E.R. Hunting, D. Ebert and M.G. Vijver. 2017. Brood pouch-mediated polystyrene nanoparticle uptake during *Daphnia magna* embryogenesis. Nanotoxicology 11: 1059–1069.

Bruneau, A., M. Fortier, F. Gagne, C. Gagnon, P. Turcotte, A. Tayabali et al. 2013. Size distribution effects of cadmium tellurium quantum dots (CdS/CdTe) immunotoxicity on aquatic organisms. Environ. Sci. Proc. Imp. 15: 596–607.

Buffet, P.E., O.F. Tankoua, J.F. Pan, D. Berhanu, C. Herrenknecht, L. Poirier et al. 2011. Behavioural and biochemical responses of two marine invertebrates *Scrobicularia plana* and *Hediste diversicolor* to copper oxide nanoparticles. Chemosphere 84: 166–174.

Buffet, P.E., J.F. Pan, L. Poirier, C. Amiard-Triquet, J.C. Amiard, P. Gaudin et al. 2013. Biochemical and behavioural responses of the endobenthic bivalve *Scrobicularia plana* to silver nanoparticles in seawater and microalgal food. Ecotoxicol. Environ. Safe 89: 117–124.

Buffet, P.E., L. Poirier, A. Zalouk-Vergnoux, C. Lopes, J.C. Amiard, P. Gaudin et al. 2014. Biochemical and behavioural responses of the marine polychaete *Hediste diversicolor* to cadmium sulfide quantum dots (CdS QDs): Waterborne and dietary exposure. Chemosphere 100: 63–70.

Burić, P., Ž. Jakšić, L. Štajner, M. Dutour Sikirić, D. Jurašin, C. Cascio et al. 2015. Effect of silver nanoparticles on Mediterranean sea urchin embryonal development is species specific and depends on moment of first exposure. Mar. Environ. Res. 111: 50–59.

Canesi, L., C. Ciacci, R. Fabbri, A. Marcomini, G. Pojana and G. Gallo. 2012. Bivalve molluscs as a unique target group for nanoparticle toxicity. Mar. Environ. Res. 76: 16–21.

Canesi, L. and P. Procházová. 2013. The invertebrate immune system as a model for investigating the environmental impact of nanoparticles. pp. 91–112. *In*: D. Boraschi and A. Duschl (eds.). Nanoparticles and the Immune System. Oxford Academic Press.

Canesi, L. and I. Corsi. 2016. Effects of nanomaterials on marine invertebrates. Sci. Total Environ. 565: 933–940.

Canesi, L., T. Balbi, R. Fabbri, A. Salis, G. Damonte, M. Volland et al. 2017. Biomolecular coronas in invertebrate species: Implications in the environmental impact of nanoparticles. NanoImpact. 8: 89–98.

Cappello, T., V. Vitale, S. Oliva, V. Villari, A. Mauceri, S. Fasulo et al. 2017. Alteration of neurotransmission and skeletogenesis in sea urchin *Arbacia lixula* embryos exposed to copper oxide nanoparticles. Comp. Biochem. Phys. C 199: 20–27.

Cid, A., A. Picado, J.B. Correia, R. Chaves, H. Silva, J. Caldeira et al. 2015. Oxidative stress and histological changes following exposure to diamond nanoparticles in the freshwater Asian clam *Corbicula fluminea* (Müller, 1774). J. Hazard Mater. 284: 27–34.

Cole, M. and T.S. Galloway. 2015. Ingestion of nanoplastics and microplastics by pacific oyster larvae. Environ. Sci. Technol. 49: 14625–14632.

Corsi, I., G.N. Cherr, H.S. Lenihan, J. Labille, M. Hassellov, L. Canesi et al. 2014. Common strategies and technologies for the ecosafety assessment and design of nanomaterials entering the marine environment. ACS Nano 8: 9694–9709.

Dai, L., K. Syberg, G.T. Banta, H. Selck and V.E. Forbes. 2013. Effects, uptake, and depuration kinetics of silver oxide and copper oxide nanoparticles in a marine deposit feeder, *Macoma balthica*. ACS Sustain. Chem. Eng. 1: 760–767.

De Marchi, L., V. Neto, C. Pretti, E. Figueira, L. Brambilla, M.J. Rodriguez-Douton et al. 2017a. Physiological and biochemical impacts of graphene oxide in polychaetes: The case of *Diopatra neapolitana*. Comp. Biochem. Phys. C 193: 50–60.

De Marchi, L., V. Neto, C. Pretti, E. Figueira, F. Chiellini, A.M.V.M. Soares et al. 2017b. The impacts of emergent pollutants on *Ruditapes philippinarum*: Biochemical responses to carbon nanoparticles exposure. Aquat. Toxicol. 187: 38–47.

De Volder, M.F., S.H. Tawfick, R.H. Baughman and A.J. Hart. 2013. Carbon nanotubes: Present and future commercial applications. Science 339: 535–539.

Della Torre, C., E. Bergami, A. Salvati, C. Faleri, P. Cirino, K.A. Dawson et al. 2014. Accumulation and embryotoxicity of polystyrene nanoparticles at early stage of development of sea urchin embryos *Paracentrotus lividus*. Environ. Sci. Technol. 48: 12302–12311.

Di, Y., Y. Aminot, D.C. Schroeder, J.W Readman and A.N. Jha. 2017. Integrated biological responses and tissue-specific expression of *p53* and *ras* genes in marine mussels following exposure to benzo(α)pyrene and C60 fullerenes, either alone or in combination. Mutagenesis 32: 77–90.

Doyle, J.J., J.E. Ward and R. Mason. 2015. An examination of the ingestion, bioaccumulation, and depuration of titanium dioxide nanoparticles by the blue mussel (*Mytilus edulis*) and the eastern oyster (*Crassostrea virginica*). Mar. Environ. Res. 110: 45–52.

Ekimov, A.I. and A.A. Onushchenko. 1982. Quantum size effect in the optical-spectra of semiconductor micro-crystals. Sov. Phys. Semicond. 16: 775–778.

Eriksen, M., L.C.M. Lebreton, H.S. Carson, M. Thiel, C.J. Moore, J.C. Borerro et al. 2014. Plastic pollution in the world's oceans: More than 5 trillion plastic pieces weighing over 250,000 tons afloat at sea. Plos One 9: e111913.

Feswick, A., R.J. Griffitt, K. Siebein and D.S. Barber. 2013. Uptake, retention and internalization of quantum dots in *Daphnia* is influenced by particle surface functionalization. Aquat. Toxicol. 130-131: 210–8.

Gagné, F., J. Auclair, P. Turcotte, M. Fournier, C. Gagnon, S. Sauvé et al. 2008. Ecotoxicity of CdTe quantum dots to freshwater mussels: impacts on immune system, oxidative stress and genotoxicity. Aquat. Toxicol. 86: 333–340.

Gambardella, C., M.G. Aluigi, S. Ferrando, L. Gallus, P. Ramoino, A.M. Gatti et al. 2013. Developmental abnormalities and changes in cholinesterase activity in sea urchin embryos and larvae from sperm exposed to engineered nanoparticles. Aquat. Toxicol. 130-131: 77–85.

García-Alonso, J., F.R. Khan, S.K. Misra, M. Turmaine, B.D.Smith, P.S. Rainbow et al. 2011. Cellular internalization of silver nanoparticles in gut epithelia of the estuarine polychaete *Nereis diversicolor*. Environ. Sci. Technol. 45: 4630–4636.

García-Negrete, C.A., M.C. Jiménez de Haro, J. Blasco, M. Soto and A. Fernández. 2015. STEM-in-SEM high resolution imaging of gold nanoparticles and bivalve tissues in bioaccumulation experiments. The Analyst 140: 3082–3089.

Girardello, F., C.C. Leite, C.S. Branco, M. Roesch-Ely, A.N. Fernandes, M. Salvador et al. 2016. Antioxidant defences and haemocyte internalization in *Limnoperna fortunei* exposed to TiO_2 nanoparticles. Aquat. Toxicol. 176: 190–196.

Gomes, T., C.G. Pereira, C. Cardoso, J.P. Pinheiro, I. Cancio and M.J. Bebianno. 2012. Accumulation and toxicity of copper oxide nanoparticles in the digestive gland of *Mytilus galloprovincialis*. Aquat. Toxicol. 118-119: 72–79.

Gomes, T., O. Araújo, R. Pereira, A.C. Almeida, A. Cravo and M.J. Bebianno. 2013a. Genotoxicity of copper oxide and silver nanoparticles in the mussel *Mytilus galloprovincialis*. Mar. Environ. Res. 84: 51–59.

Gomes, T., C.G. Pereira, C. Cardoso and M.J. Bebianno. 2013b. Differential protein expression in mussels *Mytilus galloprovincialis* exposed to nano and ionic Ag. Aquat. Toxicol. 136-137: 79–90.

Gomes, T., S. Chora, C.G. Pereira, C. Cardoso and M.J. Bebianno. 2014a. Proteomic response of mussels *Mytilus galloprovincialis* exposed to CuO NPs and Cu^{2+}: An exploratory biomarker discovery. Aquat. Toxicol. 155: 327–336.

Gomes, T., C.G. Pereira, C. Cardoso, V.S. Sousa, M.R. Teixeira, J.P. Pinheiro et al. 2014b. Effects of silver nanoparticles exposure in the mussel *Mytilus galloprovincialis*. Mar. Environ. Res. 101: 208–214.

Gottschalk, F., T. Sun and B. Nowack. 2013. Environmental concentrations of engineered nanomaterials: Review of modeling and analytical studies. Environ. Pollut. 181: 287–300.

Gui, X., J. Wei, K. Wang, A. Cao, H. Zhu, Y. Jia et al. 2010. Carbon nanotube sponges. Adv. Mater. 22: 617–621.

Hartmann, N.B., L.M. Skjolding, T. Nolte and A. Baun. 2016. Aquatic ecotoxicity testing of nanoplastics-lessons learned from nanoecotoxicology. pp. 43–44. *In*: SETAC Europe 26th Annual Meeting—abstract book. Nantes, France: SETAC Europe.

Hayashi, Y. and P. Engelmann. 2013. Earthworm's immunity in the nanomaterial world: New room, future challenges. Invertebr Surviv. J. 10: 69–76.

Hsu, P.C.L., M. O'Callaghan, N. Al-Salim and M.R.H. Hurst. 2012. Quantum dot nanoparticles affect the reproductive system of *Caenorhabditis elegans*. Environl. Toxicol. Chem. 31: 2366–74.

Hu, W., S. Culloty, G. Darmody, S. Lynch, J. Davenport, S. Ramirez-Garcia et al. 2014. Toxicity of copper oxide nanoparticles in the blue mussel, *Mytilus edulis*: a redox proteomic investigation. Chemosphere 108: 289–99.

Hull, M.S., P. Chaurand, J. Rose, M. Auffan, J.Y. Bottero, J.C. Jones et al. 2011. Filter-feeding bivalves store and biodeposit colloidally stable gold nanoparticles. Environ. Sci. Technol. 45: 6592–6599.

Hund-Rinke, K., A. Baun, D. Cupi, T.F. Fernandes, R. Handy, J.H. Kinross et al. 2016. Regulatory ecotoxicity testing of nanomaterials—proposed modifications of OECD test guidelines based on laboratory experience with silver and titanium dioxide nanoparticles. Nanotoxicology 10: 1442–1447.

Jackson, B.P., D. Bugge, J.F. Ranville and C.Y. Chen. 2012. Bioavailability, toxicity, and bioaccumulation of quantum dot nanoparticles to the amphipod *Leptocheirus plumulosus*. Environ. Sci. Technol. 46: 5550–6.

Jeong, C.B., E.J. Won, H.M. Kang, M.C. Lee, D.S. Hwang, U.K Hwang et al. 2016. Microplastic size-dependent toxicity, oxidative stress induction, and p-JNK and p-p38 activation in the monogonont rotifer (*Brachionus koreanus*). Environl. Sci. Technol. 50: 8849–8857.

Kádár, E., D.M. Lowe, M. Solé, A.S. Fisher, A.N. Jha, J.W. Readman et al. 2010. Uptake and biological responses to nano-Fe versus soluble $FeCl_3$ in excised mussel gills. Anal. Bioanal. Chem. 396: 657–666.

Kádár, E., G.A. Tarran, A.N. Jha and S.N. Al-Subiai. 2011. Stabilization of engineered zero-valent nanoiron with Na-acrylic copolymer enhances spermiotoxicity. Environ. Sci. Technol. 45: 3245–3251.

Katsumiti, A., D. Gilliland, I. Arostegui and M.P. Cajaraville. 2014. Cytotoxicity and cellular mechanisms involved in the toxicity of CdS quantum dots in hemocytes and gill cells of the mussel *Mytilus galloprovincialis*. Aquat. Toxicol. 153: 39–52.

Katsumiti, A., I. Arostegui, M. Oron, D. Gilliland, E. Valsami-Jones and M.P. Cajaraville. 2015a. Cytotoxicity of Au, ZnO and SiO_2 NPs using *in vitro* assays with mussel hemocytes and gill cells: Relevance of size, shape and additives. Nanotoxicology 1–9.

Katsumiti, A., D. Gilliland, I. Arostegui and M.P. Cajaraville. 2015b. Mechanisms of toxicity of Ag nanoparticles in comparison to bulk and ionic Ag on mussel hemocytes and gill cells. Plos One 10: e0129039.

Katsumiti, A., R. Tomovska and M.P. Cajaraville. 2017. Intracellular localization and toxicity of graphene oxide and reduced graphene oxide nanoplatelets to mussel hemocytes *in vitro*. Aquat. Toxicol. 188: 138–147.

Kim, D.-H., J. Puthumana, H.M. Kang, M.C. Lee, C.B. Jeong, J. Han et al. 2016. Adverse effects of MWCNTs on life parameters, antioxidant systems, and activation of MAPK signaling pathways in the copepod *Paracyclopina nana*. Aquat. Toxicol. 179: 115–124.

Lambert, S. and M. Wagner. 2016. Characterisation of nanoplastics during the degradation of polystyrene. Chemosphere 145: 265–268.

Langston, W.J., M.J. Bebianno and G.R. Burt. 1998. Metal handling strategies in molluscs. pp. 219–283. *In*: W.J. Langston and M.J. Bebianno (eds.). Metal Metabolism in Aquatic Environments. Chapman and Hall, London.

Lapresta-Fernández, A., A. Fernández and J. Blasco. 2012. Nanoecotoxicity effects of engineered silver and gold nanoparticles in aquatic organisms. Trends. Anal. Chem. 32: 40–59.

Lee, W.M. and Y.J. An. 2015. Evidence of three-level trophic transfer of quantum dots in an aquatic food chain by using bioimaging. Nanotoxicology 9: 407–412.

Lee, J.W., H.M. Kang, E.J. Won, D.S. Hwang, D.H. Kim, S.J. Lee et al. 2016a. Multi-walled carbon nanotubes (MWCNTs) lead to growth retardation, antioxidant depletion, and activation of the ERK signaling pathway but decrease copper bioavailability in the monogonont rotifer (*Brachionus koreanus*). Aquat. Toxicol. 172: 67–79.

Lee, J.W., E.J. Won, H.M. Kang, D.S. Hwang, D.H. Kim, R.K. Kim et al. 2016b. Effects of multi-walled carbon nanotube (MWCNT) on antioxidant depletion, the ERK signaling pathway, and copper bioavailability in the copepod (*Tigriopus japonicus*). Aquat. Toxicol. 171: 9–19.

Limbach, L.K., P. Manser, R.N. Grass, A. Bruinik and W.J. Stark. 2007. Exposure of engineered nanoparticles to human lung epithelial cells: Influence of chemical composition and catalytic activity on oxidative stress. Environ. Sci. Technol. 41: 4158–4163.

Libralato, G., D. Minetto, S. Totaro, I. Mičetić, A. Pigozzo, E. Sabbioni et al. 2013. Embryotoxicity of TiO_2 nanoparticles to *Mytilus galloprovincialis* (Lmk). Mar. Environ. Res. 92: 71–78.

Madhav, M.R., S. Einstein Mariya David, R.S. Suresh Kumar, J.S. Swathy, M. Bhuvaneshwari, Amitava Mukherjee et al. 2017. Toxicity and accumulation of Copper oxide (CuO) nanoparticles in different life stages of Artemia salina. Environ. Toxicol. Pharmacol. 52: 227–238.

Magesky, A. and E. Pelletier. 2015. Toxicity mechanisms of ionic silver and polymer-coated silver nanoparticles with interactions of functionalized carbon nanotubes on early development stages of sea urchin. Aquat. Toxicol. 167: 106–123.

Magesky, A., C.A.O. Ribeiro and E. Pelletier. 2016. Physiological effects and cellular responses of metamorphic larvae and juveniles of sea urchin exposed to ionic and nanoparticulate silver. Aquat. Toxicol. 174: 208–227.

Magesky, A., C.A.O. Ribeiro, L. Beaulieu and E. Pelletier. 2017. Silver nanoparticles and dissolved silver activate contrasting immune responses and stress-induced heat shock protein expression in sea urchin. Environ. Toxicol. Chem. 9999: 1–15.

Mahapatra, I., T.Y. Sun, J.R.A. Clark, P.J. Dobson, K. Hungerbuehler, R. Owen et al. 2015. Probabilistic modelling of prospective environmental concentrations of gold nanoparticles from medical applications as a basis for risk assessment. J. Nanobiotechnol. 13.

Maisano, M., T. Cappello, E. Catanese, V. Vitale, A. Natalotto, A. Giannetto et al. 2015. Developmental abnormalities and neurotoxicological effects of CuO NPs on the black sea urchin *Arbacia lixula* by embryotoxicity assay. Mar. Environ. Res. 111: 121–127.

Manfra, L., A. Rotini, E. Bergami, G. Grassi, C. Faleri and I. Corsi. 2017. Comparative ecotoxicity of polystyrene nanoparticles in natural seawater and reconstituted seawater using the rotifer *Brachionus plicatilis*. Ecotox. Environ. Safe 145: 557–563.

Manzo, S., M.L. Miglietta, G. Rametta, S. Buono and G. Di Francia. 2013. Embryotoxicity and spermiotoxicity of nanosized ZnO for Mediterranean sea urchin *Paracentrotus lividus*. J. Hazard. Mater. 254-255: 1–9.

Marisa, I., M.G. Marin, F. Caicci, E. Franceschinis, A. Martucci and V. Matozzo. 2015. *In vitro* exposure of haemocytes of the clam *Ruditapes philippinarum* to titanium dioxide (TiO_2) nanoparticles: Nanoparticle characterisation, effects on phagocytic activity and internalisation of nanoparticles into haemocytes. Mar. Environ. Res. 103: 11–17.

Matranga, V. and I. Corsi. 2012. Toxic effects of engineered nanoparticles in the marine environment: Model organisms and molecular approaches. Mar. Environ. Res. 76: 32–40.

McGillicuddy, E., I. Murray, S. Kavanagh, L. Morrison, A. Fogarty, M. Cormican et al. 2017. Silver nanoparticles in the environment: Sources, detection and ecotoxicology. Sci. Total Environ. 575: 231–246.

Munari, M., J. Sturve, G. Frenzilli, M.B. Sanders, P. Christian, M. Nigro et al. 2014. Genotoxic effects of Ag2S and CdS nanoparticles in blue mussel (*Mytilus edulis*) haemocytes. J. Chem. Ecol. 1–7.

Nasser, F. and I. Lynch. 2016. Secreted protein eco-corona mediates uptake and impacts of polystyrene nanoparticles on *Daphnia magna*. J. Proteomics 137: 45–51.

Navarro, E., A. Baun, R. Behra, N.B. Hartmann, J. Filser, A. Miao et al. 2008. Environmental behavior and ecotoxicity of engineered nanoparticles to algae, plants and fungi. Ecotoxicology 17: 372–386.

Pan, J.F., P.E. Buffet, L. Poirier, C. Amiard–Triquet, D. Gilliland, Y. Joubert et al. 2012. Size dependent bioaccumulation and ecotoxicity of gold nanoparticles in an endobenthic invertebrate: The Tellinid clam *Scrobicularia plana*. Environ. Pollut. 168: 37–43.

Petersen, E.J., S.A. Diamond, A.J. Kennedy, G.G. Goss, K. Ho, J.R. Lead et al. 2015. Adapting OECD aquatic toxicity tests for use with manufactured nanomaterials: key issues and consensus recommendations. Environ. Sci. Technol. 49: 9532–47.

Peyrot, C., C. Gagnon, F. Gagné, K.J. Wilkinson, P. Turcotte and S. Sauvé. 2009. Effects of cadmium telluride quantum dots on cadmium bioaccumulation and metallothionein production to the freshwater mussel, *Elliptio complanata*. Comp. Biochem. Physiol. C 150: 246–251.

Pinsino, A., R. Russo, R. Bonaventura, A. Brunelli, A. Marcomini and V. Matranga. 2015. Titanium dioxide nanoparticles stimulate sea urchin immune cell phagocytic activity involving TLR/p38 MAPK-mediated signalling pathway. Sci. Rep. 5.

Pinsino, A., E. Bergami, C. Della Torre, M.L. Vannuccini, P. Addis, M. Secci et al. 2017. Amino-modified polystyrene nanoparticles affect signalling pathways of the sea urchin (*Paracentrotus lividus*) embryos. Nanotoxicology 11: 201–209.

Qu, Y., W. Li, Y. Zhou, X. Liu, L. Zhang, L. Wang et al. 2011. Full assessment of fate and physiological behavior of quantum dots utilizing *Caenorhabditis elegans* as a model organism. Nano Lett. 11: 3174–3183.

Renault, S., M. Baudrimont, N. Mesmer-Dudons, S. Mornet and A. Brisson. 2008. Impact of nanoparticle exposure on two freshwater species: a phytoplanktonic alga (*Scenedesmus subspicatus*) and a benthic bivalve (*Corbicula fluminea*). Gold Bull. 41: 116–126.

Ringwood, A.H., M. McCarthy, T.C. Bates and D.L. Carroll. 2010. The effects of silver nanoparticles on oyster embryos. Mar. Environ. Res. 69: S49–S51.

Rist, S., A. Baun and N.B. Hartmann. 2017. Ingestion of micro- and nanoplastics in *Daphnia magna*—Quantification of body burdens and assessment of feeding rates and reproduction. Environ. Pollut. 228: 398–407.

Rocha, T.L., T. Gomes, V.S. Sousa, N.C. Mestre and M.J. Bebianno. 2015a. Ecotoxicological impact of engineered nanomaterials in bivalve molluscs: An overview. Mar. Environ. Res. 111: 74–88.

Rocha, T.L., T. Gomes, J.P. Pinheiro, V.S. Sousa, L.M. Nunes, M.R. Teixeira et al. 2015b. Toxicokinetics and tissue distribution of cadmium-based quantum dots in the marine mussel *Mytilus galloprovincialis*. Environ. Pollut. 204: 207–214.

Rocha, T.L., T. Gomes, E. Giuliani and M.J. Bebianno. 2016. Subcellular partitioning kinetics, metallothionein response and oxidative damage in the marine mussel *Mytilus galloprovincialis* exposed to cadmium-based quantum dots. Sci. Total Environ. 554-555: 130–141.

Rocha, T.L., N.C. Mestre, S.M.T. Sabóia-Morais and M.J. Bebianno. 2017. Environmental behaviour and ecotoxicity of quantum dots at various trophic levels: A review. Environ. Int. 98: 1–17.

Rotini, A., A. Gallo, I. Parlapiano, M.T. Berducci, R. Boni, E. Tosti et al. 2018. Insights into the CuO nanoparticle ecotoxicity with suitable marine model species. Ecotox. Environ. Safe 147: 852–860.

Santillán-Urquiza, E., F. Arteaga-Cardona, C. Torres-Duarte, B. Cole, B. Wu, M.A. Méndez-Rojas et al. 2017. Facilitation of trace metal uptake in cells by inulin coating of metallic nanoparticles. Roy. Soc. Open. Sci. 4: 170480.

Selck, H., R.D. Handy, T.F. Fernandes, S.J. Klaine and E.J. Petersen. 2016. Nanomaterials in the aquatic environment: A European Union-United States perspective on the status of ecotoxicity testing, research priorities, and challenges ahead: Nanomaterials in the aquatic environment. Environ. Toxicol. Chem. 35: 1055–1067.

Shi, W., Y. Han, C. Guo, X. Zhao, S. Liu, W. Su et al. 2017. Immunotoxicity of nanoparticle $nTiO_2$ to a commercial marine bivalve species, *Tegillarca granosa*. Fish Shellfish Immun. 66: 300–306.

Šiller, L., M.L. Lemloh, S. Piticharoenphun, B.G. Mendis, B.R. Horrocks, F. Brümmer et al. 2013. Silver nanoparticle toxicity in sea urchin *Paracentrotus lividus*. Environ. Pollut. 178: 498–502.

Smith, A.M., H. Duan, A.M. Mohs and S. Nie. 2008. Bioconjugated quantum dots for *in vivo* molecular and cellular imaging. Adv. Drug. Deliv. Rev. 60: 1226–1240.

Sugantharaj David, E.M.D., M. Madurantakam Royam, S.K. Rajamani Sekar, B. Manivannan, S. Jalaja Soman, A. Mukherjee et al. 2017. Toxicity, uptake, and accumulation of nano and bulk cerium oxide particles in *Artemia salina*. Environ. Sci. Pollut. Res. 24: 24187–24200.

Tedesco, S., H. Doyle, J. Blasco, G. Redmond and D. Sheehan. 2010. Oxidative stress and toxicity of gold nanoparticles in *Mytilus edulis*. Aquat. Toxicol. 100: 1781.

Tiwari, P., K. Vig, V. Dennis and S. Singh. 2011. Functionalized gold nanoparticles and their biomedical applications. Nanomaterials 1: 31–63.

Torres-Duarte, C., A.S. Adeleye, S. Pokhrel, L. Mädler, A.A. Keller and G.N. Cherr. 2016. Developmental effects of two different copper oxide nanomaterials in sea urchin (*Lytechinus pictus*) embryos. Nanotoxicology 10: 671–679.

Torres-Duarte, C., K.M. Ramos-Torres, R. Rahimoff and G.N. Cherr. 2017. Stage specific effects of soluble copper and copper oxide nanoparticles during sea urchin embryo development and their relation to intracellular copper uptake. Aquat. Toxicol. 189: 134–141.

Wang, C., H. Jia, L. Zhu, H. Zhang and Y. Wang. 2017. Toxicity of α-Fe_2O_3 nanoparticles to *Artemia salina* cysts and three stages of larvae. Sci. Total Environ. 598: 847–855.

Ward, J.E. and D.J. Kach. 2009. Marine aggregates facilitate ingestion of nanoparticles by suspension-feeding bivalves. Mar. Environ. Res. 68: 137–142.

Wu, B., C. Torres-Duarte, B.J. Cole and G.N. Cherr. 2015. Copper oxide and zinc oxide nanomaterials act as inhibitors of multidrug resistance transport in sea urchin embryos: Their role as chemosensitizers. Environ. Sci. Technol. 49: 5760–5770.

Zhao, Y., X. Wang, Q. Wu, Y. Li and D. Wang. 2015. Translocation and neurotoxicity of CdTe quantum dots in RMEs motor neurons in nematode *Caenorhabditis elegans*. J. Hazard. Mater. 283: 480–489.

Zhang, P., H. Selck, S.R. Tangaa, C. Pang and B. Zhao. 2017. Bioaccumulation and effects of sediment-associated gold- and graphene oxide nanoparticles on *Tubifex tubifex*. J. Environ. Sci. 51: 138–145.

Zhou, Y., Q. Wang, B. Song, S. Wu, Y. Su, H. Zhang et al. 2015. A real-time documentation and mechanistic investigation of quantum dots-induced autophagy in live *Caenorhabditis elegans*. Biomaterials 72: 38–48.

Zhu, S., F. Luo, W. Chen, B. Zhu and G. Wang. 2017a. Toxicity evaluation of graphene oxide on cysts and three larval stages of *Artemia salina*. Sci. Total Environ. 595: 101–109.

Zhu, S., F. Luo, X. Tu, W.C. Chen, B. Zhu and G.X. Wang. 2017b. Developmental toxicity of oxidized multi-walled carbon nanotubes on *Artemia salina* cysts and larvae: Uptake, accumulation, excretion and toxic responses. Environ. Pollut. 229: 679–687.

Zhu, S., M.Y. Xue, F. Luo, W.C. Chen, B. Zhu and G.X. Wang. 2017c. Developmental toxicity of Fe_3O_4 nanoparticles on cysts and three larval stages of *Artemia salina*. Environ. Pollut. 230: 683–691.

Zuykov, M., E. Pelletier, C. Belzile and S. Demers. 2011a. Alteration of shell nacre micromorphology in blue mussel *Mytilus edulis* after exposure to free-ionic silver and silver nanoparticles. Chemosphere 84: 701–706.

Zuykov, M., E. Pelletier and S. Demers. 2011b. Colloidal complexed silver and silver nanoparticles in extrapallial fluid of *Mytilus edulis*. Mar. Environ. Res. 71: 17–21.

3

In Vitro Testing

In Vitro Toxicity Testing with Bivalve Mollusc and Fish Cells for the Risk Assessment of Nanoparticles in the Aquatic Environment

Alberto Katsumiti and *Miren P. Cajaraville**

Introduction: Use of *In Vitro* Testing for Risk Assessment of Nanoparticles

In the last decades, *in vitro* approaches have been considered to provide an important additional data for the 'Hazard identification' in the risk assessment paradigm. These techniques have emerged following the 3R philosophy (replacement, reduction and refinement), responding to an increasing public demand for alternative methods to minimise the use of animals in research. *In vitro* techniques have risen as a cost-effective alternative that provide fast and reproducible results, avoiding external confounding factors.

New concepts to increase efficiency, minimise costs and improve evidence-based strategy have been proposed. New methods based on the use of primary cells and cell lines to assess the risk of chemicals to human and environmental health were developed and standardised. In the last years, the Organisation for Economic Cooperation and Development (OECD) has published a series of guidelines with standardised *in vitro* toxicity test methods. Among others, these standardised tests covered endpoints to address dermal absorption (OECD 2004), skin and eye irritation (OECD 2013, OECD 2014a, OECD 2015b,c), genotoxicity (OECD 2015a, OECD 2014b,c,d, OECD 2015e), endocrine disruption (OECD 2015d) and more recently published, *in vitro* methods for fish hepatic clearance (OECD 2018a,b). Launching of the EU regulation for the Registration, Evaluation, Authorisation and Restriction of Chemicals (REACH) (European Commission 2006), has been of upmost importance since this regulation encourages the use of *in vitro* approaches for toxicity testing.

CBET Research Group, Dept. Zoology and Animal Cell Biology; Faculty of Science and Technology and Research Centre for Experimental Marine Biology and Biotechnology PIE, University of the Basque Country UPV/EHU, Sarriena Auzoa z/g, E-48940 Leioa, Basque Country (Spain).

* Corresponding author: mirenp.cajaraville@ehu.eus

Currently, there is no specific standardised method to evaluate the toxicity of nanomaterials (NMs). The great variety of types of NMs with distinct chemical composition, size, shape and coatings, and the diversity of exposure scenarios that could influence NMs behaviour and stability, make the establishment of standard protocols very difficult. However, great efforts have been done by the OECD (http://www.oecd.org/env/ehs/nanosafety/publications-series-safety-manufactured-nanomaterials.htm) and regulatory bodies such as the European Chemicals Agency (ECHA, https://echa.europa.eu/regulations/nanomaterials) in order to establish systematic approaches to evaluate NMs toxicity.

A wide variety of *in vitro* tests have been employed to identify effects of NMs on cell viability, proinflammatory responses, oxidative stress and genotoxicity, among others. Primary cells and immortalised cell lines of several mammalian cell types including epithelial, endothelial, phagocytic, neuronal and hepatic cells have been extensively used in *in vitro* toxicity testing of NMs and have shown great sensitivity (Jones and Grainger 2009). Selection of the most relevant cell type for *in vitro* toxicity testing with NMs is often influenced by the exposure route. In airborne exposure studies, pulmonary alveolar cells have been used to assess the toxicity of multiwalled carbon nanotubes (Sweeney et al. 2014, Ruenraroengsak et al. 2016), Ag NPs (Sweeney et al. 2015a), Au NPs (Di Bucchianico et al. 2014), TiO_2 NPs (Sweeney et al. 2015b) and CuO NPs (Di Bucchianico et al. 2013, Misra et al. 2013, Katsumiti et al. 2018). For dermal toxicity assessment, skin fibroblast and keratinocytes cell lines have been successfully employed to address the toxicity of TiO_2 NPs (Sayes et al. 2006), ZnO NPs (Zvyagin et al. 2008) and Ag NPs (Samberg et al. 2010) among others. Alternatively, skin organ culture has also been successfully used to assess dermal toxicity caused by CuO NPs (Cohen et al. 2013).

A great number of studies have addressed *in vitro* toxicity testing using mammalian cells in order to assess NMs toxicity (reviewed in Jones and Grainger 2009, Park et al. 2009, Drasler et al. 2017). These studies have shown that differences in size, shape and crystal structure of NMs lead to different degrees of cellular uptake and toxicity (Brunner et al. 2006, Sayes et al. 2006, Duffin et al. 2007, Yang et al. 2009). Other properties such as surface charge have also been demonstrated to influence NMs toxicity *in vitro*. Harush-Frenkel et al. (2008) showed that positively charged nanoparticles (NPs) more easily penetrate epithelial Madin-Darby canine (MDCK) cells membrane compared to negatively charged particles. Specific mechanisms for NMs internalisation may differ according to their characteristics and cell type used in the assay (Mahmoudi et al. 2012, Unfried et al. 2007). Generally, NPs can be internalised into mammalian cells mainly through endocytic mechanisms such as pinocytosis, including macropinocytosis, clathrin- and caveolin-mediated endocytosis and clathrin- and caveolin-independent endocytosis (small particles < 150 nm), and via phagocytosis (particles or aggregates > 250 nm) (Conner and Schmid 2003). Other internalisation pathways such as passive diffusion through cell membrane pores and passive uptake have also been reported since some NMs have been found in the cells' cytosol and not in membrane-bound organelles (Rimai et al. 2000, Geiser et al. 2005, Mu et al. 2012, Lesniak et al. 2005, Rothen-Rutishauser et al. 2006). Cellular responses related with the mode of action (MOA) of NMs have been identified in *in vitro* studies with mammalian cells. Several evidences indicate that for many types of NPs toxicity occurs via oxidative stress leading to pro-inflammatory responses (Donaldson et al. 2003, Jones and Grainger 2009, Park et al. 2009, Fu et al. 2014, Wang and Tang 2018) and via the fibre paradigm (Dorger et al. 2001, Donaldson and Tran 2004) leading to lipid peroxidation (Panessa-Warren et al. 2008, Fu et al. 2014) and genotoxicity (Schins and Knaapen 2007, Park et al. 2009, Wang and Tang 2018). These findings on the MOA of NMs at cellular level could potentially be used as predictors of *in vivo* toxicity; thus, *in vitro* toxicity tests represent an important tool for hazard identification and human risk assessment of nanomaterials.

NMs may enter into the aquatic environment either by direct application of NMs to surface waters or indirectly via industrial discharges or wastewater treatment effluents disposal, aerial deposition dumping and run-off (Batley et al. 2012, Baker et al. 2014). Once in the aquatic systems, the physico-chemical properties of NMs may be altered according to the characteristics of the media. The fate of

NMs in seawater is likely very different than in freshwater. In seawater, agglomeration and aggregation will occur, affecting the behaviour of NMs (Stolpe and Hassellov 2007). Aggregates may sink to the ocean floor unless they encounter significant changes in temperature, ionic strength and natural organic matter. When interacting with other compounds, NMs may form organic or inorganic complexes that will be available for filter-feeding organisms (Matranga and Corsi 2012, Canesi and Corsi 2016). NMs may also dissociate, releasing ions that will interact with biotic and abiotic entities (Ju-Nam and Lead 2008). Differences in environmental parameters such as temperature, pH, ionic strength and organic complexation can also affect the solubility of NMs and thus, their dissolution kinetics (Tiede et al. 2009). Solubility is a key factor in determining NMs toxicity, especially in the case of metal bearing NPs, as the release of soluble metal ions from the surface of the particles will determine the proportional exposure/bioavailability to organisms and increase the risk of metal toxicity (Fabrega et al. 2011, Scown et al. 2010). The extent to which NMs interact with abiotic and biotic entities, which includes the possibility of becoming carriers of other environmental pollutants, should be the focus of risk assessment in the marine environment (Hartmann and Baun 2010).

Efforts to use *in vitro* assays with cells of aquatic organisms to assess the potential toxic effects of NMs in freshwater and marine environments have recently grown. Many different aquatic organisms have been used to address the toxicity of NMs *in vitro,* among them: bacteria (Heinlaan et al. 2008, Fabrega et al. 2009, Kang et al. 2009, Li et al. 2011, Georgantzopoulou et al. 2013), protozoa (Kvitek et al. 2009, Blinova et al. 2010), algae (Hund-Rinke and Simon 2006, Baun et al. 2008, Navarro et al. 2008, Aruoja et al. 2009, Wong et al. 2010, Keller et al. 2012, Georgantzopoulou et al. 2013, Lei et al. 2016, Bergami et al. 2017, Schiavo et al. 2017, Zhao et al. 2017), bivalves (Canesi et al. 2008, 2010, 2012, 2016, Gagné et al. 2008, Moore et al. 2009, Ringwood et al. 2009, Couleau et al. 2012, Bruneau et al. 2013, Katsumiti et al. 2014, 2015a, 2015b, 2016, 2017, 2018, Volland et al. 2018, Sendra et al. 2018) and fishes (Reeves et al. 2008, Vevers and Jha 2008, Fako and Furgeson 2009, Kühnel et al. 2009, Farkas et al. 2010, Wise et al. 2010, Wang et al. 2011, Christen and Fent 2012, George et al. 2012, Christen et al. 2013, Fernández-Cruz et al. 2013, Vo et al. 2013, Lammel and Navas 2014, Munari et al. 2014, Song et al. 2014, Taju et al. 2014, Yue et al. 2014, Connolly et al. 2015, Bermejo-Nogales et al. 2017, Lammel and Sturve 2018). Although these studies have provided relevant data on the toxic effects of NMs in cells of aquatic organisms *in vitro*, there is still a lack of standardised methods to characterise NMs properties and behaviour in exposure media and to assess nanotoxicity *in vitro*. Further, a conceptual framework for the use of *in vitro* toxicity tests with cells of aquatic organisms in ERA is lacking.

In this chapter, we reviewed the literature on *in vitro* toxicity testing using cells of aquatic organisms (e.g., fish and bivalve molluscs) to assess NMs toxicity, discuss current methodologies and their gaps and suggest a conceptual framework for the risk assessment of NMs in aquatic ecosystems.

Fish Cell Models for *In Vitro* Toxicity Testing of Nanomaterials

Fish cells have been used for many decades as cell models in biomedical research and for assessing the toxicity of xenobiotics in the aquatic environment. They are also good candidates to reduce *in vivo* animal testing. *In vitro* assays using fish cells have been proposed as potential alternatives to fish *in vivo* acute toxicity tests (Castaño et al. 1996, Segner 1998, Caminada et al. 2006, Schirmer 2006, Fent 2007, OECD 2012). Several authors have reported a good correlation between *in vivo* and *in vitro* toxicological data, testing different chemicals (Babich and Borenfreund 1987, Segner and Lenz 1993, Castaño et al. 1996, Fent 2001, 2007, Na et al. 2009, Taju et al. 2012) and also NMs (Taju et al. 2014). However, as pointed out by Segner (1998), *in vitro* findings show a less satisfying correlation with *in vivo* data when the tested chemical exerts its toxicity by interference with tissue-specific biochemical or physiological functions. For non-specific toxicants, studies indicated that results obtained *in vitro* could provide valuable data to decipher mechanisms of action of different chemicals including NMs, thus allowing a reduction in the use of whole animals.

Both primary cell cultures and cell lines from fishes have been used in a wide variety of toxicological studies (Schirmer 2006). One of the main advantages in the use of primary cell cultures is that these cells generally maintain many of the important markers and cellular functions seen *in vivo*. Guillouzo (1998) reported that primary hepatocytes retain liver's specific functions and responses to toxicants for several days up to weeks. Three-dimensional (3D) fish primary liver spheroids have been used for assessment of chemicals toxicity as an alternative approach to mimic conditions *in vivo* (Baron et al. 2012, 2017, Uchea et al. 2015). However, primary cells have a finite lifespan and limited proliferation capacity, making it difficult to maintain an experimental routine. Instead, most of the *in vitro* studies based on fish cells used cell lines possibly because they are easy to handle and genetically identical, which aid to provide consistent and reproducible results.

The rainbow trout gonadal RTG-2 cell line was the first fish permanent cell line generated (Wolf and Quimby 1962). Since then, many other fish cell lines have been established from different marine and freshwater fish species. In 1980, 61 fish cell lines were listed from 36 different species (Wolf and Mann 1980). In the following decade, this number increased to 159 cell lines from 74 fish species (Fryer and Lannan 1994) and in 2011, Lakra et al. (2011) reported 124 new fish cell lines (59 cell lines from 19 freshwater species, 54 from 22 marine species and 11 from 3 brackish water species) generated only in the years between 1994 and 2010.

The most used fish cell lines in Nanotoxicology derived from tissues as liver (Christen et al. 2013, Fernández-Cruz et al. 2013, Vo et al. 2013, Song et al. 2014, Lammel and Navas 2014, Connolly et al. 2015, Bermejo-Nogales et al. 2017, Lammel and Sturve 2018), gills (Kühnel et al. 2009, George et al. 2012, Vo et al. 2013, Taju et al. 2014, Yue et al. 2014, Lammel and Sturve 2018), gonads (Vevers and Jha 2008, Vo et al. 2013, Munari et al. 2014, Connolly et al. 2015, Bermejo-Nogales et al. 2017) and skin (Reeves et al. 2008, Wise et al. 2010, Christen and Fent 2012, Vo et al. 2013). Fish cell lines have demonstrated great sensitivity to detect, among others, cytotoxicity of TiO_2 NPs (Reeves et al. 2008, Vevers and Jha 2008, Lammel and Sturve 2018), ZnO NPs (Fernández-Cruz et al. 2013, Bermejo-Nogales et al. 2017), Cu NPs (Song et al. 2014), SiO_2 NPs (Christen and Fent 2012, Vo et al. 2013, Bermejo-Nogales et al. 2017), Ag NPs (Wise et al. 2010, Christen et al. 2013, Munari et al. 2014, Taju et al. 2014, Yue et al. 2014, Connolly et al. 2015), tungsten cobalt NPs (Kühnel et al. 2009), graphene nanoplatelets (Lammel and Navas 2014) and carbon nanotubes (Bermejo-Nogales et al. 2017), and also to detect genotoxic effects of TiO_2 NPs (Reeves et al. 2008, Vevers and Jha 2008) and Ag NPs (Wise et al. 2010, Munari et al. 2014, Taju et al. 2014). Table 1 shows some recent examples of *in vitro* approaches using fish cells and their highlighted discoveries in the assessment of NMs toxicity.

In vitro tests with fish cells have been used as fast screening tools for the initial assessment of NMs toxicity. Screening tests based on cell viability have been helpful for the ranking and grouping of NMs cytotoxicity (Kühnel et al. 2009, Farkas et al. 2010, Wang et al. 2011, Munari et al. 2014, Bermejo-Nogales et al. 2017). Munari et al. (2014) found that As_2S NPs reduced RTG-2 cell viability slightly whereas CdS quantum dots were cytotoxic, possibly related to the higher dissolution of the later in cell culture media. Similarly, silver NPs were more toxic than Au NPs in rainbow trout primary hepatocytes (Farkas et al. 2010). Kühnel et al. (2009) reported that the combination of metallic Co with tungsten carbide NPs increases the toxicity of the later. Based on cell viability assays, Bermejo-Nogales et al. (2017) ranked NMs according to their cytotoxicity to RTG-2 cell line: ZnO > MWCNTs $\geq CeO_2 = SiO_2$, and reported that coating agents may influence NMs toxicity, as shown for ZnO NPs. Similarly, Wang et al. (2011) reported the overall ranking of the toxicity of metal oxides tested in catfish primary cells and in the human hepatocarcinoma cell line (HepG2): CuO > ZnO > Co_3O_4 > TiO_2. These results highlight the utility of *in vitro* cytotoxicity assays for the classification of NMs according to their hazard potential.

Another strategy for grouping NMs according to their toxicity is based on the identification of their MOA. Outcomes of *in vitro* tests on the MOA of NMs can serve as reference for designing more relevant confirmatory *in vivo* experiments. As in mammalian cells, reactive oxygen species (ROS) may be one of the main drivers of NPs toxicity in fish cells (Wang et al. 2011, Song et al. 2014). Enhanced

Table 1. Examples of *in vitro* studies using fish cells to assess NMs toxicity.

Species	**Cell Type**	**Primary Cells (PC) or Cell Line (CL)**	**Tested NM**	***In vitro* Assay/Analytical Technique**	**Tested for Nano Interference (Y: yes/N: no)**	**Highlights**	**References**
Goldfish (*Carassius auratus*)	skin	GFSk-S1 (CL)	TiO_2	• NRRT • Comet Assay	• N • N	Cyto- and genotoxicity enhanced with UVA.	Reeves et al. 2008
Rainbow trout (*Oncorhyncus mykiss*)	gonadal	RTG-2 (CL)	TiO_2	• NRRT • Comet Assay • Fpg Comet Assay • CBMN	• N • N • N • N	Cyto- and genotoxicity enhanced with UVA.	Vevers and Jha 2008
Rainbow trout (*Oncorhyncus mykiss*)	gill cell	RTgill-W1 (CL)	WC and WC-Co	• Alamar blue • CFDA-AM • TEM	• N • N	Cells uptake NPs independently on their state of aggregation; toxicity caused by anions and NPs.	Kühnel et al. 2009
Rainbow trout (*Oncorhyncus mykiss*)	hepatocytes	(PC)	Ag and Au	• Alamar blue • CFDA-AM • ROS	• N • N • N	Ag was cytotoxic but did not induce ROS; Au was not cytotoxic but induced ROS.	Farkas et al. 2010
Medaka (*Oryzias latipes*)	Skin (fin tissue)	OLHNI2 (CL)	Ag	• CF Assay • Chromosomal aberrations and aneuploidy	• N • N	Cytotoxic and genotoxic effects.	Wise et al. 2010
Catfish (*Ictalurus punctatus*)	hepatocytes	(PC)	ZnO, TiO_2, CuO and Co_3O_4	• MTT • ROS • Apoptosis	• N • N • N	Primary fish hepatocytes were less sensitive than HepG2 cell line; cytotoxicity was partially due to ROS.	Wang et al. 2011
Fathead minnow (*Pimephales promelas*)	fibroblasts	FHM (CL)	SiO_2 and Ag-dopped SiO_2	• MTT • qPCR • CYP1A • MXR • ROS • TEM	• N • N • N • N • N	Ag-dopped SiO_2 was more toxic than only SiO_2; Ag-dopped SiO_2 induced ROS; SiO_2 and Ag-dopped SiO_2 induced ER stress response and altered cytochrome P4501A activity.	Christen and Fent 2012
Rainbow trout (*Oncorhyncus mykiss*)	gill cells	RTgill-W1 (CL)	Ag	• Cell membrane integrity	• N	Shape influenced Ag NMs toxicity; correlation between *in vitro* and *in vivo*.	George et al. 2012
Zebrafish (*Danio rerio*)	liver	ZFL (CL)	Ag	• MTT • ROS • qPCR • Immunoblotting • TEM	• N • N • N • N	Ag NPs internalised in ZFL; induction of ROS; ER stress response, and alterations in transcriptional levels of TNF-α, p53 and Bax.	Christen et al. 2013

Topminnow fish (*Poecilipsis lucida*)	hepatoma	PLHC-1 (CL)	ZnO	• MTT • NRU • LDH • LUCS • ROS	• N • N • N • N • N	Aggregation of NPs contributed to their toxicity in fish cells, whereas in human cells toxicity was due to the dissolved fraction; ROS did not contribute to NPs toxicity.	Fernández-Cruz et al. 2013
Rainbow trout (*Oncorhyncus mykiss*)	gills	RTgill-W1 (CL)	SiO_2	• Alamar blue • Cell morphology	• Y	Toxicity was size-, time-, temperature- and dose-dependent; cell lines derived from external tissues (e.g., skin, gills) were more sensitive than cells derived from internal tissues (liver, brain, intestine, gonads) or embryos.	Vo et al. 2013
	intestine	RTgut-GC (CL)					
	liver	RTL-W1 (CL)					
	brain	RTBrain (CL)					
Fathead minnow (*Pimephales promelas*)	liver	FHML2-6 (CL)					
	testis	FHMT-W1 (CL)					
Goldfish (*Carassius auratus*)	brain	GFB3C (CL)					
	skin	GFSk-S1 (CL)					
Zebrafish (*Danio rerio*)	abdomen	GloFish (CL)					
	blastula	ZEB2J (CL)					
American Eel (*Anguilla rostrata*)	brain	EelBrain (CL)					
Haddock (*Melanogrammusaeglefinius*)	embryo	HEW (CL)					
Topminnow fish (*Poecilipsis lucida*)	hepatoma	PLHC-1 (CL)	GO and CXYG	• Alamar blue • CFDA-AM • NRU • Fluorescamine Assay • ROS • MMP • SEM • TEM	• Y • Y • Y • Y • N • N	Graphene nanoplatelets spontaneously penetrated plasma membrane and accumulated in the cytosol; GO and CXYG reduced MMP, increased ROS but barely decreased cell viability.	Lammel and Navas 2014
Rainbow trout (*Oncorhyncus mykiss*)	gonadal	RTG-2 (CL)	CdS and Ag_2S	• LDH • XTT • Comet Assay	• N • N • N	CdS highly cytotoxic and induced genotoxicity; Ag_2S showed neither cytotoxic nor genotoxic effects.	Munari et al. 2014
Topminnow fish (*Poecilipsis lucida*)	hepatoma	PLHC-1 (CL)	Cu	• Resazurin reduction • ROS	• Y • Y	NPs' physico-chemical properties influenced their toxicity; comparing with human cell lines, fish cell lines were less sensitive.	Song et al. 2014
Rainbow trout (*Oncorhyncus mykiss*)	hepatoma	RTH-149 (CL)					

Table 1 contd. ...

... Table 1 contd.

Species	Cell Type	Primary Cells (PC) or Cell Line (CL)	Tested NM	*In vitro* Assay/Analytical Technique	Tested for Nano Interference (Y: yes/N: no)	Highlights	References
Indian carp (*Catla catla*)	heart gills	SICH (CL) ICG (CL)	Ag	• MTT • NRU • Cell morphology • Comet Assay • Nuclear fragmentation • SOD • GSH • CAT • Lipid peroxidation	• N • N • N • N • N • N • N • N	Linear correlation between *in vitro* and *in vivo* results; activation of the antioxidant system in the three cell lines exposed to Ag NPs.	Taju et al. 2014
Indian carp (*Labeorohita*)	gills	LRG (CL)					
Rainbow trout (*Oncorhyncus mykiss*)	gills	ICG (CL)	Ag	• Alamar blue • CFDA-AM • NRU	• N • N • N	Cell culture media influenced NPs stability; toxicity due to the presence of Ag ions but also due to the particulate form.	Yue et al. 2014
Rainbow trout (*Oncorhyncus mykiss*)	hepatocytes	(PC)	Ag	• Alamar blue • CFDA-AM • NRU	• Y • Y • Y	Cell culture media influenced NPs toxicity; significant differences in sensitivity between primary cells and cell lines and also among cell lines.	Connolly et al. 2015
	hepatoma	RTH-149 (CL)					
	liver	RTL-W1 (CL)					
	gonadal	RTG-2 (CL)					
Rainbow trout (*Oncorhyncus mykiss*)	gonadal	RTG-2 (CL)	ZnO, MWCNT, SiO_2 and CeO_2	• Alamar blue • CFDA-AM • NRU	• Y • Y • Y	PLHC-1 cells exhibited higher sensitivity than RTG-2 cells; SiO_2 and CeO_2 showed low cytotoxicity; ZnO-NM exerted cytotoxicity mainly by altering lysosome function and metabolic activity, while MWCNTs by plasma membrane disruption.	Bermejo-Nogales et al. 2017
Topminnow fish (*Poecilipsis lucida*)	hepatoma	PLHC-1 (CL)					
Rainbow trout (*Oncorhyncus mykiss*)	liver	RTL-W1 (CL)	TiO_2	• Alamar blue • CFDA-AM • CLSM • TEM	• Y • Y	Slight cytotoxicity; TiO_2 internalised in intracellular vesicles.	Lammel and Sturve 2018
	gills	RTgill-W1 (CL)					

CAT: catalase assay; CBMN: cytokinesis-blocked micronucleus assay; CF: colony forming; CFDA-AM: 5-carboxyfluorescein diacetate acetoxymethyl ester; CLSM: confocal laser scanning microscopy; CXYG: carboxyl graphene; ER: endoplasmic reticulum; Fpg: formamidopyrimidine DNA glycosylase; GO: graphene oxide; GSH: reduced glutathione assay; LDH: lactate dehydrogenase; LUCS: light-up cell signal; MMP: mitochondrial membrane potential; MTT: thiazolyl blue tetrazolium bromide assay; MXR: multixenobiotic resistance transport activity; MWCNT: multi-walled carbon nanotubes; NRRT: neutral red retention time assay; NRU: neutral red uptake; ROS: reactive oxygen species assay; SEM: scanning electron microscopy; SOD: superoxide dismutase assay; TEM: transmission electron microscopy; WC: tungsten carbide; WC-Co: tungsten carbide combined with cobalt.

levels of ROS were found in fish cells exposed to Au NPs (Farkas et al. 2010), Cu NPs (Song et al. 2014), ZnO, TiO_2, CuO and Co_3O_4 NPs (Wang et al. 2011) and graphene nanoplatelets (Lammel and Navas 2014). Increases in ROS production may lead to a cascade of events involving oxidative stress, lipid peroxidation and genotoxicity. Taju et al. (2014) reported that Ag NPs significantly increased lipid peroxidation and decreased GSH, SOD and CAT levels in fish cell lines in a dose-dependent manner caused possibly by the increase in ROS levels. Genotoxic effects found in fish cells exposed to TiO_2 NPs may also be caused by the increase in ROS levels (Reeves et al. 2008, Vevers and Jha 2008). In addition, studies in fish cells have identified other targets for NPs toxicity. Yue et al. (2014) reported that lysosomal membrane integrity in RTgill-W1 cells was significantly more sensitive to citrate-coated Ag NPs than cellular metabolic activity or cell membrane integrity, indicating that lysosomes are a particular target of toxicity of Ag NPs. In PLHC-1 and RTG-2 cell lines, Bermejo-Nogales et al. (2017) reported that ZnO NPs exerted toxicity mainly by altering lysosomal function and metabolic activity whereas MWCNTs induced toxicity via plasma membrane disruption.

Even though *in vitro* toxicity assays have demonstrated usefulness in assessing NMs' toxicity, they should be used with caution as some NMs are known to interfere with assay chromophores/fluorochromes or with absorbance/fluorescence spectra, giving rise to false positive or negative results. As shown by Lammel and Sturve (2018), TiO_2 NPs interfered with spectrophotometric and fluorescence readout, causing a reduction in absorbance/fluorescence values. Other authors have reported similar results in other *in vitro* tests with fish cells (Connolly et al. 2015). As shown in Table 1, not all current *in vitro* assays in fish cells have been tested for NMs interference. Several strategies involving modification of the test protocols, use of cell-free assays, or the selection of a more suitable test have been suggested to avoid NMs' interference. This subject is addressed in more detail in the Section 'Risk Assessment of Metal and Metal-Bearing Nanoparticles Using Mussel Cells *In Vitro*' of this Chapter.

Bivalve Mollusc Cell Models for *In Vitro* Toxicity Testing of Nanomaterials

Attempts to develop cell cultures from aquatic invertebrates were made quite early in the history of tissue culture (Gomot 1971, Rannou 1968, 1971). Molluscs are probably the most intensively studied group of aquatic invertebrates (Keller and Zam 1990, Rinkevich 1999, Yoshino et al. 2013). Due to the high commercial interest and extensive use of bivalve molluscs as sentinel species in environmental risk assessment, there has been a considerable interest in the development of *in vitro* models in this group of animals (Robledo and Cajaraville 1997, Elston 2000, Olabarrieta et al. 2001, Gómez-Mendikute and Cajaraville 2002, 2003, Gómez-Mendikute et al. 2005, Cajaraville 2009, Yoshino et al. 2013).

Despite great efforts to generate cell lines from molluscs, only a single cell line has been established. The Bge cell line was generated from embryonic cells of the freshwater gastropod mollusc *Biomphalaria glabrata* (Hansen 1976). Because *B. glabrata* is a snail host of the trematode parasite *Schistosoma mansoni*, which causes the disease called schistosomiasis, the Bge cell line has been used mainly in studies related with host-parasite interactions in schistosomiasis, and not in risk assessment of contaminants in aquatic environments. Notwithstanding, primary cell cultures from molluscan species have been used in toxicological studies.

The number of mollusc species used in *in vitro* toxicology experiments is relatively small (Yoshino et al. 2013), and not surprisingly, the ones used are those of high commercial interest (e.g., mussels, oysters, clams). In *in vitro* toxicity testing of NMs, the most used primary cells from molluscs are haemocytes and gill cells (Canesi et al. 2008, 2010, 2014, 2016, Gagné et al. 2008, Ciacci et al. 2012, Couleau et al. 2012, Katsumiti et al. 2014, 2015a, 2015b, 2016, 2017, 2018, Sun et al. 2017, Volland et al. 2018, Sendra et al. 2018).

Haemocytes or immunocytes comprise the internal defence system in bivalve molluscs. Haemocytes are also involved in digestion, shell repair, respiration, osmoregulation, transportand excretion (Cheng 1981). In mussels, two main types of haemocytes have been described: granulocytes and agranulocytes or hyalinocytes (Cheng 1981, Cajaraville and Pal 1995, Carballal et al. 1997, 1998). Granulocytes are large cells (10–20 μm in diameter) with a small nucleus, characterised by containing numerous cytoplasmic granules, most of which are lysosomes (Cajaraville and Pal 1995). Granulocytes are the main haemocyte type involved in phagocytosis, intracellular digestion and immune defence (Carballal et al. 1997). Hyalinocytes are agranular small cells (4–6 μm in diameter) that present characteristics of undifferentiated cells: a small volume of cytoplasm that contains few organelles, abundant free ribosomes, and a central large nucleus (Cajaraville and Pal 1995). Although less active than granulocytes, these cells are also involved in phagocytosis, intracellular digestion and immune defence (Cajaraville and Pal 1995, Carballal et al. 1997). Hyalinocytes are probably intermediate forms that give rise to granulocytes, since they seem to have an important role in protein synthesis and the granules contain abundant enzymes (Carballal et al. 1997). Gill cells are epithelial cells responsible for uptake and transport of nutrients, gas exchange and ionic osmoregulation (Owen 1974, 1978, Owen and McCrae 1976). Gill cells are the first target and uptake site of many contaminants in the aquatic environment (Livingstone and Pipe 1992, Gómez-Mendikute et al. 2005).

Mussels have been shown to be sensitive model species to detect the toxicity of NPs both *in vivo* and *in vitro* (Canesi et al. 2012, Matranga and Corsi 2012). Mussel haemocytes and gill cells have been successfully used to determine the toxicity of NMs. Gagné et al. (2008) found significant effects of CdTe NPs and also of ionic Cd ($CdSO_4$) on the immune competence of the freshwater mussel *Elliptio complanata*. In the zebra mussel *Dreissena polymorpha*, Couleau et al. (2012) found that TiO_2 NPs inhibit haemocytes phagocytosis but do not affect their cell viability, in contrast to our findings in the Mediterranean marine mussel *Mytilus galloprovincialis* (Katsumiti et al 2015a). In the same species, *in vitro* exposure of haemocytes to C60 fullerene, and TiO_2 and SiO_2 NPs did not cause effects on haemocytes lysosomal membrane stability but the three NMs induced lysozyme release, extracellular oxyradical generation and nitric oxide (NO) production (Canesi et al. 2010). Overall, inflammatory and cell-mediated immune responses have been observed in mussels' haemocytes after exposure to different NMs (Canesi et al. 2010, Ciacci et al. 2012, Bruneau et al. 2013, Katsumiti et al. 2014, 2015a, 2015b, 2016, 2017, 2018). Gill cells have been also demonstrated to be a suitable model to assess the potential effects of NMs (Katsumiti et al. 2012, 2014, 2015a, 2015b, 2016, 2018). Table 2 shows some recent examples of *in vitro* approaches using mollusc cells to assess NMs toxicity.

Methodological Aspects

There are a number of methodological aspects to take into account when performing *in vitro* toxicity tests with NMs. Conventional *in vitro* toxicological assays have been originally designed to test soluble compounds and not NMs. For this reason, prior evaluation and adaptation of these assays are recommended. In addition, a detailed physico-chemical characterisation of the primary features and behaviour of NMs in the cell culture media of interest is essential.

NP Interference and Modifications of the In Vitro Tests to be Used for NP Toxicity Assessment

Due to the special properties of NMs such as high adsorption capacity, optical properties, surface charge and catalytic activities, it is worth expecting that they might interfere with conventional *in vitro* assays, especially those based on absorbance or fluorescence detection (Stone et al. 2009, Kroll et al. 2012). Many authors have already reported interference of NMs in several assays (Kroll et al. 2012, Lammel et al. 2013, Lupu and Popescu 2013, Ong et al. 2014, Guadagnini et al. 2015). Carbon black NPs were found to oxidise 2',7' dichlorodihydrofluorescein diacetate (H_2DCF-DA) in a cell-free system, whereas ZnO NPs decreased lactate dehydrogenase (LDH) activity and TiO_2 NPs reduced interleukin 8 (IL-8) levels (Kroll et al. 2012). Lammel et al. (2013) reported up to 34% of fluorescence quenching caused by the interaction of graphene nanoplatelets with the fluorophores 5-carboxyfluorescein (5-CF), rezasurin and neutral red (NR). Lupu and Popescu (2013) found a

Table 2. Examples of *in vitro* studies using bivalve mollusc cells to assess NMs toxicity. All studies listed were performed in primary cells.

Species	Cell Type	Tested NM	*In vitro* Assay/Analytical Technique	Tested for Nano Interference (Y: yes/N: no)	Highlights	References
Mussels (*Mytilus galloprovincialis*)	haemocytes	Nanosized carbon black (NCB)	• NRRT • NCB uptake • Lysozyme activity • NO • ROS • PI • MMP • Western blots (p38 and JNK MAPK)	• N • N • N • N • N • N	Concentration-dependent uptake of NCB; increase in extracellular lysozyme, ROS and NO; activation of p38 and JNK.	Canesi et al. 2008
Freshwater mussels (*Elliptioco mplanata*)	haemocytes	CdTe	• Phagocytic activity	• N	CdTe exposure decreased phagocytic activity.	Gagné et al. 2008
Oysters (*Crassostrea virginica*)	hepatopancreas cells	Fullerene	• NRU • Confocal microscopy	• N	Fullerene aggregates were found into lysosomes and promoted lysosomal destabilisation; endocytotic and lysosomal pathways may be major targets of fullerenes.	Ringwood et al. 2009
Mussels (*Mytilus galloprovincialis*)	haemocytes	C60 fullerene, TiO_2, SiO_2	• NRRT • Lysozyme activity • NO • ROS • Western blots (p38 MAPK)	• N • N • N • N	None of the NMs affected lysosomal membrane stability; however, all of them increased lysozyme release, NO and ROS; inflammatory response mediated by p38.	Canesi et al. 2010
Mussels (*Mytilus galloprovincialis*)	haemocytes	TiO_2, SiO_2, ZnO, CeO_2	• NRRT • Lysozyme activity • Phagocytic activity • NO • ROS • PI • MMP • CL • TEM	• N • N • N • N • N • N • N • N	All metal oxide NPs moderately decreased lysosomal membrane stability and increased NO and ROS; only TiO_2 NPs induced lysozyme release; SiO_2 NPs increased phagocytosis whereas CeO_2 NPs decreased it.	Ciacci et al. 2012

Table 2 contd. ...

... Table 2 contd.

Species	Cell Type	Tested NM	*In vitro* Assay/Analytical Technique	Tested for Nano Interference (Y: yes/N: no)	Highlights	References
Zebra mussel (*Dreissena polymorpha*)	haemocytes	TiO_2	• MTT • Phagocytic activity • Western blots (ERK1/2, p38 MAPK) • CLSM • TEM	• N • N	Inhibition of phagocytosis; internalisation of TiO_2 NPs; increased phosphorylation levels of ERC1/2 and p38.	Couleau et al. 2012
Mussels (*Mytilus galloprovincialis*)	haemocytes	TiO_2combinedwith2,3,7,8-TCDD	• NRRT • Phagocytic activity	• N • N	TiO_2 and 2,3,7,8-TCDD can exert synergistic or antagonistic effects; trojan-horse effect of TiO_2.	Canesi et al. 2014
Mussels (*Mytilus galloprovincialis*)	haemocytes and gill cells	CdS	• NR • MTT • ROS • CAT • Comet Assay • AcP • MXR • Actin cytoskeleton (only in haemocytes) • Phagocytic activity (only in haemocytes) • Confocal microscopy • Na-K-ATPase activity (only in gill cells)	• Y • Y • Y • N • N • Y • N • N • Y • Y	ROS, DNA damage, AcP activity and MXR increased in both cells; CAT increased in haemocytes; particle-specific stimulation of haemocytes phagocytosis; CdS quantum dots uptake in endolysosomal vesicles.	Katsumiti et al. 2014
Mussels (*Mytilus galloprovincialis*)	haemocytes and gill cells	TiO_2	• NR • MTT	• Y • Y	Low dose-dependent cytotoxicity of TiO_2 NPs; toxicity varied with mode of synthesis, crystalline structure and size of NPs; additives influenced TiO_2 NPs toxicity.	Katsumiti et al. 2015a
Mussels (*Mytilus galloprovincialis*)	haemocytes and gill cells	Ag	• NR • MTT • ROS • CAT • Comet Assay • AcP • MXR	• Y • Y • Y • N • N • Y • N	Dose- and size-dependent cytotoxicity of Ag NPs; main mechanisms of toxicity of Ag NPs involved oxidative stress and genotoxicity in both cells, activation of lysosomal AcP activity, disruption of actin	Katsumiti et al. 2015b

			• Actin cytoskeleton (only in haemocytes) • Phagocytic activity (only in haemocytes) • Na-K-ATPase activity (only in gill cells)	• N • Y • Y	cytoskeleton and stimulation of phagocytosis in hemocytes, increase of MXR and inhibition of Na-K-ATPase activity in gill cells.	
Mussels (*Mytilus galloprovincialis*)	haemocytes	PS-NH_2	• NRRT • Phagocytic activity • ROS • Western blots (p38 MAPK, PKC) • Isolation of PS-NH_2-protein complexes • TEM	• N • N • N • Y	In haemolymph, PS-NH_2 increased cellular damage and ROS production with respect to seawater; effects were apparently mediated by deregulation of p38 MAPK signaling; formation of a NP bio-corona.	Canesi et al. 2016
Mussels (*Mytilus galloprovincialis*)	haemocytes and gill cells	Au, ZnO and SiO_2	• NR • MTT • Confocal microscopy	• Y • Y	Dose- and size-dependent cytotoxicity; toxicity was: ZnO > Au > SiO_2; shape and solubility of ZnO NPs influenced their toxicity; the additives increased toxicity of Au and ZnO NPs; SiO_2 NPs were not internalised in haemocytes.	Katsumiti et al. 2016
Mussels (*Mytilus galloprovincialis*)	haemocytes	GO and rGO	• MTT • Cell membrane integrity • ROS • TEM	• Y • N • Y	Low dose-dependent cytotoxicity; additive increased bioavailability of rGO; GO and rGO caused invaginations and perforations in the plasma membrane and decreased cell membrane integrity; GO and rGO were found in the cytosol and in endolysosomal vesicles.	Katsumiti et al. 2017
Marine scallops (*Chlamys farreri*)	haemocytes	CuO	• Cell membrane integrity • Lysosomal content • ROS • Comet Assay • TEM • SEM	• N • N • N • N	NPs induced membrane damage and increased ROS; particles rather than the dissolved ions were the dominant source of NP toxicity.	Sun et al. 2017

Table 2 contd. ...

... Table 2 contd.

Species	**Cell Type**	**Tested NM**	***In vitro* Assay/Analytical Technique**	**Tested for Nano Interference (Y: yes/N: no)**	**Highlights**	**References**
Mussels (*Mytilus galloprovincialis*)	haemocytes and gill cells	CuO	• NR • MTT • ROS • CAT • Comet Assay • AcP • MXR • Actin cytoskeleton (only in haemocytes) • Phagocytic activity (only in haemocytes) • Na-K-ATPase activity (only in gill cells)	• Y • Y • Y • N • N • Y • N • N • Y • N	Dose-dependent cytotoxicity; CuO NPs increased CAT and AcP activities, produced DNA damage and induced MXR in both cells; particle-specific stimulation of phagocytosis; mussel cells showed similar sensitivity compared to the TT1 human pulmonary cell line.	Katsumiti et al. 2018
Clams (*Ruditapes philippinarum*)	haemocytes	CeO_2	• LMS • Phagocytic activity • ROS	• N • N • N	Shape, zeta potential and biocorona formation influences CeO_2 NPs toxicity. Negatively charged and rounded CeO_2 NPs provoked loss of LMS and decrease in haemocytes phagocytic activity.	Sendra et al. 2018
Clams (*Ruditapes philippinarum*)	haemocytes	CuO	• MTT • Comet Assay • Gene expression (AIF, BCL2, CathD, GST, MT, SOD, TXNDC9, TNF, XGadd45G)	• Y • N	Dose-dependent cytotoxicity; CuO NPs increased DNA damage, decrease XGadd45G expression and increase in CathD expression; synthesis methods influenced characteristics (aggregation and dissolution), behaviour, and toxicity of bare CuO NPs. Dissolved copper plays a key role in CuO NPs toxicity.	Volland et al. 2018

2, 3, 7, 8-TCDD: 2,3,7,8-tetrachlorodibenzo-p-dioxins; AcP: acid phosphatase assay; AIF: allograft inflammatory factor; BCL2: B cell lymphoma 2 – like 1; CAT: catalase assay; CathD: cathepsin D; CL: mitochondrial cardiolipin oxidation assay; CLSM: confocal laser scanning microscopy; ERK: 1/2, extracellular signal regulating kinase 1 and 2; GO: graphene oxide; GST: glutathione S-transferase; JNK: cJun N-terminal kinase; LMS: lysosomal membrane stability; MAPK: mitogen activated protein kinase; MMP: mitochondrial membrane potential; MT: metallothionein; MTT: thiazolyl blue tetrazolium bromide assay; MXR: multixenobiotic resistance transport activity; NO: nitric oxide assay; NRRT: neutral red retention time assay; NRU: neutral red uptake; PI: propidium iodide; PKC: Protein kinase C; $PS\text{-}NH_2$: amino-modified polystyrene NPs; rGO: reduced graphene oxide; ROS: reactive oxygen species assay; SEM: scanning electron microscopy; SOD: superoxide dismutase; TEM: transmission electron microscopy; TXNDC9: thioredoxin domain containing 9; TNF: tumor necrosis factor – like; XGadd45G: growth arrest and DNA-damage-inducible, gamma.

photocatalytic interaction between TiO_2 NPs and the thiazolyl blue tetrazolium bromide (MTT) resulting in the reduction of MTT to formazan under biological relevant conditions. Ong et al. (2014) reported that Si, CdSe and TiO_2 NPs and helical rosette nanotubes interfered with at least one of the *in vitro* assays tested [3-(4,5-dimethylthiazol-2-yl)-5-(3-carboxymethoxyphenyl)-2-(4-sulfophenyl)-2H-tetrazolium MTS, Alamar blue, catalase, LDH, bicinchoninic BCA or Bradford protein quantification], resulting in either under- or overestimation of toxicity. Guadagnini et al. (2015) found that the medically relevant poly-lactic-co-glycolic acid-polyethylene oxide copolymer (PLGA-PEO), TiO_2, SiO_2, uncoated and oleic acid coated Fe_3O_4 NPs interfered with cytotoxicity, pro-inflammatory and oxidative stress detection assays, and interference was intrinsically related with NP characteristics (composition, size, coatings and agglomeration).

A detailed physico-chemical characterisation of the NPs may allow the identification of potential interferences and avoid assays' misinterpretations. Interferences with spectrophotometric or fluorimetric measurements depend on the optical properties of the NPs. Surface properties and composition, determine the affinity of NPs with assays reagents. Coatings may also significantly influence NPs reactivity. Light scattering properties of, for example, TiO_2 NPs could also interfere with optical density measurements and fluorescence detection (Guadagnini et al. 2015). NPs aggregation may be problematic in cell counting or flow cytometry techniques and could also interfere with spectrophotometric and fluorimetric assays by blocking the beam light from passing (Guadagnini et al. 2015).

Some possible solutions are available to avoid these interference processes. In general, after *in vitro* exposure, NPs should be eliminated from the assays as much as possible by several washes. Testing 24 different types of NPs, Kroll et al. (2012) demonstrated that at concentrations above 50 mg/cm^2, NPs interfered with optical measurements of 2',7' dichlorofluorescein (DCF), MTT and LDH assays; however, except for carbon black NPs, interference could be prevented by lowering particle concentrations to 10 mg/cm^2. In some toxicity tests, NPs can also be eliminated from the assays by centrifugation and reading absorption/fluorescence in supernatants (Sweeney et al. 2014, Katsumiti et al. 2014, 2015a, 2015b, 2016, 2017, 2018).

Cell-free assays and the use of positive and negative particle controls are also strongly recommended in order to test NMs' reactivity with media and assays components (Drasler et al. 2017). In all assays, it would be important to include: (1) NMs-only suspended in culture media in order to verify positive or negative interferences with absorbance or fluorescence; (2) NMs plus assay reagents in order to verify the reactivity of NMs with assays reagents; (3) inhibition/enhancement controls in order to identify false-positives or negatives caused by the interaction of for example, NMs with inflammatory mediators such as cytokines and endotoxins (Dobrovolskaia et al. 2009).

Selection of Cell Culture Media and the Protein Corona Paradigm

One important factor to take into account when evaluating the toxicity *in vitro* of NMs is their interaction with different components of the cell culture media. A standard complete medium is generally composed by a buffered solution containing proteins, vitamins, amino acids and ionic salts, supplemented with antibiotics and serum (e.g., newborn calf serum, NCS; foetal calf or bovine serum, FCS/FBS). Commercial media such as Dulbelco's modified Eagle medium (DMEM), Roswell Park Memorial Institute medium (RPMI1640), Basal Medium Eagle (BME) and Leibovitz L-15 medium are widely used for mammalian cell culture and for the culture of cells from aquatic organisms such as mussels and fish. It is recommended to avoid the use of media with a high protein content for NMs exposure (Drasler et al. 2017) since proteins represent one of the most important medium components that affect the stability of NMs in the cell culture media. Thus, the selection of a low protein content medium such as the minimum essential medium (MEM) could be a worthy option. The use of serum may also influence NMs' properties and availability to the cells. Tedja et al. (2012) reported that FBS reduced aggregate size of TiO_2 NPs increasing cellular uptake of NPs in A549 and H1299 lung cell lines. Serum proteins adsorbed to gold nanorods' surface switched their positive charge to negative, affecting significantly their uptake and toxicity (Alkilany et al. 2009). In order to avoid interference of serum with the NMs, *in vitro* toxicity tests can be performed in serum-free medium or in a medium with a reduced serum content (in case it is essential for cells maintenance) as the amount and source of serum proteins can strongly affect NMs interaction with the cells (Lesniak et al. 2010, Drasler et al. 2017).

On the other hand, the use of proteins and serum in culture media represents a more realistic exposure scenario. When using cell culture media with proteins and/or serum or hemolymph, NMs interact with proteins forming a protein corona which may affect NMs identity, thus affecting particles interactions with target cells. This phenomenon is well-known as the "protein corona paradigm" (Monopoli et al. 2011, Monopoli et al. 2012, Docter et al. 2015, Canesi et al. 2017). Several authors have been paying special attention to the influence of protein corona in the biological reactivity of different NMs. Studies have been performed using a pre-incubation step of NMs with biological fluids in order to reproduce a more realistic exposure scenario. In experiments using human lung cells *in vitro*, the incubation of NMs with pulmonary surfactants prior to *in vitro* exposures is recommended, as *in vivo* NMs are trapped in the surfactant of the lungs prior to cross the alveolar barrier reaching the blood system where they will be delivered to different tissues. In mussels and other filter-feeding organisms, NMs suspended in water would enter the organism through the gills and reach the haemolymph, where interaction of NMs with protein corona would occur. Then, due to the open circulatory system in these molluscs (Gosling 2003), haemolymph would deliver these NMs to different tissues. In order to reproduce a more realistic scenario, Canesi et al. (2016) exposed haemocytes *in vitro* to amino-modified polystyrene NPs (PS-NH_2) suspended in haemolymph serum (HS) and compared their toxicity with PS-NH_2 suspended in artificial seawater (ASW). The authors of this study found that PS-NH_2 in HS were more toxic than PS-NH_2 suspended in ASW, inducing higher lysosomal and plasma membrane damage, ROS production, and dysregulation of p38 MAPK signalling compared to cells exposed to PS-NH_2 in ASW (Canesi et al. 2016). Thus, a thorough characterisation of protein corona is encouraged for a better understanding of NMs toxicity.

Dosimetry

In order to assess cytotoxicity and MOA of NMs, unrealistically high NM concentrations are sometimes required for identification of toxic and non-toxic doses. However, for assessment of potential NMs hazard to aquatic organisms, NM doses employed in the experiments should be based on realistic and ecologically relevant concentrations. Currently, data regarding production quantities and the relative amounts of NMs released from production plants is rare and the occupational and environmental exposure values are hard to estimate. However, a few studies regarding the life cycle assessment of industrial NMs are available. These studies estimated significant releases of NMs from production sites into the atmosphere and the environment (Mueller and Nowack 2008, Gottschalk et al. 2009, Tiede et al. 2009, Chio et al. 2012, Dumont et al. 2015). Gottschalk et al. (2013) reviewed the available literature concerning environmental concentrations of six NMs (TiO_2, ZnO, Ag, fullerenes, carbon nanotubes and CeO_2) in different compartments. Tiede et al. (2009) estimated the concentrations of several NMs arising from consumer products that will reach wastewaters and finally, the aquatic environment. Using predictive computational models, environmental concentrations of different NPs were estimated in Taiwanese rivers (Chio et al. 2012) and in Europe (Dumont et al. 2015). More recently, researchers were able to determine NP concentrations in the environment as shown by Li et al. (2016) who measured Ag NPs concentrations in wastewater treatment plant effluents in Germany. These predictions and the determination of NM environmental concentrations can support the selection of NMs concentrations to be used in *in vitro* toxicity testing. Since many times realistic doses are significantly different to cytotoxic doses, *in vitro* exposures can be performed at a wide range of concentrations, including ecologically relevant doses and cytotoxic ones. Starting from the same range of concentrations, parallel exposures to materials with similar chemical composition but varying in size or other physico-chemical properties (e.g., ionic and bulk forms of metal-bearing NPs) is recommended in order to compare toxicity dose ranges. Additives present in the NM preparations should be also tested at the equivalent concentrations present in NP suspensions in order to evaluate their contribution to NMs' toxicity.

Battery of In Vitro Toxicity Tests

As addressed by many authors, toxicity of NMs varies significantly according to the cell type used, type of NMs tested, cell culture system utilised, properties of the media, and toxicity tests employed. Individual cytotoxicity methods are currently not sufficient for evaluation of toxic potential of NMs. A battery of tests including multiple *in vitro* toxicity assays for different endpoints (including dye-free approaches) and if possible, more than one representative cell type should be employed for a reliable nanotoxicological assessment (Kroll et al. 2011). Also, different tests should be employed for the same endpoint as toxic effects may occur by different routes. For example, using the three cell viability assays Alamar blue, 5-carboxyfluorescein diacetate acetoxymethyl ester (CFDA-AM), and neutral red uptake (NRU), Bermejo-Nogales et al. (2017) found that cytotoxicity of ZnO NPs, multi-walled carbon nanotubes (MWCNTs), SiO_2 NPs and CeO_2 NPs in RTG-2 and PLHC-1 fish cell lines occurred mainly by lysosomal disruption as results of the NRU test were more marked than those of Alamar blue and CFDA-AM. Similar results were found by Connolly et al. (2015) and Yue et al. (2014) for Ag NPs. For genotoxicity assessment, Doak et al. (2012) recommended a battery of assays addressing different genotoxic and mutagenic endpoints since no single method is able to detect all different genome damage which includes DNA damage, chromosome aberrations and mutations. In order to decipher MOAs of NMs, a battery of tests covering uptake mechanisms and cellular functions is highly recommended. Using a battery of *in vitro* toxicity assays, many authors successfully identified specific MOAs of NMs in fish cells (Christen and Fent 2012, Lammel and Navas 2014, Taju et al. 2014) and bivalve mollusc cells (Canesi et al. 2008, 2010, 2016, Ciacci et al. 2012, Couleau et al. 2012, Katsumiti et al. 2014, 2015b, 2018, Sun et al. 2017, Volland et al. 2018, Sendra et al. 2018).

Risk Assessment of Metal and Metal-Bearing Nanoparticles Using Mussel Cells *In Vitro*

In this section, we focus on the cytotoxicity and MOAs of metal and metal-bearing NPs in cells of marine mussels, in the framework of the risk assessment of this important group of NMs.

Toxicity Ranking of Metal-Bearing Nanoparticles Based on Cell Viability Assays

The present discussion on the use of rapid screening tests for toxicity ranking of metal-bearing NPs is based on cell viability of mussel haemocytes and gill cells exposed to CdS quantum dots (QDs) (Katsumiti et al. 2014), and NPs of TiO_2 (Katsumiti et al. 2015a), Ag (Katsumiti et al. 2015b), Au, ZnO, SiO_2 (Katsumiti et al. 2016) and CuO (Katsumiti et al. 2018). In all these studies, cells were exposed *in vitro* for 24 h to different NPs, in parallel with exposures to their respective bulk and ionic counterparts. Toxicity ranking was established based on the lethal concentration (LC50) values obtained in the cell viability tests (NR and MTT assays) with haemocytes and gill cells. Selected NPs varied in chemical composition, size, shape, crystal structure, mode of synthesis and presence of additives (Table 3). NPs were characterised by TEM or SEM for particle size and by DLS for particle size distribution, aggregation and the measurement of the Z potential (Table 3). Besides of that, chemical stability of selected NPs in seawater was studied and is shown in Table 3.

Based on the results of the two cell viability assays and LC50 values obtained, toxicity ranking of metal-bearing NPs was established as: Ag NPs > CuO NPs > CdS QDs > ZnO NPs > TiO_2 NPs > Au NPs > SiO_2 NPs (Fig. 1). The type of metal (chemical composition) was found to be the main

Table 3. Summary table of the selected NPs, their physico-chemical properties and LC50 values obtained from the cell viability assay (MTT) in mussel haemocytes. Characterisation data of CuO NPs was published by Buffet et al. (2011) and data on size and shape of Au NPs were published by Pan et al. (2012)

Metal type	NPs/ODs	Additives	Mode of synthesis	Size (nm)	Crystal structure	Shape	Z potential (mV)	Aggregation (in seawater)	Chemical stability (in seawater)	LC50 values (mg/L) Haemocytes (MTT assay)
Ag	Mal-Ag20	Maltose	Wet chemistry	20	nd	Spherical	–30/–35	Aggregate	Release ions	5.8
	Mal-Ag40	Maltose	Wet chemistry	40	nd	Spherical	–30/–35	Aggregate	Release ions	8.43
	Mal-Ag100	Maltose	Wet chemistry	100	nd	Spherical	–30/–35	Aggregate	Release ions	9.5
	Ag20	Unknown	Unknown[a]	20	nd	Spherical	–50 ± 0.5	nd	nd	22.7
	Ag80	Unknown	Unknown[a]	80	nd	Spherical	–44 ± 1	nd	nd	21.7
CuO	CuO	-	Plasma	100	Tenorite	nd	+26.3	Aggregate	Release ions	9.37
CdS	CdS	-	Wet chemistry	5	nd	nd	nd	nd	Release ions	10.4
ZnO	ZnO<130-EcodisP90	Ecodis P90	Milling	20–70	nd	Rods	–39.9	Aggregate	Release ions	13.4
	ZnO<280-EcodisP90	Ecodis P90	Milling	500x260x10–100	nd	plates	–54.5		Release ions	39.4
TiO_2	WtC10TiO_2	-	Wet chemistry	10	Rutile	Rods	–49.1	Aggregate	nd	25.5
	WtC40TiO_2	-	Wet chemistry	40	Rutile	Rods	–42.07	Aggregate	nd	34.4
	WtC60TiO_2	-	Wet chemistry	60	Rutile	Rods	–43.6	Aggregate	nd	37.6
	WtC60TiO_2 P1100TiO_2	-	Plasma	100	55% Rutile/45% Anatase	Spherical and polyhedral	–11.9	nd	nd	54.9
	Mi60 TiO_2	DSLS (b)	Milling	60	Rutile	Sharp edges isometric[d]	–24.6	Aggregate	nd	19.8
	P25 TiO_2	-	Unknown[a]	21 (c)	10% Rutile/70% Anatase (20% other materials)	nd	–25.1[e]	Aggregate	nd	30.8
Au	Au5-Cit	Na-citrate	Wet chemistry	5	nd	Rods	–35	Aggregate	Stable	76.3
	Au15-Cit	Na-citrate	Wet chemistry	15	nd	Rods	–35	Aggregate	Stable	81.6
	Au40-Cit	Na-citrate	Wet chemistry	40	nd	Rods	–35	Aggregate	Stable	83.4
SiO_2	$SiO_2$15-Larg	L-arginine	Wet chemistry	15	nd	Spherical	nd	Aggregate	Stable	>100
	$SiO_2$30-Larg	L-arginine	Wet chemistry	30	nd	Spherical	nd	Aggregate	Stable	>100
	$SiO_2$70-Larg	L-arginine	Wet chemistry	70	nd	Spherical	nd	Aggregate	Stable	>100
	Fluorescent $SiO_2$27	L-arginine	Wet chemistry	27	nd	Spherical	nd	Aggregate	Stable	>100

nd: no data; (a) commercial NPs; (b) DSLS: dissodium laureth sulfoccinate; (c) according to the data available in the manufacturer's webpage (http://www.aerosil.com/product/aerosil/ja/effects/photocatalyst/pages/default.aspx); (d) according with Deiana et al. (2013); (e) according with Griffitt et al. (2008).

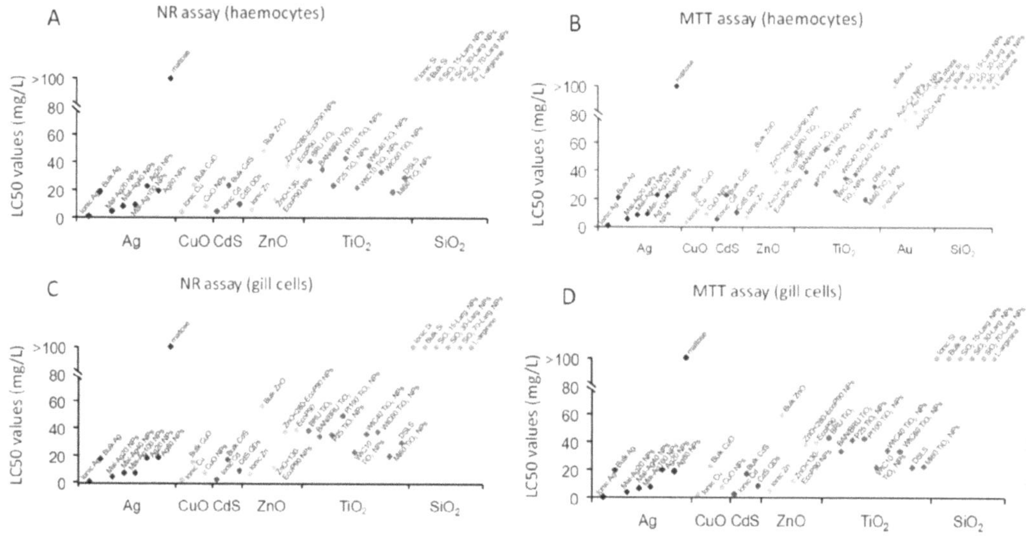

Figure 1. Cytotoxicity results based on the LC50 values obtained in the NR and MTT assays in haemocytes (A and B) and gill cells (C and D) exposed to ionic and bulk forms and to different NPs and additives (Table 3). Cytotoxicity of Au forms was tested only in haemocytes with the MTT assay. The different metals were ordered from the most to the least toxic. BRU: Bulk rutile TiO_2; BAN: Bulk anatase TiO_2.

factor explaining NPs toxicity. In the same cell type, Ciacci et al. (2012) reported a similar ranking of toxicity based on the NRRT assay for the following metal-bearing NPs: ZnO > TiO_2 = SiO_2 > CeO_2. According to the authors, chemical composition and behaviour of NPs in cell culture media could explain toxicity differences (Ciacci et al. 2012). Similar results were also reported in other species. In zebrafish, embryos exposed to the same metal-bearing NPs of Ag, CdS, CuO, ZnO, Au, TiO_2 and SiO_2 shown in Table 3, toxicity ranking was: CuO > ZnO > TiO_2 (Vicario-Parés et al. 2014) and Ag > ZnO > CdS > Au > SiO_2 (Lacave et al. 2016) in line with the degree of solubility of the different NPs.

Solubility seems to play an important role in the toxicity of metal-bearing NPs. Comparing solubility of NPs, insoluble Au and SiO_2 NPs (Katsumiti et al. 2016) were less toxic than soluble CdS QDs, ZnO, Ag and CuO NPs (Katsumiti et al. 2014, 2015b, 2016, 2018) (Fig. 1). Dissolution of ions from metal-bearing NPs such as ZnO, Ag and CdS QDs appears to be a major driver of toxicity in many aquatic organisms. The toxicity of ZnO NPs has been previously attributed to the dissolution of ionic Zn in freshwater (Aruoja et al. 2009, Franklin et al. 2007, Heinlaan et al. 2008), seawater (Wong et al. 2010, Miller et al. 2010, Bondarenko et al. 2013, Canesi and Corsi 2016) and cell culture media (Xia et al. 2008, Song et al. 2010, Bondarenko et al. 2013). CdS quantum dots have also been reported as soluble in different media (Ringwood et al. 2006, Gagné et al. 2008, Rocha et al. 2017). Ringwood et al. (2006) reported cytotoxicity to digestive gland cells of oysters *Crassostrea virginica* related with a partial degradation of CdS quantum dots. Gagné et al. (2008) found that cytotoxicity of CdTe quantum dots to haemocytes of mussels (*E. complanata*) were associated with particle dissolution. Studies have suggested that dissolution of Ag NPs is responsible for their toxicity (Lok et al. 2007, Navarro et al. 2008, Bondarenko et al. 2013, Canesi and Corsi 2016); however, others suggest that the toxicity is unlikely to be attributable to particle dissolution only (Chae et al. 2009, Griffitt et al. 2009, Yeo and Yoon 2009, Gomes et al. 2013b, Jimeno-Romero et al. 2017a). In mussels exposed for 15 days to CuO NPs, Gomes et al. (2012) found an increasing bioacumulation of NPs in parallel with a progressive enhancement of metallothionein levels along the time, whereas in exposures to ionic Cu, elimination was significant by day 15 and the increase in metallothionein levels occurred only at the end of the exposure period (day 15). Since only a small fraction of ionic Cu was released from CuO NPs (< 1% of total mass), the effects were attributed to the nano form of Cu. Similar results were obtained for CdS quantum dots (Katsumiti et al. 2014) and for CuO NPs (Katsumiti et al. 2018) in *in vitro* toxicity tests with mussel cells, toxicity being caused not just due to the release of ions from NPs. That is to say, several lines of evidence suggest that, related with the type of metal, NPs solubility is one of the main factors that explain the toxicity ranking of metal-bearing NPs in mussel haemocytes and gill cells. Nevertheless, other factors specifically related with their nanoform could also contribute to their toxicity.

In agreement with the previous idea, comparing NPs with their corresponding ionic and bulk forms, toxicity was: ionic > NPs > bulk (Fig. 1). These results appear to indicate that toxicity could be related with the dissolution rate of each metal form, because higher the metal dissolution, the greater the toxicity. Many NPs are actually designed to be more soluble than their bulk counterparts. In seawater, for example, the solubility of ZnO NPs can be more than twice of that of bulk ZnO (Wong et al. 2010). Karlsson et al. (2008) showed that CuO NPs are highly toxic when compared to the bulk form of CuO, in human alveolar epithelial cell line A549, especially in terms of oxidative damage and ROS production, and this was related with metal dissolution. Other *in vitro* studies have also shown that ionic forms of metals are more toxic than metal-bearing NPs due to the higher bioavailability of ions in the former (Miura and Shinohara 2009, Li and Hurt 2010, Farkas et al. 2011, Taju et al. 2014). There are relatively few studies comparing NP, ionic and bulk forms but generally bulk forms are the least toxic form, possibly related to their lower ability to release metal ions when compared to ionic and NP forms (Liu et al. 2011, Angel et al. 2013, Katsumiti et al. 2014, 2015a, 2015b, 2016, 2018).

As shown in Fig. 1, smaller NPs were generally more toxic than the larger ones of the same chemical composition. Size-dependent toxicity was found for Au NPs (Au5-Cit > Au15-Cit and Au40-Cit, Katsumiti et al. 2016), TiO_2 NPs (WtC10 > WtC40 ~ WtC60, Katsumiti et al. 2015a) and Ag NPs (Mal-Ag20 > Mal-Ag40 ~ Mal-Ag100, Katsumiti et al. 2015b). Other studies have also described that smaller NPs are more cytotoxic than larger ones for TiO_2 NPs (Oberdörster et al. 2005, L'Azou et al. 2008) and for Ag NPs (Carlson et al. 2008, Jiang et al. 2008a, Zhang et al. 2014, Gliga et al. 2014). Auffan et al. (2009) suggested that there is a critical size of about 30 nm below which NPs exhibit properties different from bulk materials. Below this size, the ratio of surface area to total atoms increases exponentially with decreasing NP size, resulting in excess of energy at the NPs surface making them thermodynamically less stable and more reactive. However, NPs with rough or porous surface may not exhibit a size dependent increase in reactivity, as well as size dependent properties may become hesitant when particles aggregate (Luyts et al. 2013). Thus, aggregation of NPs, especially in the case of metal-based NPs, could also play a critical role in toxicity. Aggregation of NPs in aqueous media is not only driven by NPs primary size and surface charge, but also media properties such as pH, presence of organic matter and ionic strength (Petosa et al. 2010, Corsi et al. 2014, Canesi and Corsi 2016). NPs are known to aggregate and precipitate in high ionic strength media such as seawater (Canesi et al. 2010, Petosa et al. 2010, Brunelli et al. 2013). Considering that high ionic strength media (similar to seawater) is used for mussel cells culture (Katsumiti et al. 2014, 2015a, 2015b, 2016, 2018), it is expected that most of the metal-bearing NPs were delivered as aggregates to mussel cells. Filter-feeding animals such as mussels, clams, scallops and oysters have shown to capture and ingest NP aggregates more efficiently than individual NPs (Kach and Ward 2008, Ward and Kach 2009). However *in vitro*, cells might uptake NPs and small aggregates via endocytic and phagocytic pathways (Conner and Schmid 2003), being larger aggregates less efficiently internalised (Champion et al. 2008). This might explain the observation in mussel cells exposed to metal-bearing NPs that larger NPs are not more toxic than smaller ones above certain size threshold.

Shape may also influence the toxicity of NPs although this is a factor much less studied than size (Fig. 1). Comparing the two types of ZnO NPs tested (Fig. 1), smaller rod-shaped NPs (20–70 nm) were more cytotoxic than larger plate-shaped ones (500 × 260 × 10–100 nm) (Katsumiti et al. 2016). Similar results were found in zebrafish embryos exposed to the same ZnO NPs (Lacave et al. 2016). Misra et al. (2013) found differences in the cytotoxicity of CuO NPs of different shapes (spheres, rods and spindles) to human alveolar lung cells. Comparing the dissolution ability and toxicity of different shapes of CuO NPs, spherical NPs were more toxic followed by rods and spindles, partly explained by the higher dissolution rate of the former (Misra et al. 2013). Thus, it seems that shape of NPs may influence their dissolution rate and hence, determine their cellular uptake, fate and toxicity.

Further, additives present in the preparation of NPs can contribute to their toxicity as shown in Fig. 1 for Na-citrate (Au NPs, Katsumiti et al. 2016) and DSLS (TiO_2 NPs, Katsumiti et al. 2015a), and to a lesser extent for Ecodis P90 (ZnO NPs, Katsumiti et al. 2016) whereas others such as maltose (Ag NPs, Katsumiti et al. 2015b) and L-arginine (SiO_2 NPs, Katsumiti et al. 2016) were not toxic to mussel cells *in vitro*. Na-citrate is a common additive used to disperse and stabilize many types of NPs such as Au or Ag NPs but at high concentrations it can be toxic. Testing the same Au NPs, Jimeno-Romero et al. (2017b) found a reduction of lysosomal membrane stability in mussels exposed to Na-citrate alone and Lacave et al. (2016) reported that Na-citrate alone decreased survival rate of zebrafish embryos. Sathishkumar et al. (2014) also found that the presence of Na-citrate in the Au NP preparations contributed to their toxicity. Authors reported that Na-citrate Au NPs were 5-fold more cytotoxic to human epithelial lung cells (A549) than Au NPs synthesised with star anise (*Illicium verum*; a commercially available spice). Concerning the toxicity of the surfactants Ecodis P90 and DSLS, Jimeno-Romero et al. (2016) reported that DSLS decreased lysosomal membrane stability of mussels. In zebrafish embryos, DSLS caused mortality and decrease in hatching rate (Vicario-Parés et al. 2014) and Ecodis P90 caused mortality and increase in malformation prevalence (Lacave et al. 2016). Thus, certain additives present in NP formulations appear to contribute significantly to NPs toxicity.

In addition to factors already discussed such as size, shape and the presence of additives, mode of synthesis and crystalline structure could also influence NPs toxicity. Among tested TiO_2 NPs, TiO_2 NPs produced by milling (Mi60) showed the highest toxicity (partly due to the presence of DSLS, as discussed before), followed by the smaller TiO_2 NPs produced by wet chemistry and TiO_2 NPs produced by plasma (Katsumiti et al. 2015a). Results suggest that the method of synthesis influences the cytotoxicity possibly by producing NPs of different physico-chemical characteristics. This is consistent considering that TiO_2 NPs produced by wet chemistry were relatively monodispersed nanorods showing the highest zeta potential values, corresponding to the most stable samples, whereas those produced by plasma contained NPs of a variety of shape and size ranges and were the least stable (Table 3).

TiO_2 with different crystalline structure showed different cytotoxicity to mussel haemocytes and gill cells (Fig. 1). Bulk anatase/rutile TiO_2 was more stable and more toxic than pure rutile TiO_2 (Katsumiti et al. 2015a). In cells exposed to TiO_2 NPs, no clear relationship was found between crystal structure and cytotoxicity (Table 3). Rutile TiO_2 NPs (Mi60, WtC10, WtC40 and WtC60) and P25 TiO_2 NPs (10% rutile; 70% anatase) were more cytotoxic than Pl100 TiO_2 NPs (55% rutile; 45% anatase). Differences in toxicity between Pl100 and P25 TiO_2 NPs could be due to the different ratios of rutile/anatase crystal forms or even to other factors including size of NPs. Generally, TiO_2 NPs with a higher content of the anatase form are more toxic than TiO_2 NPs with a higher content of the rutile form, at least partly because anatase generates more free radicals than rutile (Hidaka et al. 2005, Hsiao and Huang 2011, Jiang et al. 2008b, Sayes et al. 2006, Uchino et al. 2002, Warheit et al. 2007).

Comparing the two cell viability assays used (NR and MTT assays), both tests generally showed similar sensitivity in all exposures (Fig. 1). NR assay is based on the incorporation of the dye into lysosomes of living cells (Borenfreund and Puerner 1984) and MTT is based on the conversion of the water soluble MTT into non-soluble purple formazan by actively respiring cells (Mosmann 1983). Although the two tests measure different physiological endpoints, a good correlation between the two cytotoxicity assays has been reported (Borenfreund et al. 1988, Triglia et al. 1991, Vian et al. 1995, Chiba et al. 1998). A strong correlation ($r = 0.939$) was found following exposure of BALB/c mouse 3T3 fibroblast cell line to 28 different pharmaceuticals and cancer chemotherapeutic drugs (Borenfreund et al. 1988). Comparing the sensitivity of NR, MTT and LDH assays, Fotakis and Timbrell (2006) reported that the first two assays were more sensitive than the last one in hepatoma cell lines exposed to cadmium chloride. In agreement, similar results in terms of LC50 and NOEC values were obtained for NR and MTT assays in mussel cells exposed to CdS QDs (Katsumiti et al. 2014), TiO_2 NPs (Katsumiti et al. 2015a), Ag NPs (Katsumiti et al. 2015b), ZnO and SiO_2 NPs (Katsumiti et al. 2016) and CuO NPs (Katsumiti et al. 2018).

Comparing the sensitivity of mussel haemocytes versus gill cells using the two cytotoxicity NR and MTT assays, generally the two cell types responded similarly to NP exposures (Fig. 1). This is not surprising considering that the two tests target general cellular functions such as membrane integrity and lysosomal activity and mitochondrial activity, respectively. Besides this, both haemocytes (Cajaraville and Pal 1995) and gill cells (Gómez-Mendikute et al. 2005) are known to contain a well-developed endolysosomal system. In fact, NPs were found internalised into endocytic vesicles of both gill cells (Koehler et al. 2008) and haemocytes (Ciacci et al. 2012, Katsumiti et al. 2014, 2017).

Studies comparing the sensitivity of bivalve and mammalian cells in response to toxicants are scarce. In exposure to Cd, Olabarrieta et al. (2001) reported that the pig renal LLC-PK1 cell line was more sensitive than mussel haemocytes. The higher resistance of mussel cells to Cd could be related to the synthesis of specific inducible Cd binding metallothioneins (Zorita et al. 2007) and to the ability for lysosomal accumulation of Cd (Marigómez et al. 1995, Marigómez et al. 2002). In CuO NP exposures, mussel haemocytes and gill cells showed a similar sensitivity compared to human TT1 cells (Katsumiti et al. 2018) and thus could provide alternative models for screening the toxicity of CuO NPs and other nanomaterials.

Summing up, toxicity ranking based on cell viability assays with mussel haemocytes and gill cells were able to demonstrate the influence of different factors, including chemical composition, dissolution rate, size, shape, additives, mode of synthesis and crystalline structure on the toxicity of studied metal and metal-bearing NPs. Overall, screening *in vitro* assays with mussel cells were able to classify NPs with different physico-chemical properties and behaviour in a comprehensive toxicity ranking that generally agreed with results of *in vivo* studies.

In an attempt to provide tools for the risk assessment of metal-bearing NPs in marine ecosystems, a heat map based on the results obtained in the screening cell viability assays was constructed (Table 4). The heat map represents the dose/effect response for each studied NP. The numbers in the table indicate the doses where the first statistically significant response was observed. Colours indicate the severity of the response according to the arbitrary scale proposed: $>$ 100 mg/L (dark green: not toxic), 100 to $>$ 10 mg/L (light green: low toxicity), 10 to $>$ 1 mg/L (yellow: moderate toxicity), 1 to $>$ 0.1 mg/L (orange: toxic) and $\leq$ 0.1 mg/L (red: very toxic). According to the data shown in Table 4, Ag NPs, CuO NPs, CdS QDs and TiO_2 NPs and the corresponding ionic and bulk forms were the most cytotoxic materials tested (Table 4). ZnO NPs and the corresponding ionic and bulk forms were moderately toxic, and Au and SiO_2 NPs and the corresponding ionic and bulk forms were in general not cytotoxic to mussel cells (Table 4). These data indicate that concentrations below 100 mg/L of Au and SiO_2 NPs were in the range of *non-effect* concentrations. According to the results shown in Table 4, ionic forms were the most toxic metal forms followed by NPs and lastly, bulk forms. Note that the ranking of toxicity based on the first statistically significant response observed (Table 4) is similar to that discussed before based on LC50 values (Table 3, Fig. 1). Exceptions are found in haemocytes (MTT assay) exposed to ionic Cd and to maltose alone, and both haemocytes (MTT assay) and gill cells (NR assay) exposed to WtC60 TiO_2 NPs where a slight but significant decrease in cell viability was observed at low concentrations. Some chemicals can induce toxic effects at low concentrations that can be compensated later on by regulatory mechanisms such as the induction of metallothioneins for metal detoxification.

Toxicity Ranking Based on the MOA of Metal-Bearing NPs

As an alternative to the use of data from cell viability assays for toxicity ranking, NMs can also be grouped according to their MOA (Drasler et al. 2017). For this, a battery of functional or mechanistic tests should be carefully selected and exposures should be performed at sublethal doses in order to avoid acute cell responses. In this section, data published in Katsumiti et al. (2014, 2015b, 2018) on the MOA of CdS QDs, Ag NPs (Ag20-Mal) and CuO NPs and their corresponding ionic and bulk

Table 4. Heat map relating doses (mg/L) with effects (colours) of NPs, additives and corresponding ionic and bulk forms on haemocytes and gill cells based on cell viability assays (NR and MTT assays). - : not tested; o: no effect.

	Haemocytes		Gill Cells	
	NR	MTT	NR	MTT
Ionic Ag	1	0.1	1	0.1
Bulk Ag	10	10	25	10
Mal-Ag20 NPs	10	10	0.1	1
Mal-Ag40 NPs	10	10	10	10
Mal-Ag100 NPs	10	10	10	10
Maltose	o	0.01	o	o
Ag20 NPs	25	25	25	10
Ag80 NPs	25	25	25	25
Ionic Cu	1	0.5	0.5	0.5
Bulk CuO	10	10	10	10
CuO NPs	1	1	0.5	1
Ionic Cd	1	0.01	1	1
Bulk CdS	10	10	10	10
CdS QDs	1	10	10	10
Ionic Zn	10	10	10	10
Bulk ZnO	25	50	25	50
ZnO<130-EcodisP90 NPs	10	10	10	10
ZnO<280-EcodisP90 NPs	10	10	10	10
Ecodis P90	25	10	10	10
TiO_2 BRU	10	10	50	50
TiO_2 RUAN	10	10	1	1
WtC10 TiO_2 NPs	10	1	10	1
WtC40 TiO_2 NPs	10	1	0.1	10
WtC60 TiO_2 NPs	50	0.1	1	50
Pl100 TiO_2 NPs	10	0.1	50	10
Mi60 TiO_2 NPs	10	1	10	10
DSLS	50	1	50	10
P25 TiO_2 NPs	10	1	10	10
Ionic Au	-	25	-	-
Bulk Au	-	o	-	-
Au5-Cit NPs	-	50	-	-
Au15-Cit NPs	-	50	-	-
Au40-Cit NPs	-	50	-	-
Na-citrate	-	50	-	-
Ionic Si	100	25	o	25
Bulk Si	o	100	o	o
$SiO_2$15-Larg NPs	50	50	100	25
$SiO_2$30-Larg NPs	100	50	100	50
$SiO_2$70-Larg NPs	100	100	o	50
L-arginine	o	o	o	o

>100	100->10	10->1	1->0.1	≤0.1	mg/L
not toxic	**low toxicity**	**moderate**	**toxic**	**very toxic**	

Color version at the end of the book

forms is discussed. Characterisation of the three selected NPs is shown in Table 3. In these studies, cells were exposed for 24 hours to sublethal concentrations (< LC25 values) and the following array of mechanistic tests was used in mussel haemocytes and gill cells to assess effects related with oxidative stress and immune responses (ROS production assay, catalase -CAT- activity, acid phosphatase -AcP- activity), genotoxicity (Comet assay), and detoxification (multixenobiotic resistance -MXR- activity). Additionally, Na-K-ATPase activity was measured only in gill cells as this enzyme activity is prominent in epithelial cells. Actin cytoskeleton integrity and phagocytic activity were assessed only in haemocytes, as well-known phagocytic cells.

Results obtained in Katsumiti et al. (2014, 2015b, 2018) are summarised in Table 5. Main mechanisms of toxicity of CdS QDs (5 nm) in mussel haemocytes and gill cells involved increased ROS production, DNA damage, CAT, AcP and MXR transport activities, and stimulation of hemocyte phagocytic activity (Katsumiti et al. 2014). Similarly, main mechanisms of toxicity of maltose-coated Ag NPs (20 nm) involved the increase of ROS production, CAT activity and DNA damage in both cell types (Katsumiti et al. 2015b). Ag NP exposure caused activation of AcP and phagocytic activities and disrupted the actin cytoskeleton in haemocytes. In gill cells, Ag NPs increased MXR activity and decreased Na-K-ATPase activity. Finally, sensitivity of mussel cells and human alveolar cells towards CuO NPs (100 nm) were compared and cellular mechanisms involved in the toxicity of CuO NPs in mussel cells were studied (Katsumiti et al. 2018). Results showed that mussel cells were more sensitive than human alveolar TT1 cells based on the ROS production assay. Main mechanisms of toxicity of CuO NPs in mussel haemocytes and gill cells involved increased ROS production, induction of CAT, AcP and MXR activities and genotoxicity. A significant increase in haemocytes phagocytic activity and inhibition of gill cells Na-K-ATPase activity were also observed. It is remarkable that for all studied endpoints, with the exception of haemocytes phagocytosis, similar effect trends were found for ionic, bulk or NP forms, indicating similar mechanisms of action. Overall, the stimulatory effect on haemocytes phagocytosis caused by CdS QDs, bulk CdS, Ag NPs and CuO NPs indicate that cell-mediated immunity represents a key target of nano- and micro-sized particles.

As previously reported for a number of different NPs, the toxicity of CdS QDs, Ag NPs and CuO NPs to mussel cells appeared to be driven by their capacity to promote oxidative stress (Cho et al. 2009), which then could cause genotoxic (Asharani et al. 2009) and cytotoxic (Hussain et al. 2005) damages. Thus, it would appear that studied NPs exerted their toxicity in a sequential manner, so that increases in ROS production stimulated oxidant-sensitive signalling pathways to eventually

Table 5. Summary of the overall trends of the effects of CdS QD, Ag NPs and CuO NPs and their corresponding ionic and bulk forms on different cellular processes of mussel haemocytes and gill cells.

			CdS						Ag						CuO			
	Ionic		Bulk		QDs		Ionic		Bulk		NPs		Ionic		Bulk		NPs	
	H	G	H	G	H	G	H	G	H	G	H	G	H	G	H	G	H	G
ROS production	↑	↑	↑	↑	↑	↑	↑	↑	-	-	↑	↑	↑	↑	-	-	↑	↑
CAT activity	↑	↑	↑	↑	↑	-	↑	↑	↑	↑	↑	↑	↑	↑	-	-	↑	↑
DNA damage	↑	↑	↑	↑	↑	↑	↑	↑	↑	-	↑	↑	↑	↑	-	-	↑	↑
AcP activity	↑	↑	↑	-	↑	↑	↑	↑	↑	-	↑	-	↑	↑	-	-	↑	↑
MXR activity	↑	↑	↑	↑	↑	↑	↑	↑	-	↑	-	↑	↑	↑	-	↑	↑	↑
Na-K-ATPase activity		↓		-		-		↓		↓		↓		↓		-		↓
Phagocytic activity	-		↑		↑		-		-		↑		-		-		↑	
Damage to the actin cytoskeleton	-		-		-		↑		↑		↑		↑		-		-	

H: haemocytes; G: gill cells; ↑: increased; ↓: decreased; -: not altered.

culminate in genotoxic and cytotoxic consequences. Increased ROS production occurred not only in exposures to NPs but also in exposures to ionic Cd and bulk CdS, to ionic Ag and to ionic Cu (Table 5).

Many *in vitro* studies have shown that exposure to NPs elevates cellular oxidative stress (Limbach et al. 2007, Fahmy and Cormier 2009, Canesi et al. 2010, Ciacci et al. 2012, Christen and Fent 2012, Christen et al. 2013, Fernández-Cruz et al. 2013, Song et al. 2014, Lammel and Navas 2014, Taju et al. 2014, Drasler et al. 2017, Akter et al. 2018). To cope with elevated ROS levels, cells display various protective responses, including activation of enzymatic and non-enzymatic antioxidant defence mechanisms. Catalase, besides other antioxidant enzymes such as glutathione peroxidase and superoxide dismutase, and phase II biotransformation enzymes, plays essential roles in returning cells to a normal redox state (Sheehan 2000). When oxidative stress overwhelms defence mechanisms, cellular macromolecules such as proteins, lipids and DNA are subject to damage. DNA damage includes chromosome deletions, mutations and single- and double-strand breakages. In mussels, several studies have confirmed the formation of DNA adducts and oxidation-induced DNA fragmentation following exposure to metal-bearing NPs such as CuO and Ag NPs (Gomes et al. 2013a, Ruiz et al. 2015, Munari et al. 2014, Chelomin et al. 2017, Mahaye et al. 2017). In response to DNA damage, cells trigger mechanisms to repair the damaged DNA. In the case of severe damage to DNA, cells may die by either necrosis or apoptosis.

In exposures to CdS QDs, Ag NPs and CuO NPs and the corresponding ionic and bulk forms, differences in sensitivity of haemocytes and gill cells were found when evaluating AcP and MXR activities (Table 5). In general, haemocytes were more sensitive in the AcP activity assay whereas gill cells were more sensitive in the MXR activity assay (Table 5). All forms of Ag increased AcP activity in haemocytes whereas in gill cells only ionic Ag produced the same effect. In exposures to Cu forms, ionic Cu and CuO NPs increased AcP activity in the two cell types, although at lower concentrations in the haemocytes. Similar results were found in Cd exposures, and bulk CdS increased AcP activity only in haemocytes. It has been previously reported that *in vivo* exposure of molluscs to metal ions might trigger the hypersynthesis of lysosomal AcP, especially in this cell model (Suresh et al. 1993, Gómez-Mendikute and Cajaraville 2003). Haemocytes are involved in detoxification through the accumulation of metallic and organic xenobiotics in their well-developed endolysosomal system (Cajaraville and Pal 1995, Marigómez et al. 1995, 2002).

MXR activity is mediated mainly by P-glycoprotein (P-gp) and P-gp-like proteins, which actively transport a wide variety of structurally and functionally diverse compounds across the cell membrane out of the cell (Bard 2000). Induction of MXR activity has been already described in mussel gill cells exposed to the ionic form of metals such as Cd, Cu and Hg (Eufemia and Epel 2000, Achard et al. 2004) and the same effect is reported for haemocytes and gill cells exposed to CdS QDs, Ag NPs and CuO NPs and their respective ionic and bulk forms (Katsumiti et al. 2014, 2015b, 2018 and Table 5). This mechanism appears to represent an important detoxification pathway of metal-bearing NPs in gill cells. As epithelial polarized cells, mussel gill cells display a MXR transport activity in order to protect themselves from xenobiotics (Cornwall et al. 1995, Bard 2000). Induction of the MXR transport has also been described in mussel haemocytes exposed to xenobiotics (Svensson et al. 2003, Marin et al. 2004).

Na-K-ATPase activity in mussel gill cells was inhibited in the presence of Ag NPs and CuO NPs. Ionic forms of Cd, Ag and Cu and bulk Ag also inhibited gill cells Na-K-ATPase activity. Cu based NPs and Ag NPs are known to cause ion-regulatory impairment in fish gills due to their inhibitory effect on the Na-K-ATPase activity, even at a low concentration (Griffitt et al. 2009, Shaw et al. 2012).

In Cd, Ag and Cu exposures, micro- and NP-specific stimulation of phagocytosis was found in mussel haemocytes. At tested concentrations, haemocytes phagocytosis increased in Ag and CuO NP exposures and in exposures to bulk CdS and CdS QDs but not in the exposures to the ionic form of the same metals. It has already been described that phagocytosis is involved in the uptake of many NPs or NP aggregates. Uptake of graphene oxide and reduced graphene oxide in mussel haemocytes occurred partially through endophagocytic mechanisms (Katsumiti et al. 2017). Exposure of mussel haemocytes to carbon black NPs was associated with rapid and concentration-dependent cellular

Table 6. Heat map relating doses (mg/L) with effects (colours) of NPs and corresponding ionic and bulk forms on different cellular processes of haemocytes and gill cells. o: no effect. ROS: Reactive oxygen species production assay; CAT: Catalase activity assay; DNA: DNA damage by Comet assay; AcP: Acid phosphatase activity assay; MXR: Multixenobiotic resistance transport activity assay; PHAGO: Phagocytic activity assay; Na-K-ATPase: Na-K-ATPase activity assay.

	Haemocytes						**Gill cells**					
	ROS	**CAT**	**DNA**	**AcP**	**MXR**	**PHAGO**	**ROS**	**CAT**	**DNA**	**AcP**	**MXR**	**Na-K-ATPase**
Ionic Ag	0.03	0.03	0.06	0.06	0.25	o	0.03	0.06	0.12	0.5	0.06	0.06
Bulk Ag	o	5	10	10	o	o	o	10	o	o	0.62	5
Mal-Ag20 NPs	0.62	1.25	1.25	0.15	o	1.25	1.25	1.25	2.5	o	0.31	1.25
Ionic Cu	0.2	0.2	0.2	0.12	2	o	0.2	1	1.5	0.25	0.02	0.25
Bulk CuO	7.5	o	10	o	o	o	10	o	o	o	10	o
CuO NPs	0.05	2.75	2.75	2.75	2.75	2.75	0.5	5	2.75	5	2.75	1.25
Ionic Cd	0.1	0.1	1	0.5	2	o	0.25	2	1	2	2	0.25
Bulk CdS	1.25	1.25	10	10	5	2.5	2.5	5	10	o	5	o
CdS QDs	2.5	5	2.5	1.25	0.31	1.25	1.25	o	5	2.5	2.5	o

> 100	100->10	10->1	1->0.1	≤ 0.1	mg/L
not toxic	**low toxicity**	**moderate**	**toxic**	**very toxic**	

Color version at the end of the book

uptake by phagocytosis (Canesi et al. 2008). At low concentrations, Ciacci et al. (2012) found that TiO_2 and ZnO NPs increased haemocytes phagocytic activity. This is not surprising since haemocytes possess a high capacity for internalisation of micro- and nano-scale particles through processes as endocytosis and phagocytosis (Moore 2006). Since phagocytosis is a proxy for immunocompetence state in bivalves (Ellis et al. 2011), an increase in phagocytic activity may reflect the activation of the immune system by bulk CdS and by CdS QDs, Ag NPs and CuO NPs. This effect on haemocytes phagocytosis could also be taken as a specific mechanism of action of micro- and nano-scale particles caused by the interaction of these nanomaterials with the cell plasma membrane. Based on published nanotoxicity studies, it is apparent that the attachment of NPs to the plasma membrane can be a critical initial process that precedes the toxicity pathways (Lesniak et al. 2013). For example, greater NP adhesion has been observed to correlate with greater cellular internalisation by mechanisms as phagocytosis (Peetla and Labhasetwar 2009, Leroueil et al. 2007, 2008, Yuan et al. 2010).

As in the Subsection 'Toxicity Ranking of Metal-Bearing Nanoparticles Based on Cell Viability Assays' of this Chapter, a dose/response heat map based on the MOA of CdS QDs, Ag NPs and CuO NPs and their respective bulk and ionic forms in mussel haemocytes and gill cells was constructed (Table 6) using the same approach explained before. Briefly, numbers in the table indicate the doses where the first statistically significant response was observed and colours indicate the severity of the response. As shown in Table 6, toxicity varied from "not toxic" to "very toxic" depending on the endpoint and cell type evaluated. Comparing NPs with their respective ionic and bulk forms, toxicity was: ionic > NPs > bulk, in agreement with the classification obtained using data from cytotoxicity assays (Table 4). Comparing the different types of metals, toxicity could be ranked as follows: Ag > Cd > Cu (Table 6). As expected, haemocytes and gill cells showed relative differences in sensitivity according to the functional tests performed. Haemocytes were generally more sensitive than gill cells for CAT and AcP activities, possibly due to their role in immune defence. MXR transport activity seemed to play an important role in detoxification of Cd, Ag and Cu mainly in gill cells. Clearly, cell-mediated immunity and gill cell function represent significant targets for metal and metal-bearing NP toxicity in mussels. Thus, the selected *in vitro* assays could provide a relevant tool in the environmental risk assessment of these NPs.

Conceptual Framework for the Use of *In Vitro* Tests in Risk Assessment of Nanomaterials in Aquatic Systems

Ecological risk assessors face an increasing demand to assess hazard potential of NMs with greater speed and accuracy. *In vitro* assays have the potential to cover these needs by providing fast and reliable supporting data for the risk assessment of NMs. Nevertheless, at the moment there are no standardised tier testing approaches for the use of *in vitro* assays in environmental risk assessment of NMs.

Here we propose a systematic conceptual framework for the use of *in vitro* tests with aquatic organisms in the risk assessment of NMs (Fig. 2). The conceptual framework is divided into three tiers that are integrated at the end providing an adverse outcome pathway. In the tier I, cell models from aquatic organisms are selected taking into account different factors such as source of cells (organism of interest) from one or more aquatic species, use of primary cells or cell lines, cell type of interest (e.g., epithelial, immune, etc.), in monoculture or co-culture, 2D or 3D system, amongst others. Exposure concentrations are selected and the suitability of performing parallel experiments with for example, ionic and bulk forms of NMs and additives alone is considered. Primary properties and behaviour of NMs in cell culture media are characterised and exposure conditions are set up by adjusting medium components (e.g., salinity, pH, presence of serum) or changing to a more suitable media with similar composition if necessary. Based on the characterisation of NMs in cell culture media, potential NM properties that could contribute to toxicity are identified. Then, in the tier II of cytotoxicity assessment, potential interference of NMs with the absorbance or fluorescence spectra in the cell viability assays are initially assessed using cell-free assays. Cell viability tests that show no interference with the presence of NMs are selected and cytotoxicity of NMs is assessed at a wide range of concentrations covering sublethal and lethal doses. LC50 values are calculated and sublethal concentrations ($\leq$ LC25 values) are selected for the tier III. Based on the results of the cell viability assays, a heat map identifying the most and least toxic NMs or NM concentrations is built.

In the tier III of MOA identification, potential interference of NMs with absorbance or fluorescence spectra of functional or mechanistic tests is initially assessed using cell-free assays. Functional tests that show no interference with the presence of NMs are selected and mechanisms addressing NMs uptake, intracellular handling, elimination and toxicity are performed in order to assess MOA of NMs. Based on the results of functional tests, a heat map identifying the most and least reactive NMs or NM concentrations is built. Finally, results of the characterisation of NMs and identification of NMs properties that potentially contribute to toxicity are integrated with cellular responses (cytotoxicity and MOA of NMs) in order to generate an adverse outcome pathway that could indicate possible adverse effects expected *in vivo*.

Future Prospects

The goal of *in vitro* studies is to determine NMs toxicity and to unravel NMs MOAs to finally predict potential for toxic effects *in vivo*. Thus, it is anticipated that obtained information on the MOAs of NMs can be later used in weight of evidence analysis or tier testing schemes to predict toxic effects *in vivo*, leading to a reduction in the use of animals in toxicity testing (Oberdörster and Kuhlbusch 2018). For this, it is essential to use environmentally relevant concentrations, as discussed before, and to be able to relate cellular responses *in vitro* with toxic effects *in vivo*. As far as we know, in molluscs there are no nanotoxicological studies published so far investigating the correlation between *in vitro* and *in vivo* responses. However, similar responses were observed in mussels exposed *in vivo* and in

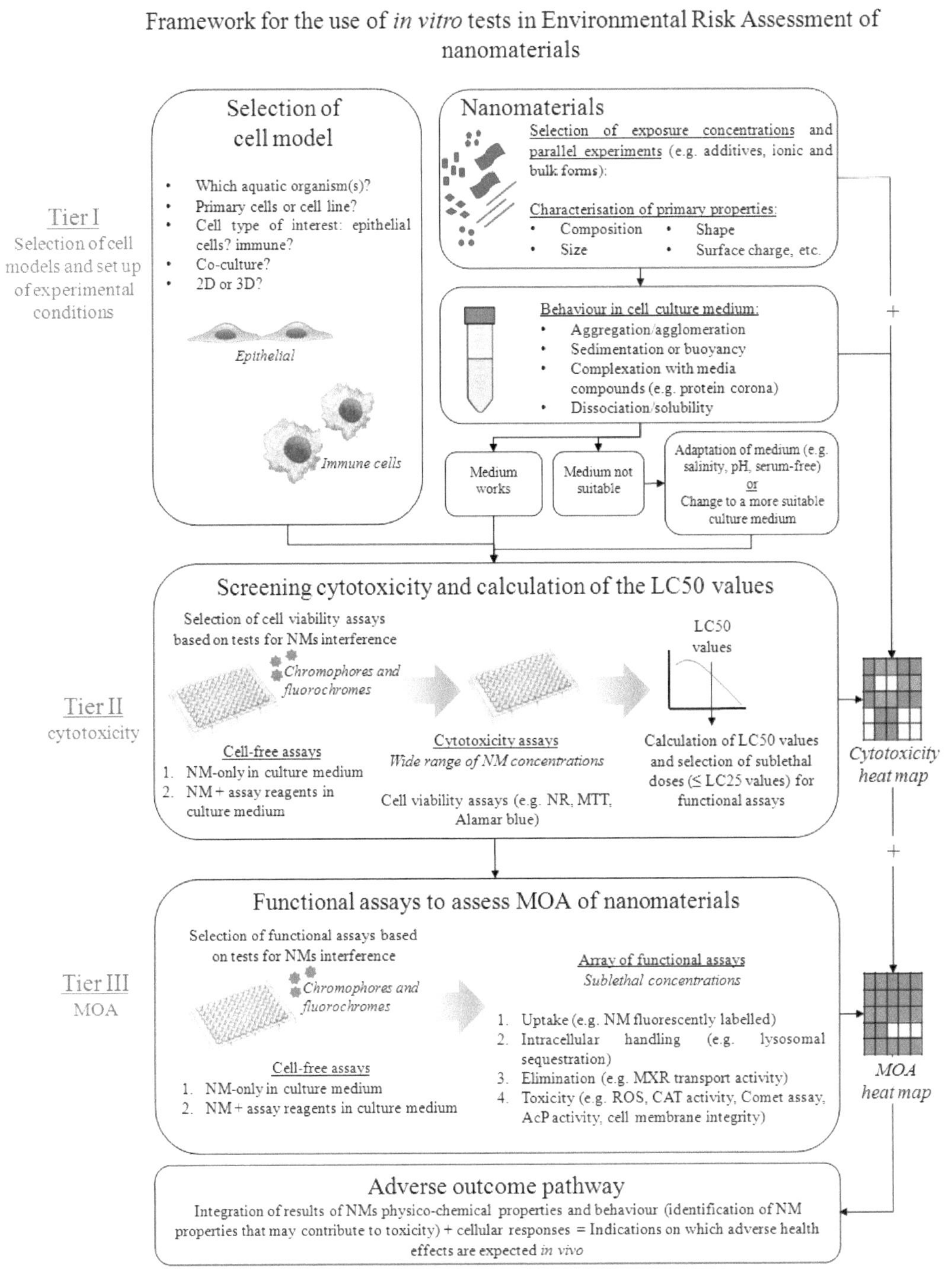

Figure 2. Conceptual framework for the use of *in vitro* tests with cells of aquatic organisms in environmental risk assessment of NMs.

mussel cells exposed *in vitro* to the same NPs. For instance, genotoxic effects were found in mussel haemocytes and gill cells exposed *in vitro* to CuO NPs (increase in DNA strand breaks, Katsumiti et al. 2018) and in haemocytes of mussels exposed *in vivo* to the same NPs (increased prevalence of micronucleus, Ruiz et al. 2015). In exposure to the same Ag NPs, lysosomal responses were found in mussel haemocytes *in vitro* (Katsumiti et al. 2015b) and in digestive glands of mussels *in vivo*

(Jimeno-Romero et al. 2017a). In fish, significant correlations were found between responses in cell lines *in vitro* and fish lethality *in vivo* after exposure to different chemicals (Segner and Lenz 1993, Castaño et al. 1996, Fent 2001, Na et al. 2009, Taju et al. 2012, 2013), including NPs (Taju et al. 2014). Using two fish species (*Catla catla* and *Labeo rohita*), Taju et al. (2014) found a linear correlation between EC50 values of Ag NPs in fish cell lines and LC50 values in whole fish exposed to the same Ag NPs. However, *in vitro* and *in vivo* responses do not always show a good correlation (Landsiedel et al. 2014). A possible reason for that could be related with the exposure media. NMs could show significant differences in physico-chemical properties and behaviour in the exposure media (cell culture media versus seawater or freshwater) and these differences may partly explain differences in NMs toxicity *in vitro* versus *in vivo*. The future direction in *in vitro* experiments should be driven towards the validation of exposure conditions that mimic as much as possible the *in vivo* scenario.

In vitro exposures should be performed as close as possible to a realistic scenario, but it is well-known that traditional 2D cell monocultures do not adequately represent a 3D tissue, especially the complex cross-talks between different cell types and between cells and the extracellular matrix. 3D culture models have been considered as more predictive *in vitro* systems, filling the gap between two-dimensional (2D) systems and *in vivo* experiments (Alepee et al. 2014). In this context, Lee et al. (2009) compared the sensitivity of 2D and 3D cell cultures of HepG2 cell line and found that toxicity of anionic and cationic gold NPs was significantly reduced in the 3D environment compared to the 2D. Similarly, Movia et al. (2011) found a significantly lower toxicity of single-walled carbon nanotubes (SWCNTs) in 3D cultures of human monocytic leukaemia THP-1 cell line compared to the same cells in 2D. For hazard identification of irritant substances, 3D human epidermis-based cell models (OECD 2015c) have been used because it closely mimics the physiological properties of human skin. This assay has been successfully applied to assess NMs' irritant potential (Kim et al. 2016). In aquatic species, 3D cell cultures of primary hepatocytes have been already established in fish species (Baron et al. 2012); however, as far as we know, it has not yet been used to assess toxicity of NMs. Thus, 3D cell culture systems with isolated cells of aquatic organisms represent an interesting field to be explored in future nanotoxicological studies. Additionally, co-cultures of different cell types from aquatic organisms such as mussel digestive gland cells (Robledo and Cajaraville 1997) or epithelial gill cells (Gómez-Mendikute et al. 2005) with mussel haemocytes could also offer an interesting *in vitro* model to study the cell-cell interactions in nanotoxicological studies.

Acknowledgements

This work was funded by EU 7th Framework Programme (NanoReTox project, CP-FP 214478-2), Spanish Ministry of Economy and Competitiveness (NanoCancer project CTM2009-13477, NanoSilverOmics project MAT2012-39372 and NanoCarrierERA project CTM2016-81130-R), Basque Government (consolidated research group IT810-13 and postdoctoral contract to AK) and University of the Basque Country (UFI 11/37).

References

Achard, M., M. Baudrimont, A. Boudou and J.P. Bourdineaud. 2004. Induction of a multixenobiotic resistance protein (MXR) in the Asiatic clam *Corbicula fluminea* after heavy metals exposure. Aquat. Toxicol. 67: 347–357.

Alepee, N., A. Bahinski, M. Daneshian, B. De Wever, E. Fritsche, A. Goldberg et al. 2014. State-of-the-art of 3D cultures (organs-on-a-chip) in safety testing and pathophysiology. ALTEX 31: 441–477.

Alkilany, A.M., P.K. Nagaria, C.R. Hexel, T.J. Shaw, C.J. Murphy and M.D. Wyatt. 2009. Cellular uptake and cytotoxicity of gold nanorods: Molecular origin of cytotoxicity and surface effects. Small 5: 701–708.

Angel, B.M., G.E. Batley, C.V. Jarolimek and N.J. Rogers. 2013. The impact of size on the fate and toxicity of nanoparticulate silver in aquatic systems. Chemosphere 93: 359–365.

Aruoja, V., H.C. Dubourguier, K. Kasemets and A. Kahru. 2009. Toxicity of nanoparticles of CuO, ZnO and TiO_2 to microalgae *Pseudokirchneriella subcapitata*. Sci. Total Environ. 407: 1461–1468.

Asharani, P.V., M.G. Low Kah, M.P. Hande and S. Valiyaveettil. 2009. Cytotoxicity and genotoxicity of silver nanoparticles in human cells. ACS Nano 3: 279–290.

Auffan, M., J. Rose, J.-Y. Bottero, G.V. Lowry, J.P. Jolivet and M.R. Wiesner. 2009. Towards a definition of inorganic nanoparticles from an environmental, health and safety perspective. Nat. Nanotechnol. 4: 634–641.

Babich, H. and E. Borenfreund. 1987. *In vitro* cytotoxicity of organic pollutants to bluegill sunfish (BF-2) cells. Environ. Res. 42: 229–237.

Baker, T.J., C.R. Tyler and T.S. Galloway. 2014. Impacts of metal and metal oxide nanoparticles on marine organisms. Environ. Pollut. 186: 257–271.

Bard, S.M. 2000. Multixenobiotic resistance as a cellular defense mechanism in aquatic organisms. Aquat. Toxicol. 48: 357–389.

Baron, M.G., W.M. Purcell, S.K. Jackson, S.F. Owen and A.N. Jha. 2012. Towards a more representative *in vitro* method for fish ecotoxicology: Morphological and biochemical characterisation of three-dimensional spheroidal hepatocytes. Ecotoxicology 21: 2419–2429.

Baron, M.G., K.S. Mintram, S.F. Owen, M.J. Hetheridge, A.J. Moody, W.M. Purcell et al. 2017. Pharmaceutical metabolism in fish: using a 3-D hepatic *in vitro* model to assess clearance. PLoS ONE 12(1): e0168837.

Batley, G.E., J.K. Kirby and M.J. McLaughlin. 2012. Fate and risk of nanomaterials in aquatic and terrestrial environments. Acc. Chem. Res. 46: 854–862.

Baun, A., S.N. Sørensen, R.F. Rasmussen, N.B. Hartmann and C.B. Koch. 2008. Toxicity and bioaccumulation of organic compounds in the presence of aqueous suspensions of aggregates of nano-C_{60}. Aquat. Toxicol. 86: 379–387.

Bergami, E., S. Pugnalini, M.L. Vannuccini, L. Manfra, C. Faleri, F. Savorelli et al. 2017. Long-term toxicity of surface-charged polystyrene nanoplastics to marine planktonic species *Dunaliella tertiolecta* and *Artemia franciscana*. Aquat. Toxicol. 189: 159–169.

Bermejo-Nogales, A., M.L. Fernández-Cruz and J.M. Navas. 2017. Fish cell lines as a tool for the ecotoxicity assessment and ranking of engineered nanomaterials. Reg. Toxicol. Pharmacol. 90: 297–307.

Blinova, I., A. Ivask, M. Heinlaan, M. Mortimer and A. Kahru. 2010. Ecotoxicity of nanoparticles of CuO and ZnO in natural water. Environ. Pollut. 158: 41–47.

Bondarenko, O., K. Juganson, A. Ivask, K. Kasemets, M. Mortimer and A. Kahru. 2013. Toxicity of Ag, CuO and ZnO nanoparticles to selected environmentally relevant test organisms and mammalian cells *in vitro*: a critical review. Arch. Toxicol. 87: 1181.

Borenfreund, E., H. Babich and N. Martin-Alguacil. 1984. Comparisons of two *in vitro* cytotoxicity assays—The neutral red and tetrazolium MTT tests. Toxicol. *In Vitro* 2: 1–6.

Borenfreund, E. and J.A. Puerner. 1984. Toxicity determined *in vitro* by morphological alterations and neutral red absorption. Toxicol. Lett. 24: 119–124.

Bruneau, A., M. Fortier, F. Gagne, C. Gagnon, P. Turcotte, A. Tayabali et al. 2013. Size distribution effects of cadmium tellurium quantum dots (CdS/CdTe) immunotoxicity on aquatic organisms. Environ. Sci. Process Impacts 15: 596–607.

Brunelli, A., G. Pojana, S. Callegaro and A. Marcomini. 2013. Agglomeration and sedimentation of titanium dioxide nanoparticles (n-TiO_2) in synthetic and real waters. J. Nanopart. Res. 15: 1684–1694.

Brunner, T.J., P. Wick, P. Manser, P. Spohn, R.N. Grass, L.K. Limbach et al. 2006. *In vitro* cytotoxicity of oxide nanoparticles: comparison to asbestos, silica, and the effect of particle solubility. Environ. Sci. Technol. 40: 4374–4381.

Cajaraville, M.P. and S.G. Pal. 1995. Morphofunctional study of the haemocytes of the bivalve mollusc *Mytilus galloprovincialis* with emphasis on the endolysosomal compartment. Cell Sturct. Funct. 20: 355–367.

Cajaraville, M.P. 2009. Potential of mussel gill cell primary cultures for toxicity testing of conventional and emerging environmental pollutants. Comp. Biochem. Physiol. 153: S86.

Caminada, D., C. Escher and K. Fent. 2006. Cytotoxicity of pharmaceuticals found in aquatic systems: Comparison of PLHC-1 and RTG-2 fish cell lines. Aquat. Toxicol. 79: 114–123.

Canesi, L., C. Ciacci, M. Betti, R. Fabbri, B. Canonico, A. Fantinati et al. 2008. Immunotoxicity of carbon black nanoparticles to blue mussel hemocytes. Environ. Int. 34: 1114–1119.

Canesi, L., C. Ciacci, D. Vallotto, G. Gallo, A. Marcomini and G. Pojana. 2010. *In vitro* effects of suspensions of selected nanoparticles (C60 fullerene, TiO_2, SiO_2) on Mytilus hemocytes. Aquat. Toxicol. 96: 151–158.

Canesi, L., C. Ciacci, R. Fabbri, A. Marcomini, G. Pojana and G. Gallo. 2012. Bivalve molluscs as a unique target group for nanoparticle toxicity. Mar. Environ. Res. 76: 16–21.

Canesi, L., G. Frenzilli, T. Balbi, M. Bernardeschi, C. Ciacci, S. Corsolini et al. 2014. Interactive effects of n-TiO_2 and 2,3,7,8-TCDD on the marine bivalve *Mytilus galloprovincialis*. Aquat. Toxicol. 153: 53–65.

Canesi, L., C. Ciacci, E. Bergami, M.P. Monopoli, K.A. Dawson, S. Papa et al. 2015. Evidence for immunomodulation and apoptotic processes induced by cationic polystyrene nanoparticles in the hemocytes of the marine bivalve Mytilus. Mar. Environ. Res. 111: 34–40.

Canesi, L., C. Ciacci, R. Fabbri, T. Balbi, A. Salis, G. Damonte et al. 2016. Interaction of cationic polystyrene nanoparticles with marine bivalve hemocytes in a physiological environment: Role of soluble hemolymph protein. Environ. Res. 150: 73–81.

Canesi, L. and I. Corsi. 2016. Effects of nanomaterials on marine invertebrates. Sci. Total Environ. 565: 933–940.

Canesi, L., T. Balbi, R. Fabbri, A. Salis, G. Damonte, M. Volland et al. 2017. Biomolecular coronas in invertebrate species: Implication in the environmental impact of nanoparticles. NanoImpact. 8: 89–98.

Carballal, M.J., C. López, C. Azevedo and A. Villalba. 1997. Enzymes involved in defense functions of hemocytes of mussel *Mytilus galloprovincialis*. J. Invertebr. Pathol. 70: 96–105.

Carballal, Villalba, A., M.J. and C. López. 1997. Seasonal variation and effects of age, food availability, size, gonadal development, and parasitism on the hemogram of *Mytilus galloprovincialis*. J. Invertebr. Pathol. 72: 304–312.

Carlson, C., S.M. Hussain, A.M. Schrand, L.K. Braydich-Stolle, K.L. Hess, R.L. Jones et al. 2008. Unique cellular interaction of silver nanoparticles: size-dependent generation of reactive oxygen species. J. Phys. Chem. B 112: 13608–13619.
Castaño, A., M.J. Cantarino, P. Castillo and J.V. Tarazona. 1996. Correlations between the RTG-2 cytotoxicity test EC50 and *in vivo* LC50 rainbow trout bioassay. Chemosphere 11: 2141–2157.
Chae, Y.J., C.H. Pham, J. Lee, E. Bae, J. Yi and M.B. Gu. 2009. Evaluation of the toxic impact of silver nanoparticles on Japanese medaka (*Oryzias latipes*). Aquat. Toxicol. 94: 320–327.
Champion, J.A., A. Walker and S. Mitragotri. 2008. Role of particle size in phagocytosis of polymeric microspheres. Pharm. Res. 25: 1815–1821.
Chelomin, V.P., V.V. Slobodskova, M. Zakhartsev and S. Kukla. 2017. Genotoxic potential of copper oxide nanoparticles in the bivalve mollusk *Mytilus trossulus*. J. Ocean Univ. China (Oceanic and Coastal Sea Research) 16: 10–20.
Cheng, T.C. 1981. Bivalves. pp. 233–300. *In*: N.A. Ratcliffe and A.F. Rowley (eds.). Invertebrate Blood Cells. Academic Press, London, UK.
Chiba, K., K. Kawakami and K. Tohyama. 1998. Simultaneous evaluation of cell viability by neutral red, MTT and cristal violet staining assays of the same cells. Toxicol. *In Vitro* 12: 251–258.
Chio, C.-P., W.-Y. Chen, W.-C. Chou, N.-H. Hsieh, M.-P. Ling and C.-M. Li. 2012. Assessing the potential risks to zebrafish posed by environmentally relevant copper and silver nanoparticles. Sci. Total Environ. 420: 111–118.
Cho, W.-S, M. Cho, J. Jeong, M. Choi, H.-Y. Cho, B.S. Han et al. 2009. Acute toxicity and pharmacokinetics of 13 nm-sized PEG-coated gold nanoparticles. Toxicol. Appl. Pharmacol. 236: 16–24.
Christen, V. and K. Fent. 2012. Silica nanoparticles and silver-doped silica nanoparticles induce endoplasmatic reticulum stress and alter cytochrome P4501A activity. Chemosphere 87: 423–434.
Christen, V., M. Capelle and K. Fent. 2013. Silver nanoparticles induce endoplasmatic reticulum stress response in zebrafish. Toxicol. Appl. Pharmacol. 272: 519–528.
Ciacci, C., B. Canonico, D. Bilanicova, R. Fabbri, K. Cortese, G. Gallo et al. 2012. Immunomodulation by different types of N-oxides in the hemocytes of the marine bivalve *Mytilus galloprovincialis*. Plos One 7: e36937.
Cohen, D., Y. Soroka, Z. Ma'or, M. Oron, M. Portugal-Cohen, F.M., Brégégère et al. 2013. Evaluation of topically applied copper(II) oxide nanoparticle cytotoxicity in human skin organ culture. Toxicol. *In Vitro* 27: 292–298.
Conner, S.D. and S.L. Schmid. 2003. Regulated portals of entry into the cell. Nature 422: 37–44.
Connolly, M., M.-L. Fernández-Cruz, A. Quesada-Garcia, L. Alte, H. Segner and J. Navas. 2015. Comparative cytotoxicity study of silver nanoparticles (AgNPs) in a variety of rainbow trout cell lines (RTL-W1, RTH-149, RTG-2) and primary hepatocytes. Int. J. Envion. Res. Publ. Health 12: 5386–5405.
Cornwall, R., B.H. Toomey, S. Bard, C. Bacon, W.M. Jarman and D. Epel. 1995. Characterization of multixenobiotic/multidrug transport in the gills of the mussel *Mytilus californianus* and identification of environmental substrates. Aquat. Toxicol. 31: 277–296.
Couleau, N., D. Techer, C. Pagnout, S. Jomini, L. Foucaud, P. Laval-Gilly et al. 2012. Hemocyte responses of *Dreissena polymorpha* following a short-term *in vivo* exposure to titanium dioxide nanoparticles: Preliminary investigations. Sci. Total Environ. 438: 490–497.
Di Bucchianico, S., M.R. Fabbrizi, S.K. Misra, E. Valsami-Jones, D. Berhanu, P. Reip et al. 2013. Multiple cytotoxic and genotoxic effects induced *in vitro* by differently shaped copper oxide nanomaterials. Mutagenesis 28: 287–299.
Di Bucchianico, S., M.R. Fabbrizi, S. Cirillo, C. Uboldi, D. Gilliland, E. Valsami-Jones et al. 2014. Aneuploidogenic effects and DNA oxidation induced *in vitro* by differently sized gold nanoparticles. Int. J. Nanomedicine 9: 2191–2204.
Doak, S.H., B. Manshian, G.J.S. Jenkins and N. Singh. 2012. *In vitro* genotoxicity testing strategy for nanomaterials and the adaptation of current OECD guidelines. Mutat. Res. 745: 104–111.
Dobrovolskaia, M.A., D.R. Germolec and J.L Weaver. 2009. Evaluation of nanoparticle immunotoxicity. Nat. Nanotechnol. 4: 411–414.
Docter, D., D. Westmeier, M. Markiewicz, S. Stolte, S.K. Knauer and R.H. Stauber. 2015. The nanoparticle biomolecule corona: lessons learned—challenge accepted? Chem. Soc. Rev. 44: 6094–6121.
Donaldson, K., V. Stone, P.J.A. Borm, L.A. Jimenez, P.S. Gilmour, P.F. Schins et al. 2003. Oxidative stress and calcium signalling in the adverse effects of environmental particles (PM10). Free Radic. Biol. Med. 34: 1369–1382.
Donaldson, K. and Tran, C.L. 2004. An introduction to the short-term toxicology of respirable industrial fibres. Mutat. Res. 553: 5–9.
Dorger, M., S. Munzing, A.M. Allmeling, K. Messmer and F. Krombach. 2001. Differential responses of rat alveolar and peritoneal macrophages to man-made vitreous fibers *in vitro*. Environ. Res. 85: 207–214.
Drasler, B., P. Sayre, K.G. Steinhäuser, A. Petri-Fink and B. Rothen-Rutishauser. 2017. *In vitro* approaches to assess the hazard of nanomaterials. NanoImpact 8: 99–116.
Duffin, R., L. Tran, D. Brown, V. Stone and K. Donaldson. 2007. Proinflammogenic effects of low-toxicity and metal nanoparticles *in vivo* and *in vitro*: Highlighting the role of particle surface area and surface reactivity. Inhal. Toxicol. 19: 849–856.
Dumont, E., A.C. Johnson, V.D.J. Keller and R.J. Williams. 2015. Nano silver and nano zinc-oxide in surface waters-exposure estimation for Europe at high spatial and temporal resolution. Environ. Pollut. 196: 341–349.
Ellis, R.P., H. Parry, J.I. Spicer, T.H. Hutchinson, R.K. Pipe and S. Widdicombe. 2011. Immunological function in marine invertebrates: responses to environmental perturbation. Fish Shellfish Immunol. 30: 1209–1222.
Elston, R.A. 2000. Molluscan diseases: a tissue-culture perspective. pp. 183–203. *In*: C. Mothersill and B. Austin (eds.). Aquatic Invertebrate Cell Culture. Praxis Publishing Ltd., Chichester, UK.

Eufemia, N.A. and D. Epel. 2000. Induction of the multixenobiotic defense mechanism (MXR), p-glycoprotein, in the mussel *Mytilus californianus* as a general cellular response to environmental stresses. Aquat. Toxicol. 49: 89–100.

European Commission. 2006. Regulation No. 1907/2006 of the European Parliament and of the Council of 18 December 2006 concerning the Registration, Evaluation, Authorisation and Restriction of Chemicals (REACH), establishing a European Chemicals Agency, amending Directive 1999/45/EC and repealing Council Regulation (EEC) No. 793/93 and Commission Regulation (EC) No 1488/94 as well as Council Directive 76/769/EEC and Commission Directives 91/155/EEC, 93/105/EC and 2000/21/EC. Available at: http://eurlex.europa.eu/LexUriServ/LexUriServ.do?uri.CELEX:32006R1907:EN.

Fabrega, J., S.R. Fawcett, J.C. Renshaw and J.R. Lead. 2009. Silver nanoparticle impact on bacterial growth: effect of pH, concentration, and organic matter. Environ. Sci. Technol. 43: 7285–7290.

Fabrega, J., S.N. Luoma, C.R. Tyler, T.S. Galloway and J.R. Lead. 2011. Silver nanoparticles: Behaviour and effects in the aquatic environment. Environ. Int. 37: 517–531.

Fahmy, B. and S.A. Cormier. 2009. Copper oxide nanoparticles induce oxidative stress and cytotoxicity in airway epithelial cells. Toxicol. *In Vitro* 23: 1365–1371.

Fako, V.E. and D.Y. Furgeson. 2009. Zebrafish as a correlative and predictive model for assessing biomaterial nanotoxicity. Adv. Drug Deliv. Rev. 61: 478–486.

Farkas, J., P. Christian, J.A. Urrea, N. Roos, M. Hassellöv, K.E. Tollefsen et al. 2010. Effects of silver and gold nanoparticles on rainbow trout (*Oncorhynchus mykiss*) hepatocytes. Aquat. Toxicol. 96: 44–52.

Fent, K. 2001. Fish cell lines as versatile tools in ecotoxicology: Assessment of cytotoxicity, cytochrome P4501A induction potential and estrogenic activity of chemicals and environmental samples. Toxicol. *In Vitro* 15: 477–488.

Fent, K. 2007. Permanent fish cell culture as important tools in Ecotoxicology. ALTEX 24: 26–28.

Fernández-Cruz, M.L., T. Lammel, M. Connolly, E. Conde, A. Isabel Barrado, S. Derick et al. 2013. Comparative cytotoxicity induced by bulk and nanoparticulate ZnO in the fish and human hepatoma cell lines PLHC-1 and Hep G2. Nanotoxicology 7: 935–952.

Fotakis, G. and J.A. Timbrell. 2006. *In vitro* cytotoxicity assays: comparison of LDH, neutral red, MTT, and protein assay in hepatoma cell lines following exposure to cadmium chloride. Toxicol. Lett. 160: 171–177.

Franklin, N.M., N.J. Rogers, S.C. Apte, G.E. Batley, G.E. Gadd and P.S. Casey. 2007. Comparative toxicity of nanoparticulate ZnO, bulk ZnO. and $ZnCl_2$ to a freshwater microalga (*Pseudokirchneriella subcapitata*): The importance of particle solubility. Environ. Sci. Technol. 41: 8484–8490.

Fryer, J.L. and C.N. Lannan. 1994. Three decades of fish cell culture: a current listing of cell lines derived from fishes. J. Tissue Cult. Meth. 16: 87–94.

Fu, P.P., Q. Xia, H.-M. Hwang, P.C. Ray and H. Yu. 2014. Mechanisms of nanotoxicity: Generation of reactive oxygen species. J. Food Drug. Anal. 22: 64–75.

Gagné, F., J. Auclair, P. Turcotte, M. Fournier, C. Gagnon, S. Sauve et al. 2008. Ecotoxicity of CdTe quantum dots to freshwater mussels: impacts on immune system, oxidative stress and genotoxicity. Aquat. Toxicol. 86: 333–340.

Geiser, M., B. Rothen-Rutishauser, N. Kapp, S. Schürch, W. Kreyling, H. Schulz et al. 2005. Ultrafine particles cross cellular membranes by nonphagocytic mechanisms in lungs and in cultured cells. Environ. Health Perspect. 13: 1555–1560.

Georgantzopoulou, A., Y.L. Balachandran, P. Rosenkranz, M. Dusinska, A. Lankoff, M. Wojewodzka et al. 2013. Ag nanoparticles: Size- and surface-dependent effects on model aquatic organisms and uptake evaluation with NanoSIMS. Nanotoxicology 7: 1168–1178.

George, S., S. Lin, Z. Ji, C.R. Thomas, L. Li, M. Mecklenburg et al. 2012. Surface defects on plate-shaped silver nanoparticles contribute to its hazard potential in a fish gill cell line and zebrafish embryos. ACS Nano 6: 3745–3759.

Gliga, A.R., S. Skoglund, I.O. Wallinder, B. Fadeel and H.L. Karlsson. 2014. Size-dependent cytotoxicity of silver nanoparticles in human lung cells: The role of cellular uptake, agglomeration and Ag release. Part. Fibre Toxicol. 11: 11.

Gomes, T., C.G. Pereira, C. Cardoso, J.P. Pinheiro, I. Cancio and M.J. Bebianno. 2012. Accumulation and toxicity of copper oxide nanoparticles in the digestive gland of *Mytilus galloprovincialis*. Aquat. Toxicol. 118-119: 72–79.

Gomes, T., O. Arouja, R. Pereira, A.C. Almeida, A. Cravo and M.J. Bebianno. 2013a. Genotoxicity of copper oxide and silver nanoparticles in the mussel *Mytilus galloprovincialis*. Mar. Environ. Res. 84: 51–59.

Gomes, T., C.G. Pereira, C. Cardoso and M.J. Bebianno. 2013b. Differential protein expression in mussel *Mytilus galloprovincialis* exposed to nano and ionic Ag. Aquat. Toxicol. 136-137: 79–90.

Gómez-Mendikute, A., A. Etxeberria, I. Olabarrieta and M.P. Cajaraville. 2002. Oxygen radicals production and actin filament disruption in bivalve haemocytes treated with benzo(a)pyrene. Mar. Environ. Res. 54: 431–436.

Gómez-Mendikute, A. and M.P. Cajaraville. 2003. Comparative effects of cadmium, copper, paraquat and benzo[a]pyrene on the actin cytoskeleton and production of reactive oxygen species (ROS) in mussel haemocytesToxicol. *in vitro* 17: 539–546.

Gómez-Mendikute, A., M. Elizondo, P. Venier and M.P. Cajaraville. 2005. Characterization of mussel gill cells *in vivo* and *in vitro*. Cell Tissue Res. 321: 131–140.

Gomot, L. 1971. The organotypic culture of invertebrates other than insects. pp. 41–136. *In*: C. Vago (ed.). Invertebrate Tissue Culture. Academic Press, New York, USA.

Gosling, E. 2003. Bivalve molluscs: Biology, ecology and culture. Fishing new books. Blackwell Publishing. 443p.

Gottschalk, F., T. Sonderer, R.W. Scholz and B. Nowack. 2009. Modelled environmental concentrations of engineered nanomaterials (TiO_2, ZnO, Ag, CNT, fullerenes) for different regions. Environ. Sci. Technol. 43: 9216–9222.

Gottschalk, F., T. Sun and B. Nowack. 2013. Environmental concentrations of engineered nanomaterials: Review of modelling and analytical studies. Environ. Pollut. 181: 287–300.

Griffitt, R.J., K. Hyndman, N.D. Denslow and D.S. Barber. 2009. Comparison of molecular and histological changes in zebrafish gills exposed to metallic nanoparticles. Toxicol. Sci. 107: 404–415.

Guadagnini, R., B. HalamodaKenzaoui, L. Walker, G. Pojana, Z. Magdolenova, D. Bilanicova et al. 2015. Toxicity screenings of nanomaterials: challenges due to interference with assay processes and components of classic *in vitro* tests. Nanotoxicology 9: 13–24.

Guillouzo, A. 1998. Liver cell models in *In Vitro* toxicology. Environ. Health Perspect. 106: 511–32.

Hansen, E.L. 1976. A cell line from embryos of *Biomphalaria glabrata* (Pulmonata): Establishment and characteristics. pp. 75–97. *In*: K. Maramorosch (ed.). Invertebrate Tissue Culture: Research Applications. Academic Press, New York, USA.

Hartmann, N.B. and A. Baun. 2010. The nano cocktail: Ecotoxicological effects of engineered nanoparticles in chemical mixtures. Integrat. Environ. Assess. Manag. 6: 311–314.

Harush-Frenkel, O., E. Rozentur, S. Benita and Y. Altschuler. 2008. Surface charge of nanoparticles determines their endocytic and transcytotic pathway in polarized MDCK cells. Biomacromolecules 9: 435–443.

Heinlaan, M., A. Ivask, I. Blinova, H.C. Dubourguier and A. Kahru. 2008. Toxicity of nanosized and bulk ZnO, CuO and TiO_2 to bacteria *Vibrio fischeri* and crustaceans *Daphnia magna* and *Thamnocephalus platyurus*. Chemosphere 71: 1308–1316.

Hidaka, H., H. Kobayshi, M. Kuga and T. Koike. 2005. Photoinduced characteristics of metal-oxide cosmetic pigments by agarose gel electrophoresis of DNA plasmids *in vitro* under UV-illumination. J. Oleo. Sci. 54: 487–494.

Hsiao, I.-L. and Y.-J. Huang. 2011. Effects of various physicochemical characteristics on the toxicities of ZnO and TiO_2 nanoparticles toward human lung epithelial cells. Sci. Total Environ. 409: 1219–1228.

Hund-Rinke, K. and M. Simon. 2006. Ecotoxic effect of photocatalytic active nanoparticles (TiO_2) on algae and daphnids. Environ. Sci. Pollut. Res. Int. 13: 225–232.

Hussain, S.M., K.L. Hess, J.M. Gearhart, K.T. Geiss and J.J. Schlager. 2005. *In vitro* toxicity of nanoparticles in BRL 3A rat liver cells. Toxicol. *in vitro* 19: 975–983.

Jiang, J., G. Oberdörster, A. Elder, R. Gelein, P. Mercer and P. Biswas. 2008b. Does nanoparticle activity depend upon size and crystal phase? Nanotoxicology 2: 33–42.

Jiang, W., B.Y. Kim, J.T. Rutka and W.C. Chan. 2008a. Nanoparticle-mediated cellular response is size-dependent. Nat. Nanotechnol. 3: 145–150.

Jimeno-Romero, A., M. Oron, M.P. Cajaraville, M. Soto and I. Marigómez. 2016. Nanoparticle size and combined toxicity of TiO_2 and DSLS (surfactant) contribute to lysosomal responses in digestive cells of mussels exposed to TiO_2 nanoparticles. Nanotoxicology 10: 1168–1176.

Jimeno-Romero, A., E. Bilbao, U. Izagirre, M.P. Cajaraville, I. Marigómez and M. Soto. 2017a. Digestive cell lysosomes as main targets for Ag accumulation and toxicity in marine mussels, *Mytilus galloprovincialis*, exposed to maltose-stabilised Ag nanoparticles of different sizes. Nanotoxicology 11: 168–183.

Jimeno-Romero, A., U. Izagirre, D. Gilliland, A. Warley, M.P. Cajaraville, I. Marigómez et al. 2017b. Lysosomal responses to different gold forms (nanoparticles, aqueous, bulk) in mussel digestive cells: A trade-off between the toxicity of the capping agent and form, size and exposure concentration. Nanotoxicology 11: 658–670.

Jones, C.F. and D.W. Grainger. 2009. *In vitro* assessment of nanomaterial toxicity. Adv. Drug Del. Rev. 61: 438–456.

Ju-Nam, Y. and J.R. Lead. 2008. Manufactured nanoparticles: An overview of their chemistry, interactions and potential environmental problems. Sci. Total Environ. 400: 396–414.

Kach, D.J. and J.E. Ward. 2008. The role of marine aggregates in the ingestion of picoplankton-size particles by suspension-feeding molluscs. Mar. Biol. 153: 797–805.

Kang, S., M.S. Mauter and M. Elimelech. 2009. Microbial cytotoxicity of carbon-based nanomaterials: Implications for river water and wastewater effluent. Environ. Sci. Technol. 43: 2648–2653.

Karlsson, H.L., P. Cronholm, J. Gustafsson and L. Moller. 2008. Copper oxide nanoparticles are highly toxic: A comparison between metal oxide nanoparticles and carbon nanotubes. Chem. Res. Toxicol. 21: 1726–1732.

Katsumiti, A., D. Berhanu, E. Valsami-Jones, D. Gilliland, M. Oron, P. Reip et al. 2012. Screening of cytotoxicity effects of different metal bearing nanoparticles on mussel hemocytes and gill cells *in vitro*. Comp. Biochem. Physiol. 163: S25.

Katsumiti, A., D. Gilliland, I. Arostegui and M.P. Cajaraville. 2014. Cytotoxicity and cellular mechanisms involved in the toxicity of CdS quantum dots in hemocytes and gill cells of the mussel *Mytilus galloprovincialis*. Aquat. Toxicol. 153: 39–52.

Katsumiti, A., D. Berhanu, K.T. Howard, I. Arostegui, M. Oron, P. Reip et al. 2015a. Cytotoxicity of TiO_2 nanoparticles to mussel hemocytes and gill cells *in vitro*: Influence of synthesis method, crystalline structure, size and additive. Nanotoxicology 9: 543–553.

Katsumiti, A., D. Gilliland, I. Arostegui and M.P. Cajaraville. 2015b. Mechanisms of toxicity of Ag nanoparticles in comparison to bulk and ionic Ag on mussel hemocytes and gill cells. Plos One 10: e0129039.

Katsumiti, A., I. Arostegui, M. Oron. D. Gilliland, E. Valsami-Jones and M.P. Cajaraville. 2016. Cytotoxicity of Au, ZnO and SiO_2 NPs using *in vitro* assays with mussel hemocytes and gill cells: Relevance of size, shape and additives. Nanotoxicology 10: 185–193.

Katsumiti, A., R. Tomovska and M.P. Cajaraville. 2017. Intracellular localization and toxicity of graphene oxide and reduced graphene oxide nanoplatelets to mussel hemocytes *in vitro*. Aquat. Toxicol. 188: 138–147.

Katsumiti, A., A.J. Thorley, I. Arostegui, P. Reip, E. Valsami-Jones, T.D. Tetley et al. 2018. Cytotoxicity and cellular mechanisms of toxicity of CuO NPs in mussel cells *in vitro* and comparative sensitivity with human cells. Toxicol. *In Vitro*. 48: 146–158.

Keller, A.A., K. Garner, R.J. Miller and H.S. Lenihan. 2012. Toxicity of nano-zero valent iron to freshwater and marine organisms. Plos One 7: e43983.
Keller, A.E. and S.G. Zam. 1990. Simplification of *in vitro* culture techniques for freshwater mussels. Environ. Toxicol. Chem. 9: 1291–1296.
Kim, H., J. Choi, H. Lee, J. Park, B.-I. Yoon, S.M. Jin et al. 2016. Skin corrosion and irritation test of nanoparticles using reconstructed three-dimensional human skin model, EpiDerm(TM). Toxicol. Res. 32: 311–316.
Koehler, A., U. Marx, K. Broeg, S. Bahns and J. Bressling. 2008. Effects of nanoparticles in *Mytilus edulis* gills and hepatopancreas—A new threat to marine life? Mar. Environ. Res. 66: 12–14.
Kroll, A., C. Dierker, C. Rommel, D. Hahn, W. Wohlleben, C. Schulze-Isfort et al. 2011. Cytotoxicity screening of 23 engineered nanomaterials using a test matrix of ten cell lines and three different assays. Part. FibreToxicol. 8: 1–19.
Kroll, A., M.H. Pillukat, D. Hahn and J. Schnekenburger. 2012. Interference of engineered nanoparticles with *in vitro* toxicity assays. Arch. Toxicol. 86: 1123–1136.
Kühnel, D., W. Busch, T. Meibner, A. Springer, A. Potthoff, V. Richter et al. 2009. Agglomeration of tungsten carbide nanoparticles in exposure medium does not prevent uptake and toxicity toward a rainbow trout gill cell line. Aquat. Toxicol. 93: 91–99.
Kvitek, L., M. Vanickova, A. Panacek, J. Soukupova, M. Dittrich, E. Valentova et al. 2009. Initial study on the toxicity of silver nanoparticles (NPs) against *Paramecium caudatum*. J. Phys. Chem. C 113: 4296–4300.
L'Azou, B., J. Jorly, D. On, E. Sellier, F. Moisan, J. Fleury-Feith et al. 2008. *In vitro* effects of nanoparticles on renal cells. Part. FibreToxicol. 5: 22.
Lacave, J.M., A. Retuerto, U. Vicario-Parés, D. Gilliland, M. Oron, M.P. Cajaraville and A. Orbea. 2016. Effects of metal-bearing nanoparticles (Ag, Au, CdS, ZnO, SiO_2) on developing zebrafish embryos. Nanotechnol. 27: 325102.
Lakra, W.S., T.R. Swaminathan and K.P. Joy. 2011. Development, characterization, conservation and storage of fish cell lines: A review. Fish Physiol. Biochem. 37: 1–20.
Lammel, T., P. Boisseaux, M.L. Fernandez-Cruz and J.M. Navas. 2013. Internalization and cytotoxicity of graphene oxide and carboxyl graphene nanoplatelets in the human hepatocellular carcinoma cell line Hep G2. Part. FibreToxicol. 10: 27.
Lammel. T. and J.M. Navas. 2014. Graphene nanoplatelets spontaneously translocate into the cytosol and physically interact with cellular organelles in the fish cell line PLHC-1. Aquat. Toxicol. 150: 55–65.
Lammel, T. and J. Sturve. 2018. Assessment of titanium dioxide nanoparticle toxicity in the rainbow trout (*Onchorynchus mykiss*) liver and gill cell lines RTL-W1 and RTgill-W1 under particular consideration of nanoparticle stability and interference with fluorometric assays. NanoImpact 11: 1–19.
Landsiedel, R., U.G. Sauer, L. Ma-Hock, J. Schnekenburger and M. Wiemann. 2014. Pulmonary toxicity of nanomaterials: A critical comparison of published *in vitro* assays and *in vivo* inhalation or instillation studies. Nanomedicine-UK 9: 2557–2585.
Lee, J., G.D. Lilly, R.C. Doty, P. Podsiadlo and N.A. Kotov. 2009. *In vitro* toxicity testing of nanoparticles in 3D cell culture. Small 5: 1213–1221.
Lei, C., L. Zhang, K. Yang, L. Zhu and D. Lin. 2016. Toxicity of iron-based nanoparticles to green algae: Effects of particle size, crystal phase, oxidation state and environmental aging. Environ. Poll. 218: 505–512.
Leroueil, P.R., S. Hong, A. Mecke, J.R. Baker, Jr., B.G. Orr and M.M. Banaszak Holl. 2007. Nanoparticle interaction with biological membranes: Does nanotechnology present a Janus face? Acc. Chem. Res. 40: 335–342.
Leroueil, P.R., S.A. Berry, K. Duthie, G. Han, V.M. Rotello, D.Q. McNerny et al. 2008. Wide varieties of cationic nanoparticles induce defects in supported lipid bilayers. Nano Lett. 8: 420–424.
Lesniak, A., A. Campbell, M.P. Monopoli, I. Lynch, A. Salvati and K.A. Dawson. 2010. Serum heat inactivation affects protein corona composition and nanoparticle uptake. Biomaterials 31: 9511–9518.
Lesniak, A., A. Salvati, M.J. Santos-Martinez, M.W. Radomski, K.A. Dawson and C. Aberg. 2013. Nanoparticle adhesion to the cell membrane and its effect on nanoparticle uptake efficiency. J. Am. Chem. Soc. 135: 1438–1444.
Lesniak, W., A.U. Bielinska, K. Sun, K.W. Janczak, X. Shi, J.R. Baker et al. 2005. Silver/dendrimer nanocomposites as biomarkers: Fabrication, characterization, *In Vitro* toxicity, and intracellular detection. Nano Lett. 5: 2123–2130.
Li, L., M. Stoiber, A. Wimmer, Z. Xu., C. Lindenblatt, B. Helmreich et al. 2016. To what extent can full-scale wastewater treatment plant effluent influence the occurrence of silver-based nanoparticles in surface waters? Environ. Sci. Technol. 50: 6327–6333.
Li, M., S. Pokhrel, X. Jin, L. Mädler, R. Damoiseaux and E.M. Hoek. 2011. Stability, bioavailability, and bacterial toxicity of ZnO and iron-doped ZnO nanoparticles in aquatic media. Environ. Sci. Technol. 45: 755–761.
Limbach, L.K., P. Wick, P. Manser, R.N. Grass and W.J. Stark. 2007. Exposure of engineered nanoparticles to human lung epithelial cells: Influence of chemical composition and catalytic activity on oxidative stress. Environ. Sci. Technol. 41: 4158–4163.
Liu, J.G. and R.H. Hurt. 2010. Ion release kinetics and particle persistence in aqueous nanosilver colloids. Environ. Sci. Technol. 44: 2169–2175.
Liu, P., R. Guan, X. Ye, J. Jiang, M. Liu, G. Huang et al. 2011. Toxicity of nano- and microsized silver particles in human hepatocyte cell line L02. J. Physiol. Conf. Ser. 304: 012036.
Livingstone, D.R. and R.K. Pipe. 1992. Mussel and environmental contaminants: Molecular and cellular aspects. pp. 425–464. *In*: E. Gosling (ed.). The Mussel Mytilus: Ecology, Physiology, Genetics and Culture. Elsevier, Amsterdam, NL.
Lok, C.N., C.M. Ho, R. Chen, Q.Y. He, W.Y. Yu, H. Sun et al. 2007. Silver nanoparticles: Partial oxidation and antibacterial activities. J. Biol. Inorg. Chem. 12: 527–534.

Lupu, A.R. and T. Popescu. 2013. The noncellular reduction of MTT tetrazolium salt by TiO_2 nanoparticles and its implications for cytotoxicity assays. Toxicol. *in vitro* 27: 1445–1450.
Luyts, K., D. Napierska, B. Nemery and P.H.M. Hoet. 2013. How physico-chemical characteristics of nanoparticles cause their toxicity: Complex and unresolved interrelations. Environ. Sci.: Processes Impacts. 15: 23–38.
Mahaye, N., M. Thwala, D.A. Cowan and N. Mussee. 2017. Genotoxicity of metal based engineered nanoparticles in aquatic organisms: A review. Mutat. Res. 773: 134–160.
Mahmoudi, M., S.N. Saeedi-Eslami, M.A. Shokrgozar, K. Azadmanesh, M. Hassanlou, H.R. Kalhor et al. 2012. Cell "vision": complementary factor of protein corona in nanotoxicology. Nano 4: 5461–5468.
Marigómez, I., M. Soto and M.P. Cajaraville. 1995. Morphofunctional patterns of cell and tissue systems involved in metal handling and metabolism. pp. 89–134. *In*: M.P. Cajaraville (ed.). Cell Biology in Environmental Toxicology. University of the Basque Country Press, Bilbao, ES.
Marigómez, I., M. Soto, M.P. Cajaraville, E. Angulo and L. Giamberini. 2002. Cellular and subcellular distribution of metals in molluscs. Micro. Res. Tech. 56: 358–392.
Marin, M., H. Legros, A. Poret, F. Leboulenger and F. Le Foll. 2004. Cell responses to xenobiotics: Comparison of MCF7 multidrug- and mussel blood cell multi-xenobiotic defense mechanisms. Mar. Environ. Res. 58: 209–213.
Matranga, V. and I. Corsi. 2012. Toxic effects of engineered nanoparticles in the marine environment: Model organisms and molecular approaches. Mar. Environ. Res. 76: 32–40.
Miller, R.J., H.S. Lenihan, E.B. Muller, N. Tseng, S.K. Hanna and A.A. Keller. 2010. Impacts of metal oxide nanoparticles on marine phytoplankton. Environ. Sci. Technol. 44: 7329–7334.
Misra, S.K., S. Nuseibeh, A. Dybowska, D. Berhanu, T.D. Tetley and E. Valsami-Jones. 2013. Comparative study using spheres, rods and spindle-shaped nanoplatelets on dispersion stability, dissolution and toxicity of CuO nanomaterials. Nanotoxicology 8: 422–432.
Miura, N. and Y. Shinohara. 2009. Cytotoxic effect and apoptosis induction by silver nanoparticles in HeLa cells. Biochem. Biophys. Res. Commun. 390: 733–737.
Monopoli, M.P., C. Aberg, A. Salvati and K.A. Dawson. 2012. Biomolecular coronas provide the biological identity of nanosized materials. Nat. Nanotechnol. 7: 779–786.
Monopoli, M.P., D. Walczyk, A. Campbell, G. Elia, I. Lynch, F.B. Bombelli et al. 2011. Physical-chemical aspects of protein corona: relevance to *in vitro* and *in vivo* biological impacts of nanoparticles. J. Am. Chem. Soc. 133: 2525–2534.
Moore, M.N. 2006. Do nanoparticles present ecotoxicological risks for the health of the aquatic environment? Environ. Int. 32: 967–976.
Moore, M.N., J.A.J. Readman, J.W. Readman, D.M. Lowe, P.E. Frickers and A. Beesley. 2009. Lysosomal cytotoxicity of carbon nanoparticles in cells of the molluscan immune system: an *in vitro* study. Nanotoxicology 3: 40–45.
Mosmann, T. 1983. Rapid colorimetric assay for cellular growth and survival: Application to proliferation and cytotoxicity assays. J. Immunol. Methods 65: 55–63.
Movia, D., A. Prina-Mello, D. Bazou, Y. Volkov and S. Giordani. 2011. Screening the cytotoxicity of single-walled carbon nanotubes using novel 3D tissue-mimetic models. ACSNano 5: 9278–9290.
Mu, Q.S., N.S. Hondow, L. Krzeminski, A.P. Brown, L.J.C. Jeuken and M.N. Routledge. 2012. Mechanism of cellular uptake of genotoxic silica nanoparticles. Part. FibreToxicol. 9: 1–11.
Mueller, N.C. and B. Nowack. 2008. Exposure modeling of engineered nanoparticles in the environment. Environ. Sci. Technol. 42: 4447–453.
Munari, M., J. Sturve, G. Frenzilli, M.B. Sanders, A. Brunelli, A. Marcomini et al. 2014. Genotoxic effects of CdS quantum dots and Ag_2S nanoparticles in fish cell lines (RTG-2). Mutat. Res. 775-776: 89–93.
Na, N., H. Guo, S. Zhang, Z. Li and L. Yin. 2009. *In vitro* and *in vivo* acute toxicity of fenpyroximate to flounder *Paralichthys olivaceus* and its gill cell line FG. Aquat. Toxicol. 92: 76–85.
Navarro, E., F. Piccapietra, B. Wagner, F. Marconi, R. Kaegi, N. Odzak et al. 2008. Toxicity of silver nanoparticles to *Chlamydomonas reinhardtii*. Environ. Sci. Technol. 42: 8959–8964.
Oberdörster, G. and T.A.J. Kuhlbusch. 2018. *In vivo* effects: Methodologies and biokinetics of inhaled nanomaterials. NanoImpact 10: 38–60.
Oberdörster, G., A. Maynard, K. Donaldson, V. Castranova, J. Fitzpatrick, K. Ausman et al. 2005. Principles for characterizing the potential human health effects from exposure to nanomaterials: Elements of a screening strategy. Part. FibreToxicol. 2: 8.
OECD (Organisation for Economic Cooperation and Development), 2004. Test No. 428: Skin absorption: *in vitro* method. OECD Publishing. http:// www.oecd-ilibrary.org/environment/test-no-428-skin-absorption-in-vitro-method_9789264071087-en.
OECD (Organisation for Economic Cooperation and Development). 2012. Fish toxicity testing framework. Series on Testing and Assessment No 171, Paris, FR.
OECD (Organisation for Economic Cooperation and Development). 2013. Test No. 437: Bovine corneal opacity and permeability test method for identifying ocular corrosives and severe irritants. OECD Publishing. http://www.oecd-ilibrary.org/environment/test-no-437-bovine-corneal-opacity-andpermeability-testmethod-for-identifying-i-chemicals-inducing-serious-eye-damageand-ii-chemicals-not-requiring-classification-for-eye-irritation-or-serious-eyedamage_9789264203846-en.
OECD (Organisation for Economic Cooperation and Development). 2014a. *In vitro* skin corrosion: reconstructed human epidermis (Rhe) test method. OECD Publishing. http://www.oecd-ilibrary.org/environment/test-no-431-in-vitro-skin-corrosion-reconstructed-human-epidermis-rhe-test-method_9789264224193-en.

OECD (Organisation for Economic Cooperation and Development). 2014b. Test No. 487: *In vitro* mammalian cell micronucleus test. OECD Publishing. http://www.oecd-ilibrary.org/environment/test-no-487-*in-vitro* mammalian-cellmicronucleus-test_9789264224438-en.

OECD (Organisation for Economic Cooperation and Development). 2014c. Test No. 473: *In vitro* mammalian chromosome aberration test. OECD Publishing. http://www.oecd-ilibrary.org/environment/test-no-473-in-vitromammalian-chromosomal-aberration-test_9789264224223-en.

OECD (Organisation for Economic Cooperation and Development). 2014d. Test No. 475: Mammalian bone marrow chromosomal aberration test. OECD Publishing. http://www.oecd.org/env/test-no-475-mammalian-bonemarrow-chromosomal-aberration-test-9789264224407-en.htm.

OECD (Organisation for Economic Cooperation and Development). 2015a. Test No. 476: *In vitro* mammalian cell gene mutation tests using the HPRT and XPRT genes. OECD Publishing. http://www.oecd-ilibrary.org/environment/test-no476-in-vitro-mammalian-cell-gene-mutation-tests-using-thehprt-and-xprtgenes_9789264243088-en;jsessionid=1h0nxo8t44eay.x-oecd-live-03.

OECD (Organisation for Economic Cooperation and Development). 2015b. Test No. 492: Reconstructed human cornea-like epithelium (RhCE) test method for identifying chemicals not requiring classification and labelling for eye irritation or serious eye damage. OECD Publishing. http://www.oecd-ilibrary.org/environment/test-no-492-reconstructed-human-cornea-like-epithelium-rhce-testmethod-for-identifying-chemicals-not-requiring-classification-and-labelling-for-eyeirritation-or-serious-eye-damage_9789264242548-en.

OECD (Organisation for Economic Cooperation and Development). 2015c. Test No. 439: *In vitro* skin irritation: reconstructed human epidermis test method. OECD Publishing. http://www.oecd.org/env/test-no-439-in-vitro-skinirritation-reconstructed-human-epidermis-test-method-9789264242845-en.htm.

OECD (Organisation for Economic Cooperation and Development). 2015d. Test No. 493: Performance-based test guideline for human recombinant estrogen eeceptor (hrER) *in vitro* assays to detect chemicals with ER binding affinity. OECD Publishing. http://www.oecd-ilibrary.org/environment/test-no-493-performance-based-test-guideline-for-human-recombinant-estrogen-receptor-hrer-invitro-assays-to-detect-chemicals-with-er-binding-affinity_9789264242623-en.

OECD (Organisation for Economic Cooperation and Development). 2015e. Guidance document on revisions to OECD genetic toxicology test guidelines. http://www.oecd.org/env/ehs/testing/section4-health-effects.htm.

OECD (Organisation for Economic Cooperation and Development). 2018a. Test No. 319A: Determination of *in vitro* intrinsic clearance using cryopreserved rainbow trout hepatocytes (RT-HEP). OECD Publishing. https://www.oecd-ilibrary.org/docserver/9789264303218-en.pdf?expires=1531729643&id=id&accname=guest&checksum=65A79001D30A53849 4AA6DF6782EB126.

OECD (Organisation for Economic Cooperation and Development). 2018b. Test No. 319B: Determination of *in vitro* intrinsic clearance using rainbow trout liver S9 subcellular fraction (RT-S9). OECD Publishing. https://www.oecd-ilibrary.org/docserver/9789264303232en.pdf?expires=1531475391&id=id&accname=guest&checksum=8DEC2F9494E9D3F54 9E50DC8035F6923.

Olabarrieta, I., B. L'Azou, S. Yuric, J. Cambar and M.P. Cajaraville. 2001. *In vitro* effects of cadmium on two different animal cell models. Toxicol. *In Vitro* 15: 511–517.

Ong, K.J., T.J. Maccormack, R.J. Clark and J.D. Ede. 2014. Widespread nanoparticle-assay interference: implications for nanotoxicity testing. Plos One 9: e90650.

Owen, G. 1974. Feeding and digestion in Bivalvia. Adv. Comp. Physiol. Biochem. 5: 1–35.

Owen, G. and J.M. McCrae. 1976. Further studies on the latero-frontal tract of bivalves. Proc. R. Soc. Lond. B 194: 527–544.

Owen, G. 1978. Classification and the bivalve gill. Phil. Trans. R. Soc. Lond. B 284: 377–385.

Panessa-Warren, B.J., J.B. Warren, M.M. Maye, D. Van der Lelie, O. Gang, S.S. Wong et al. 2008. Human epithelial cell processing of carbon and gold nanoparticles. Int. J. Nanotechnol. 5: 55–91.

Park, M.V.D.Z., D.P.K. Lankveld, H. Van Loveren and W.H. De Jong. 2009. The status of *in vitro* toxicity studies in the risk assessment of nanomaterials. Nanomedicine 4: 669–685.

Peetla, C. and V. Labhasetwar. 2009. Effect of molecular structure of cationic surfactants on biophysical interactions of surfactant-modified nanoparticles with a model membrane and cellular uptake. Langmuir 25: 2369–2377.

Petosa, A.R., D.P. Jaisi, I.R. Quevedo, M. Elimelech and N. Tufenkji. 2010. Aggregation and deposition of engineered nanomaterials in aquatic environments: Role of physicochemical interactions. Environ. Sci. Technol. 44: 6532–6549.

Rannou, M. 1968. Formation de spicules dans des cultures cellulaires de cnidaire (gorgone). Vie Millieu 19: 53–57.

Rannou, M. 1971. Cell culture of invertebrates other than molluscs and arthropods. *In*: C. Vago (Ed.). Invertebrate Tissue Culture. Academic Press, New York, USA.

Reeves, J.F., S.J. Davies, N.J.F. Dodd and A.N. Jha. 2008. Hydroxyl radicals (OH) are associated with titanium dioxide (TiO_2) nanoparticle- induced cytotoxicity and oxidative DNA damage in fish cells. Mutat. Res. 640: 113–122.

Rimai, D.S., D.J. Quesnel and A.A. Busnaina. 2000. The adhesion of dry particles in the nanometer to micrometer-size range. Colloids Surf. A 165: 3–10.

Ringwood, A.H., S. Khambhammettu, P. Santiago, E. Bealer, M. Stogner, J. Collins et al. 2006. Characterization, imaging and degradation studies of quantum dots in aquatic organisms. Mater. Res. Soc. Symp. Proc. 895: 0895-G04-06-S04-06.1.

Ringwood, A.H., N. Levi-Polyachenko and D.L. Carroll. 2009. Fullerene exposures with oysters: Embryonic, adult, and cellular responses. Environ. Sci. Technol. 43: 7136–7141.

Rinkevich, B. 1999. Cell cultures from marine invertebrates. Obstacles, new approaches and recent improvements. J. Biotechnol. 70: 133–153.

Robledo, Y. and M.P. Cajaraville. 1997. Isolation and morphofunctional characterization of mussel digestive gland cells *in vitro*. Eur. J. Cell. Biol. 72: 362–369.
Rocha, T.L., N. Mestre, S.M.T. Sabóia-Morais and M.J. Bebianno. 2017. Environmental behaviour and ecotoxicity of quantum dots at various trophic levels: A review. Environ. Int. 98: 1–17.
Rothen-Rutishauser, B.M., S. Schürch, B. Haenni, N. Kapp and P. Gehr. 2006. Interaction of fine particles and nanoparticles with red blood cells visualized with advanced microscopic techniques. Environ. Sci. Technol. 40: 4353–4359.
Ruenraroengsak, P., S. Chen, S. Hu, J. Melbourne, S. Sweeney, A.J. Thorley et al. 2016. Translocation of functionalized multi-walled carbon nanotubes across human pulmonary alveolar epithelium: Dominant role of epithelial type 1 cells. ACS Nano 10: 5070–5085.
Ruiz, P., A. Katsumiti, J.A. Nieto, J. Bori, A. Jimeno-Romero, P. Reip et al. 2015. Short-term effects on antioxidant enzymes and long-term genotoxic and carcinogenic potential of CuO nanoparticles compared to bulk CuO and ionic copper in mussels *Mytilus galloprovincialis*. Mar. Environ. Res. 111: 107–120.
Samberg, M.E., S.J. Oldenburg and N.A. Monteiro-Riviere. 2010. Evaluation of silver nanoparticle toxicity in skin *in vivo* and keratinocytes *in vitro*. Environ Health Perspect. 118: 407–413.
Sathishkumar, M., S. Pavagadhi, A. Mahadevan and R. Balasubramanian. 2014. Biosynthesis of gold nanoparticles and related cytotoxicity evaluation using A549 cells. Ecotoxicol. Environ. Saf. 114: 232–240.
Sayes, C.M., R. Wahi, P.A. Kurian, Y. Liu, J.L. West, K.D. Ausman et al. 2006. Correlating nanoscale titania structure with toxicity: a cytotoxicity and inflammatory response study with human dermal fibroblasts and human lung epithelial cells. Toxicol. Sci. 92: 174–185.
Sendra, M., M. Volland, T. Balbi, R. Fabbri, M.P. Yeste, J.M. Gatica et al. 2018. Cytotoxicity of CeO_2 nanoparticles using *in vitro* assay with *Mytilus galloprovincialis* hemocytes: Relevance of zeta potential, shape and biocorona formation. Aquat. Toxicol. 200: 13–20.
Schiavo, S., N. Duroudier, E. Bilbao, M. Mikolaczyk, J. Schäfer, M.P. Cajaraville et al. 2017. Effects of PVP/PEI coated and uncoated silver NPs and PVP/PEI coating agent on three species of marine microalgae. Sci. Total Environ. 577: 45–53.
Schins, R.P.F. and A.M. Knaapen. 2007. Genotoxicity of poorly soluble particles. Inhal. Toxicol. 19: 189–198.
Schirmer, K. 2006. Proposal to improve vertebrate cell cultures to establish them as substitutes for the regulatory testing of chemicals and effluents using fish. Toxicology 224: 163–183.
Scown, T.M., E.M. Santos, B.D. Johnston, B. Gaiser, M. Baalousha, S. Mitov et al. 2010. Effects of aqueous exposure to silver nanoparticles of different sizes in rainbow trout. Toxicol. Sci. 115: 521–534.
Segner, H. and D. Lenz. 1993. Cytotoxicity assays with the rainbow trout R1 cell line. Toxicol. *In Vitro* 7: 537–540.
Segner, H. 1998. Fish cell lines as a tool in aquatic toxicology. pp. 1–38. *In*: T. Braunbeck, D.E. Hinton and B. Streit (eds.). Fish Ecotoxicology. Springer Basel AG, Berlin, Germany.
Shaw, B.J., G. Al-Bairuty and R.D. Handy. 2012. Effects of waterborne copper nanoparticles and copper sulphate on rainbow trout, (*Oncorhynchus mykiss)*: Physiology and accumulation. Aquat. Toxicol. 116-117: 90–101.
Sheehan, D. 2000. Applications of invertebrate cell culture in studies of biomarkers and ecotoxicology. pp. 337–352. *In*: C. Mothersill and B. Austin (eds.). Aquatic Invertebrate Cell Culture. Praxis Publishing Ltd., Chichester, UK.
Song, L., M. Connolly, M.L. Fernández-Cruz, M.G. Vijver, M. Fernández, E. Conde et al. 2014. Species-specific toxicity of copper nanoparticles among mammalian and piscine cell lines. Nanotoxicology 8: 383–393.
Song, W., J. Zhang, J. Guo, J. Zhang, F. Ding, L. Li et al. 2010. Role of the dissolved zinc ion and reactive oxygen species in cytotoxicity of ZnO nanoparticles. Toxicol. Lett. 199: 389–397.
Stolpe, B. and M. Hassellov. 2007. Changes in size distribution of fresh water nanoscale colloidal matter and associated elements on mixing in seawater. Geochim. Cosmochim. Acta 71: 3292–3301.
Stone, V., H. Johnston and R.P.F. Schins. 2009. Development of *in vitro* systems for nanotoxicology: Methodological considerations. Crit. Rev. Toxicol. 39: 613–626.
Sun, X., B. Chen, B. Xia, Q. Han, L. Zhu and K. Qu. 2017. Are CuO nanoparticles effects on hemocytes of the marine scallop (*Chlamys farreri*) caused by particles and/or corresponding released ions? Ecotoxicol. Environ. Saf. 139: 65–72.
Suresh, P.G., M.K. Reju and A. Mohandas. 1993. Haemolymph phosphatase activity levels in two freshwater gastropods exposed to copper. Sci. Total Environ. 1: 1265–1277.
Svensson, S., A. Särngren and L. Förlin. 2003. Mussel blood cells, resistant to the cytotoxic effects of okaidaic acid, do not express cell membrane p-glycoprotein activity (multixenobiotic resistance). Aquat. Toxicol. 65: 27–37.
Sweeney, S., D. Berhanu, S. Misra, A.J. Thorley, E. Valsami-Jones and T.D. Tetley. 2014. Multi-walled carbon nanotube length as a critical determinant of bioreactivity with primary human pulmonary alveolar cells. Carbon 78: 26–37.
Sweeney, S., I.G. Theodorou, M. Zambianchi, S. Chen, A. Gow, S. Schwander et al. 2015a. Silver nanowire interactions with primary human alveolar type-II epithelial cell secretions: contrasting bioreactivity with human alveolar type-I and type-II epithelial cells. Nanoscale 7: 10398.
Sweeney, S., D. Berhanu, P. Ruenraroengsak, A.J. Thorley, E. Valsami-Jones and T.D. Tetley. 2015b. Nano-titanium dioxide bioreactivity with human alveolar type-I-like epithelial cells: Investigating crystalline phase as a critical determinant. Nanotoxicology 4: 482–492.
Sweeney, S., D. Grandolfo, P. Ruenraroengsak and T.D. Tetley. 2015c. Functional consequences for primary human alveolar macrophages following treatment with long, but not short, multiwalled carbon nanotubes. Int. J. Nanomed. 10: 3115–3129.
Taju, G., S. Abdul Majeed, K.S.N. Nambi, V. Sarath Babu, S. Vimal, S. Kamatchiammal et al. 2012. Comparison of *in vitro* and *in vivo* acute toxicity assays in *Etroplus suratensis* (Bloch, 1790) and its three cell lines in relation to tannery effluent. Chemosphere 87: 55–61.

Taju, G., S.A. Majeed, K.S.N. Nambi and A.S.S. Hameed. 2014. *In vitro* assay for the toxicity of silver nanoparticles using heart and gill cell lines of *Catla catla* and gill cell line of *Labeorohita*. Comp. Biochem. Physiol. 161: 41–52.

Tedja, R., M. Lim, R. Amal and C. Marquis. 2012. Effects of serum adsorption on cellular uptake profile and consequent impact of titanium dioxide nanoparticles on human lung cell lines. ACS Nano 6: 4083–4093.

Thevenot, P., J. Cho, D. Wavhal, R.B. Timmons and L. Tang. 2008. Surface chemistry influence cancer killing effect of TiO_2 nanoparticles. Nanomedicine 4: 226–236.

Tiede, K., M. Hassellov, E. Breitbarth, Q. Chaudhry and A.B.A. Boxall. 2009. Considerations for environmental fate and ecotoxicity testing to support environmental risk assessment for engineered nanoparticles. J. Chromatogr. A 1216. 503–509.

Triglia, D., S. Sherard Braa, C. Yonan and G.K. Naughton. 1991. Cytotoxicity testing using neutral red and MTT assays on a three-dimensional human skin substrate. Toxicol. *In Vitro* 5: 573–578.

Uchea, C., S. Owen and K. Chipman. 2015. Functional xenobiotic metabolism and efflux transporters in trout hepatocyte spheroid cultures. Toxicol. Res. 4: 494–507.

Uchino, T., H. Tokunaga, M. Ando and H. Utsumi. 2002. Quantitative determination of OH radical generation and its cytotoxicity induced by TiO_2-UVA treatment. Toxicol. *In Vitro* 16: 629–635.

Unfried, K., C. Albrecht, L.O. Klotz, A. Von Mikecz, S. Grether-Beck and R.P.F. Schins. 2007. Cellular responses to nanoparticles: target structures and mechanisms. Nanotoxicology 1: 52–71.

Vevers, W.F. and A.N. Jha. 2008. Genotoxic and cytotoxic potential of titanium dioxide (TiO_2) nanoparticles on fish cells *in vitro*. Ecotoxicology 17: 411–421.

Vian, L., J. Vincent, J. Maurin, I. Fabre, J. Giroux and J.P. Cano. 1995. Comparison of three *in vitro* cytotoxicity assays for estimating surfactant ocular irritation. Toxicol. *In Vitro* 9: 185–190.

Vicario-Parés, U., L. Castañaga, J.M. Lacave, M. Oron, P. Reip, D. Berhanu et al. 2014. Comparative toxicity of metal oxide nanoparticles (CuO, ZnO and TiO_2) to developing zebrafish embryos. J. Nanopart. Res. 16: 2550.

Vo, N.T., M.R. Bufalino, K.D. Hartlen, V. Kitaev and L.E. Lee. 2013. Cytotoxicity evaluation of silica nanoparticles using fish cell lines. *In Vitro* Cell. Dev. Biol. Anim. 50: 427–438.

Volland, M., M. Hampel, A. Katsumiti, M.P. Yeste, J.M. Gatica, M.P. Cajaraville et al. 2018. Synthesis methods influence characteristics, behaviour and toxicity of bare CuO NPs compared to bulk CuO and ionic Cu after *in vitro* exposure of *Ruditapes philippinarum* hemocytes. Aquat. Toxicol. 199: 285–295.

Wang, Y., W.G. Aker, H.M. Hwang, C.G. Yedjou, H. Yu and P.B. Tchounwou. 2011. A study of the mechanism of *in vitro* cytotoxicity of metal oxide nanoparticles using catfish primary hepatocytes and human HepG2 cells. Sci. Total Environ. 409: 4753–4762.

Wang, Y. and M. Tang. 2018. Review of *in vitro* toxicological research of quantum dot and potentially involved mechanisms. Sci. Total Environ. 625: 940–962.

Ward, J.E. and D.J. Kach. 2009. Marine aggregates facilitate ingestion of nanoparticles by suspension-feeding bivalves. Mar. Environ. Pollut. 68: 137–142.

Warheit, D.B., R.A. Hoke, C. Finlay, E.M. Donner, K.L. Reed and C.M. Sayes. 2007. Development of a base set of toxicity tests using ultrafine TiO_2 particles as a component of nanoparticle risk management. Toxicol. Lett. 171: 99–110.

Wise, Sr, J.P., B.C. Goodale, S.S. Wise, G.A. Craig, A.F. Pongan, R.B. Walter et al. 2010. Silver nanospheres are cytotoxic and genotoxic to fish cells. Aquat. Toxicol. 97: 34–41.

Wolf, K. and M.C. Quimby. 1962. Established eurythermic line of fish cells *in vitro*. Science 135: 1065–1066.

Wolf, K. and J.A. Mann. 1980. Poikilotherm vertebrate cell lines and viruses: A current listing for fishes. *In Vitro* 16: 168–179.

Wong, S.W., P.T. Leung, A.B. Djurišić and K.M.Y. Leung. 2010. Toxicities of nano zinc oxide to five marine organisms: influences of aggregate size and ion solubility. Anal. Bioanal. Chem. 396: 609–618.

Xia, T., M. Kovochich, M. Liong, L. Madler, B. Gilbert, H. Shi et al. 2008. Comparison of the mechanism of toxicity of zinc oxide and cerium oxide nanoparticles based on dissolution and oxidative stress properties. ACS Nano 2: 2121–2134.

Yang, H., C. Liu, D. Yang, H. Zhang and Z. Xi. 2009. Comparative study of cytotoxicity, oxidative stress and genotoxicity induced by four typical nanomaterials: The role of particle size, shape and composition. J. Appl. Toxicol. 29: 69–78.

Yeo, M.K. and J.W. Yoon. 2009. Comparison of the effects of nano-silver antibacterial coatings and silver ions on zebrafish embryogenesis. Mol. Cell Toxicol. 5: 23–31.

Yoshino, T.P., U. Bickham and C.J. Bayne. 2013. Molluscan cells in culture: Primary cell cultures and cell lines. Can. J. Zool. 91: doi:10.1139/cjz-2012-0258.

Yuan, H., J. Li, G. Bao and S. Zhang. 2010. Variable nanoparticle-cell adhesion strength regulates cellular uptake. Phys. Rev. Lett. 105: 138101.

Yue, Y., R. Behra, L. Sigg, P. Fernández Freire, S. Pillai and K. Schirmer. 2014. Toxicity of silver nanoparticles to a fish gill cell line: Role of medium composition. Nanotoxicology 9: 54–63.

Zhang, T., L. Wang, Q. Chen and C. Chen. 2014. Cytotoxic potential of silver nanoparticles. Yonsei Med. J. 55: 283–291.

Zhao, J., X. Cao, Z. Wang, Y. Dai and B. Xing. 2017. Mechanistic understanding toward the toxicity of graphene-family materials to freshwater algae. Water Res. 111: 18–27.

Zorita, I., E. Bilbao, A. Schad, I. Cancio, M. Soto and M.P. Cajaraville. 2007. Tissue- and cell specific expression of metallothionein genes in cadmium- and copper-exposed mussels analyzed by *in situ* hybridization and RT–PCR. Toxicol. Appl. Pharmacol. 220: 186–196.

Zvyagin, A.V., X. Zhao, A. Gierden, W. Sanchez, J. Ross and M.S. Roberts. 2008. Imaging of zinc oxide nanoparticle penetration in human skin *in vitro* and *in vivo*. J. Biomedical Optics 13: 064031.

4

Toxicity Tests and Bioassays for Aquatic Ecotoxicology of Engineered Nanomaterials

Susanne Schmidt,[1] *Dana Kühnel*[1,*] and *Anita Jemec Kokalj*[2]

Challenges in Aquatic Toxicity Testing of ENMs

Ecotoxicological studies aim to investigate the effects of pollutants on organisms. Effects can be described as changes in the state or dynamics at the organism level or at other levels of biological organization. These may include the sub-cellular level, cellular level, tissues, individuals, populations, communities, ecosystems, landscapes and biosphere (Leeuwen and Vermeire 2007). The primary tool of ecotoxicological studies that enables the assessment of effects is the toxicity test (Leeuwen and Vermeire 2007, Rand 1995). The toxicity test is used to evaluate the concentration of the chemical and the duration of exposure required to produce the criterion effect. Depending on the duration of exposure in the toxicity test, the effects can be classified as acute, sub-chronic and chronic.

As considered with the rising of nanoecotoxicology, the assessment of particulate materials toxicity in biological test systems is a challenging task and a number of ENM-specific issues need consideration. As generally recognized, the toxicity tests that have been initially developed for the testing of organic chemicals may have shortcomings when transferred one-to-one to particulate test items such as nanomaterials. For organic chemicals, the tests are optimized to achieve constant exposure conditions. Because chemicals equally distribute in the medium, this can be monitored by measurement of the exposure concentrations over time. However, the situation in toxicity testing of ENMs is different due to the inherent dynamic of ENMs as test items, which results in many processes that warrant careful consideration and characterization during the toxicity test.

In this chapter, first the major challenges that need attention in ENM toxicity testing in general are addressed and furthermore specific interferences and modifications necessary for each of the toxicity test groups are described. Other chapters in this book discuss in more detail the actual effects that ENMs exert on different types of organisms.

[1] Helmholtz-Centre for Environmental Research – UFZ, Department of Bioanalytical Ecotoxicology, Permoserstr. 15, 04318 Leipzig, Germany.
Emails: susanne.schmidt@ufz.de; anita.jemec@bf.uni-lj.si
[2] University of Ljubljana, Biotechnical Faculty, Department of Biology, Jamnikarjeva 101, 1000 Ljubljana, Slovenia.
* Corresponding author: dana.kuehnel@ufz.de

Properties, Behaviour and Transformations of ENMs in Liquid Media

As a first step in most aquatic toxicity tests and bioassays, ENMs need to be transferred into the liquid media employed. In order to equally distribute the particles in the media, usually energy is applied in the form of shaking or ultrasound. Sonication is a widely used method for preparation of suspensions, but needs to be operated with care, as it may influence particle properties and behaviour (Kroll et al. 2013). For example, sonication may promote aging and transformation processes of ENMs (Taurozzi et al. 2011, Tantra et al. 2015). Once the ENMs are dispersed in the liquid media, in dependence on particle properties as well as the composition of the test medium, ENMs may be present as aggregates, agglomerates, single particles, partly dissolved particles, or even fully dissolved particles, meaning that only ions are present (Skjolding et al. 2016). One has to keep in mind that each organism or each assay puts specific requirements to the composition of the test medium, which implies that numerous media are in use. However, the composition of media has consequences for particle behaviour which hence, needs to be assessed individually (e.g., influence on agglomeration, complexation of ions) (Hund-Rinke et al. 2016, Kroll et al. 2013, Kühnel et al. 2009). Specific peculiarities for test media are discussed in detail in the respective organism sections. Furthermore, ENMs may interact with media constituents such as ions, chelators, proteins or organisms exudates, all of which will further influence particle dispersal. For example, the formation of coatings or so called coronas around the surface of particles may stabilize the particles in the media and will promote the presence of single particles (Docter et al. 2015, Monopoli et al. 2011). In addition, a number of transformation processes constantly change the materials surface, for example by redox reactions or changes in metal speciation.

Overall, one has to keep in mind that during the exposure, dynamic changes in the ENM properties occur and that exact knowledge on these changes over time will strongly support the final evaluation of results.

Consequences for Attachment, Bioaccumulation and Kinetics in Organisms

The behaviour of ENMs in liquid media has consequences for the exposure of organisms. While for organic chemicals the organisms are equally exposed to the test substance, irrespective of their position in an exposure vessel or their mobility, for ENMs, the distribution in the test media is mostly dissimilar. Even though there are some ENMs that form homogenous suspensions, in most cases they are not evenly distributed (heterogeneous suspensions) and undergo constant transformations during the test such as agglomeration or sedimentation, as well as changes on their surface (corona formation, redox reactions). With regard to the organisms or cells, this results in very variable exposure conditions. In addition, the contact with ENMs also depends on the organisms mobility; for example, the sessile zebrafish embryo will be strongly exposed to settled nanomaterials, whereas mobile organisms such as daphnids will be exposed via the water phase. Hence, the bioavailability of ENMs for organisms will differ. In addition, the attachment of ENMs on organism surfaces contributes to their hazard by for example, promoting physical effects (Rodea-Palomares et al. 2011).

The sedimentation of ENMs in the test media poses a common difficulty in toxicity tests/bioassays using liquid media (Brunelli et al. 2013), and different approaches have been applied to avoid sedimentation, for example, by agitation (Hund-Rinke et al. 2016). This was done to achieve maximum contact between test organism and the ENMs, but also to achieve the intended exposure concentrations for the duration of the tests. In general, the dynamic behaviour of ENMs hampers the differentiation between administered and effective dose, which however is essential to derive meaningful dose-reponse relationships and effect values of toxicity as needed for hazard assessment (Praetorius et al. 2014).

Interferences of ENMs with the Test Components or with Detection Methods

Several potential interferences of ENMs with the detection of endpoints in various tests/bioassays were described and need to be accounted for by using appropriate positive and negative controls. Such interferences include the interaction of ENMs with, for example, colorimetric or fluorimetric compounds used in cell-based bioassays. By binding such compounds, ENMs may inhibit their metabolisation or alter enzymatic activity, leading to false-negative results (e.g., Wörle-Knirsch et al. 2006, Ong et al. 2014, Bonvin et al. 2017). Optical interference may occur when ENMs have an intrinsic absorbance or fluorescence, which may lead to false-positive results. In addition, for autotroph organisms, the shading of light from the organisms is an issue (Skjolding et al. 2016). Also, intrinsic catalytic properties of ENMs may lead to the disruption of crucial test components.

Overall, due to the wide variety in ENMs and their properties, it is difficult to propose static procedures; rather ENM and organism specific peculiarities need to be taken into account when conducting a test. This is for example accounted for in the structured approach suggested by (Potthoff et al. 2015). In any case, an extensive characterization of various physical chemical particle properties is suggested, an extensive list of relevant parameters is presented in Table 1. It is crucial to perform the characterization of ENMs in the medium specific to the test or assay and ideally over the whole test duration, in order to understand the particle behaviour over time. Related to the potential interferences described here there are ongoing activities to adopt testing guidelines such as those by the OECD or provide in-depth guidance specific to handling ENMs (OECD TG 318, 2017, Hund-Rinke et al. 2016, Kühnel and Nickel 2014).

Table 1. Extensive list of parameters that need consideration for ENM toxicity testing (adopted from DaNa criteria checklist, https://www.nanopartikel.info/en/nanoinfo/methods/991-literature-criteria-checklist and Schmidt et al. 2017).

General Methodology	Source/Generation of Particles
	• Method of particle sampling (volume [m^3/h], stages, sampler, sampling duration, filter)
	• Filter preparation (before/after sampling, storage; preparation for chemical analysis-organic extracts, water-soluble fraction, whole particles)
Particle characterization	• Average PM concentration [mass per volume; number per volume] • Particle chemical composition (carbon, metals)
	• Particle size, size distribution, aggregate size
	• Surface chemistry (hydrophobic, hydrophilic)/surface reactivity and/or surface charge, surface area
	• Particle morphology (shape)
Exposure conditions	• Method for preparation of suspensions (composition of dispersion medium, preparation of stock solution or direct dosing, way of dispersal, energy input, nominal concentration)
	• Type of exposure (suspended particles, direct contact, complexity of mixtures)
Consideration of particle behavior under test conditions	• Extent of agglomeration/aggregation resp. particle size distribution under experimental conditions
	• Evaluation of particle behaviour over test duration (sedimentation of particles, floating)
	• Dosage used classified clearly to be "non-overload" or "overload conditions"
	• Interaction (particles with cells or organism)
	• Release of attached components (e.g., chemicals, metals)
Testing parameters	• Controls (positive and negative controls), check for interferences • Concentration administered: in µg/ml, µg/cm^2; N (particle)/cell or organism or pg/cell • Use of reference material

Protocols Used for Aquatic Toxicity Testing of ENMs

Toxicity Test Organisms

It is suggested that a battery of bioassays/toxicity tests on organisms from different trophic levels is used to reduce the risk of over/underestimation of the toxicity. The most common criteria for selection of organisms to be used as test species are: they should be representative in terms of the trophic level, route of exposure and morphology, and it should be easy to keep them under laboratory conditions (Leeuwen and Vermeire 2007). To overcome the need for culturing the test organisms, some of the toxicity tests have been developed as ready-to-use kits. These rely on the supply of organisms as dormant forms (cysts, seeds of plants, immobilized algae, lyophilized bacteria or yeast) (http://www.microbiotests.be, n.d.).

The most common organisms that are used in aquatic toxicity testing are algae, bacteria, yeast, crustaceans, rotifers, nematodes, plants and fish (Wieczerzak et al. 2016). Also, single-cell-based assays for evaluating environmental quality are commonly used (Neale et al. 2017). In aquatic ecotoxicology of ENMs, the same test organisms have been employed as in "traditional" aquatic toxicology. Based on nanoE-Tox database (Juganson et al. 2015, Table S5 in Supplementary information), which relies on Thomson Reuters Web of Science, the distribution of test organisms used in aquatic ecotoxicology of ENMs was analysed (Fig. 1). An extensive search (910 publications) was also performed by Chen

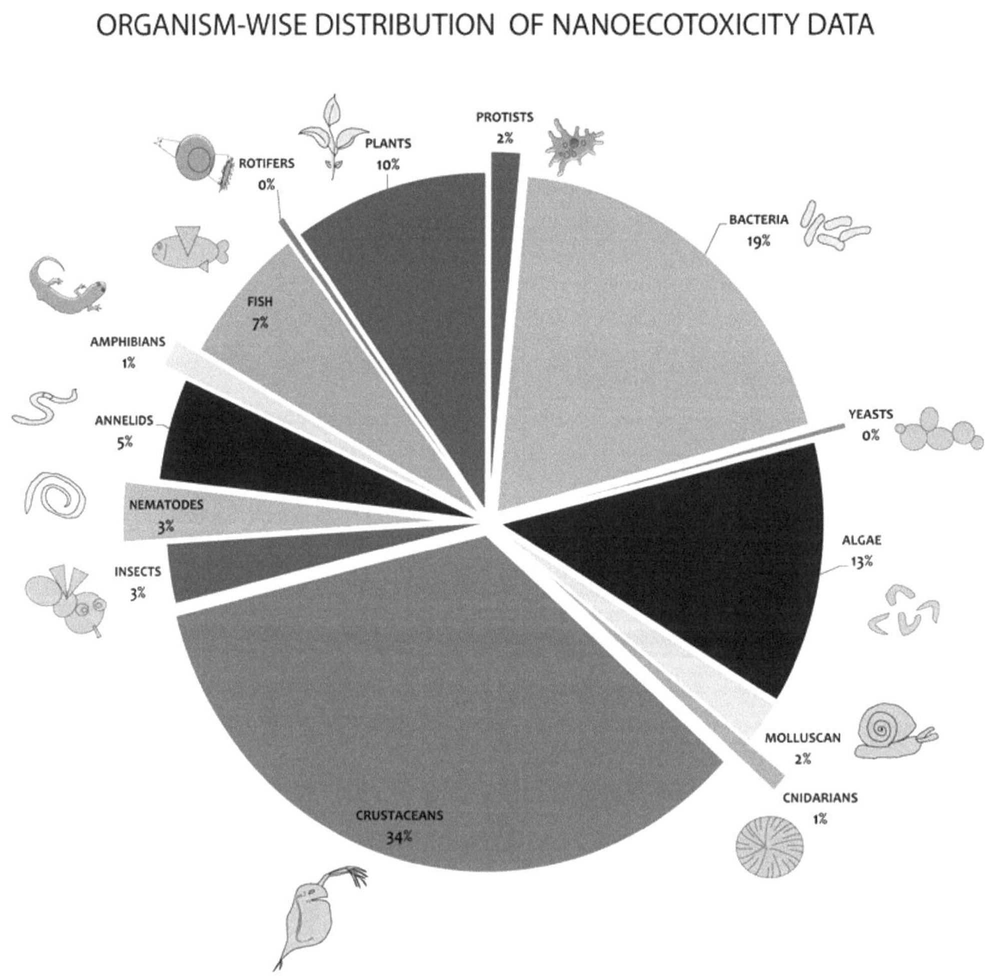

Figure 1. This figure shows the distribution of test organisms used in nanoecotoxicity studies (both aquatic and terrestrial) as listed by Juganson et al. (2015) (Table S5, Supplementary information). The total number of studies was 1432.

et al. (2015) using the same web source. Their search however resulted in a higher share of studies devoted to bacteria (44%), fish (27%) and yeast (12%), because they also included the names of genera in the search. Nevertheless, both reviews (Juganson et al. 2015, Chen et al. 2015) reveal that bacteria, algae, fish and crustaceans are among most commonly applied aquatic toxicity test models.

In this chapter toxicity tests with unicellular organisms (microbes, algae, protozoa), invertebrates (crustaceans, mollusks, insects, nematodes), vertebrates (amphibians, fish), vertebrate cells (fish cells or cell line) and higher aquatic plants are presented. For each of the test organism group, the scope of the tests, list of common species used, advantages and limitations for ENMs testing and needs for further optimization and development is presented. An informative search on the test species used in nanoecotoxicology was done by using the search terms: test group/species name (e.g., fish, mollusk*, amphibia*, crustacean*, nematode*, insect*, alga*, microbe*, etc.), AND nano* (Thomson Reuters Web of Science). In some cases, the number of publications is only informative, because publications reporting nano units (e.g., ng/L) are included as well. Furthermore, in the case of plants, the search also retrieved the use of ENMs as fertilisers and pesticides, and biosynthesis of ENMs by plants; in the case of insects, the use of ENMs as insecticides is included.

Additionally, for ENMs toxicity tests concerning fish, algae, bacteria and amphibia a more detailed search was performed, using the following search combination: Fish/algae/bacteria/reptile/mollusk/nematode/insect/plants* OR amphibi* AND nanopartic1* OR nanomaterial* OR NP AND ecotox* OR aquatic tox* AND assay or *test* NOT sediment OR soil.

As well, the nanoE-Tox database concerning ecotoxicity of nanomaterials was used to get an insight on the type and number of species used for ENM toxicity testing (Juganson et al. 2015).

Microbes and Protozoa

In the literature dealing with ecotoxicology of ENMs, several microbial species as well as protozoa *Tetrahymena thermophila* were used. As reported by Juganson et al. (2015), 21 different bacterial strains have been used in testing of ENMs. Yeast test organisms are much less diverse with two species commonly used in this group: *Saccharomyces cerevisiae* and *Candida albicans*. An overview of the different species used is given in Table 2.

General Principles of the Bioassays

In the nanoecotoxicology literature, three endpoints/effects were assessed on microbes after exposure to ENMs. In the following paragraphs, a short description of each bioassay is given. In Table 3 the effects and corresponding bioassays were summarized. The bacterial growth inhibition is analyzed via optical density measurements or determination of colony formation. For *Pseudomonas putida* and *Vibrio fischeri*, guidelines for assessing bacterial growth inhibition are available, that is, ISO 10712:1995 (ISO 10712:1995 1995) and DIN 38412-L37:1999 (DIN 1999), respectively. The exposure time is different for the different species; for example, *V. fischeri* are exposed for 6 h and the highest exposure time is examined for *Gordonia* sp. (48 h). Additionally bacterial toxicity can be assessed by the bioluminescence inhibition; typically exposing *V. fischeri* bacteria for max. 30 min. Standardized protocols are available for this assay as well (DIN EN ISO 11348-2: 2009, 2009, ISO 21338: 2010, 2010). Additionally, growth inhibition was assessed with protozoa and yeast, where no standardized protocols exist. For testing viability, different assays are available. Bacterial and yeast viability can be examined by the standard plate count assay or by analyzing the colony-forming ability. Furthermore, staining methods can be used for analyzing, for example, membrane permeability of bacterial cells or the nitrification inhibition of ammonia oxidizing bacteria. For assessing the mutagenic potential of ENMs, the Ames fluctuation test (ISO 11350:2012 2012) was used by Schiwy et al. (2016). The viability of protozoa is analyzed by measuring the ATP content using the luciferin–luciferase method.

Table 2. Overview of different microbes and protozoa used for toxicity testing of ENMs, their morphological features and the corresponding references. The number of publications as retrieved in ISI Thomson Reuters Web of Science using keywords "species name" and "nano*" are shown; number in brackets indicate the number of publications in Web of Science category 'Environmental Sciences' (last accessed 4th December 2017). The references retrieved by the detailed search as depicted in the introduction to this chapter are given in the last column. In some cases, studies were found by cross referencing.

Species	Morphology	No. of Publications	References
Aeromonas hydrophila	Gram-negative, 0.3–1.0 x 1.0–3.5 µm, motile by single polar flagellum	78 (5)	Tong et al. 2013
Bacillus aquimaris	Gram-variable, 0.5–0.7 x 1.2–3.5 µm, motile by peritrichous flagella	0	Kumar et al. 2014, Iswarya et al. 2016
Bacillus subtilis	Gram-positive, 2.0–3.0 x 0.7–0.8 µm, motile by peritrichous flagella	1426 (137)	Nam et al. 2014
Bacillus thuringiensis	Gram-positive, 1.1–1.2 x 3.0–5.0 µm, motile	138 (10)	Kumar et al. 2014
Escherichia coli	Gram-negative, 1.1–1.5 x 2.0–6.0 µm, motile by peritrichous flagella or non-motile	13,299 (821)	Li et al. 2011, Nam et al. 2014
Gordonia sp.	Gram-positive	2 (1)	Chen et al. 2016
Nitrosomonas europaea	Gram-negative, 1.1–1.8 µm, motile by polar flagella or non-motile	54 (28)	Radniecki et al. 2011
Pseudomonas fluorescens	Gram-negative, 0.7–0.8 x 2.3–2.8 µm, motile by polar flagella	209 (24)	Ivask et al. 2014
Pseudomonas putida	Gram-negative, 0.7–1.1 x 2–4 µm, motile by polar multitrichous flagella	242 (61)	Matzke et al. 2014, Picado et al. 2015
Salmonella typhimurium (*TA98* and *TA100*)	Gram-negative, 2–5/0.7–1.5 µm, motile	786 (21)	Schiwy et al. 2016
Vibrio fischeri	Gram-negative, motile by polar flagella	153 (73)	Cerrillo et al. 2015, Rossetto et al. 2014, Rossetto et al. 2014, Sanchis et al. 2016, Casado et al. 2013, Choi et al. 2014, Jemec et al. 2016, Svartz et al. 2017, Andreani et al. 2017, Aruoja et al. 2015, Mohmood et al. 2016, Bondarenko et al. 2016, Picado et al. 2015
Saccharomyces cerevisiae	Yeast, 5–6 µm	1262 (43)	Kasemets et al. 2009, Otero-González et al. 2013, Ivask et al. 2014, Bondarenko et al. 2016
Candida albicans	Yeast, 4–10 µm	1137 (12)	Lipovsky et al. 2011
Tetrahymena thermophila	Ciliated protozoan, 30–60 µm, motile by ciliates	67 (19)	Jemec et al. 2016, Aruoja et al. 2015, Bondarenko et al. 2016

Advantages and Limitations for ENM Testing

Advantages

In general, bioassays involving microorganisms are easy to conduct and rapid. Both prokaryotic and eukaryotic cells are available for testing. With regard to bacteria, different pathogenic or environmentally relevant strains can be used. Instead of media rich in organic substance, deionized water can be used for some species also, reducing the potential transformation of metallic ENMs related to speciation as well as to avoid corona formation on the ENMs (Bondarenko et al. 2016).

Table 3. Overview of the different bioassays used for toxicity testing of ENMs, with details on the observed endpoints, the determined test parameter and the references which used the bioassay.

Endpoint	Bioassay	Test Parameter	References
Growth inhibition	Growth inhibition (ISO 10712:1995, DIN 38412-L37:1999)	Optical density, Colony formation	Chen et al. 2013, Matzke et al. 2014, Picado et al. 2015, Li et al. 2011, Nam et al. 2014
	Bioluminescence inhibition test (DIN EN ISO 11348-2:2009, 2009), ISO 21338:2010, 2010)	Bioluminescence inhibition	Jung et al. 2015, Aruoja et al. 2015, Picado et al. 2015, Choi et al. 2014, Cerrillo et al. 2015, Jemec et al. 2016, Rossetto et al. 2014, Rossetto et al. 2014, Sanchis et al. 2016, Svartz et al. 2017, Andreani et al. 2017, Mohmood et al. 2016, Casado et al. 2013, Bondarenko et al. 2016
	Growth inhibition/acute toxicity (Protozoa)		Hu et al. 2015
	Growth inhibition assay (yeast)	Colony formation	Kasemets et al. 2009, Otero-González et al. 2013
Viability	Bacterial viability assay	Colony formation	Aruoja et al. 2015, Ivask et al. 2014, Iswarya et al. 2016, Kumar et al. 2014
	Protozoan viability assay	Colony formation	Jemec et al. 2016, Aruoja et al. 2015, Bondarenko et al. 2016
	Yeast viability assay	Colony formation	Lipovsky et al. 2011, Kasemets et al. 2009, Ivask et al. 2014, Bondarenko et al. 2016
	Life/death staining assay (Ethidium bromide, Acridine orange, etc.)	Measurement of membrane integrity or metabolic activity of the cell	Binh et al. 2014, Tong et al. 2013, Tong et al. 2013
	Nitrification inhibition assay	NO^{2-} production	Radniecki et al. 2011
DNA damage	Ames fluctuation assay (ISO 11350:2012)	Count of revertants	Schiwy et al. 2016

Limitations, Needs for Optimization and Future Challenges

Despite being simple and rapid, a number of issues need consideration when assessing hazard of ENMs with microorganisms. First, most growth inhibition assays are conducted in rich media in order to provide optimal growth conditions for the cells. Hence, sorption or interference of media constituents with the ENMs may occur, influencing, for example, ENM agglomeration by corona formation or the dissolution behaviour (Bondarenko et al. 2016).

As the Microtox test depends on the detection of visible light that is produced by the luminescent bacteria, the ENMs may lead to false-positive results because of turbidity of suspensions (Svartz et al. 2017, Rossetto et al. 2014, Rossetto et al. 2014). In addition, the Microtox Test is sensitive to pH changes induced by particles because the test medium is not buffered. With regard to the above mentioned replacement of rich media by deionized water (Bondarenko et al. 2016), one has to keep in mind that deionized water is not a natural environment and may trigger osmotic stress in cells. This may change the susceptibility of cells to ENMs and lead to false-positive results. Further, the different compositions of cell walls of bacteria strains need consideration, as they influence the permeability of bacteria cells for ENMs as well as attachment of ENMs to their surface. Accordingly, ENM induced physical damage may differ substantially between different species (Handy et al. 2012).

Algae

Species

In the literature dealing with ecotoxicology of ENMs, several algae species were used for assessing the toxicity of ENMs. An overview of the different algae species used is given in Table 4. Juganson et al. (2015) identified nine different green algal and four red algal species in their nanoE-Tox database. Most commonly applied species are *Raphidocelis subcapitata* (formerly known as *Pseudokirchneriella subcapitata*), *Chlamydomonas reinhardtii*, *Chlorella vulgaris* and *Desmodesmus subspicatus* (formerly known as *Scenedesmus subspicatus*).

General Principles of the Bioassays

In the literature, several algae specific endpoints/effects were assessed after ENM exposure. In Table 5,the effects and corresponding bioassays used in nanoecotoxicology were summarized. In most cases, the algae growth inhibition was used to assess ENM toxicity (OECD TG 201, 2011, ISO 8692:2012, 2017). The purpose of the growth inhibition test is to determine the effects of a substance on the growth of freshwater microalgae and/or cyanobacteria. Exponentially growing test organisms are exposed to the test particles in batch cultures over a period of normally 72 hours. In spite of the relatively brief test duration, effects over several generations can be assessed. Growth is quantified from measurements of the algal biomass as a function of time. Algal biomass is defined as the dry weight per volume, for example, mg algae per liter test solution. However, dry weight is difficult to measure and therefore surrogate parameters are used. Of these surrogates, cell counts are most often used. Other surrogate parameters include cell volume, pigment fluorescence, optical density, etc. (OECD 2010).

In the algal viability assay ('spot test'), the ability of the toxicant exposed algae to form colonies on toxicant-free nutrient agar is determined (Suppi et al. 2015). The advantage of this test design is the avoidance of nutrient removal effects. In addition the viability of the algae cells can be determined via colorimetric methods, for example, the MTT Assay. Furthermore, morphological changes of algae after ENMs exposure can be assessed by using microscopy, for example, Scanning Electron Microscopy (SEM). Additionally, these methods allow analyzing potential interactions/attachment between algae and ENMs. Another typically assessed endpoint is the determination of inhibition effects on Photosystem II (PS II). The fluorescence emitted by PS II complexes is measured to

Table 4. Overview of different algae species used for toxicity testing of ENMs, their morphological features and the corresponding references. The number of publications as retrieved in ISI Thomson Reuters Web of Science using keywords "species name" and "nano*" are shown, number in brackets indicate number of publications in Web of Science category 'Environmental Sciences' (accessed December 4th 2017). The references retrieved by the detailed search as depicted in the introduction to this chapter are given in the last column.

Species	**Morphology**	**No. of Publications**	**References**
Chlamydomonas reinhardtii	Oval/egg-shaped, about 10 µm in diameter, single-cell algae	360 (107)	Sendra et al. 2017
Chlorella spirulina	Spherical-shaped, about 2 to 10 µm in diameter, single-cell algae	15 (2)	Iswarya et al. 2016
Chlorella vulgaris	Spherical-shaped, about 4 to 10 µm in diameter, single-cell algae	196 (45)	Clement et al. 2013a, 2013b, Gong et al. 2011, Polonini et al. 2015, Schwab et al. 201, Suman et al. 2015
Desmodesmus subspicatus (formerly known as *Scenedesmus subspicatus*)	Elliptic- or spindle-shaped, length: 7–15 µm, width: 312 µm, formation of colonies consisting of simple rows with 4 to 16 cells	80 (45)	García-Cambero et al. 2013, Hund-Rinke and Simon 2006, Schivvy et al. 2016
Dunaliella salina	Oval-, spherical-, spindle-, elliptic- or cylindric-shaped, length: 5–29 µm, width: 3.8–20.3 µM, single-cell algae	34 (6)	Shirazi et al. 2015
Dunaliella tertiolecta	Rod- to ovoid-shaped, 9–11 µm, single-cell algae	61 (20)	Schiavo et al. 2016
Phaeodactylum tricornutum	Fusiform-, triradiate-, or oval-shaped, length: 20–35 µm, width: 3–4 µm	97 (19)	Clement et al. 2013a, 2013b, Minetto et al. 2017, Mohmood et al. 2016, Sendra et al. 2017
Picochlorum sp.	small unicellular algae, some genera are halotolerant	6 (4)	Hazeem et al. 2016
Raphidocelis subcapitata (formerly known as *Pseudokirchneriella subcapitata*)	Curved -, twisted-Shaped, length: 8–14 µm, width: 2–3 µm, single-cell algae	347 (207)	Andreani et al. 2017, Aruoja et al. 2015, Bondarenko et al. 2016, Booth et al. 2015, Bouldin et al. 2008, Casado et al. 2013, Choi et al. 2014, Clar et al. 2015, Franklin et al. 2007, Ivask et al. 2014, Jemec et al. 2016, Joonas et al. 2017, Kennedy et al. 2010, Malysheva et al. 2016, Manier et al. 2013, Nam et al. 2014, Neale et al. 2015, Pereira et al. 2017, Picado et al. 2015, Ribeiro et al. 2014, Schwab et al. 2011, Hoecke et al. 2011, 2009, 2008, Zhang et al. 2016
Scenedesmus obliquus	Elliptic- or spindle-shaped, formation of colonies	44 (24)	Wang et al. 2016, Zhang et al. 2016
Euglena gracilis	35–65 x 5–15 µm, motile by two flagella	26 (5)	Hu et al. 2015

determine the photosynthetic activity. This can be done through the pulsed amplitude modulation (PAM) method or the commercially available LuminoTox Assay (Lab-Bell Inc., Québec, Canada). In one study, the genotoxicity of three particles on algae was also determined by using the Comet Assay, which detects DNA strand breaks, alkali-labile sites, and incomplete excision repair events. It involves the encapsulation of cells in a low-melting-point agarose suspension, lysis of the cells in neutral or alkaline (pH > 13) conditions, and electrophoresis of the suspended lysed cells. The term "comet" refers to the pattern of DNA migration through the electrophoresis gel, which often resembles a comet.

Table 5. Overview of the different bioassays used for toxicity testing of ENMs, with details on the observed endpoints, the determined test parameter and the references which used the bioassay.

Endpoints	Bioassay	Test Parameter	References
Growth inhibition	Algae growth inhibition assay (ISO 8692:2012, OECD TG 201)	Cell density via cell counting or determination of cell volume	Hazeem et al. 2016, Shirazi et al. 2015, Kennedy et al. 2010, Andreani et al. 2017, Choi et al. 2014, Nam et al. 2014, Bouldin et al. 2008, Casado et al. 2013, Franklin et al. 2007, Joonas et al. 2017, Pereira et al. 2017, Schwab et al. 2011, Hoecke et al. 2011, 2008, Zhang et al. 2016, Picado et al. 2015, Wang et al. 2016, Iswarya et al. 2016, Polonini et al. 2015, Mohmood et al. 2016, Minetto et al.2017, Sendra et al. 2017, Bondarenko et al. 2016
		Cell density via determination of pigment fluorescence (Chlorophyll a, carotenoids)	Bennett et al. 2013, Hoecke et al. 2009, Hazeem et al. 2016, Shirazi et al. 2015, Jemec et al. 2016, Booth et al. 2015, Clar et al. 2015, Malysheva et al. 2016, Manier et al. 2013, Neale et al. 2015, Schiwy et al. 2016, Hoecke et al. 2011, Aruoja et al. 2015, Ivask et al. 2014, Hund-Rinke and Simon 2006, García-Cambero et al. 2013, Gong et al. 2011
		Optical density	Clement et al. 2013, Clement et al. 2013, Ribeiro et al. 2014
	Algaltoxkit FTM	Optical density, pigment fluorescence	Ksiazyk et al. 2015
Viability	Algal viability assay	Colony formation	Aruoja et al. 2015, Joonas et al. 2017
	Life/death staining (MTT-Assay, PI, Acridine orange, etc.)	Measurement of membrane integrity or metabolic activity of the cell	Suman et al. 2015, Sendra et al. 2017, Schiavo et al. 2016
Morphological changes	Morphological analysis of algal cells	Microscopic observation of shape and cell integrity	Pereira et al. 2017
PS II inhibition	LuminoTox assay	Fluorescence	Manusadzianas et al. 2012, Clement 2013, Clement et al. 2013
	Imaging-PAM	Fluorescence	Wang et al. 2012, 2015, Neale et al. 2015, Polonini et al. 2015, Schwab et al. 2011
	Phyto-PAM	Fluorescence	Sendra et al. 2017, Zhang et al. 2016
DNA damage	Comet assay	Analysis of DNA strand breaks via gel electrophoresis	Schiavo et al. 2016

Advantages and Limitations for ENM Testing

Advantages

The great advantages of using algae for toxicity testing are the ease of cultivation and the relatively brief test duration, where effects over several generations can be assessed. Furthermore, shaking of algae suspensions is required also during ENM exposure; this prevents sedimentation of ENMs and enables a close contact of ENMs and algae (Hund-Rinke et al. 2016).

Limitations, Needs for Optimization and Future Challenges

On the other hand, some facts should be considered when using algae for ENM toxicity testing. As already mentioned in section regarding challenges in toxicity testing of ENMs, differences in the recipes of test media are a critical point (Handy et al. 2012, Hund-Rinke et al. 2016). In the case of ENM testing with algae, the chelating agent EDTA can interfere with metal ENMs and thus have an influence on the test results. Additionally, it should be kept in mind that the algae growth medium contains $MgSO_4$ and $CaCl_2$ and these cations will promote agglomeration (Handy et al. 2012). Furthermore, algae produce exudates, which may also promote stability or agglomeration of the ENMs. Further, ENMs as well as their agglomerates may influence light penetration (shading) and thus, photosynthesis, and therefore algal growth will be affected in an indirect manner (Handy et al. 2012). With regard to the shading issue, the shaking of the test vessels may be an advantage, because algae will have access to light when situated close to the test tube walls; but could be also a critical point due to orthokinetic (shear) aggregation and the altered nature and frequency of ENM collision with the algae (Handy et al. 2012). Another aspect is the quality and quantity of light provided, especially for certain photo-reactive ENMs (e.g., Nanoceria and Nanotitania), where light, which is essential for algal growth, can promote the generation of reactive oxygen species (ROS) (Handy et al. 2012). Regarding the test parameters for the growth inhibition, the cell count should only be used when ENMs and algae differ substantially in size and the ENMs are stable, that is, no agglomerates are formed. Only then can ENMs and algae be clearly distinguished from each other. Especially for ENM forming cell-sized agglomerates and/or adhering to cell surface, it is difficult to distinguish cells from particle agglomerates (Hartmann et al. 2012).

In order to adequately consider these interferences in future testing, the following modifications to the OECD TG 201 (OECD TG 201 2011) and for algae testing in general when assessing effects of ENMs were suggested (see also (ECHA 2017). Potential shading effects induced by ENMs should be considered by methods such as those previously developed (Sörensen et al. 2015, Hund-Rinke and Simon 2006, Hjorth et al. 2016). For example a 'sandwich test' is a test design that physically separates algae and ENMs. Also, mechanical effects of ENMs and attachment to the organisms should be assessed (e.g., Geitner et al. 2016). In order to prevent the sedimentation of the ENMs, controlled stirring/shaking regimes are suggested (Hund-Rinke et al. 2016); however, these may also exert physical effects on cells. Because the ENMs may interfere with the determination of the biomass by cell counting, it is recommended to check different methods, for example, hemocytometer, measurement of cell volume, etc. (Hund-Rinke et al. 2016). As alternative measure of biomass, the determination of *in vitro* Chlorophyll A (Chla) or other pigments such as carotenoids (Hartmann et al. 2012, Hjorth et al. 2016, Hoecke et al. 2009, Hund-Rinke et al. 2016, Shirazi et al. 2015) is suggested, if the tested nanomaterial is checked for auto fluorescence when using chlorophyll extracts (Hartmann et al. 2012). For specific ENMs, potential effects of their photoactivity or catalytic properties on their toxicity need to be considered (Hund-Rinke et al. 2016).

Crustaceans

Species

Crustaceans are among the most represented invertebrate species used for ENM toxicity testing (Juganson et al. 2015). As estimated in this latter study, studies involving crustaceans constituted approximately one third (500/1518) of all studies. Among these, studies using water flea *Daphnia magna* by far exceed other crustaceans (Table 6).

Table 6. Overview of different crustacean species used for toxicity testing of ENMs. The number of publications as retrieved in ISI Thomson Reuters Web of Science using keywords "species name" and "nano*" are shown, number in brackets indicate number of publications in Web of Science category 'Environmental Sciences' (accessed December 4th 2017 4.12.17). The most cited references of studies with nanomaterials are listed in the last column.

Species	Order, Habitat, Life-Stage Used	No. of Publications	References
Daphnia magna	Cladocera, waterflea, freshwaters, mostly lakes, rivers, and temporary pools; neonates less than 24 h old	975 (275)	Cui et al. 2017, Heinlaan et al. 2008, Wiench et al. 2009, Lovern and Klaper 2006, Blinova et al. 2010
Daphnia similis	The same as above	12 (6)	Artells et al. 2013, Clemente et al. 2014
Daphnia pulex	The same as above	28 (8)	Hall et al. 2009, Klaper et al. 2009, Auffan et al. 2013, Artells et al. 2013
Ceriodaphnia dubia	The same as above	63 (20)	Bhuvaneshwari et al. 2017, Hall et al. 2009, Wang et al. 2011, Dalai et al. 2013
Ceriodaphnia cornuta	The same as above	9 (8)	Ishwarya et al. 2017
Thamnocephalus platyurus	Anostraca, freshwaters, mostly lakes, rivers, and temporary pools; neonates less than 24 h old	22 (15)	Heinlaan et al. 2008, Blinova et al. 2010, Casado et al. 2013
Heterocypris incongruens	Ostracoda, temporary ponds or the littoral zone of lakes; neonates less than 24 h old	5 (0)	Manzo et al. 2011
Hyalella azteca	Amphipoda, benthic freshwater; adults or juveniles	21 (6)	Oberdörster et al. 2006, Mwangi et al. 2012
Gammarus fossarum	Amphipoda, benthic freshwater; adults or juveniles	12 (3)	Bundschuh et al. 2011, Mehennaoui et al. 2016
Corophium volutator	Amphipoda, benthic freshwater species, adults	5 (0)	Fabrega et al. 2011, Dogra et al. 2016

General Principles of the Toxicity Tests

Daphnia magna toxicity test is widely accepted in ecotoxicology community and is internationally standardized (OECD TG 211, 2008, OECD TG 202, 2004, ISO 6341:2012, 2012, ISO 10706:2000, 2000). Beside *D. magna*, *D. similis*, *D. pulex* and *Ceriodaphnia dubia* are also commonly used (ISO 20665:2008 2008). International standard exists for ostracods *Heterocypris incongruens* (ISO 14371:2012 2012), *Thamnocephalus platyurus* (ISO 14380:2011 2011) and *Hylella azteca* (EPA, 2009, ISO 16303:2013 2013). Some of the crustacean bioassays are also available commercially as Toxkit Microbiotests (http://www.microbiotests.be, n.d.). These are the: *Thamnocephalus* Toxicity Test (THAMNOTOXKIT F), *Daphnia magna* test (DAPHTOXKIT F magna), *Ceriodaphnia* test (CERIODAPHTOXKIT F) and Ostracod (*Heterocypris incongruens*) Crustacean Toxicity Test (OSTRACODTOXKIT F). The accessability to these kits also increases their wide applicability in toxicity studies.

In the case of *Daphnia* sp. acute immobilization test, young daphnids less than 24 h old (neonates) are exposed to the test substance at a range of concentrations for a period of 48 h. Immobilization is recorded at 24 h and 48 h and compared with control values. Immobility of daphnids is defined as the lack of swimming after gentle agitation of the liquid for 15 s (ISO 6341:2012 2012). The animals are considered immobile even if they still move antennae. Commonly, the authors refer to immobility of daphnids as their mortality, but the latter needs to be confirmed by the absence of heart beat. A 24 h LC50 bioassay with *T. platyurus* is performed in a multiwell test plate using instar II-III larvae, which are hatched from cysts (Heinlaan et al. 2008). After 24 h of exposure, the mortality of larvae is assessed as the absence of movement during 10 seconds of observation. Water sediment toxicity test with ostracod crustacean *H. incongruens* is a "direct sediment contact" bioassay and is performed in multiwell test plates using neonates hatched from cysts (ISO 14371:2012). After 6 days of exposure, the percentage mortality and the growth of the crustaceans are determined and compared to the results obtained in control uncontaminated exposure (Chial and Persoone 2002) (Table 7).

Long term toxicity of substances to *D. magna* is assessed via the reproduction of adult daphnids (OECD TG 211, 2008, ISO 10706:2000). *D. magna* less than 24 h old are exposed for a period of 21 days. The survival of parents and the number of live offspring produced per live parent at the end of the test is recorded (Wiench et al. 2009). In addition to the reproduction endpoint (number of offspring) as suggested by standards, additional endpoints such as average brood size, time to first brood, time between broods, number of broods per female and length of animals are also very informative (Adam et al. 2015). Chronic toxicity to *dubia* is based on reproduction inhibition after (7 ± 1) days (ISO 20665:2008 2008). Also here the number of offspring is assessed.

In the case of *H. azteca*, *Gammarus fossarum* and *Corophium volutator*, the exposure protocols are not standardised and therefore various experimental set-up and endpoints have been employed (Mwangi et al. 2012, Oberdörster et al. 2006, Bundschuh et al. 2011).

Table 7. Overview of the different toxicity tests for ENMs, with details on the observed endpoints, the determined test parameter and the references which used the test.

Endpoint	Toxicity Test*	Test Parameter	References
Immobility	*Daphnia* sp. acute immobilisation toxicity test (OECD 202, ISO 6341:2012) (48 h)	Swimming observation, daphnia is considered immobile if it does not swim after 15 s of gentle agitation.	Artells et al. 2013, Clemente et al. 2014, Heinlaan et al. 2008, Wiench et al. 2009, Lovern and Klaper 2006
Mortality	*Thamnocephalus* acute toxicity test (ISO 14380:2011) (24 h)	The larvae are considered dead if they do not show any movement during 10 seconds of observation.	Heinlaan et al. 2008, Casado et al. 2013
Mortality/growth	*Heterocypris incongruens* toxicity test (ISO 14380:2011) (6 d)	Direct contact toxicity test for Freshwater Sediments. Percentage mortality and the length of ostracods are assessed.	Manzo et al. 2011
Reproduction	*Daphnia magna* reproduction toxicity test (ISO 10706:2000; OECD 201) (21 d)	Number of living offspring produced per surviving parent animal at the end of the test and the survival of parent animals are reported.	Wiench et al. 2009
	Ceriodaphnia dubia chronic toxicity (ISO 20665:2008) (7 d)		
Moult rate	*Daphnia magna* acute toxicity test (96 h)	Moult rate is estimated.	Dabrunz et al. 2011
Behaviour	*Daphnia pulex* and *Daphnia similis* acute toxicity test (48 h)	Swimming velocity.	Artells et al. 2013

*different generic names apply to these tests. The most descriptive one is used.

Advantages and Limitations for ENM Testing

Advantages

There are many advantages of using crustacean tests for toxicity testing of ENMs. Firstly, some of them are easily accessible because they are commercially available as kits (DAPHTOXKIT F™ *magna*; DAPHTOXKIT F™ *pulex*; OSTRACODTOXKIT F™ *Heterocypris*; RAPIDTOXKIT F™ *Thamnocephalus* and CERIODAPHTOXKIT F™ *Ceriodaphnia*) (http://www.microbiotests.be, n.d.). The protocols are internationally standardised which enables the comparison of data across laboratories. They have been estimated as user-friendly and cost-effective (Wieczerzak et al. 2016). Advantage of some tests is that the organisms are hatched from dormant cysts which enables their constant availability and reduces the cost of maintenance. Daphnids and thamnocephalus are non-selective particle-feeding organism and thus, a very relevant model for ENM exposure. Also, amphipods are potentially exposed to ENMs deposited on the sediment or plants (*H. azteca, G. fossarum*). In particular *C. volutator* is exposed when burrowing in the sediment. In terms of sensitivity of crustacean toxicity tests to different ENMs, the acute *D. magna* toxicity tests was the most sensitive among 15 toxicity tests investigated (Bondarenko et al. 2016). These authors tested seven well-characterized ENMs in 15 toxicity tests covering bacteria, yeast, algae, protozoa, crustacean and fish. Daphnids are in particular sensitive to metal nanomaterials (Hund-Rinke et al. 2018).

Limitations, Needs for Optimization and Future Challenges

Among crustacean tests, most of the effort in scientific community was to adapt the existing *D. magna* protocols (OECD TG 202, 2004, ISO 6341:2012, 2012). In the acute (48 h) protocol, the water is not renewed and therefore ENM sedimentation is expected. Medium replacement during the experiment may increase stress and lead to additional mortality. Also, aeration could result in introducing additional mechanical stress for daphnids. It has been suggested to redesign the aeration in the test vessels and protect the animals from the stream of bubbles by introducing a mesh (Handy et al. 2012, Sørensen et al. 2015). One of the proposed modifications of *D. magna* test was to adapt the exposure medium to such an extent that the dispersion is more stable. This could be achieved by lowering the ionic strength and pH (Hund-Rinke et al. 2016). Daphnia exposures are usually done in waters of greater water hardness, but this leads to greater ENM agglomerations. Choosing an alternative test species adapted to softer water (e.g., *D. pulex*) may be an alternative (Petersen et al. 2015).

Also some issues were raised about whether immobilisation of daphnids is a relevant end-point in the case of ENM testing (Handy et al. 2012). It has been shown that ENMs adsorb onto the carapace and appendages of *D. magna* and physically restrict the movement (Rosenkranz et al. 2009). Another "particle control" for the mechanical effects on ENMs was suggested, but the problem that arises here is the choice of such appropriate positive control. Even though the primary particle size of particle control would match the testing ENM size, it would never behave precisely the same as the test material. Such control would only indicate whether the toxicity is mainly mechanical or chemical in origin (Handy et al. 2012, Handy et al. 2012). In the case of daphnids, reduced moult rate was also suggested as a relevant end-point (Dabrunz et al. 2011). This was evidenced in the case of TiO_2 which absorbed on the body surface of daphnids and affected the moult (Dabrunz et al. 2011). Both of these latter studies pointed out that although no effect of TiO_2 on daphnids was observed after a standard 48 h exposure, significant effects were found after prolongation to 72 and 96 h. Also, when daphnids were incubated in a clean medium for 24 h after 48 h exposure to TiO_2, they exhibited evident mortality. Due to these data, it would be worth considering modifying the exposure period.

Another concern is the introduction of food during chronic exposures, because it will interfere with ENM measurements of size during the test. Also, introducing the food into the test system may affect the actual uptake of ENMs by daphnids (Handy et al. 2012). This could be avoided by adapting the time of organism feeding and test medium change. Daphnids could be fed prior to transfer into new medium. Although most of the limitations and modifications were presented for daphnids, at least some of them also apply for other crustacean toxicity tests. The stability of ENMs in all aquatic media is a problem; therefore their stability should be ensured in all toxicity tests. Also, the issue of ENM adsorption onto body surface and potential interference with moult is probably an issue in all crustaceans.

Mollusks

Species

Freshwater mollusks are among those organisms which are less commonly used for ENM toxicity testing (Juganson et al. 2015). Much more studies exist for marine mollusks (Canesi et al. 2012). According to Thomson Reuters ISI Web of Science, the highest share of studies with molluscs ENMs was done using two bivalve species *Corbicula fluminea* and *Dreissena polymorpha* and two gastropods *Lymnaea stagnalis* and *Potamopyrgus antipodarum* (Table 8).

Table 8. Overview of different mollusk species used for toxicity testing of ENMs. The number of publications as retrieved in ISI Thomson Reuters Web of Science using keywords "species name" and "nano*" are shown (accessed 20.11.2017), number in brackets indicate number of publications in Web of Science category 'Environmental Sciences' (accessed December 4th 2017). Some representative references of studies with ENMs are listed.

Species	Class, Habitat, Life-Stage Used	No. of Publications	References
Corbicula fluminea	Bivalvia; rivers, lakes, ditches and pools of fresh or brackish waters, attached to mud or sand; adults	21 (4)	Canesi et al. 2012, Hull et al. 2011
Dreissena polymorpha	Bivalvia; sessile bivalve mollusk, forming dense colonies on various hard substrates in fresh and slightly brackishwaters; adults	32 (4)	Couleau et al. 2012, Garaud et al. 2016
Elliptio complanata	Bivalvia; shoals of lakes or river-lakes, most abundant in substrates composed of clay and sand; adults	9 (1)	Gagné et al. 2013, Peyrot et al. 2009
Sphaerium corneum	Bivalvia; shallow, freshwater habitats with slow moving waters, including freshwater lakes, rivers and creeks; adults	2 (0)	Völker et al. 2015
Lymnaea stagnalis	Gastropoda; slowly running water, and standing water bodies; adults	32 (2)	Oliver et al. 2014, Croteau et al. 2014
Potamopyrgus antipodarum	Gastropoda; freshwater and brackish streams and lakes; wide variety of habitats including reservoirs, post-industrial ponds, geothermal streams, polder marsh, coastal lakes, estuaries and the sea; adults	19 (0)	Pang et al. 2012, Ramskov et al. 2015, Völker et al. 2015
Lymnaea luteola L.	Gastropoda; temporary water bodies, which dry up in summer, and it tides over the unfavourable conditions by burying itself in the mud, adults	4 (3)	Ali et al. 2012
Physa acuta	Gastropoda; freshwater rivers, streams, lakes, ponds, and swamps; anthropogenic reservoirs, occurring in warm water discharges from power stations and in some rivers; adults	6 (1)	Justice and Bernot 2014

General Principles of the Toxicity Tests

Two OECD TGs exist for mollusks: OECD TG 243 (2016): *Lymnaea stagnalis* Reproduction Test and OECD TG 242 (2016): *Potamopyrgus antipodarum* Reproduction Test; however *L. stagnalis* standard reproduction test has so far not been used to test ENMs (Table 9). The objective of the *P. antipodarum* reproduction test is to assess the effect of chemicals on reproductive output by evaluating embryo

Table 9. Overview of the different toxicity tests for ENMs, with details on the observed endpoints, the determined test parameter and the references which used the test.

Endpoint	**Toxicity Test**	**Test Parameter**	**References**
Mortality	*Potamopyrgus antipodarum* chronic toxicity test (8 weeks) *Potamopyrgus antipodarum* chronic toxicity test (14 d) *Sphaerium corneum* chronic test (28 d)	No movement after a gentle agitation.	Pang et al. 2012, Völker et al. 2015, Ramskov et al. 2015
	Physa acuta acute test (96 h)	Adult and juvenile snails exposed; juvenile mortality is estimated as the absence of hearth beat.	Gonçalves et al. 2017
Growth	*Potamopyrgus antipodarum* chronic toxicity test (8 weeks)	Shell length.	Pang et al. 2012
Feeding rate	*Potamopyrgus antipodarum* chronic toxicity test (8 weeks)	Dry weight of fecal pellets produced per day.	Pang et al. 2012
Behaviour	*Dreissena polymorpha* chronic toxicity test (21 d)	Filtration rate by assessing the removal of particles or food.	Garaud et al. 2016
Reproduction	*Potamopyrgus antipodarum* reproduction toxicity test (OECD TG 242) (28 d)	Embryo numbers in the brood pouch.	Ramskov et al. 2014
	Sphaerium corneum reproduction toxicity test (28 d)	No. of embryos per adult.	Völker et al. 2015
	Physa acuta reproduction toxicity test (13 d)	No. of eggs (embryos) per egg mass; hatching success.	Gonçalves et al. 2017

numbers in the brood pouch at the end of 28 days exposure. The survival of adults and reproduction over the 28 days and the end exposure period are examined. Acute endpoints in mollusks include standard observations of mortality, growth of shell, feeding rate and filtering rate. The acute exposure periods vary considerably among authors, most probably due to the fact that there is no standardised acute protocol for mollusks available. Recently, Gonçalves et al. (2017) presented an interesting acute (96 h) nanotoxicity study exposing egg mass, juvenile and adult snails *Physa acuta*.

Advantages and Limitations for ENM Testing

Advantages

There are not many studies addressing the effects of ENMs on mollusks. Therefore, the advantages of their use are also rarely assessed. However, in the past there was great effort to standardise the mollusk-based tests for the purpose of testing endocrine disruptors. Both *L. stagnalis* and *P. antipodarum* OECD TGs tests proved to be reliable, reproducible and provide important information regarding the effects of substance on the reproduction of organism (Ducrot et al. 2014, Ruppert et al. 2017).

Limitations, Needs for Optimization and Future Challenges

No specific proposals regarding the adaptation of molluskan OECD tests have previously been suggested (Hund-Rinke et al. 2016, Handy et al. 2012, Handy et al. 2012, Hjorth et al. 2016, Petersen et al. 2015). Following an example of other aquatic tests (e.g., crustaceans), the molluskan test could also be modified in order to increase the stability of ENMs in test medium. In the crustacean test, the medium renewal during the toxicity test is not an option since this may physically affect sensitive crustaceans. It remains to be discussed whether this could be an option for mollusks. Also, some modification of the test medium composition would be possible, as has been suggested for the *Physa acuta* test (Gonçalves et al. 2017). In this particular case, the authors suggested to use less chloride ions which change the speciation of some metals, for example, Ag. However, some other modifications in terms of decreasing the ionic strength of the medium and adapting the pH are also possible as suggested for crustaceans (Hund-Rinke et al. 2016, Handy et al. 2012, Handy et al. 2012, Petersen et al. 2015) and plants (Gubbins et al. 2011).

Aquatic Insects

Species

Nine different insect species were identified in a detailed search on ENM toxicity related studies (Juganson et al. 2015), but out of these, only two species, *Chironomus riparius* and *Chironomus dilutes*, are aquatic species (Lee et al. 2009, Waissi-Leinonen et al. 2012, Bour et al. 2015, Oberholster et al. 2011). Also, in other nanoecotoxicity reviews, insects are rarely mentioned (Kwak et al. 2016, Bondarenko et al. 2016). A part of literature related to ENMs and insects is concerned with the use of nanomaterials as potential insecticides, either directly or as carriers of other active substances (Rai and Ingle 2012, Benelli 2016). When applying "nanoparticles" and "insects" as keywords on ISI web of Science, 77 hits were retrieved (24.11.17), but only five of them were in relation to ecotoxicity. None of these was done on aquatic insects. This is an indication that these organisms are indeed rarely applied in nanoecotoxicity studies. Other aquatic insects that are used in ecotoxicity studies with conventional chemicals are larvae of mayflies *Hexagenia limbata, H. bilineata* (Harwood et al. 2014), *Baetis tricaudatus* (Irving 2000), *Cloeon dipterum* and *Caenis miliaria* (Peterson et al. 1994), and larvae of caddisfly *Hydropsychidae* (Siegfried 1993), and damselfly larvae (*Lestes sponsa* and *Cordulia aenea*) (Peterson et al. 1994). All these are potential aquatic test species for ENMs testing.

General Principles of the Toxicity Tests

The Chironomus toxicity test is standardised according to various international organisations (US EPA; ASTM, AFNOR) and OECD (OECD TG 235 2011): *Chironomus* sp., Acute Immobilisation Test and OECD TG 233 (2010): Sediment-Water Chironomid Life-Cycle Toxicity Test Using Spiked Water or Spiked Sediment. *Chironomus* immobilisation test (OECD TG 235 2011) is based on OECD TG 202 (2004). First instar larvae of *C. riparius* are used in this test, since they have been shown to be the most sensitive larval stage. Further, this instar is free swimming and therefore not stressed by the absence of sediment. Although *C. riparius* is the preferred species, *C. dilutus* or *C. yoshimatsui* may also be used. Immobilisation of larvae is recorded at 24 and 48 h.

The sediment-water chironomid life-cycle toxicity test (OECD TG 233 2010) is designed to assess the effects of prolonged exposure of chemicals to the life-cycle of the sediment-dwelling freshwater dipteran *Chironomus* sp. First instar chironomid larvae are exposed to five concentrations of the test chemical in sediment-water systems. The test substance is spiked into the water or alternatively, the sediment, and first instar larvae are subsequently introduced into test beakers in which the sediment and water concentrations have been stabilised. Chironomid emergence, time to emergence, and sex ratio of the fully emerged and alive midges are assessed. The viability of the 2nd generation is also assessed. No defined period of the test is defined, but the emergence period of 20 to 65 days is acceptable. In ENM related studies, other acute end-points beside immobility and mortality were also applied. For example, these are growth (Waissi-Leinonen et al. 2015, Waissi et al. 2017) and avoidance behaviour (Oberholster et al. 2011) (Table 10).

Table 10. Overview of the different toxicity tests used for testing of ENMs, with details on the observed endpoints, the determined test parameter and the references which used the test.

Endpoint	Toxicity Tests	Test Parameter	References
Immobility	*Chironomus riparius* acute toxicity test (48 h) (OECD TG 235)	No movement after gentle agitation	Lee et al. 2016
Mortality	*Chironomus tentans* chronic toxicity test (10 d)	No movement after gentle agitation	Oberholster et al. 2011
Growth rate	*Chironomus riparius* chronic toxicity test (10 d)	Body length; head capsule length and head capsule width	Waissi et al. 2017, Waissi-Leinonen et al. 2015
Avoidance behaviour	*Chironomus tentans* chronic toxicity test (10 d)	Test organisms entering the sediment were used as a measure of avoidance behaviour	Oberholster et al. 2011
Reproduction	*Chironomus riparius* chronic toxicity test (42 d)	Time to first emergence, number of adults emerged from larvae	Waissi et al. 2017, Waissi-Leinonen et al. 2015
	Chironomus riparius chronic toxicity test (25 d) (OECD TG 233)	Time to first emergence, number of adults emerged from larvae	Lee et al. 2016

Advantages and Limitations for ENM Testing

Advantages

C. riparius is a non-biting sediment-dwelling, detritus-feeding organism, and its larvae are widely used in sediment toxicity experiments of conventional chemicals, but not yet so commonly for ENMs (Meregalli et al. 2000, Taenzler et al. 2007, Waissi-Leinonen et al. 2015). Chironomids are ecologically relevant because of their widespread distribution, numerical abundance and ecological importance (Taenzler et al. 2007). They can easily be cultured under laboratory conditions, and have a relatively short life cycle. *C. riparius* is a midge whose first three life stages take place under aquatic conditions (egg stage, four larval stages and pupal stage). The larval stage is of special interest due to its complete ongoing metamorphosis during the instar phases, which plays an important role in the life cycle. Likewise, larvae are in contact with the sediment during the entire larval stage (Taenzler et al. 2007). This habitat specific characteristic makes them appropriate for sediment testing of ENMs, because ENMs commonly settle out of the suspension and sediment on the ground. Currently, the number of acute ecotoxicity studies far outnumbers the chronic exposures. Using this insect toxicity test could provide necessary information regarding the chronic aquatic toxicity of ENMs.

Limitations, Needs for Optimization and Future Challenges

Some suggestions regarding the sediment testing with *Chironomus* sp. were given (Handy et al. 2012). The authors suggest using artificial sediment instead of natural sediment to standardise the sediment composition which affects the ENMs' properties. It was also suggested that the mode of dosing the sediment with ENMs should be harmonised. In the chironomid test, it is also possible to spike the overlying water with the test chemical. In the case of ENMs, this scenario may be more environmentally relevant (Waissi-Leinonen et al. 2012, Waissi-Leinonen et al. 2015). Nanomaterials will most probably settle out of the suspension resulting in the deposition on the sediment surface. The renewal of sediment to prevent ENM modification was not estimated as a feasible option. Also, if food is added during the test, it is suggested that this is done prior to water changes; although it cannot be prevented that food will settle on the sediment. No specific proposals regarding the adaptation of *Chironomus* OECD tests have previously been suggested by Hund-Rinke et al. (2016).

Aquatic Nematodes

Species

Caenorhabditis elegans is the most commonly used nematode species used in nanoecotoxicology (Juganson et al. 2015). It is a free-living soil nematode mainly found in the interstitial waters of soil particles, for example, pore water. This species is in principle considered as soil test organism (Donkin and Dusenbery 1993, Roh et al. 2009, Khare et al. 2011). It is also considered as such by ASTM (ASTM 2012). However, most commonly, these organisms are experimentally exposed to aqueous solution of ENMs and indeed many times suggested also as aquatic test species (Williams and Dusenbery 1990, Boyd and Williams 2003, Clavijo et al. 2016). Also, the international standard ISO 10872:2010 (2010) on *C. elegans* considers it both an aquatic and soil test organism. As per our knowledge, the most commonly applied genuine aquatic nematode toxicity test model for ENMs is the benthic sediment-dwelling organism *Lumbriculus variegatus*.

General Principles of the Toxicity Tests

ISO 10872:2010 (2010) specifies a method for determining the toxicity of environmental samples on growth, fertility and reproduction of *C. elegans*. First juvenile stage nematodes are exposed to aqueous suspension/dispersion of test chemical for 96 h in the dark. At the end of the test, the growth of animals is assessed by measuring their length. Reproduction is assessed as the number of young nematodes per adult hermaphrodite. Details regarding the feeding and obtaining the required developmental stage of *C. elegans* is described in Hanna et al. (2016). Although the ISO method for *C. elegans* toxicity testing is available, it appears to only have been used for testing nano FeOx (Höss et al. 2015) and TiO_2 (Angelstorf et al. 2014), but the majority of other ENM related studies with *C. elegans* have used non-standardised protocols with variable exposure set-ups and durations (Hanna et al. 2016). Only the endpoints tested according to the standard are described here (Table 11).

The *Lumbriculus* toxicity test is standardised according to OECD (OECD TG 225 2007), but no ISO standard exists. Adult worms of similar weight are exposed to a series of toxicant concentrations applied to the sediment phase of a sediment-water system. Artificial sediment and reconstituted water

Table 11. Overview of the different toxicity tests used for testing of ENMs, with details on the observed endpoints, the determined test parameter and the references, which used the test.

Endpoint	Toxicity Tests	Test Parameter	References
Growth	*Caenorhabditis elegans* chronic toxicity test (96 h) (ISO 10872:2010)	Body length.	Höss et al. 2015, Angelstorf et al. 2014, Pakarinen et al. 2011
	Chronic *Lumbriculus* toxicity test (28 d) (OECD TG 225)	Dry body weight.	Pakarinen et al. 2011
	Chronic *Lumbriculus* toxicity test (28 d) (OECD TG 225)	Total biomass including adult and young worms.	Pakarinen et al. 2011, Rajala et al. 2016
Mortality	Chronic *Lumbriculus* toxicity test (28 d) (OECD TG 225)	No reaction after a gentle mechanical stimulus.	Pakarinen et al. 2011, Rajala et al. 2016
	Lumbriculus acute toxicity test (24 h)	No reaction after a gentle mechanical stimulus.	Khan et al. 2015
Feeding	Chronic *Lumbriculus* toxicity test (28 d) (OECD TG 225)	Egestion rate as mass of fecal pellets.	Pakarinen et al. 2011, Rajala et al. 2016
Reproduction	*Caenorhabditis elegans* chronic toxicity test (96 h) (ISO 10872:2010)	Number of juvenile offspring.	Höss et al. 2015, Angelstorf et al. 2014, Hanna et al. 2016
	Chronic *Lumbriculus* toxicity test (28 d) (OECD TG 225)	An increase in worm number.	Pakarinen et al. 2011

should be used as media. The test animals are exposed to the sediment-water systems for a period of 28 days. The sediment is amended with a food source to ensure that the worms will grow and reproduce under control conditions. In this way it is ensured that the test animals are exposed through the water and sediment as well as by their food. After the exposure, the reproduction and animal biomass are assessed (Table 11). Some studies also report acute (24 h) exposure of *L. variegatus* to assess the mortality (Khan et al. 2015).

Advantages and Limitations for ENM Testing

Advantages

C. elegans is already an established model for environmental and developmental toxicological research of conventional chemicals (Höss et al. 2015, Clavijo et al. 2016). It has been recently suggested as a great model to test the toxicity of ENMs (Hanna et al. 2016). These nematodes feed on bacteria and are mainly found in microbe-rich environments. It is well suited for toxicity testing due to easy culturing, short generation time, transparent body and well known physiology. An additional advantage is that it grows both in terrestrial and aquatic environments (Höss et al. 2015). It can therefore be used to test sediment, soil and water. Apart from toxicity studies, it is an useful model to study the body distribution of ENMs (Qu et al. 2011).

Aquatic oligochaetes, such as *L.variegatus*, play an important role in the sediments of aquatic systems, both as a consumers of small benthic food particles (such as algae, detritus, bacteria and fungi) and as a prey item for other species (Xie et al. 2016). This species has been used to study the trophic transfer of contaminants and is well suited for such work due to its ability to accumulate contaminants (Ankley et al. 1994). The test organism is exposed to the test substance via all possible uptake routes (e.g., contact with, and ingestion of contaminated sediment particles, but also via pore water and overlying water). Sedimentary environments typically contain high organic matter content and microbial cells attached as biofilms. These efficiently sorb ENMs, therefore the sediment surface is an environment with potentially high bioavailability of ENMs (Handy et al. 2012). This exposure mode is therefore relevant for *L. variegatus*. Because they ingest small food particles they may regularly come in contact with ENMs.

Limitations, Needs for Optimization and Future Challenges

Sediment toxicity tests are most relevant for those ENMs that are unstable in test medium (Petersen et al. 2015). The test animal should be selected in line with its type of feeding behaviour and accessibility of ENMs (Handy et al. 2012).

The following modifications of the OECD TG 225 *Lumbriculus* sp. toxicity test were suggested (Handy et al. 2012, Handy et al. 2012, Hund-Rinke et al. 2016). The discussion was about how to prevent the sorption of ENMs to organic material in the sediment. The suggestion was to decrease the peat content from 5 to 2 percentage. In such a way, the bioavailability of ENMs to organisms would be increased. Upon such alteration of organic matter content, no alterations in the behaviour and reproduction of control group of *Lumbriculus* was observed (Hund-Rinke et al. 2016). Another issue was raised about the type of sediment used. Handy et al. (2012) suggested that artificial sediment should be used instead of natural sediment to standardise the sediment composition which affects the ENMs' properties. It was suggested that ENM aging during the test exposure should be considered. However the renewal of test medium was not advised. The food should be added to the sediment prior to water changes. The need for harmonised mixing protocol of ENMs into the sediment was pointed out.

As most of the existing standards, the *C. elegans* toxicity test is designed for testing of dissolved chemicals and at the moment there is no guidance on how to use this test for ENMs (Hanna et al. 2016). It has been suggested that some modifications of the existing ISO 10872:2010 (ISO 10872:2010

2010) could lead to more reproducible results. For example, it was reported that the viability and concentration of bacteria *E. coli* which is used to feed the nematodes plays a role in nematode growth. A more precise and robust method for the quantification of the bacteria concentration could help decrease the variability of the assay. Also, media composition affects the toxicity outcome. It is suggested that additional testing of the robustness of this test with different ENMs may reveal other important biases or limits to the applicability of this test which should also be considered (Hanna et al. 2016).

Amphibians

Species

Generally, amphibians are among less commonly used test organisms in ecotoxicology and nanoecotoxicology. Only three different species were reported in nanoE-Tox database (Juganson et al. 2015). The most commonly used is African clawed frog *Xenopus laevis*. Most probably this is due to the fact that the test is standardized (ASTM E1439) – 98; Standard Guide for Conducting the Frog Embryo Teratogenesis Assay-Xenopus (FETAX) (ASTM E1439 - 12 2012). Also, one potential reason is that these are vertebrates and thus subjected to strict laws regarding the use as experimental animals. Therefore, the AMPHITOX, A Customized Set of Toxicity Tests Employing Amphibian Embryos, was also developed as an alternative *in vitro* test. An overview of the different species used is given in Table 12.

Table 12. Overview of different amphibian species used for toxicity testing of ENMs, their common name, habitat and lifes-stage used, and the corresponding references. The number of publications as retrieved in ISI Thomson Reuters Web of Science using keywords "species name" and "nano*" are shown, number in brackets indicate number of publications in Web of Science category 'Environmental Sciences' (accessed December 4th 2017). The references retrieved by the detailed search as depicted in the introduction to this chapter are given in the last column.

Species	Common Name, Habitat, Life-Stage Used	No. of Publications	References
Ambystoma mexicanum	Axolotl, Mexican canals and wetlands, larvae	3 (–)	Mouchet et al. 2007
Lithobates sylvaticus	Wood frog, North America, larvae	1 (1)	Fong et al. 2016
Pleurodeles waltl	Sharped-ripped salamander, Western Europe, larvae	2 (–)	Bour et al. 2017
Rhinella arenarum	Argentine common toad, South America, larvae	5 (3)	Svartz et al. 2017
Xenopus laevis	African clawed frog, Sub-Saharan Africa (introduced: North America, South America, and Europe), embryo/larvae	240 (20)	Mouchet et al. 2008, Nations et al. 2015, 2011, Perelshtein et al. 2015

General Principles of the Toxicity Tests

Mortality, development and DNA damage are the most commonly assessed endpoints in amphibians after ENM exposure (Table 13). The protocol with *Xenopus laevis* is standardized (FETAX- Frog Embryo Teratogenesis Assay Xenopus) (ASTM E1439 - 12 2012). However, most of the authors used non-standardized toxicity tests with different amphibian species. Most often, lethal or sublethal effects on amphibian larvae were analyzed. Additionally, morphological effects, behaviour and also, neurotoxic endpoints can be assessed. Thereby the amphibian larvae can be exposed over different time spans, that is, acute (96 h) to chronic. In one study, the metamorphosis after ENM exposure was analyzed. Therefore the time point of metamorphosis, as well as the weight of the larvae were assessed (Fong et al. 2016). A Micronucleus (MN) assay is also adopted by OECD (OECD 1997, ISO 21427-1:2006, 2006). Blood samples of the exposed individuals were obtained and the number of erythrocytes containing one MN or more in a total sample of 1,000 erythrocytes per larva was

Table 13. Overview of the different toxicity tests used for testing of ENMs, with details on the observed endpoints, the determined test parameter and the references, which used the test.

Endpoint	Toxicity Test	Test Parameter	References
Mortality	Customized set of toxicity tests employing amphibian embryos (AMPHITOX test)	Lethal and sublethal effect, behavior	Svartz et al. 2017
	Frog embryo teratogenesis toxicity test Xenopus (FETAX Assay) (ASTM E1439 - 12, 2012)	Mortality, malformations, stage, snout vent length, and total body length	Nations et al. 2011, Perelshtein et al. 2015
	Toxicity assessment on amphibian larvae	Mortality, growth inhibition, behavior	Bour et al. 2017, Mouchet et al. 2007, Mouchet et al. 2008, Nations et al. 2015
Development	Larval metamorphosis	Metamorphosis (time, mass)	Fong et al. 2016
DNA damage	Micronucleus assay (OECD 1997, ISO 21427-1: 2006, 2006)	Formation of micronuclei (Microscopy)	Bour et al. 2017, Mouchet et al. 2008, Mouchet et al. 2007

determined. MN are formed during the anaphase of mitotic cell divisions from chromosomal fragments, or whole chromosomes, that are "left behind" when the nucleus divides. After the telophase, these fragments may not be included in the nuclei of daughter cells, and form single or multiple micronuclei in the cytoplasm (Bour et al. 2017, Mouchet et al. 2007, 2008).

Advantages and Limitations for ENM Testing

Advantages

The general amphibian eggs and tadpoles are easy to obtain in high numbers. They are easy to observe with regard to mobility and malformations. Furthermore, they have a biphasic life-cycle which allows to evaluate effects on metamorphosis. Amphibia are vertebrates, and due to genetic homology with humans, pathways of toxicity are conserved. Hence, results obtained for amphibian allow a good prediction of ENM mode-of-action in humans as well. The larvae are mobile and should hence get into contact with ENMs under test conditions.

Limitations, Needs for Optimization and Future Challenges

Most studies involving amphibians were performed in tap water or distilled water amended with physiologically important salts. A couple of studies described the agglomeration of ENMs with subsequent sedimentation in a time dependent manner and hence, transient exposure conditions (Mouchet et al. 2008, Nations et al. 2011, Fong et al. 2016, Svartz et al. 2017, Hyung et al. 2007). Hence, the exposure conditions in relation to agglomeration and sedimentation of ENMs should be monitored. To prevent sedimentation of ENMs following the agglomeration, one study placed an air diffuser in the water to promote ENM dispersal (Mouchet et al. 2007). Also, daily renewal of the medium was considered to allow working in more controlled conditions (Hyung et al. 2007). In addition, one study observed the stabilization of CNTs in suspension by the addition of food and the progressive release of natural organic matter by the animals (Hyung et al. 2007).

Attachment of ENMs to larvae gills as well as uptake into the gut was observed (Mouchet et al. 2007). It was suspected that the blockage of gills leads to anoxia in *Xenopus*, pointing to the fact that physical or mechanical effects of ENMs occur in amphibians. Low doses of metals (i.e., copper) were reported to facilitate metamorphosis and growth (Nations et al. 2015), hence potentially falsifying test results.

In a microcosm experiment involving several organisms, an increase in toxicity of CeO_2 ENM was observed, compared to single-exposure of larvae only (Bour et al. 2015). The reasons for this observation were not fully elucidated but it was suspected that trophic exposure may differ from direct exposure. In addition, transformations of ENMs may occur in a microcosm due to the presence of bacteria. Also the interaction with the other organisms (i.e., corona formation by biomolecules) may have an impact. However, similar effects were observed in microcosms for chemicals such as flame retardants and may hence be more system-imminent issues than test item-imminent. A general limitation in the micronucleus assay is the risk to falsely identify aggregates of ENMs as micronuclei fragments (Handy et al. 2012).

Fish

For the assessment of ENMs toxicity towards fish, whole organisms and fish embryos as well as fish cells or cell lines have been used. Due to the different aspects that need specific consideration for toxicity testing of ENMs, the organism tests and embryo tests (hereafter referred to as whole-organism fish test) are discussed separately from the tests involving fish cells.

Whole-Organism Fish Toxicity Tests

Species

Many different fish species were used in the literature for ENM toxicity testing. A summary is given in Table 14. Seven different fish species were identified in a detailed search on ENM toxicity related studies published in nanoE-Tox database (Juganson et al. 2015). Most often zebrafish *Danio rerio*, rainbow trout *Oncorhynchus mykiss* or carp *Cyprius carpio* at different life stages are used for ENM toxicological studies.

General Principles of the Toxicity Tests

For the assessment of lethal and/or sublethal effects in fish, different life stages are used, that is, adult fish, larvae or embryos. For all of these, toxicity tests standardized protocols are available (OECD TG 203, 1992, OECD TG 212, 1998, OECD TG 236, 2013). The most commonly assessed endpoints are mortality and behavioural changes. Additionally, in the fish embryo toxicity test several other, especially sublethal, effects can be analyzed, for example, hatching, edema, pigmentation, etc. In addition, there is also a specific toxicity test for swimming behaviour analysis of fish embryos available. Chen and colleagues analyzed the movement of zebrafish embryos after TiO_2 exposure in 24-well microplates with a video camera after mechanical stimulation. Afterwards, locomotion parameters (average velocity, maximum velocity and percent time active) were analyzed and used for toxicity assessment (Chen et al. 2011). A summary is presented in Table 15. Additionally, biomarkers were often used for further analysis of biochemical processes after ENM exposure, for example, enzyme activity.

Advantages and Limitations for ENM Testing

Advantages

Fish are vertebrates and due to genetic homology with humans pathways of toxicity are conserved. Hence, results obtained for fish allow a good prediction of ENM mode-of-action in humans as well. Fish tests and fish embryo tests are *in vivo* tests, and the use of whole organisms allows for the detection of effects which are specific to certain organs as well as changes to behaviour. In addition, the assessment of chronic effects is possible.

Table 14. Overview of different fish species used for toxicity testing of ENMs. The number of publications as retrieved in ISI Thomson Reuters Web of Science using keywords "species name" and "nano*" are shown, number in brackets indicate number of publications in Web of Science category 'Environmental Sciences' (accessed December 4th 2017). The references retrieved by the detailed search as depicted in the introduction to this chapter are given in the last column.

Species	Common Name, geographic region, Life-Stage Used	No. of Publications	References
Acipenser baerii	Siberian sturgeon, Siberian river basins, larvae	2 (1)	Ostaszewska et al. 2016
Carassius auratus	Goldfish, East Asia, adult fish	76 (25)	Ates et al. 2015, Benavides et al. 2016, Picado et al. 2015
Chapalichthys pardalis	Polka-dot splitfin, Mexico, adult fish	1 (1)	Valerio-Garcia et al. 2017
Cyprinodon variegatus	Sheepshead minnow, North and Central America, larvae	2 (1)	Griffitt et al. 2012
Cyprinus carpio	Common carp, Europe and Asia, adult fish	142 (65)	Gaiser et al. 2012, 2009, Jang et al. 2014, Lee et al. 2014, Mansouri et al. 2016, Hao et al. 2009
Cyprius carpio	Common carp, Europe and Asia, larvae	142 (65)	Hao and Chen 2012, Oprsal et al. 2015, Zhu et al. 2008
Danio rerio	Zebrafish, Himalayan region, adult fish	545 (205)	Bacchetta et al. 2016, Caceres-Velez et al. 2016, Li et al. 2017, Rocco et al. 2015, Weil et al. 2015, García-Cambero et al. 2013)
Danio rerio	Zebrafish, Himalayan region, larvae/embryo	545 (205)	Asharani et al. 2008, Bondarenko et al. 2016, Chen et al. 2011, Fouqueray et al. 2013, Henry et al. 2007, Jemec et al. 2016, Nam et al. 2014, Ribeiro et al. 2014, Schiwy et al. 2016, Song et al. 2015, Hoecke et al. 2009, Wang et al. 2015, Wang et al. 2012, Weil et al. 2015
Fundulus heteroclitus	Mummichog, North America, larvae/embryos	23 (5)	Bone et al. 2015
Gasterosteus aculeatus	Three-spined stickleback, Northern Hemisphere, adult fish	6 (3)	Sanders et al. 2008
Oncorhynchus mykiss	Rainbow trout, Asia and North America, adult fish	391 (139)	Connolly et al. 2016
Oncorhynchus mykiss	Rainbow trout, Asia and North America, larvae	391 (139)	Song et al. 2015
Oreochromis niloticus	Nile tilapia, Africa, adult fish	64 (22)	Kaya et al. 2017
Oryzias melastigma	Japanese rice fish, East and South Asia, adult fish	12 (4)	Wang and Wang 2014, Wu et al. 2010
Oryzias latipes	Japanese rice fish, East and South Asia, embryos	154 (69)	Li et al. 2009, Chen et al. 2013, Wu et al. 2010, Nam et al. 2014
Perca flavescens	Yellow perch, North America, larvae	7 (3)	Martin et al. 2017
Piaractus mesopotamicus	Paraná River pacu, South America, larvae	3 (2)	Clemente et al. 2013
Pimephales promelas	Fathead minnow, North America, adult fish	57 (42)	Bisesi et al. 2014, Zhu et al. 2006
Pimephales promelas	Fathead minnow, North America, embryos	57 (42)	Hoheisel et al. 2012, Laban et al. 2010, Kennedy et al. 2010
Pimephales promelas	Fathead minnow, North America, larvae	57 (42)	Song et al. 2015, Hoheisel et al. 2012
Poecilia reticulate	Guppy, South America, adult fish	2 (1)	Qualhato et al. 2017

Table 15. Overview of the different toxicity tests used for testing of ENMs, with details on the observed endpoint, the determined test parameter and the references, which used the test.

Endpoint	Toxicity Test	Test Parameter	References
Lethal and sublethal effects	Fish toxicity test (OCED TG 203)	Mortality, Behavioral changes	Ates et al. 2015, Benavides et al. 2016, Picado et al. 2015, Bacchetta et al. 2016, Caceres-Velez et al. 2016, Li et al. 2017, Rocco et al. 2015, Weil et al. 2015, Bisesi et al. 2014, Connolly et al. 2016, Gaiser et al. 2012, Jang et al. 2014, Lee et al. 2014, Mansouri et al. 2016, Kaya et al. 2017, Qualhato et al. 2017, Valerio-Garcia et al. 2017, Wu et al. 2010, Wang and Wang 2014, Hao et al. 2009, Zhu et al. 2006, Sanders et al. 2008, García-Cambero et al. 2013
	Fish early life stage test (OECD TG 212)	Mortality, Growth, Behavior	Henry et al. 2007, Zhu et al. 2008, Griffitt et al. 2012, Hao and Chen 2012, Hoheisel et al. 2012, Clemente et al. 2013, Fouqueray et al. 2013, Bone et al. 2015, Song et al. 2015, Wang et al. 2015, Weil et al. 2015, Oprsal et al. 2015, Ostaszewska et al. 2016, Martin et al. 2017, Boyle et al. 2015
	Fish embryo toxicity test (OECD TG 236)	Mortality, Movement, heartbeat and blood circulation, hatching, sublethal effects, heart rate, pigmentation	Asharani et al. 2008, Li et al. 2009, Hoecke et al. 2009, Wu et al. 2010, Kennedy et al. 2010, Laban et al. 2010, Chen et al. 2011, Wang et al. 2012, 2015, Hoheisel et al. 2012, Chen et al. 2013, Nam et al. 2014, Ribeiro et al. 2014, Weil et al. 2015, Bone et al. 2015, Bondarenko et al. 2016, Jemec et al. 2016, Boyle et al. 2015, Schiwy et al. 2016
Behavior	Swimming behavior test	Movement speed or distance	Chen et al. 2011

Limitations, Needs for Optimization and Future Challenges

The maintenance of fish requires, depending on species, comparable high volumes of water. Hence, for ENM exposures, large amounts of ENM are needed. For that reason, the flow-through method is often replaced by semi-static exposures (Handy et al. 2012). Particle sedimentation is an often observed phenomenon in fish studies. In addition, ENM agglomeration was facilitated by mucus secreted by the fish. Any deposited material needs to be removed from the tank during water changes in order to avoid up-concentration of ENM (Handy et al. 2012). To circumvent sedimentation, ENMs were also applied trophically (Fouqueray et al. 2013). The exposure durations are relatively long.

Because fish are vertebrates, for ethical reasons, often the use of fish embryos as replacement for animal testing is favoured. There are established tests available, for example, for zebrafish embryos and medaka. The use of embryos requires less sample volume. When working with embryos, one has to keep in mind that they are not mobile and reside at the bottom of the culture vessel, due to ENM sedimentation; the real concentration the embryos are exposed to may be much higher than the one nominally applied. Also, the embryos are protected by the chorion, which consist of a complex structure traversed by pores with a diameter of about 200 nm (Hart and Donovan 1983). Despite some ENMs being smaller than the pores, the chorion was shown to effectively sorb different types of nanomaterials and act as a barrier that protects the embryo (Böhme et al. 2017, Böhme et al. 2015). Hence, some experimenters considered dechorionation, that is, removal of the chorion before the onset of exposure to ENMs (Bodewein et al. 2016); however this is not a realistic exposure scenario.

With regard to the fish embryo test, it was suggested to adopt a stirring setup to avoid the sedimentation of ENMs on the embryos (Hund-Rinke et al. 2016). The embryos are placed in inserts which are connected to the exposure solution. The exposure solution is constantly agitated by a magnetic stirrer (Hund-Rinke et al. 2016). In addition, dechorionation was suggested, in order to avoid the protective role of the embryo and allow for direct exposure of the embryo.

Table 16. Overview of different fish cell lines used for toxicity testing of ENMs. The number. of publications as retrieved in ISI Thomson Reuters Web of Science using keywords "species name" and "nano*" are shown, number in brackets indicate number of publications in Web of Science category 'Environmental Sciences' (accessed December 4th 2017). The references retrieved by the detailed search as depicted in the introduction to this chapter are given in the last column.

Species	Common Name, Geographic region, Life-Stage Used	Cell Line	No. of Publications	References
Oncorhynchus mykiss	Rainbow trout, Asia and North America, hepatoma	RTH-149	4 (3)	Bermejo-Nogales et al. 2017, Connolly et al. 2015, Fernandez et al. 2013
Oncorhynchus mykiss	Rainbow trout, Asia and North America, gonade	RTG-2	13 (10)	Casado et al. 2013, Bermejo-Nogales et al. 2017, Connolly et al. 2015, Fernandez et al. 2013
Oncorhynchus mykiss	Rainbow trout, Asia and North America, gut	RTgutGC	3 (1)	Jemec et al. 2016
Oncorhynchus mykiss	Rainbow trout, Asia and North America, gill	RTgill	13 (2)	Kühnel et al. 2009
Oncorhynchus mykiss	Rainbow trout, Asia and North America, liver	RTL-W1	5 (4)	Bermejo-Nogales et al. 2017, Connolly et al. 2015, Fernandez et al. 2013
Oncorhynchus tshawytscha	Chinook salmon, North America and Asia, embryo	CHSE-214	5 (2)	Srikanth et al. 2015
Poeciliopsis lucida	Clearfin livebearer, Mexico, hepatoma	PLHC-1	9 (3)	Hernandez-Moreno et al. 2016, Bermejo-Nogales et al. 2017

Fish Cell-based Assays

Species

In addition to 'whole organism'- toxicity test, some fish cell lines were used for analyzing ENM toxicity. The applied cell lines are summarized in Table 16. Most of the cell lines were derived from rainbow trout organs. Furthermore, primary hepatocytes of rainbow trout are used to analyze ENM toxicity (Bermejo-Nogales et al. 2017, Connolly et al. 2015, Gaiser et al. 2012, 2009).

General Principles of the Bioassays

Overall, fish cell-based bioassays are used for determination of cytotoxic effects after ENM exposure. Therefore different colorimetric assays are available for analyzing cell viability. For example, the metabolic activity (Alamar Blue Assay), the plasma membrane integrity (CFDA-AM Assay) or the lysosomal uptake of the dye (Neutral Red Uptake Assay) are analyzed. In the MTT-Assay, the mitochondrial function, that is, the ability of viable cells to reduce 3-(4,5-dimethylthiazol-2-yl)-2,5 diphenyltetrazolium bromide into the blue formazan product, is evaluated photometrically.

Morphological changes of fish cell after ENM exposure, as well as particle attachment and uptake can be assessed by using microscopy, for example, Scanning Electron Microscopy (SEM). For determination of genotoxic/mutagenic effects, blood samples of fish were analyzed by the Comet Assay or the Micronucles Assay. An overview of different fish cell-based bioassays used for ENM testing is given in Table 17.

Table 17. Overview of the different bioassays used for toxicity testing of ENMs, with details on the observed endpoint, the determined test parameter and the references which used the bioassay.

Endpoint	Bioassay	Test Parameter	References
Cytotoxicity	Alamar blue assay	Metabolic activity via fluorescence	Fernandez et al. 2013, Bermejo-Nogales et al. 2017, Connolly et al. 2015, Jemec et al. 2016, Kühnel et al. 2009
	CFDA-AM assay	Plasma membrane integrity via fluorescence	Hernandez-Moreno et al. 2016, Bermejo-Nogales et al. 2017, Connolly et al. 2015, Fernandez et al. 2013, Jemec et al. 2016, Kühnel et al. 2009
	Kenacid Blue Protein (KBP) assay	Total protein via fluorescence	Fernandez et al. 2013
	LDH release assay	Cell damage via absorbance	Gaiser et al. 2009, 2012
	MTT Assay	Mitochondrial function via absorbance	Srikanth et al. 2015, Bermejo-Nogales et al. 2017, Fernandez et al. 2013, Hernandez-Moreno et al. 2016
	Neutral red uptake assay	Lysosomal uptake via fluorescence	Casado et al. 2013, Bermejo-Nogales et al. 2017, Hernandez-Moreno et al. 2016, Connolly et al. 2015, Jemec et al. 2016, Srikanth et al. 2015, Kühnel et al. 2009
DNA damage	Micronucleus assay	Formation of micronuclei	Caceres-Velez et al. 2016, Clemente et al. 2013, Qualhato et al. 2017
	Comet assay	Analysis of DNA strand breaks via gel electrophoresis	Caceres-Velez et al. 2016, Clemente et al. 2013, Qualhato et al. 2017, Rocco et al. 2015

Advantages and Limitations for ENM Testing

Advantages

Primary cells or cell lines derived from different organs of fish are an ethical favourable option compared to the use of fish. Cell cultivation is relatively easy but requires working under sterile conditions; depending on preferred temperature, fish cells grow slower compared to cell lines derived from mammals. The test durations are relatively short. In principle, bioassays involving cell lines are applicable to high-throughput testing.

Limitations, Needs for Optimization and Future Challenges

Some cell types grow in suspension, but most reside at the bottom of the culture vessels. Hence, ENMs that undergo sedimentation will deposit on top of the cells, and the real dose that cells are exposed to may be much higher than the nominal concentration. There are a couple of potential interferences of ENMs with cells as well as with the detection of specific endpoints. This applies to assays that use colorimetric and fluorimetric readouts, because ENMs may interfere with the respective compounds. In order to avoid false-negative or false-positive results by such interferences, appropriate positive and negative controls should be employed for each assay (Ong et al. 2014). In addition, optical interference may occur when ENMs have an intrinsic absorbance or fluorescence (Schultz et al. 2014).

Aquatic Macrophytes

Species

NanoE-Tox database reports 26 different plant species used in ENMs' ecotoxicity studies (Juganson et al. 2015) (data until 2015). Out of these, only four grow in an aquatic environment: freshwater

Table 18. Overview of different aquatic macrophytes used for toxicity testing of ENMs. The number of publications as retrieved in ISI Thomson Reuters Web of Science using keywords "species name" and "nano*" are shown, number in brackets indicate number of publications in Web of Science category 'Environmental Sciences' (accessed December 4th 2017). Some representative references of studies with nanomaterials are listed. The numbers of publications retrieved by the detailed search are given in brackets.

Species	Common Name, Habitat	No. Publications	References
Lemna minor	Duckweed; common duckweed or lesser duckweed; floating plant; growing in dense colonies, forming a mat on the water surface	56 (20)	Gubbins et al. 2011, Pereira et al. 2017
Lemna gibba	Duckweed; gibbous duckweed, swollen duckweed, or fat duckweed; habitat same as above	20 (4)	Oukarroum et al. 2013
Spirodela polyrhiza	Duckweed; common duckmeat, greater duckweed, or common duckweed, habitat same as above	15 (1)	Jiang et al. 2012
Oryza sativa	Asian rice; fields that are flooded for part of the growing season	200* (1)	Rico et al. 2013
Schoenoplectus tabernaemontani	Bulrush; moist and wet habitat, and sometimes in shallow water	3 (0)	Zhang et al. 2015
Elodea Canadensis	Pondweed; submergent macrophyte	5 (0)	Johnson et al. 2011

*This count is overestimated because it mostly includes the use of nanomaterials in agriculture.

duckweeds *Lemna gibba* and *Lemna minor,* wetland rice *Oryza sativa* and bulrush *Schoenoplectus tabernaemontani.* Beside these species, ENM studies using aquatic plants *Elodea Canadensis* (Koetsem et al. 2016), *Spirodela polyrhiza* (Jiang et al. 2014) and *Lemna paucicostata* (Kim et al. 2011) were also found. According to the Thomson Reuters Web of Science (WoS), there are currently 83 studies retrieved when using the search term "nano*" and lemna*. Most of the ecotoxicity studies use *L. minor* (Gubbins et al. 2011, Pereira et al. 2017), followed by *L. gibba* (Oukarroum et al. 2013). Generally, it has been noted previously that aquatic macrophytes are not commonly being used in ENM testing (Handy et al. 2012) (Table 18).

In many scientific studies, terrestrial species are grown hydroponically, meaning without soil using mineral nutrient solutions in water (Stampoulis et al. 2009, Hawthorne et al. 2012). Still, toxicity tests with *Allium cepa*, *Zea mays* and *Vicia faba* are considered terrestrial tests by international organisations (OECD TG 208, OECD TG 208 2006, OECD TG 227, OECD TG 227 2006) and they will not be addressed here. In some studies, plant seeds of different terrestrial plants are used as a test system and exposed on soaked filter paper (Lin and Xing 2007). Likewise, these toxicity tests are not considered as aquatic.

General Principles of the Lemna Growth Toxicity Test

Only one plant aquatic toxicity test is used routinely and is recommended by many international organisations: the duckweed *Lemna* sp. growth test (ISO 20079:2005, ISO 20079:2005, 2005, OECD TG 221, 2006, ASTM: E1415, 2012, EPA, 2012). It is also commercially available (Carolina Investigations®). Commercial kit also exists for another duckweed species, the *Spirodela polyrhiza* (SPIRODELA DUCKWEED TOXKIT^{TM}). Also records of ISO standard for *Myriophyllum aquaticum* (ISO 16191:2013 2013) were found, but no studies using this species to test ENM toxicity were found.

According to the OECD TG 221 (2006), the plant cultures of the genus *Lemna* are allowed to grow as monocultures in different concentrations of the test substance over a period of seven days. The principle is to assess the effect of a pollutant on the growth. Frond number is the primary measurement variable. At least one other measurement variable (total frond area, dry weight or fresh weight) is also measured, since some substances may affect other measurement variables much more than frond numbers. Commonly, the root growth and chlorophyll content are also commonly assessed as end-points (Kalčíková et al. 2016). The organism used for this test is either *L. gibba* or *L. minor*. Plant material may be obtained from a culture collection, another laboratory or from the field.

According to ISO 20079:2005 (2005) the duckweed species *L. minor* is used as model organism for higher aquatic plants. This standard specifies this method for the determination of the growth-inhibiting response of duckweed (*Lemna minor*) specifically to substances and mixtures contained in water, treated municipal wastewater and industrial effluents. The subsequent inhibition of growth is calculated from the observation of frond number, frond area, chlorophyll and dry weight.

Advantages and Limitations for ENM Testing

Advantages

Duckweeds are floating freshwater plants, meaning that they are exposed to ENMs present in upper layers of the water body. They are primary producers and are widely distributed across the globe. *Lemna* sp. have gained broad acceptance as bioassays in ecotoxicological research (Park et al. 2013). They have simple structure and small size, rapid rates of growth, they are easy to culture and handle and are sensitive to a wide range of pollutants (Wang 1990, Park et al. 2013). Direct observation of all end-points is possible which enables easy handling and observation (Wieczerzak et al. 2016). The organisms are commercially available as kits and hence, easily accessible.

Limitations, Needs for Optimization and Future Challenges

Although the *Lemna* sp. growth inhibition test is an internationally standardised guideline, surprisingly little effort has been done to adapt the existing protocols for ENM testing. For example, this protocol was not among the OECD TGs that were considered within the MARINA project (Hund-Rinke et al. 2016). Also no proposals regarding this toxicity test was presented in related scientific literature (Handy et al. 2012, Handy et al. 2012, Hjorth et al. 2016, Petersen et al. 2015).

For this toxicity test, some general modifications that were suggested for other aquatic bioassays could apply. For example, the stability of ENMs during testing should be ensured. The use of low ionic strength medium was suggested to prevent sedimentation and hence ensure exposure of lemna situated in the upper layer of the water (Gubbins et al. 2011). Also, it has been shown that ENMs adsorb onto plants which may result in physical impact (Pereira et al. 2017). Similarly, as in the case of algae, ENMs may adsorb to leave surface and obstruct the light (physical shading), causing decays in light absorption and consequently, inhibiting the growth of plants (Handy et al. 2012). So, here some steps could also be undertaken to consider shading effect as has been done for algae (Sørensen et al. 2015).

Also some previous efforts were already undertaken to adapt the standard *Lemna* growth toxicity test (AFNOR 1996, ISO 20079:2005, 2005, OECD TG 221, 2006, ASTM: E1415, 2012, EPA, 2012) for conventional chemicals. Park et al. (2013) suggested a simple, rapid, cost-effective, sensitive and precise 48 h toxicity test where re-growth of roots is used as an endpoint. In this bioassay, the roots are excised prior to exposure and subsequently, newly developed roots are measured after toxicant exposure. These authors showed that root growth is a more sensitive endpoint than frond number. Since roots are the main *Lemna* organ exposed to ENM testing, the root growth is a relevant endpoint.

Bioassays and Toxicity Tests to Assess Hazard of ENMs—Summary & Conclusion

As overviewed in this chapter, there are various bioassays and toxicity tests that have been used to assess the potential hazardous effect of ENMs. Species from different trophic levels, habitats and with different feeding behaviours were employed. In general, none of these tests are to be considered as inappropriate for ENM hazard assessment. However, as illustrated in this chapter, almost all of these tests are prone to certain interferences with ENMs at some point in the test procedure. Currently, however, numerous approaches to deal with these interferences or circumvent them when conducting the test are reported in the scientific literature.

These novel insights are now also incorporated into harmonized and finally, standardized protocols. Currently, it is in the hands of individual experimenters to select out of many recommendations an appropriate experimental design. With regard to regulatory testing of ENMs, however, harmonized approaches are needed. The current practice it to use test guidelines established for chemicals and adopt them based on guidance provided by regulatory agencies as well as the scientific literature. There is an ongoing debate on OECD level, whether test guidelines specifically tailored to ENMs are needed, or whether the ENM-specific peculiarities that need consideration are summarized in overarching guidance documents (e.g., OCED 2012, OECD 2002). Such guidance was also provided by ECHA (ECHA 2017) in order to allow the appropriate assessment of ENMs under REACH legislation. For most existing organism tests, such guidance will be sufficient, but needs for novel test guidelines were identified in the field of environmental fate and behaviour of ENMs (e.g., new TG318, OECD 2017). However, specific attention should be paid to organism groups that are exposed to ENMs with high likelihood of exposure, for example, soil and sediment organisms. At the moment, standards exist only for *Chironomus* sp. and *Lumbriculus*. These tests may be complemented by tests employing, for example, amphipods and mollusks, which are not commonly applied, but may yield valuable additional information on the effects of ENMs towards environmental organisms.

Currently, various efforts for standardization of test protocols for ENM specific toxicity tests and bioassays employing various organisms are under way (ISO, CEN, etc.). Such harmonization efforts will also support modelling approaches to improve the prediction of ENM toxicity (nano-QSAR approaches). So far, the variations in test design hamper the generalization of test results for such approaches (Chen et al. 2015).

Acknowledgements

The authors acknowledge the support by the DaNa2.0 project (Data and knowledge on nanomaterials—Evaluation of socially relevant scientific facts) funded by the German Federal Ministry for Education and Research (BMBF), grant no. 03X0131. AJK is financed by research programme P1-0184 (Slovenian Research Agency).

References

Adam, N., A. Vakurov, D. Knapen and R. Blust. 2015. The chronic toxicity of CuO nanoparticles and copper salt to Daphnia magna. J. Hazard. Mater. 283: 416–422.

Ali, D., S. Alarifi, S. Kumar, M. Ahamed and M.A. Siddiqui. 2012. Oxidative stress and genotoxic effect of zinc oxide nanoparticles in freshwater snail *Lymnaea luteola* L. Aquat. Toxicol. 124: 83–90.

Andreani, T., V. Nogueira, V.V. Pinto, M.J. Ferreira, M.G. Rasteiro, A.M. Silva et al. 2017. Influence of the stabilizers on the toxicity of metallic nanomaterials in aquatic organisms and human cell lines. Sci. Total Environ. 607: 1264–1277.

Angelstorf, J.S., W. Ahlf, F. von der Kammer and S. Heise. 2014. Impact of particle size and light exposure on the effects of TiO_2 nanoparticles on Caenorhabditis elegans. Environ. Toxicol. Chem. 33(10): 2288–2296.

Ankley, G.T., E.N. Leonard and V.R. Mattson. 1994. Prediction of bioaccumulation of metals from contaminated sediments by the oligochaete, Lumbriculus variegatus. Water Res. 28(5): 1071–1076.

Artells, E., J. Issartel, M. Auffan, D. Borschneck, A. Thill, M. Tella et al. 2013. Exposure to cerium dioxide nanoparticles differently affect swimming performance and survival in two daphnid species. PLoS One. 8(8): e71260.

Aruoja, V., S. Pokhrel, M. Sihtmäe, M. Mortimer, L. Mädler and A. Kahru. 2015. Toxicity of 12 metal-based nanoparticles to algae, bacteria and protozoa. Environ. Sci. Nano. 2(6): 630–644.

Asharani, P.V., Y. Lian Wu, Z. Gong andS. Valiyaveettil. 2008. Toxicity of silver nanoparticles in zebrafish models. Nanotechnology. 19(25): 255102.

ASTM. 2012. Standard Guide for Conducting the Frog Embryo Teratogenesis Assay-Xenopus (FETAX).

ASTM: E1415. 2012. Standard Guide for Conducting Static Toxicity Tests With Lemna gibbaNo Title.

ASTM E1439 - 12. 2012. Standard Guide for Conducting the Frog Embryo Teratogenesis Assay-Xenopus (FETAX).

Ates, M., Z. Arslan, V. Demir, J. Daniels and I.O. Farah. 2015. Accumulation and Toxicity of CuO and ZnO Nanoparticles Through Waterborne and Dietary Exposure of Goldfish (Carassius auratus). Environ. Toxicol. 30(1): 119–128.

Auffan, M., D. Bertin, P. Chaurand, C. Pailles, C. Dominici, J. Rose et al. 2013. Role of molting on the biodistribution of CeO_2 nanoparticles within Daphnia pulex. Water Res. 47(12): 3921–3930.
Bacchetta, C., G. Lopez, G. Pagano, D.T. Muratt, L.M. de Carvalho and J.M. Monserrat. 2016. Toxicological Effects Induced by Silver Nanoparticles in Zebra Fish (Danio Rerio) and in the Bacteria Communities Living at Their Surface. Bull. Environ. Contam. Toxicol. 97(4): 456–462.
Benavides, M., J. Fernandez-Lodeiro, P. Coelho, C. Lodeiro and M.S. Diniz. 2016. Single and combined effects of aluminum (Al_2O_3) and zinc (ZnO) oxide nanoparticles in a freshwater fish, Carassius auratus. Environ. Sci. Pollut. Res. 23(24): 24578–24591.
Benelli, G. 2016. Plant-mediated biosynthesis of nanoparticles as an emerging tool against mosquitoes of medical and veterinary importance: a review. Parasitol. Res. 115(1): 23–34.
Bennett, S.W., A. Adeleye, Z.X. Ji and A.A. Keller. 2013. Stability, metal leaching, photoactivity and toxicity in freshwater systems of commercial single wall carbon nanotubes. Water Res. 47(12): 4074–4085.
Bermejo-Nogales, A., M. Connolly, P. Rosenkranz, M.L. Fernandez-Cruz and J.M. Navas. 2017. Negligible cytotoxicity induced by different titanium dioxide nanoparticles in fish cell lines. Ecotoxicol. Environ. Saf. 138: 309–319.
Bhuvaneshwari, M., D. Kumar, R. Roy, S. Chakraborty, A. Parashar, A. Mukherjee et al. 2017. Toxicity, accumulation, and trophic transfer of chemically and biologically synthesized nano zero valent iron in a two species freshwater food chain. Aquat. Toxicol. 183: 63–75.
Binh, C.T.T., T. Tong, J.F. Gaillard, K.A. Gray and J.J. Kelly. 2014. Acute effects of TiO_2 nanomaterials on the viability and taxonomic composition of aquatic bacterial communities assessed via high-throughput screening and next generation sequencing. PLoS One. 9(8).
Bisesi, J.H., J. Merten, K. Liu, A.N. Parks, A. Afrooz, J.B. Glenn et al. 2014. Tracking and Quantification of Single-Walled Carbon Nanotubes in Fish Using Near Infrared Fluorescence. Environ. Sci. Technol. 48(3): 1973–1983.
Blinova, I., A. Ivask, M. Heinlaan, M. Mortimer and A. Kahru. 2010. Ecotoxicity of nanoparticles of CuO and ZnO in natural water. Environ. Pollut. 158(1): 41–47.
Bodewein, L., F. Schmelter, S. Di Fiore, H. Hollert, R. Fischer and M. Fenske. 2016. Differences in toxicity of anionic and cationic PAMAM and PPI dendrimers in zebrafish embryos and cancer cell lines. Toxicol. Appl. Pharmacol. 305: 83–92.
Böhme, S., M. Baccaro, M. Schmidt, A. Potthoff, H.-J. Stärk, T. Reemtsma et al. 2017. Metal uptake and distribution in the zebrafish (Danio rerio) embryo: differences between nanoparticles and metal ions. Environ. Sci. Nano. 4(5): 1005–1015.
Böhme, S., H.-J. Stärk, T. Reemtsma and D. Kühnel. 2015. Effect propagation after silver nanoparticle exposure in zebrafish (Danio rerio) embryos: a correlation to internal concentration and distribution patterns. Environ. Sci. Nano. 2(6): 603–614.
Bondarenko, O.M., M. Heinlaan, M. Sihtmäe, A. Ivask, I. Kurvet, E. Joonas et al. 2016. Multilaboratory evaluation of 15 bioassays for (eco)toxicity screening and hazard ranking of engineered nanomaterials: FP7 project NANOVALID. Nanotoxicology. 10(9): 1229–1242.
Bone, A.J., C.W. Matson, B.P. Colman, X.Y. Yang, J.N. Meyer and R.T. Di Giulio. 2015. Silver nanoparticle toxicity to atlantic killifish (fundulus heteroclitus) and caenorhabditis elegans: a comparison of mesocosm, microcosm, and conventional laboratory studies. Environ. Toxicol. Chem. 34(2): 275–282.
Bonvin, D., H. Hofmann and M. Mionić Ebersold. 2017. Assessment of nanoparticles' safety: corrected absorbance-based toxicity test. Analyst. 142(13): 2338–2342.
Booth, A., T. Storseth, D. Altin, A. Fornara, A. Ahniyaz, H. Jungnickel et al. 2015. Freshwater dispersion stability of PAA-stabilised-cerium-oxide nanoparticles and toxicity towards Pseudokirchneriella subcapitata. Sci. Total Environ. 505: 596–605.
Bouldin, J.L., T.M. Ingle, A. Sengupta, R. Alexander, R.E. Hannigan and R.A. Buchanan. 2008. Aqueous toxicity and food chain transfer of quantum Dots (TM) in freshwater algae and Ceriodaphnia dubia. Environ. Toxicol. Chem. 27(9): 1958–1963.
Bour, A., F. Mouchet, S. Cadarsi, J. Silvestre, D. Baque, L. Gauthier et al. 2017. CeO_2 nanoparticle fate in environmental conditions and toxicity on a freshwater predator species: a microcosm study. Environ. Sci. Pollut. Res. 24(20): 17081–17089.
Bour, A., F. Mouchet, L. Verneuil, L. Evariste, J. Silvestre, E. Pinelli et al. 2015. Toxicity of CeO_2 nanoparticles at different trophic levels–effects on diatoms, chironomids and amphibians. Chemosphere. 120: 230–236.
Boyd, W.A. and P.L. Williams. 2003. Comparison of the sensitivity of three nematode species to copper and their utility in aquatic and soil toxicity tests. Environ. Toxicol. Chem. 22(11): 2768–2774.
Boyle, D., H. Boran, A.J. Atfield and T.B. Henry. 2015. Use of an exposure chamber to maintain aqueous phase nanoparticle dispersions for improved toxicity testing in fish. Environ. Toxicol. Chem. 34(3): 583–588.
Bundschuh, M., J.P. Zubrod, D. Englert, F. Seitz, R.R. Rosenfeldt and R. Schulz. 2011. Effects of nano-TiO 2 in combination with ambient UV-irradiation on a leaf shredding amphipod. Chemosphere. 85(10): 1563–1567.
Caceres-Velez, P.R., M.L. Fascineli, C.K. Grisolia, E.C.D. Lima, M.H. Sousa, P.C. de Morais et al. 2016. Genotoxic and histopathological biomarkers for assessing the effects of magnetic exfoliated vermiculite and exfoliated vermiculite in Danio rerio. Sci. Total Environ. 551: 228–237.
Canesi, L., C. Ciacci, R. Fabbri, A. Marcomini, G. Pojana and G. Gallo. 2012. Bivalve molluscs as a unique target group for nanoparticle toxicity. Mar. Environ. Res. 76: 16–21.
Casado, M.P., A. Macken and H.J. Byrne. 2013b. Ecotoxicological assessment of silica and polystyrene nanoparticles assessed by a multitrophic test battery. Environ. Int. 51: 97–105.

Cerrillo, C., G. Barandika, A. Igartua, O. Areitioaurtena, A. Marcaide and G. Mendoza. 2015. Ecotoxicity of multiwalled carbon nanotubes: Standardization of the dispersion methods and concentration measurements. Environ. Toxicol. Chem. 34(8): 1854–1862.

Chen, D., X. Li, T. Soule, F. Yorio and L. Orr. 2016. Effects of solution chemistry on antimicrobial activities of silver nanoparticles against *Gordonia* sp. Sci. Total Environ. 566: 360–367.

Chen, P.J., W.L. Wu and K.C.W. Wu. 2013. The zerovalent iron nanoparticle causes higher developmental toxicity than its oxidation products in early life stages of medaka fish. Water Res. 47(12): 3899–3909.

Chen, T.-H., C.-Y. Lin and M.-C. Tseng. 2011. Behavioral effects of titanium dioxide nanoparticles on larval zebrafish (Danio rerio). Mar. Pollut. Bull. 63(5–12): 303–308.

Chial, B. and G. Persoone. 2002. Cyst-based toxicity tests XIII—Development of a short chronic sediment toxicity test with the ostracod crustacean Heterocypris incongruens: Methodology and precision. Environ. Toxicol. 17(6): 528–532.

Choi, M.H., Y. Hwang, H.U. Lee, B. Kim, G.W. Lee, Y.K. Oh et al. 2014. Aquatic ecotoxicity effect of engineered aminoclay nanoparticles. Ecotoxicol. Environ. Saf. 102: 34–41.

Clar, J.G., S.A. Gustitus, S. Youn, C.A.S. Batista, K.J. Ziegler and J.C.J. Bonzongo. 2015. Unique Toxicological Behavior from Single-Wall Carbon Nanotubes Separated via Selective Adsorption on Hydrogels. Environ. Sci. Technol. 49(6): 3913–3921.

Clavijo, A., M.F. Kronberg, A. Rossen, A. Moya, D. Calvo, S.E. Salatino et al. 2016. The nematode Caenorhabditis elegans as an integrated toxicological tool to assess water quality and pollution. Sci. Total Environ. 569: 252–261.

Clement, L., C. Hurel and N. Marmier. 2013. Toxicity of TiO_2 nanoparticles to cladocerans, algae, rotifers and plants - Effects of size and crystalline structure. Chemosphere. 90(3): 1083–1090.

Clement, L., A. Zenerino, C. Hurel, S. Amigoni, E.T. de Givenchy, F. Guittard et al. 2013. Toxicity assessment of silica nanoparticles, functionalised silica nanoparticles, and HASE-grafted silica nanoparticles. Sci. Total Environ. 450: 120–128.

Clemente, Z., V.L. Castro, L.O. Feitosa, R. Lima, C.M. Jonsson, A.H.N. Maia et al. 2013. Fish exposure to nano-TiO_2 under different experimental conditions: Methodological aspects for nanoecotoxicology investigations. Sci. Total Environ. 463: 647–656.

Clemente, Z., V.L. Castro, C.M. Jonsson and L.F. Fraceto. 2014. Minimal levels of ultraviolet light enhance the toxicity of TiO_2 nanoparticles to two representative organisms of aquatic systems. J. Nanoparticle Res. 16(8): 2559.

Connolly, M., M.L. Fernandez-Cruz, A. Quesada-Garcia, L. Alte, H. Segner and J.M. Navas. 2015. Comparative Cytotoxicity Study of Silver Nanoparticles (AgNPs) in a Variety of Rainbow Trout Cell Lines (RTL-W1, RTH-149, RTG-2) and Primary Hepatocytes. Int. J. Environ. Res. Public Health. 12(5): 5386–5405.

Connolly, M., M. Fernandez, E. Conde, F. Torrent, J.M. Navas and M.L. Fernandez-Cruz. 2016. Tissue distribution of zinc and subtle oxidative stress effects after dietary administration of ZnO nanoparticles to rainbow trout. Sci. Total Environ. 551: 334–343.

Couleau, N., D. Techer, C. Pagnout, S. Jomini, L. Foucaud, P. Laval-Gilly et al. 2012. Hemocyte responses of Dreissena polymorpha following a short-term *in vivo* exposure to titanium dioxide nanoparticles: preliminary investigations. Sci. Total Environ. 438: 490–497.

Croteau, M.-N., S.K. Misra, S.N. Luoma and E. Valsami-Jones. 2014. Bioaccumulation and toxicity of CuO nanoparticles by a freshwater invertebrate after waterborne and dietborne exposures. Environ. Sci. Technol. 48(18): 10929–10937.

Cui, R., Y. Chae and Y.-J. An. 2017. Dimension-dependent toxicity of silver nanomaterials on the cladocerans Daphnia magna and Daphnia galeata. Chemosphere. 185: 205–212.

Dabrunz, A., L. Duester, C. Prasse, F. Seitz, R. Rosenfeldt, C. Schilde et al. 2011. Biological surface coating and molting inhibition as mechanisms of TiO_2 nanoparticle toxicity in Daphnia magna. PLoS One. 6(5): e20112.

Dalai, S., S. Pakrashi, N. Chandrasekaran and A. Mukherjee. 2013. Acute toxicity of TiO_2 nanoparticles to Ceriodaphnia dubia under visible light and dark conditions in a freshwater system. PLoS One. 8(4): e62970.

DIN. 1999. Part 37: Determination of the inhibitory effect of water on the growth of bacteria (Photobacterium phosphoreum cell multiplication inhibition test) (L 37). Bio-assays (gr. L). German sta.

Docter, D., D. Westmeier, M. Markiewicz, S. Stolte, S.K. Knauer and R.H. Stauber. 2015. The nanoparticle biomolecule corona: lessons learned – challenge accepted? Chem. Soc. Rev. 44(17): 6094–6121.

Dogra, Y., K.P. Arkill, C. Elgy, B. Stolpe, J. Lead, E. Valsami-Jones et al. 2016. Cerium oxide nanoparticles induce oxidative stress in the sediment-dwelling amphipod Corophium volutator. Nanotoxicology. 10(4): 480–487.

Donkin, S.G. and D.B. Dusenbery. 1993. A soil toxicity test using the nematode Caenorhabditis elegans and an effective method of recovery. Arch. Environ. Contam. Toxicol. 25(2): 145–151.

Ducrot, V., C. Askem, D. Azam, D. Brettschneider, R. Brown, S. Charles et al. 2014. Development and validation of an OECD reproductive toxicity test guideline with the pond snail Lymnaea stagnalis (Mollusca, Gastropoda). Regul. Toxicol. Pharmacol. 70(3): 605–614.

ECHA. 2017. https://echa.europa.eu/guidance-documents/guidance-on-information-requirements-and-chemical-safety-assessment [WWW Document].

EPA. 2009. OCSPP 850.1735. 2009. Spiked Whole Sediment 10-Day Toxicity Test, Freshwater Invertebrates.

EPA, 2012. 2012. Ecological Effects Test Guidelines OCSPP 850.4400: Aquatic Plant Toxicity Test Using Lemna spp.Title.

Fabrega, J., R. Tantra, A. Amer, B. Stolpe, J. Tomkins, T. Fry et al. 2011. Sequestration of zinc from zinc oxide nanoparticles and life cycle effects in the sediment dweller amphipod Corophium volutator. Environ. Sci. Technol. 46(2): 1128–1135.

Fernandez, D., C. Garcia-Gomez and M. Babin. 2013. *In vitro* evaluation of cellular responses induced by ZnO nanoparticles, zinc ions and bulk ZnO in fish cells. Sci. Total Environ. 452: 262–274.

Fong, P.P., L.B. Thompson, G.L.F. Carfagno and A.J. Sitton. 2016. Long-term exposure to gold nanoparticles accelerates larval metamorphosis without affecting mass in wood frogs (Lithobates sylvaticus) at environmentally relevant concentrations. Environ. Toxicol. Chem. 35(9): 2304–2310.

Fouqueray, M., P. Noury, L. Dherret, P. Chaurand, K. Abbaci, J. Labille et al. 2013. Exposure of juvenile Danio rerio to aged TiO_2 nanomaterial from sunscreen. Environ. Sci. Pollut. Res. 20(5): 3340–3350.

Franklin, N.M., N.J. Rogers, S.C. Apte, G.E. Batley, G.E. Gadd and P.S. Casey. 2007. Comparative toxicity of nanoparticulate ZnO, bulk ZnO, and ZnCl2 to a freshwater microalga (Pseudokirchneriella subcapitata): The importance of particle solubility. Environ. Sci. Technol. 41(24): 8484–8490.

Gagné, F., J. Auclair, M. Fortier, A. Bruneau, M. Fournier, P. Turcotte et al. 2013. Bioavailability and immunotoxicity of silver nanoparticles to the freshwater mussel Elliptio complanata. J. Toxicol. Environ. Heal. Part A. 76(13): 767–777.

Gaiser, B.K., T.F. Fernandes, M.A. Jepson, J.R. Lead, C.R. Tyler, M. Baalousha et al. 2012. Interspecies comparisons on the uptake and toxicity of silver and cerium dioxide nanoparticles. Environ. Toxicol. Chem. 31(1): 144–154.

Gaiser, B.K., T.F. Fernandes, M. Jepson, J.R. Lead, C.R. Tyler and V. Stone. 2009. Assessing exposure, uptake and toxicity of silver and cerium dioxide nanoparticles from contaminated environments. Environ. Heal. 8.

Garaud, M., M. Auffan, S. Devin, V. Felten, C. Pagnout, S. Pain-Devin et al. 2016. Integrated assessment of ceria nanoparticle impacts on the freshwater bivalve Dreissena polymorpha. Nanotoxicology. 10(7): 935–944.

Garcia-Cambero, J.P., M.N. Garcia, G.D. Lopez, A.L. Herranz, L. Cuevas, E. Perez-Pastrana et al. 2013. Converging hazard assessment of gold nanoparticles to aquatic organisms. Chemosphere. 93(6): 1194–1200.

Geitner, N.K., S.M. Marinakos, C. Guo, N. O'Brien and M.R. Wiesner. 2016. Nanoparticle Surface Affinity as a Predictor of Trophic Transfer. Environ. Sci. Technol. 50(13): 6663–6669.

Gonçalves, S., M. D Pavlaki, R. Lopes, J. Hammes, J.A. Gallego-Urrea, M. Hassellöv et al. 2017. Effects of silver nanoparticles on the freshwater snail Physa acuta: The role of test media and snails' life cycle stage. Environ. Toxicol. Chem. 36(1): 243–253.

Gong, N., K.S. Shao, W. Feng, Z.Z. Lin, C.H. Liang and Y.Q. Sun. 2011. Biotoxicity of nickel oxide nanoparticles and bio-remediation by microalgae Chlorella vulgaris. Chemosphere. 83(4): 510–516.

Griffitt, R.J., N.J. Brown-Peterson, D.A. Savin, C.S. Manning, I. Boube, R.A. Ryan et al. 2012. Effects of chronic nanoparticulate silver exposure to adult and juvenile sheepshead minnows (Cyprinodon variegatus). Environ. Toxicol. Chem. 31(1): 160–167.

Gubbins, E.J., L.C. Batty and J.R. Lead. 2011. Phytotoxicity of silver nanoparticles to Lemna minor L. Environ. Pollut. 159(6): 1551–1559.

Hall, S., T. Bradley, J.T. Moore, T. Kuykindall and L. Minella. 2009. Acute and chronic toxicity of nano-scale TiO_2 particles to freshwater fish, cladocerans, and green algae, and effects of organic and inorganic substrate on TiO_2 toxicity. Nanotoxicology. 3(2): 91–97.

Handy, R.D., G. Cornelis, T. Fernandes, O. Tsyusko, A. Decho, T. Sabo-Attwood et al. 2012. Ecotoxicity test methods for engineered nanomaterials: Practical experiences and recommendations from the bench. Environ. Toxicol. Chem. 31(1): 15–31.

Handy, R.D., N. Van Den Brink, M. Chappell, M. Mühling, R. Behra, M. Dušinská et al. 2012. Practical considerations for conducting ecotoxicity test methods with manufactured nanomaterials: What have we learnt so far? Ecotoxicology. 21(4): 933–972.

Hanna, S.K., G.A. Cooksey, S. Dong, B.C. Nelson, L. Mao, J.T. Elliott et al. 2016. Feasibility of using a standardized Caenorhabditis elegans toxicity test to assess nanomaterial toxicity. Environ. Sci. 3(5): 1080–1089.

Hao, L.H. and L. Chen. 2012. Oxidative stress responses in different organs of carp (Cyprinus carpio) with exposure to ZnO nanoparticles. Ecotoxicol. Environ. Saf. 80: 103–110.

Hao, L.H., Z.Y. Wang and B.S. Xing. 2009. Effect of sub-acute exposure to TiO2 nanoparticles on oxidative stress and histopathological changes in Juvenile Carp (Cyprinus carpio). J. Environ. Sci. 21(10): 1459–1466.

Hart, N.H. and M. Donovan. 1983. Fine structure of the chorion and site of sperm entry in the egg of Brachydanio. J. Exp. Zool. 227(2): 277–296.

Hartmann, N.B., C. Engelbrekt, J. Zhang, J. Ulstrup, K.O. Kusk and A. Baun. 2012. The challenges of testing metal and metal oxide nanoparticles in algal bioassays: titanium dioxide and gold nanoparticles as case studies. Nanotoxicology. 7(6): 1082–1094.

Harwood, A.D., A.K. Rothert and M.J. Lydy. 2014. Using Hexagenia in sediment bioassays: Methods, applicability, and relative sensitivity. Environ. Toxicol. Chem. 33(4): 868–874.

Hawthorne, J., C. Musante, S.K. Sinha and J.C. White. 2012. Accumulation and phytotoxicity of engineered nanoparticles to Cucurbita pepo. Int. J. Phytoremediation. 14(4): 429–442.

Hazeem, L.J., M. Bououdina, S. Rashdan, L. Brunet, C. Slomianny and R. Boukherroub. 2016. Cumulative effect of zinc oxide and titanium oxide nanoparticles on growth and chlorophyll a content of *Picochlorum* sp. Environ. Sci. Pollut. Res. 23(3): 2821–2830.

Heinlaan, M., A. Ivask, I. Blinova, H.-C. Dubourguier and A. Kahru. 2008. Toxicity of nanosized and bulk ZnO, CuO and TiO_2 to bacteria Vibrio fischeri and crustaceans Daphnia magna and Thamnocephalus platyurus. Chemosphere. 71(7): 1308–1316.

Henry, T.B., F.M. Menn, J.T. Fleming, J. Wilgus, R.N. Compton and G.S. Sayler. 2007. Attributing effects of aqueous C-60 nano-aggregates to tetrahydrofuran decomposition products in larval zebrafish by assessment of gene expression. Environ. Health Perspect. 115(7): 1059–1065.

Hernandez-Moreno, D., L.X.Y. Li, M. Connolly, E. Conde, M. Fernandez, M. Schuster et al. 2016. Mechanisms underlying the enhancement of toxicity caused by the coincubation of zinc oxide and copper nanoparticles in a fish hepatoma cell line. Environ. Toxicol. Chem. 35(10): 2562–2570.

Hjorth, R., S.N. Sørensen, M.E. Olsson, A. Baun and N.B. Hartmann. 2016. A certain shade of green: Can algal pigments reveal shading effects of nanoparticles? Integr. Environ. Assess. Manag. 12(1): 200–202.

Hoheisel, S.M., S. Diamond and D. Mount. 2012. Comparison of nanosilver and ionic silver toxicity in Daphnia magna and Pimephales promelas. Environ. Toxicol. Chem. 31(11): 2557–2563.

Höss, S., B. Frank-Fahle, T. Lueders and W. Traunspurger. 2015. Response of bacteria and meiofauna to iron oxide colloids in sediments of freshwater microcosms. Environ. Toxicol. Chem. 34(11): 2660–2669.

Hu, C., Q. Wang, H. Zhao, L. Wang, S. Guo and X. Li. 2015. Ecotoxicological effects of graphene oxide on the protozoan Euglena gracilis. Chemosphere. 128: 184–190.

Hull, M.S., P. Chaurand, J. Rose, M. Auffan, J.-Y. Bottero, J.C. Jones et al. 2011. Filter-feeding bivalves store and biodeposit colloidally stable gold nanoparticles. Environ. Sci. Technol. 45(15): 6592–6599.

Hund-Rinke, K., A. Baun, D. Cupi, T.F. Fernandes, R. Handy, J.H. Kinross et al. 2016. Regulatory ecotoxicity testing of nanomaterials–proposed modifications of OECD test guidelines based on laboratory experience with silver and titanium dioxide nanoparticles. Nanotoxicology. 10(10): 1442–1447.

Hund-Rinke, K., K. Schlich, D. Kühnel, B. Hellack, H. Kaminski and C. Nickel. 2018. Grouping concept for metal and metal oxide nanomaterials with regard to their ecotoxicological effects on algae, daphnids and fish embryos. NanoImpact. 9: 52–60.

Hund-Rinke, K. and M. Simon. 2006. Ecotoxic effect of photocatalytic active nanoparticles TiO_2 on algae and daphnids. Environ. Sci. Pollut. Res. 13(4): 225–232.

Hyung, H., J.D. Fortner, J.B. Hughes and J.H. Kim. 2007. Natural organic matter stabilizes carbon nanotubes in the aqueous phase. Env. Sci Technol. 41(1): 179–184.

Irving, E.C. 2000. Ecotoxicological Responses of the Mayfly Baetis Tricaudatus to Dietary and Water-Borne Cadmium.

Ishwarya, R., B. Vaseeharan, S. Shanthi, S. Ramesh, P. Manogari, K. Dhanalakshmi et al. 2017. Green Synthesized Silver Nanoparticles: Toxicity Against Poecilia reticulata Fishes and Ceriodaphnia cornuta Crustaceans. J. Clust. Sci. 28(1): 519–527.

ISO 10706:2000. 2000. Water quality—Determination of long term toxicity of substances to Daphnia magna Straus (Cladocera, Crustacea).

ISO 10712:1995. 1995. Water quality—Pseudomonas putida growth inhibition test (Pseudomonas cell multiplication inhibition test.

ISO 10872:2010. 2010. Water quality—Determination of the toxic effect of sediment and soil samples on growth, fertility and reproduction of Caenorhabditis elegans (Nematoda).

ISO 11348-2:2009. 2009. Water quality—Determination of the inhibitory effect of water samples on the light emission of Vibrio fischeri (Luminescent bacteria test)—Part 2: Method using liquid-dried bacteria.

ISO 11350:2012. 2012. Water quality—Determination of the genotoxicity of water and waste water— Salmonella/microsome fluctuation test (Ames fluctuation test).

ISO 14371:2012. 2012. Water quality—Determination of fresh water sediment toxicity to Heterocypris incongruens (Crustacea, Ostracoda).

ISO 14380:2011. 2011. Water quality—Determination of the acute toxicity to Thamnocephalus platyurus (Crustacea, Anostraca.

ISO 16191:2013. 2013. Water quality—Determination of the toxic effect of sediment on the growth behaviour of Myriophyllum aquaticum.

ISO 16303:2013. 2013. Water quality—Determination of toxicity of fresh water sediments using Hyalella azteca.

ISO 20079:2005. 2005. ISO 20079:2005 Water quality—Determination of the toxic effect of water constituents and waste water on duckweed (Lemna minor) —Duckweed growth inhibition test.

ISO 20665:2008. 2008. Water quality—Determination of chronic toxicity to Ceriodaphnia dubia.

ISO 21338:2010. 2010. Water quality—Kinetic determination of the inhibitory effects of sediment, other solids and coloured samples on the light emission of Vibrio fischeri (kinetic luminescent bacteria test).

ISO 21427-1:2006. 2006. Water quality—Evaluation of genotoxicity by measurement of the induction of micronuclei—Part 1: Evaluation of genotoxicity using amphibian larvae.

ISO 6341:2012. 2012. Water quality—Determination of the inhibition of the mobility of Daphnia magna Straus (Cladocera, Crustacea)—Acute toxicity test.

ISO 8692:2012. 2017. Water quality—Fresh water algal growth inhibition test with unicellular green algae.

Iswarya, V., J. Manivannan, A. De, S. Paul, R. Roy, J.B. Johnson et al. 2016. Surface capping and size-dependent toxicity of gold nanoparticles on different trophic levels. Environ. Sci. Pollut. Res. 23(5): 4844–4858.

Ivask, A., I. Kurvet, K. Kasemets, I. Blinova, V. Aruoja, S. Suppi et al. 2014. Size-dependent toxicity of silver nanoparticles to bacteria, yeast, algae, crustaceans and mammalian cells *in vitro*. PLoS One. 9(7).

Jang, M.H., W.K. Kim, S.K. Lee, T.B. Henry and J.W. Park. 2014. Uptake, Tissue Distribution, and Depuration of Total Silver in Common Carp (Cyprinus carpio) after Aqueous Exposure to Silver Nanoparticles. Environ. Sci. Technol. 48(19): 11568–11574.

Jemec, A., A. Kahru, A. Potthoff, D. Drobne, M. Heinlaan, S. Bohme et al. 2016. An interlaboratory comparison of nanosilver characterisation and hazard identification: Harmonising techniques for high quality data. Environ. Int. 87: 20–32.

Jiang, H., M. Li, F. Chang, W. Li and L. Yin. 2012. Physiological analysis of silver nanoparticles and $AgNO_3$ toxicity to Spirodela polyrhiza. Environ. Toxicol. Chem. 31(8): 1880–1886.
Jiang, H., X. Qiu, G. Li, W. Li and L. Yin. 2014. Silver nanoparticles induced accumulation of reactive oxygen species and alteration of antioxidant systems in the aquatic plant Spirodela polyrhiza. Environ. Toxicol. Chem. 33(6): 1398–1405.
Johnson, M.E., S.A. Ostroumov, J.F. Tyson and B. Xing. 2011. Study of the interactions between Elodea canadensis and CuO nanoparticles. Russ. J. Gen. Chem. 81(13): 2688–2693.
Joonas, E., V. Aruoja, K. Olli, G. Syvertsen-Wiig, H. Vija and A. Kahru. 2017. Potency of (doped) rare earth oxide particles and their constituent metals to inhibit algal growth and induce direct toxic effects. Sci. Total Environ. 593: 478–486.
Juganson, K., A. Ivask, I. Blinova, M. Mortimer and A. Kahru. 2015. NanoE-Tox: New and in-depth database concerning ecotoxicity of nanomaterials. Beilstein J. Nanotechnol. 6: 1788.
Jung, Y., C.B. Park, Y. Kim, S. Kim, S. Pflugmacher and S. Baik. 2015. Application of Multi-Species Microbial Bioassay to Assess the Effects of Engineered Nanoparticles in the Aquatic Environment: Potential of a Luminous Microbial Array for Toxicity Risk Assessment (LumiMARA) on Testing for Surface-Coated Silver Nanoparticle. Int. J. Environ. Res. Public Health. 12(7): 8172–8186.
Justice, J.R. and R.J. Bernot. 2014. Nanosilver inhibits freshwater gastropod (Physa acuta) ability to assess predation risk. Am. Midl. Nat. 171(2): 340–349.
Kalčíková, G., M. Zupančič, A. Jemec and A.Ž. Gotvajn. 2016. The impact of humic acid on chromium phytoextraction by aquatic macrophyte Lemna minor. Chemosphere. 147: 311–317.
Kasemets, K., A. Ivask, H.C. Dubourguier and A. Kahru. 2009. Toxicity of nanoparticles of ZnO, CuO and TiO_2 to yeast Saccharomyces cerevisiae. Toxicol. Vitr. 23(6): 1116–1122.
Kaya, H., M. Duysak, M. Akbulut, S. Yilmaz, M. Gurkan, Z. Arslan et al. 2017. Effects of subchronic exposure to zinc nanoparticles on tissue accumulation, serum biochemistry, and histopathological changes in tilapia (Oreochromis niloticus). Environ. Toxicol. 32(4): 1213–1225.
Kennedy, A.J., M.S. Hull, A.J. Bednar, J.D. Goss, J.C. Gunter, J.L. Bouldin et al. 2010. Fractionating Nanosilver: Importance for Determining Toxicity to Aquatic Test Organisms. Environ. Sci. Technol. 44(24): 9571–9577.
Khan, F.R., K.B. Paul, A.D. Dybowska, E. Valsami-Jones, J.R. Lead, V. Stone et al. 2015. Accumulation dynamics and acute toxicity of silver nanoparticles to Daphnia magna and Lumbriculus variegatus: implications for metal modeling approaches. Environ. Sci. Technol. 49(7): 4389–4397.
Khare, P., M. Sonane, R. Pandey, S. Ali, K.C. Gupta and A. Satish. 2011. Adverse effects of TiO_2 and ZnO nanoparticles in soil nematode, Caenorhabditis elegans. J. Biomed. Nanotechnol. 7(1): 116–117.
Kim, E., S.H. Kim, H.-C. Kim, S.G. Lee, S.J. Lee and S.W. Jeong. 2011. Growth inhibition of aquatic plant caused by silver and titanium oxide nanoparticles. Toxicol. Environ. Health Sci. 3(1): 1–6.
Klaper, R., J. Crago, J. Barr, D. Arndt, K. Setyowati and J. Chen. 2009. Toxicity biomarker expression in daphnids exposed to manufactured nanoparticles: changes in toxicity with functionalization. Environ. Pollut. 157(4): 1152–1156.
Kroll, A., D. Kühnel and K. Schirmer. 2013. Testing nanomaterial toxicity in unicellular eukaryotic algae and fish cell lines. Methods in Molecular Biology.
Ksiazyk, M., M. Asztemborska, R. Steborowski and G. Bystrzejewska-Piotrowska. 2015. Toxic Effect of Silver and Platinum Nanoparticles Toward the Freshwater Microalga Pseudokirchneriella subcapitata. Bull. Environ. Contam. Toxicol. 94(5): 554–558.
Kühnel, D., W. Busch, T. Meißner, A. Springer, A. Potthoff, V. Richter et al. 2009. Agglomeration of tungsten carbide nanoparticles in exposure medium does not prevent uptake and toxicity toward a rainbow trout gill cell line. Aquat. Toxicol. 93(2-3): 91–99.
Kühnel, D. and C. Nickel. 2014. The OECD expert meeting on ecotoxicology and environmental fate—Towards the development of improved OECD guidelines for the testing of nanomaterials. Sci. Total Environ. 472.
Kumar, D., J. Kumari, S. Pakrashi, S. Dalai, A.M. Raichur, T.P. Sastry et al. 2014. Qualitative toxicity assessment of silver nanoparticles on the fresh water bacterial isolates and consortium at low level of exposure concentration. Ecotoxicol. Environ. Saf. 108: 152–160.
Kwak, J. Il, R. Cui, S.-H. Nam, S.W. Kim, Y. Chae and Y.-J. An. 2016. Multispecies toxicity test for silver nanoparticles to derive hazardous concentration based on species sensitivity distribution for the protection of aquatic ecosystems. Nanotoxicology. 10(5): 521–530.
Laban, G., L.F. Nies, R.F. Turco, J.W. Bickham and M.S. Sepulveda. 2010. The effects of silver nanoparticles on fathead minnow (Pimephales promelas) embryos. Ecotoxicology. 19(1): 185–195.
Lee, J.W., J.E. Kim, Y.J. Shin, J.S. Ryu, I.C. Eom, J.S. Lee et al. 2014. Serum and ultrastructure responses of common carp (*Cyprinus carpio* L.) during long-term exposure to zinc oxide nanoparticles. *Ecotoxicol.* Environ. Saf. 104: 9–17.
Lee, S.-W., S.-M. Kim and J. Choi. 2009. Genotoxicity and ecotoxicity assays using the freshwater crustacean Daphnia magna and the larva of the aquatic midge Chironomus riparius to screen the ecological risks of nanoparticle exposure. Environ. Toxicol. Pharmacol. 28(1): 86–91.
Lee, S.-W., S.-Y. Park, Y. Kim, H. Im and J. Choi. 2016. Effect of sulfidation and dissolved organic matters on toxicity of silver nanoparticles in sediment dwelling organism, Chironomus riparius. Sci. Total Environ. 553: 565–573.
Li, H.C., Q.F. Zhou, Y. Wu, J.J. Fu, T. Wang and G.B. Jiang. 2009. Effects of waterborne nano-iron on medaka (Oryzias latipes): Antioxidant enzymatic activity, lipid peroxidation and histopathology. Ecotoxicol. Environ. Saf. 72(3): 684–692.
Li, M., L.Z. Zhu and D.H. Lin. 2011. Toxicity of ZnO Nanoparticles to Escherichia coli: Mechanism and the Influence of Medium Components. Environ. Sci. Technol. 45(5): 1977–1983.

Li, Y.X., B. Men, Y. He, H.M. Xu, M.Q. Liu and D.S. Wang. 2017. Effect of single-wall carbon nanotubes on bioconcentration and toxicity of perfluorooctane sulfonate in zebrafish (Danio rerio). Sci. Total Environ. 607: 509–518.
Lin, D. and B. Xing. 2007. Phytotoxicity of nanoparticles: Inhibition of seed germination and root growth. Environ. Pollut. 150(2): 243.
Lipovsky, A., Y. Nitzan, A. Gedanken and R. Lubart. 2011. Antifungal activity of ZnO nanoparticles—the role of ROS mediated cell injury. Nanotechnology. 22(10): 105101.
Lovern, S.B. and R. Klaper. 2006. Daphnia magna mortality when exposed to titanium dioxide and fullerene (C60) nanoparticles. Environ. Toxicol. Chem. 25(4): 1132–1137.
Malysheva, A., N. Voelcker, P.E. Holm and E. Lombi. 2016. Unraveling the Complex Behavior of AgNPs Driving NP-Cell Interactions and Toxicity to Algal Cells. Environ. Sci. Technol. 50(22): 12455–12463.
Manier, N., A. Bado-Nilles, P. Delalain, O. Aguerre-Chariol and P. Pandard. 2013. Ecotoxicity of non-aged and aged CeO_2 nanomaterials towards freshwater microalgae. Environ. Pollut. 180: 63–70.
Mansouri, B., A. Maleki, S.A. Johari, B. Shahmoradi, E. Mohammadi, S. Shahsavari et al. 2016. Copper Bioaccumulation and Depuration in Common Carp (Cyprinus carpio) Following Co-exposure to TiO_2 and CuO Nanoparticles. Arch. Environ. Contam. Toxicol. 71(4): 541–552.
Manusadzianas, L., C. Caillet, L. Fachetti, B. Gylyte, R. Grigutyte, S. Jurkoniene et al. 2012. Toxicity of copper oxide nanoparticle suspensions to aquatic biota. Environ. Toxicol. Chem. 31(1): 108–114.
Manzo, S., A. Rocco, R. Carotenuto, F.D.L. Picione, M.L. Miglietta, G. Rametta et al. 2011. Investigation of ZnO nanoparticles' ecotoxicological effects towards different soil organisms. Environ. Sci. Pollut. Res. 18(5): 756–763.
Martin, J.D., T.L.L. Colson, V.S. Langlois and C.D. Metcalfe. 2017. Biomarkers of exposure to nanosilver and silver accumulation in yellow perch (Perca Flavescens). Environ. Toxicol. Chem. 36(5): 1211–1220.
Matzke, M., K. Jurkschat and T. Backhaus. 2014. Toxicity of differently sized and coated silver nanoparticles to the bacterium Pseudomonas putida: risks for the aquatic environment? Ecotoxicology. 23(5): 818–829.
Mehennaoui, K., A. Georgantzopoulou, V. Felten, J. Andreï, M. Garaud, S. Cambier et al. 2016. Gammarus fossarum (Crustacea, Amphipoda) as a model organism to study the effects of silver nanoparticles. Sci. Total Environ. 566: 1649–1659.
Meregalli, G., A.C. Vermeulen and F. Ollevier. 2000. The use of chironomid deformation in an *in situ* test for sediment toxicity. Ecotoxicol. Environ. Saf. 47(3): 231–238.
Minetto, D., G. Libralato, A. Marcomini and A.V. Ghirardini. 2017. Potential effects of TiO_2 nanoparticles and TiCl4 in saltwater to Phaeodactylum tricornutum and Artemia franciscana. Sci. Total Environ. 579: 1379–1386.
Mohmood, I., C.B. Lopes, I. Lopes, D.S. Tavares, A. Soares, A.C. Duarte et al. 2016. Remediation of mercury contaminated saltwater with functionalized silica coated magnetite nanoparticles. Sci. Total Environ. 557: 712–721.
Monopoli, M.P., D. Walczyk, A. Campbell, G. Elia, I. Lynch, F. Baldelli Bombelli et al. 2011. Physical−chemical aspects of protein corona: relevance to *in vitro* and *in vivo* biological impacts of nanoparticles. J. Am. Chem. Soc. 133(8): 2525–2534.
Mouchet, F., P. Landois, E. Flahaut, E. Pinelli and L. Gauthier. 2007. Assessment of the potential *in vivo* ecotoxicity of Double-Walled Carbon Nanotubes (DWNTs) in water, using the amphibian Ambystoma mexicanum. Nanotoxicology. 1(2): 149–156.
Mouchet, F., P. Landois, E. Sarremejean, G. Bernard, P. Puech, E. Pinelli et al. 2008. Characterisation and *in vivo* ecotoxicity evaluation of double-wall carbon nanotubes in larvae of the amphibian Xenopus laevis. Aquat. Toxicol. 87(2): 127–137.
Mwangi, J.N., N. Wang, C.G. Ingersoll, D.K. Hardesty, E.L. Brunson, H. Li et al. 2012. Toxicity of carbon nanotubes to freshwater aquatic invertebrates. Environ. Toxicol. Chem. 31(8): 1823–1830.
Nam, S.H., W.M. Lee, Y.J. Shin, S.J. Yoon, S.W. Kim, J.I. Kwak et al. 2014. Derivation of guideline values for gold (III) ion toxicity limits to protect aquatic ecosystems. Water Res. 48: 126–136.
Nations, S., M. Long, M. Wages, J.D. Maul, C.W. Theodorakis and G.P. Cobb. 2015. Subchronic and chronic developmental effects of copper oxide (CuO) nanoparticles on Xenopus laevis. Chemosphere. 135: 166–174.
Nations, S., M. Wages, J.E. Canas, J. Maul, C. Theodorakis and G.P. Cobb. 2011. Acute effects of Fe_2O_3, TiO_2, ZnO and CuO nanomaterials on Xenopus laevis. Chemosphere. 83(8): 1053–1061.
Neale, P.A., R. Altenburger, S. Aït-Aïssa, F. Brion, W. Busch, G. de Aragão Umbuzeiro et al. 2017. Development of a bioanalytical test battery for water quality monitoring: Fingerprinting identified micropollutants and their contribution to effects in surface water. Water Res. 123: 734–750.
Neale, P.A., A.K. Jamting, E. O'Malley, J. Herrmann and B.I. Escher. 2015. Behaviour of titanium dioxide and zinc oxide nanoparticles in the presence of wastewater-derived organic matter and implications for algal toxicity. Environ. Sci. 2(1): 86–93.
Oberdörster, E., S. Zhu, T.M. Blickley, P. McClellan-Green and M.L. Haasch. 2006. Ecotoxicology of carbon-based engineered nanoparticles: effects of fullerene (C 60) on aquatic organisms. Carbon N. Y. 44(6): 1112–1120.
Oberholster, P.J., N. Musee, A.-M. Botha, P.K. Chelule, W.W. Focke and P.J. Ashton. 2011. Assessment of the effect of nanomaterials on sediment-dwelling invertebrate Chironomus tentans larvae. Ecotoxicol. Environ. Saf. 74(3): 416–423.
OCED. 2012. Guidance on sample preparation and dosimetry for the safety testing of manufactured nanomaterials. Ser. Saf. Manuf. Nanomater. No. 36. ENV/JM/MON.
OECD. 1997. Test No. 474: Mammalian Erythrocyte Micronucleus Test. OECD Publ. Paris.
OECD. 2002. Guidance Document on Aquatic Toxicity Testing of Difficult Substances and Mixtures. OECD Ser. Test. Assess.
OECD. 2010. OECD Guidelines for the Testing of Chemicals (March 2006).
OECD TG 201. 2011. Freshwater Alga and Cyanobacteria, Growth Inhibition Test. OECD Publ. Paris.
OECD TG 202. 2004. Test No. 202: Daphnia sp. Acute Immobilisation Test. OECD Publ. Paris.

OECD TG 203. 1992. Fish, Acute Toxicity Test. OECD Publ. Paris.

OECD TG 208. 2006. Terrestrial Plant Test: Seedling Emergence and Seedling Growth Test. OECD Guidelines for the Testing of Chemicals, Section, 2. OECD Publ. Paris.

OECD TG 211. 2008. Daphnia magna Reproduction Test. OECD Publ. Paris.

OECD TG 212. 1998. Fish, Short-term Toxicity Test on Embryo and Sac-Fry Stages. OECD Publ. Paris.

OECD TG 221. 2006. Lemna sp. Growth Inhibition Test. OECD Publ. Paris.

OECD TG 225. 2007. Sediment-Water Lumbriculus Toxicity Test Using Spiked Sediment. OECD Publ. Paris.

OECD TG 227. 2006. Terrestrial Plant Test: Vegetative Vigour Testle. OECD Publ. Paris.

OECD TG 233. 2010. Sediment-Water Chironomid Life-Cycle Toxicity Test Using Spiked Water or Spiked Sediment. OECD Publ. Paris.

OECD TG 235. 2011. Chironomus sp., Acute Immobilisation Test. OECD Publ. Paris.

OECD TG 236. 2013. Fish Embryo Acute Toxicity (FET) Test. OECD Publ. Paris.

OECD TG 242. 2016. Potamopyrgus antipodarum Reproduction Test. OECD Publ. Paris.

OECD TG 243. 2016. Lymnaea stagnalis Reproduction Test. OECD Publ. Paris.

OECD TG 318. 2017. Test No. 318: Dispersion Stability of Nanomaterials in Simulated Environmental Media. OECD Publ. Paris.

Oliver, A.L.-S., M.-N. Croteau, T.L. Stoiber, M. Tejamaya, I. Römer, J.R. Lead et al. 2014. Does water chemistry affect the dietary uptake and toxicity of silver nanoparticles by the freshwater snail Lymnaea stagnalis? Environ. Pollut. 189: 87–91.

Ong, K.J., T.J. MacCormack, R.J. Clark, J.D. Ede, V.A. Ortega, L.C. Felix et al. 2014. Widespread Nanoparticle-Assay Interference: Implications for Nanotoxicity Testing. PLoS One. 9(3): e90650.

Oprsal, J., L. Blaha, M. Pouzar, P. Knotek, M. Vlcek and K. Hrda. 2015. Assessment of silver nanoparticle toxicity for common carp (Cyprinus carpio) fish embryos using a novel method controlling the agglomeration in the aquatic media. Environ. Sci. Pollut. Res. 22(23): 19124–19132.

Ostaszewska, T., M. Chojnacki, M. Kamaszewski and E. Sawosz-Chwalibog. 2016. Histopathological effects of silver and copper nanoparticles on the epidermis, gills, and liver of Siberian sturgeon. Environ. Sci. Pollut. Res. 23(2): 1621–1633.

Otero-González, L., C. García-Saucedo, J.A. Field and R. Sierra-Álvarez. 2013. Toxicity of TiO_2, ZrO_2, Fe0, Fe_2O_3, and Mn_2O_3 nanoparticles to the yeast, Saccharomyces cerevisiae. Chemosphere. 93(6): 1201–1206.

Oukarroum, A., L. Barhoumi, L. Pirastru and D. Dewez. 2013. Silver nanoparticle toxicity effect on growth and cellular viability of the aquatic plant Lemna gibba. Environ. Toxicol. Chem. 32(4): 902–907.

Pakarinen, K., E.J. Petersen, M.T. Leppänen, J. Akkanen and J.V.K. Kukkonen. 2011. Adverse effects of fullerenes (nC 60) spiked to sediments on Lumbriculus variegatus (Oligochaeta). Environ. Pollut. 159(12): 3750–3756.

Pang, C., H. Selck, S.K. Misra, D. Berhanu, A. Dybowska, E. Valsami-Jones et al. 2012. Effects of sediment-associated copper to the deposit-feeding snail, Potamopyrgus antipodarum: A comparison of Cu added in aqueous form or as nano- and micro-CuO particles. Aquat. Toxicol. 106-107(0): 114–122.

Park, A., Y.-J. Kim, E.-M. Choi, M.T. Brown and T. Han. 2013. A novel bioassay using root re-growth in Lemna. Aquat. Toxicol. 140: 415–424.

Pereira, F.F., E.C. Paris, J.D. Bresolin, M.M. Foschini, M.D. Ferreira and D.S. Correa. 2017. Investigation of nanotoxicological effects of nanostructured hydroxyapatite to microalgae Pseudokirchneriella subcapitata. Ecotoxicol. Environ. Saf. 144: 138–147.

Perelshtein, I., A. Lipovsky, N. Perkas, A. Gedanken, E. Moschini and P. Mantecca. 2015. The influence of the crystalline nature of nano-metal oxides on their antibacterial and toxicity properties. Nano Res. 8(2): 695–707.

Petersen, E.J., S.A. Diamond, A.J. Kennedy, G.G. Goss, K. Ho, J. Lead et al. 2015. Adapting OECD aquatic toxicity tests for use with manufactured nanomaterials: key issues and consensus recommendations. Environ. Sci. Technol. 49(16): 9532–9547.

Peterson, H.G., C. Boutin, P.A. Martin, K.E. Freemark, N.J. Ruecker and M.J. Moody. 1994. Aquatic phyto-toxicity of 23 pesticides applied at expected environmental concentrations. Aquat. Toxicol. 28(3-4): 275–292.

Peyrot, C., C. Gagnon, F. Gagné, K.J. Willkinson, P. Turcotte and S. Sauvé. 2009. Effects of cadmium telluride quantum dots on cadmium bioaccumulation and metallothionein production to the freshwater mussel, Elliptio complanata. Comp. Biochem. Physiol. Part C Toxicol. & Pharmacol. 150(2): 246–251.

Picado, A., S.M. Paixao, L. Moita, L. Silva, M. Diniz, J. Lourenco et al. 2015. A multi-integrated approach on toxicity effects of engineered TiO_2 nanoparticles. Front. Environ. Sci. Eng. 9(5): 793–803.

Polonini, H.C., H.M. Brandao, N.R.B. Raposo, M.A.F. Brandao, L. Mouton, A. Coute et al. 2015. Size-dependent ecotoxicity of barium titanate particles: the case of Chlorella vulgaris green algae. Ecotoxicology. 24(4): 938–948.

Potthoff, A., M. Weil, T. Meißner and D. Kühnel. 2015. Towards sensible toxicity testing for nanomaterials: proposal for the specification of test design. Sci. Technol. Adv. Mater. 16(6): 65006.

Praetorius, A., N. Tufenkji, K.-U. Goss, M. Scheringer, F. von der Kammer and M. Elimelech. 2014. The road to nowhere: equilibrium partition coefficients for nanoparticles. Environ. Sci. Nano. 1(4): 317.

Qu, Y., W. Li, Y. Zhou, X. Liu, L. Zhang, L. Wang et al. 2011. Full assessment of fate and physiological behavior of quantum dots utilizing Caenorhabditis elegans as a model organism. Nano Lett. 11(8): 3174–3183.

Qualhato, G., T.L. Rocha, E.C.D. Lima, D.M.E. Silva, J.R. Cardoso, C.K. Grisolia et al. 2017. Genotoxic and mutagenic assessment of iron oxide (maghemite-gamma-Fe_2O_3) nanoparticle in the guppy Poecilia reticulata. Chemosphere. 183: 305–314.

Radniecki, T.S., D.P. Stankus, A. Neigh, J.A. Nason and L. Semprini. 2011. Influence of liberated silver from silver nanoparticles on nitrification inhibition of Nitrosomonas europaea. Chemosphere. 85(1): 43–49.

Rai, M. and A. Ingle. 2012. Role of nanotechnology in agriculture with special reference to management of insect pests. Appl. Microbiol. Biotechnol. 94(2): 287–293.

Rajala, J.E., K. Mäenpää, E.-R. Vehniäinen, A. Väisänen, J.J. Scott-Fordsmand, J. Akkanen et al. 2016. Toxicity testing of silver nanoparticles in artificial and natural sediments using the benthic organism Lumbriculus variegatus. Arch. Environ. Contam. Toxicol. 71(3): 405–414.

Ramskov, T., M.-N. Croteau, V.E. Forbes and H. Selck. 2015. Biokinetics of different-shaped copper oxide nanoparticles in the freshwater gastropod, Potamopyrgus antipodarum. Aquat. Toxicol. 163: 71–80.

Ramskov, T., H. Selck, G. Banta, S.K. Misra, D. Berhanu, E. Valsami-Jones et al. 2014. Bioaccumulation and effects of different-shaped copper oxide nanoparticles in the deposit-feeding snail Potamopyrgus antipodarum. Environ. Toxicol. Chem. 33(9): 1976–1987.

Rand, G.M. 1995. Fundamentals of Aquatic Toxicology: Effects, Environmental Fate and Risk Assessment.

Ribeiro, F., J.A. Gallego-Urrea, K. Jurkschat, A. Crossley, M. Hassellov, C. Taylor et al. 2014. Silver nanoparticles and silver nitrate induce high toxicity to Pseudokirchneriella subcapitata, Daphnia magna and Danio rerio. Sci. Total Environ. 466: 232–241.

Rico, C.M., M.I. Morales, A.C. Barrios, R. McCreary, J. Hong, W.-Y. Lee et al. 2013. Effect of cerium oxide nanoparticles on the quality of rice (*Oryza sativa* L.) grains. J. Agric. Food Chem. 61(47): 11278–11285.

Rocco, L., M. Santonastaso, F. Mottola, D. Costagliola, T. Suero, S. Pacifico et al. 2015. Genotoxicity assessment of TiO_2 nanoparticles in the teleost Danio rerio. Ecotoxicol. Environ. Saf. 113: 223–230.

Rodea-Palomares, I., K. Boltes, F. Fernández-Piñas, F. Leganés, E. García-Calvo, J. Santiago et al. 2011. Physicochemical characterization and ecotoxicological assessment of CeO_2 nanoparticles using two aquatic microorganisms. Toxicol. Sci. 119(1): 135–145.

Roh, J., S.J. Sim, J. Yi, K. Park, K.H. Chung, D. Ryu et al. 2009. Ecotoxicity of silver nanoparticles on the soil nematode Caenorhabditis elegans using functional ecotoxicogenomics. Environ. Sci. Technol. 43(10): 3933–3940.

Rosenkranz, P., Q. Chaudhry, V. Stone and T.F. Fernandes. 2009. A comparison of nanoparticle and fine particle uptake by Daphnia magna. Environ. Toxicol. Chem. 28(10): 2142–2149.

Rossetto, A., S.P. Melegari, L.C. Ouriques and W.G. Matias. 2014. Comparative evaluation of acute and chronic toxicities of CuO nanoparticles and bulk using Daphnia magna and Vibrio fischeri. Sci. Total Environ. 490: 807–814.

Rossetto, A., D.S. Vicentini, C.H. Costa, S.P. Melegari and W.G. Matias. 2014. Synthesis, characterization and toxicological evaluation of a core-shell copper oxide/polyaniline nanocomposite. Chemosphere. 108: 107–114.

Ruppert, K., C. Geiß, C. Askem, R. Benstead, R. Brown, M. Coke et al. 2017. Development and validation of an OECD reproductive toxicity test guideline with the mudsnail Potamopyrgus antipodarum (Mollusca, Gastropoda). Chemosphere. 181: 589–599.

Sanchis, J., M. Olmos, P. Vincent, M. Farre and D. Barcelo. 2016. New Insights on the Influence of Organic Co-Contaminants on the Aquatic Toxicology of Carbon Nanomaterials. Environ. Sci. Technol. 50(2): 961–969.

Sanders, M.B., M. Sebire, J. Sturve, P. Christian, I. Katsiadaki, B.P. Lyons et al. 2008. Exposure of sticklebacks (Gasterosteus aculeatus) to cadmium sulfide nanoparticles: Biological effects and the importance of experimental design. Mar. Environ. Res. 66(1): 161–163.

Schiavo, S., M. Oliviero, M. Miglietta, G. Rametta and S. Manzo. 2016. Genotoxic and cytotoxic effects of ZnO nanoparticles for Dunaliella tertiolecta and comparison with SiO_2 and TiO_2 effects at population growth inhibition levels. Sci. Total Environ. 550: 619–627.

Schiwy, A., H.M. Maes, D. Koske, M. Flecken, K.R. Schmidt, H. Schell et al. 2016. The ecotoxic potential of a new zero-valent iron nanomaterial, designed for the elimination of halogenated pollutants, and its effect on reductive dechlorinating microbial communities. Environ. Pollut. 216: 419–427.

Schmidt, S., R. Altenburger and D. Kühnel. 2017. From the air to the water phase: implication for toxicity testing of combustion-derived particles. Biomass Convers. Biorefinery. 10.

Schultz, A.G., D. Boyle, D. Chamot, K.J. Ong, K.J. Wilkinson, J.C. McGeer et al. 2014. Aquatic toxicity of manufactured nanomaterials: Challenges and recommendations for future toxicity testing. Environ. Chem. 11(3): 207–226.

Schwab, F., T.D. Bucheli, L.P. Lukhele, A. Magrez, B. Nowack, L. Sigg et al. 2011. Are carbon nanotube effects on green algae caused by shading and agglomeration? Environ. Sci. Technol. 45(14): 6136–6144.

Sendra, M., I. Moreno-Garrido, M.P. Yeste, J.M. Gatica and J. Blasco. 2017. Toxicity of TiO_2, in nanoparticle or bulk form to freshwater and marine microalgae under visible light and UV-A radiation. Environ. Pollut. 227: 39–48.

Shirazi, M.A., F. Shariati, A.K. Keshavarz and Z. Ramezanpour. 2015. Toxic effect of aluminum oxide nanoparticles on Green Micro-Algae dunaliella salina. Int. J. Environ. Res. 9(2): 585–594.

Siegfried, B.D. 1993. Comparative toxicity of pyrethroid insecticides to terrestrial and aquatic insects. Environ. Toxicol. Chem. 12(9): 1683–1689.

Skjolding, L.M., S.N. Sørensen, N.B. Hartmann, R. Hjorth, S.F. Hansen and A. Baun. 2016. Aquatic ecotoxicity testing of nanoparticles—the quest to disclose nanoparticle effects. Angew. Chemie - Int. Ed. 55(49): 15224–15239.

Song, L., M.G. Vijver, W. Peijnenburg, T.S. Galloway and C.R. Tyler. 2015. A comparative analysis on the in vivo toxicity of copper nanoparticles in three species of freshwater fish. Chemosphere. 139: 181–189.

Sørensen, S.N., R. Hjorth, C.G. Delgado, N.B. Hartmann and A. Baun. 2015. Nanoparticle ecotoxicity-physical and/or chemical effects? Integr. Environ. Assess. Manag. 11(4): 722–724.

Srikanth, K., E. Pereira, A.C. Duarte, I. Ahmad and J.V. Rao. 2015. Assessment of cytotoxicity and oxidative stress induced by titanium oxide nanoparticles on Chinook salmon cells. Environ. Sci. Pollut. Res. 22(20): 15579–15586.

Stampoulis, D., S.K. Sinha and J.C. White. 2009. Assay-dependent phytotoxicity of nanoparticles to plants. Environ. Sci. Technol. 43(24): 9473–9479.

Suman, T.Y., S.R.R. Rajasree and R. Kirubagaran. 2015. Evaluation of zinc oxide nanoparticles toxicity on marine algae chlorella vulgaris through flow cytometric, cytotoxicity and oxidative stress analysis. Ecotoxicol. Environ. Saf. 113: 23–30.

Suppi, S., K. Kasemets, A. Ivask, K. Künnis-Beres, M. Sihtmäe, I. Kurvet et al. 2015. A novel method for comparison of biocidal properties of nanomaterials to bacteria, yeasts and algae. J. Hazard. Mater. 286: 75–84.

Svartz, G., M. Papa, M. Gosatti, M. Jordan, A. Soldati, P. Samter et al. 2017. Monitoring the ecotoxicity of gamma-Al_2O_3 and Ni/gamma-Al_2O_3 nanomaterials by means of a battery of bioassays. Ecotoxicol. Environ. Saf. 144: 200–207.

Taenzler, V., E. Bruns, M. Dorgerloh, V. Pfeifle and L. Weltje. 2007. Chironomids: suitable test organisms for risk assessment investigations on the potential endocrine disrupting properties of pesticides. Ecotoxicology. 16(1): 221–230.

Tantra, R., A. Sikora, N.B. Hartmann, J.R. Sintes and K.N. Robinson. 2015. Comparison of the effects of different protocols on the particle size distribution of TiO_2 dispersions. Particuology. 19: 35–44.

Taurozzi, J.S., V.A. Hackley and M.R. Wiesner. 2011. Ultrasonic dispersion of nanoparticles for environmental, health and safety assessment – issues and recommendations. Nanotoxicology. 5(4): 711–729.

Tong, T.Z., C.T.T. Binh, J.J. Kelly, J.F. Gaillard and K.A. Gray. 2013. Cytotoxicity of commercial nano-TiO_2 to *Escherichia coli* assessed by high-throughput screening: Effects of environmental factors. Water Res. 47(7): 2352–2362.

Tong, T.Z., A. Shereef, J.S. Wu, C.T.T. Binh, J.J. Kelly, J.F. Gaillard et al. 2013. Effects of Material Morphology on the Phototoxicity of Nano-TiO_2 to Bacteria. Environ. Sci. Technol. 47(21): 12486–12495.

Valerio-Garcia, R.C., A.L. Carbajal-Hernandez, E.B. Martinez-Ruiz, V.H. Jarquin-Diaz, C. Haro-Perez and F. Martinez-Jeronimo. 2017. Exposure to silver nanoparticles produces oxidative stress and affects macromolecular and metabolic biomarkers in the goodeid fish Chapalichthys pardalis. Sci. Total Environ. 583: 308–318.

Van Hoecke, K., K.A.C. De Schamphelaere, S. Ramirez-Garcia, P. Van der Meeren, G. Smagghe and C.R. Janssen. 2011. Influence of alumina coating on characteristics and effects of SiO_2 nanoparticles in algal growth inhibition assays at various pH and organic matter contents. Environ. Int. 37(6): 1118–1125.

Van Hoecke, K., K.A.C. De Schamphelaere, P. Van der Meeren, S. Lucas and C.R. Janssen. 2008. Ecotoxicity of silica nanoparticles to the green alga Pseudokirchneriella subcapitata: Importance of surface area. Environ. Toxicol. Chem. 27(9): 1948–1957.

Van Hoecke, K., J.T.K. Quik, J. Mankiewicz-Boczek, K.A.C. De Schamphelaere, A. Elsaesser, P. Van der Meeren et al. 2009. Fate and Effects of CeO_2 Nanoparticles in Aquatic Ecotoxicity Tests. Environ. Sci. Technol. 43(12): 4537–4546.

Van Koetsem, F., Y. Xiao, Z. Luo and G. Du Laing. 2016. Impact of water composition on association of Ag and CeO_2 nanoparticles with aquatic macrophyte Elodea canadensis. Environ. Sci. Pollut. Res. 23(6): 5277–5287.

van Leeuwen, C.J. and T.G. Vermeire. 2007. Risk Assessment of Chemicals: an Introduction.

Völker, C., I. Kämpken, C. Boedicker, J. Oehlmann and M. Oetken. 2015. Toxicity of silver nanoparticles and ionic silver: comparison of adverse effects and potential toxicity mechanisms in the freshwater clam Sphaerium corneum. Nanotoxicology. 9(6): 677–685.

Waissi-Leinonen, G.C., I. Nybom, K. Pakarinen, J. Akkanen, M.T. Leppänen and J.V.K. Kukkonen. 2015. Fullerenes (nC 60) affect the growth and development of the sediment-dwelling invertebrate Chironomus riparius larvae. Environ. Pollut. 206: 17–23.

Waissi-Leinonen, G.C., E.J. Petersen, K. Pakarinen, J. Akkanen, M.T. Leppänen and J.V.K. Kukkonen. 2012. Toxicity of fullerene (C60) to sediment-dwelling invertebrate Chironomus riparius larvae. Environ. Toxicol. Chem. 31(9): 2108–2116.

Waissi, G.C., K. Väänänen, I. Nybom, K. Pakarinen, J. Akkanen, M.T. Leppänen et al. 2017. The chronic effects of fulleneneC 60-associated sediments in the midge Chironomus riparius–Responses in the first and the second generation. Environ. Pollut. 229: 423–430.

Wang, D., J. Hu, B.E. Forthaus and J. Wang. 2011. Synergistic toxic effect of nano-Al_2O_3 and As (V) on Ceriodaphnia dubia. Environ. Pollut. 159(10): 3003–3008.

Wang, J. and W.X. Wang. 2014. Low bioavailability of silver nanoparticles presents trophic toxicity to marine medaka (Oryzias melastigma). Environ. Sci. Technol. 48(14): 8152–8161.

Wang, W. 1990. Literature review on duckweed toxicity testing. Environ. Res. 52(1): 7–22.

Wang, Z., J.W. Chen, X.H. Li, J.P. Shao and W. Peijnenburg. 2012. Aquatic toxicity of nanosilver colloids to different trophic organisms: Contributions of particles and free silver ion. Environ. Toxicol. Chem. 31(10): 2408–2413.

Wang, Z., Y.C. Gao, S. Wang, H. Fang, D.F. Xu and F. Zhang. 2016. Impacts of low-molecular-weight organic acids on aquatic behavior of graphene nanoplatelets and their induced algal toxicity and antioxidant capacity. Environ. Sci. Pollut. Res. 23(11): 10938–10945.

Wang, Z., J.T.K. Quik, L. Song, E.J. Van den Brandhof, M. Wouterse and W. Peijnenburg. 2015. Humic substances alleviate the aquatic toxicity of polyvinylpyrrolidone-coated silver nanoparticles to organisms of different trophic levels. Environ. Toxicol. Chem. 34(6): 1239–1245.

Weil, M., T. Meissner, W. Busch, A. Springer, D. Kuhnel, R. Schulz et al. 2015. The oxidized state of the nanocomposite Carbo-Iron (R) causes no adverse effects on growth, survival and differential gene expression in zebrafish. Sci. Total Environ. 530: 198–208.

Wieczerzak, M., J. Namieśnik and B. Kudłak. 2016. Bioassays as one of the Green Chemistry tools for assessing environmental quality: A review. Environ. Int. 94: 341–361.

Wiench, K., W. Wohlleben, V. Hisgen, K. Radke, E. Salinas, S. Zok et al. 2009. Acute and chronic effects of nano-and non-nano-scale TiO_2 and ZnO particles on mobility and reproduction of the freshwater invertebrate Daphnia magna. Chemosphere. 76(10): 1356–1365.

Williams, P.L. and D.B. Dusenbery. 1990. Aquatic toxicity testing using the nematode, Caenorhabditis elegans. Environ. Toxicol. Chem. 9(10): 1285–1290.
Wörle-Knirsch, J.M., K. Pulskamp and H.F. Krug. 2006. Oops they did it again! carbon nanotubes hoax scientists in viability assays. Nano Lett. 6(6): 1261–1268.
Wu, Y., Q. Zhou, H. Li, W. Liu, T. Wang and G. Jiang. 2010. Effects of silver nanoparticles on the development and histopathology biomarkers of Japanese medaka (Oryzias latipes) using the partial-life test. Aquat. Toxicol. 100(2): 160–167.
Xie, L., X. Wu, H. Chen, Y. Luo, Z. Guo, J. Mu et al. 2016. The bioaccumulation and effects of selenium in the oligochaete Lumbriculus variegatus via dissolved and dietary exposure routes. Aquat. Toxicol. 178: 1–7.
Zhang, D., T. Hua, F. Xiao, C. Chen, R.M. Gersberg, Y. Liu et al. 2015. Phytotoxicity and bioaccumulation of ZnO nanoparticles in Schoenoplectus tabernaemontani. Chemosphere. 120: 211–219.
Zhang, L., Y.L. He, N. Goswami, J.P. Xie, B. Zhang and X.J. Tao. 2016. Uptake and effect of highly fluorescent silver nanoclusters on Scenedesmus obliquus. Chemosphere. 153: 322–331.
Zhang, Y.Q., R. Dringen, C. Petters, W. Rastedt, J. Koser, J. Filser et al. 2016. Toxicity of dimercaptosuccinate-coated and un-functionalized magnetic iron oxide nanoparticles towards aquatic organisms. Environ. Sci. 3(4): 754–767.
Zhu, S.Q., E. Oberdorster and M.L. Haasch. 2006. Toxicity of an engineered nanoparticle (fullerene, C-60) in two aquatic species, Daphnia and fathead minnow. Mar. Environ. Res. 62: S5–S9.
Zhu, X.S., L. Zhu, Y.P. Lang and Y.S. Chen. 2008. Oxidative stress and growth inhibition in the freshwater fish Carassius auratus induced by chronic exposure to sublethal fullerene aggregates. Environ. Toxicol. Chem. 27(9): 1979–1985.

5

Nanomaterial Transport and Ecotoxicity in Fish Embryos

Kerry J. Lee[1,*] and *Lauren M. Browning*[2]

Nanomaterials (NMs) possess unique physicochemical properties, providing a vast amount of applications, such as design of high quality consumer products, effective disease diagnosis and treatment therapies (Murphy et al. 2008, Sun et al. 2005, Tiwari and Amiji 2006, Xu and Patel 2005). The increased use of nanomaterials can potentially lead to contamination of the aquatic environments. Their small sizes give them the ability to enter into living cells and organisms and, given their unique properties, may potentially cause toxicity and damage to *in vivo* systems (Agrawal et al. 2007, Lee et al. 2007, Xu et al. 2004). Even though the mechanisms of nanomaterial reactivity are still not completely understood, their interactions with live cells and embryos have been studied extensively and the toxic effects related to concentration dependence have been reported (Agrawal et al. 2007, Lee et al. 2007, Xu et al. 2004). With increasing use of nanomaterials in commonly used products, the potential of these materials to come in contact with our water sources is amplified and may pose the risk of polluting natural bodies of water and contaminating the aquatic environments. This increases the chances of interaction with aquatic organisms in their natural environment and risk for development to be affected.

There are many different possible routes for nanomaterials to come in contact with our water sources. The obvious mass contamination causes are in manufacturing and potential large accidental spills by industries using nanomaterials in processing; however, there are many other ways for nanomaterials to enter our water sources (Boxall et al. 2007, Wiesner et al. 2006). Increased human use of nanomaterial-based products raise the risk of exposure to our water sources and have hitherto gone unnoticed. Nanoparticle-based pharmaceuticals are becoming a reality and once ingested by the patient, the nanomaterial will be excreted and passed to the sewer systems (Boxall et al. 2007). For example, new drug treatments based on nanoparticle delivery vehicles are already being used in Alzheimer's, Wilson's, and Parkinson's diseases (Cui et al. 2005, Dobson 2001). In addition, nanoparticle-based diagnostic tools and treatments are being used for targeting cancers (Brigger et al. 2002). Specifically, nanoparticles have been used to target breast cancer cells for the purpose of drug delivery (Yezhelyev et al. 2006). Sewer water contamination can also come from human use of

[1] Department of Biological Sciences, Florida Gulf Coast University, Fort Myers, Florida 33965.
[2] Department of Biological Sciences, Old Dominion University, Norfolk, Virginia 23529.
Email: lbrownin@odu.edu
* Corresponding author: kerlee@fgcu.edu

products containing nanomaterials, such as tires, sports equipment, video screens, cosmetics, lotions, sunscreens and toothpastes in personal hygiene care (Boxall et al. 2007, Wiesner et al. 2006). The problem is that once nanomaterials are passed to the sewer systems, it is possible that they may be passed to surface waters. Direct contact with the surface water is also possible from products that are directly applied to the human skin surface, such as sunscreens and cosmetics, and when humans go swimming or bathe, the nanomaterials come in contact with water directly (Boxall et al. 2007, Wiesner et al. 2006). Another possible contamination source is when products containing nanomaterials are disposed of in the household and then sent to the landfill (Boxall et al. 2007, Wiesner et al. 2006). This offers direct contact with ground water contamination from run-off in the landfill. Paints now also contain nanomaterials and provide another route of contamination of sewer water through direct run-off from the painted surface, either industrial or domestic (Boxall et al. 2007, Wiesner et al. 2006). Also, if paint is applied to structures that will primarily be underwater, such as, docks and boats or ships, there is a possibility the nanomaterials will leach into the natural bodies of water directly (Boxall et al. 2007). Nanomaterials are being used more frequently in fuels and as catalysts in motor vehicles and their exhaust emission can result in nanoparticles being dispersed into the air, which can additionally result in contamination of surface waters (Boxall et al. 2007, Dahl et al. 2006). It is important to consider that as nanomaterials are becoming increasingly used, there is clear risk to contamination of the environment and our natural water sources.

There are many dynamics influencing the transport and ecotoxicity of nanomaterials in an aquatic environment. One factor to evaluate is when engineered and manufactured nanomaterials are deposited into an unknown water environment, there is potential for the nanomaterial to undergo unpredictable changes, such as aggregation (Boxall et al. 2007, Wiesner et al. 2006). The ability for nanomaterials to have different sizes can make their toxicity variable in aquatic environments, which is important to consider as well (Brown et al. 2001, Hoet et al. 2004, Howard 2004). Another factor for risk of increased toxicity is that many engineered nanoparticles also contain stabilizers on the surface to resist aggregation and surface chemistry modifications will further make their changes in the natural water environments haphazard (Boxall et al. 2007, Sayes et al. 2004, Zhu et al. 2010). This could have unknown consequences for the aquatic environment and organisms. The type and concentration of the nanomaterial used also poses a risk for varying and unforeseen levels of toxicity. Different types of nanoparticles are being used in various applications and products previously mentioned that could widely affect the severity in toxicity to aquatic embryo organism development. For instance, silver (Ag), titanium dioxide (TiO_2), zinc oxide (ZnO) and fullerene nanomaterials are being used in sunscreens, cosmetics, personal care products, food packaging and paints (Boxall et al. 2007, Tran and Salmon 2011, Wiesner et al. 2006). Fluorescent silica nanoparticles, because of their desired qualities for imaging, are being used as diagnostic tools and drug and gene delivery vehicles for treatment of diseases (Burns et al. 2008, Yezhelyev et al. 2006). Quantum dots (QDs) are also being used more in biological and medical imaging and have promise for use in alternative nanoparticle research and medicine (Dahan et al. 2003, Moronne et al. 1998). Gold (Au) nanoparticles have been used in a range of applications for many years due to their inert activity with biological tissues and have been considered for use in diagnosis and treatment for disease as mediums for targeting cancer cells and drug delivery, respectively (Chen et al. 2007, Daniel and Astruc 2004, Handley 1989, Murphy et al. 2008, Perrault et al. 2009). Magnetic nanoparticles in the form of iron oxide (Fe_3O_4) and cobalt ferrite ($CoFe_2O_4$) are being used in biomedical magnetic resonance imaging (MRI), cancer therapy, drug delivery and labeling of cells for diagnosis (Guglielmo et al. 2010, Hong et al. 2013, Pankhurst et al. 2003, Sincai et al. 2001). Copper (Cu) and copper-based (CuO) nanoparticles are being used in micro-electronic devices, magnetic recording equipment, and in catalysis (Cattaruzza et al. 2007, Gawande et al. 2016, Khanna et al. 2007, Lin et al. 2005). As just described, nanomaterials are synthesized using many types of elements and modified to suite their potential application; thus it remains difficult to foresee their potential toxicity in fish embryo development. By using fish embryos as a model system for toxicity, researchers can predict relative biocompatibility of different types of nanomaterials. Fish embryos are desirable as an aquatic organism model system because they share the same genetic material as humans. The following sections briefly highlight these different nanomaterials and their effect on development of fish embryos.

The toxicity of various types of nanomaterials on aquatic embryonic development has been widely investigated for many of the factors mentioned above. Ag nanoparticle transport and toxicity has been studied in embryonic development of several different types of fish species. In zebrafish (*Danio rerio*), Ag nanoparticles of small size (~ 12 nm) were determined to be able to transport by simple diffusion into developing embryos (Fig. 1) and invoke toxicity and death in a concentration dependent manner, producing a wide variety of deformities (Fig. 2) (Lee et al. 2007). Another study using zebrafish, found that Ag nanoparticles (~ 13 nm) were able to diffuse into the developing embryo at various developmental stages and caused stage dependent deformities during development in a concentration dependent mode (Fig. 3) (Lee et al. 2013a). An additional study using small Ag nanoparticles, ~ 12 nm in size, with surface charge modifications demonstrated charge dependent toxicity in zebrafish embryonic development, with the positively charged Ag nanoparticles being the least toxic to embryo development and the negatively charged being the most toxic (Lee et al. 2013b). Further studies done with different sizes of Ag nanoparticles (~ 40 nm and ~ 95 nm) indicate that as the size of the nanoparticle increases, there is an increase in the effect of the toxicity on zebrafish embryo development (Lee et al. 2012a, Lee et al. 2012b, Browning et al. 2013a). It was found that ~ 40 nm Ag nanoparticles were able to enter embryos by simple diffusion and cause deformity and death at a much lower concentration, acquainting toxicity to size dependence (Fig. 4) (Lee et al. 2012a). Larger Ag nanoparticles (~ 95 nm) were able to transport into developing zebrafish embryos and cause toxicity at lower doses than smaller Ag nanoparticles (Fig. 5) (Lee et al. 2012b, Browning et al. 2013a). In addition to zebrafish, fathead minnow (*Pimephalespromelas*) embryos were also exposed to varying sizes of Ag nanoparticles that

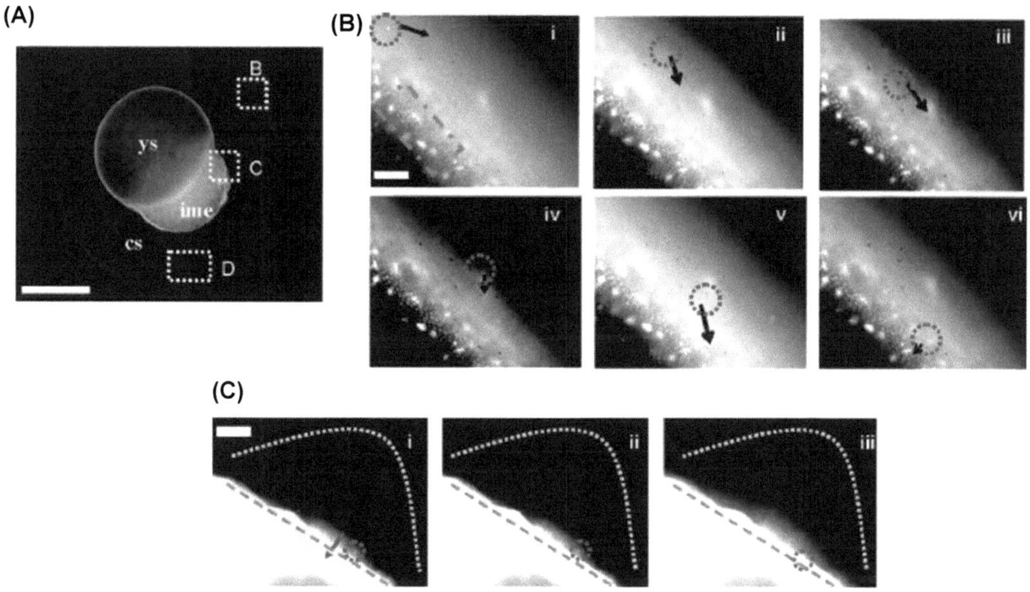

Figure 1. Real-time monitoring and characterization of transport of individual Ag nanoparticles in a cleavage-stage living embryo (64-cell stage; 2–2.25 hpf). (A) Optical image of a cleavage-stage embryo, showing chorionic space (cs), yolk sac (ys), and inner mass of the embroy (ime). The transport of single Ag nanoparticles at the interface of egg water with chorionic space, interface of chorionic space with inner mass of the embryo, water/chorionic space, as marked by B, C, respectively. Snapshots of transport of single nanoparticles at the interfaces of egg were used to distinguish them from tissue debris or vesicles in embroy are illustrated in (B) and (C). LSPR spectra (color) of individual nanoparticles were used to distinguish them from tissue debris or vesicles in embryos. (B) Sequential dark-flied optical images, illustrating the transport of single Ag nanoparticles, as indicated by the circle, from the egg water (extra-embryo) into the chorionoc space via an array of chorion pore canals (CPCs), highlighted by a rectangle, (C) Sequential dark-field optical images, illustrating the transport of single Ag nanoparticles, as indicated by the circle, from chorionic space into inner mass of the embryo. The time interval between sequential images in (B) and (C) is 25 s. Scale bar 400 μm (A) and 15 μm (B, C).

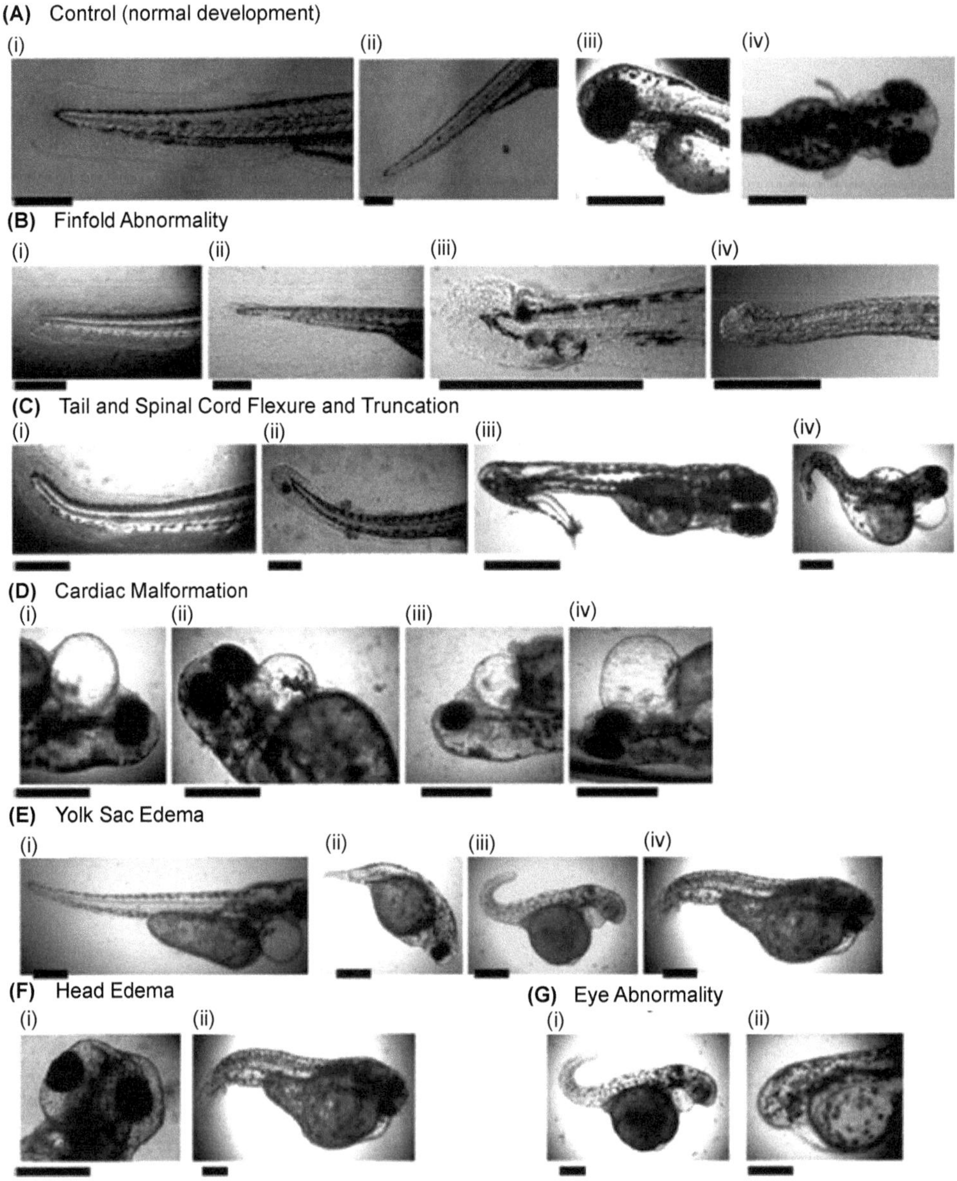

Figure 2. Representative optical images of (A) normally developed and (B–G) deformed zebrafish. (A) Normal development of (i) finfold, (ii) tail/spinal cord, (iii) cardiac, (iii, iv) yolk sac, cardiac, head, and eye. (B–G) Deformed zebrafish: (B) finfold abnormality; (C) tail and spinal cord flexure and truncation; (D) cadiac malformation; (E) yolk sac edema; (F) head edema, showing both (i) head edema and (ii) head edema and eye abnormality; (G) eye abnormality, showing both (i) eye abnormality and (ii) eyeless. Scale bar = 500 μm.

were able to transport and cause different deformities during the embryos' development (Fig. 6) (Laban et al. 2010). Furthermore, Japanese medaka fish (*Oryzias latipes*) embryos exposed to Ag nanoparticles, having an average diametre of ~ 30 nm, exhibited various morphological deformities as well (Fig. 7) (Wu and Zhou 2012). These studies indicate that varying sizes of Ag nanoparticles can certainly affect the toxicity and development of different aquatic organisms.

The influence of titanium dioxide (TiO_2) nanoparticle ecotoxicity on embryonic development of several different types of fish species has also been studied. TiO_2 nanoparticles (~ 25 nm) either 100 percent anatase or 20 percent/80 percent rutile/anatase were able to invoke low toxicity and death in zebrafish embryos, having a mortality rate of only 5 percent in anatase and a much higher rate for

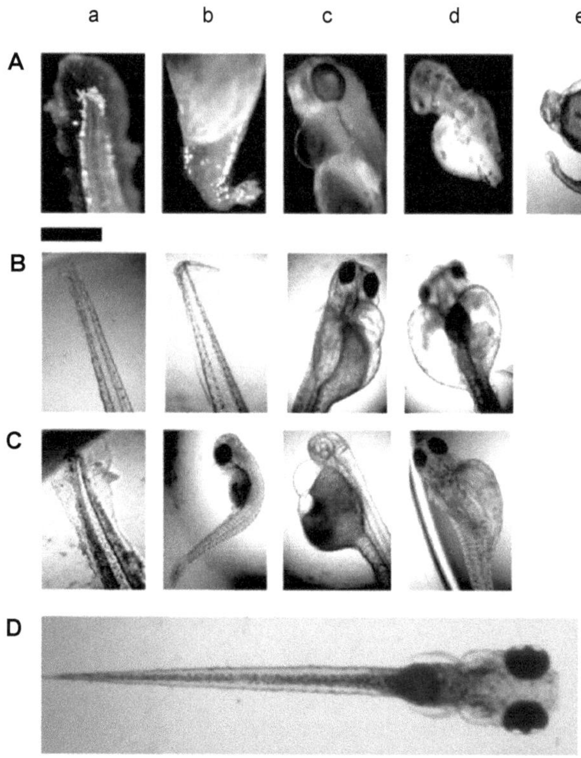

Figure 3. Optical images of (A–C) deformed and (D) normally developed zebrafish. (A–C) Deformed zebrafish are observed as (A) stage-I, (B) stage-III, and (C) stage-IV embryos have been incubated with the Ag NPs for 2 h (acute treatment), and develop in egg water over 120 hpf, which show (a) finfold abnormality; (b) tail/spinal cord flexure; (c) cardiac malformation/edema; (d) yolk sac edema, and (e*) and (f*) acephaly (*the severest and rare deformation with no-head, but beating heart). Scale bar is 500 μm for all images.

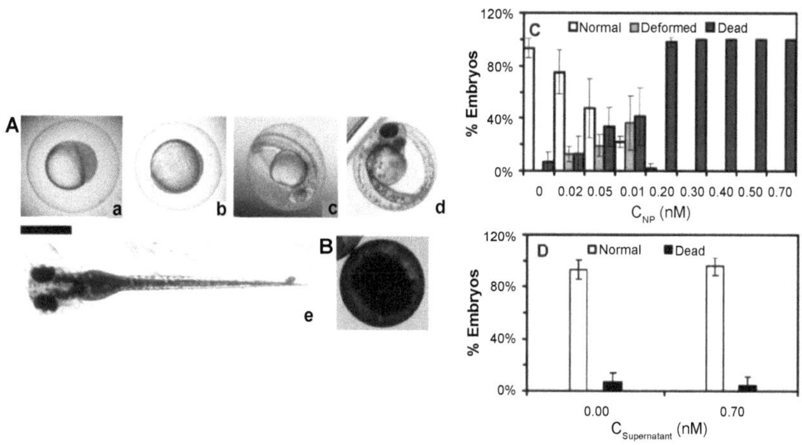

Figure 4. Study of dose-dependent nanotoxicity of Ag NPs (41.6 ± 9.1 nm) using zebrafish embryos as ultrasensitive *in vivo* assays. (A) Optical images of normally developing embryos at (a) cleavage stage (0.75–2.25 hpf); (b) late gastrula stage (10 hpf); (c) late segmentation stage (24 hpf); (d) hatching stage (48 hpf); and (e) fully developed larvae (120 hpf); (B) Dead embryo. (C) Histograms of distributions of embryos that developed into normal and deformed zebrafish or became dead versus NP concentration. (D) Control experiments: histograms of the distributions of embryos that developed into normal zebrafish or became dead either in egg water alone or versus supernatant concentration, which was collected from washing NPs with DI water. A total of 36 embryos were studied for each NP concentration and control in C–D. The percentage of embryos that developed into normal and deformed zebrafish or became dead was calculated by dividing its number by the total number of embryos used at each concentration. The means and standard deviations (error bars) of each given percentage at each given concentration from each of 3 replicates are presented. The scale bar is 500 μm for all images in A.

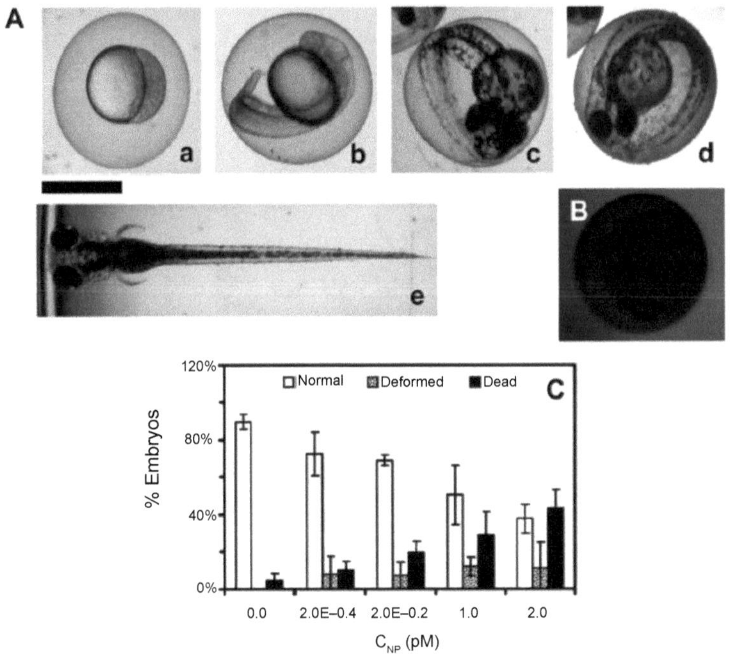

Figure 5. Study of toxic effects of Ag NPs (~ 95) on embryonic development. (A) Optical images of normally developing embryos at (a) cleavage-stage (0.75–2.25 hpf); (b) late segmentation stage (24 hpf); (c) early hatching stage (48 hpf); (d) late hatching stage (72 hpf); (e) fully developed larvae (120 hpf). (B) Dead embryo. (C) Histograms of distributions of embryos that developed to normal or deformed zebrafish or became dead versus NP concentration. Scale bar is 500 μm for all images in (A) and (B).

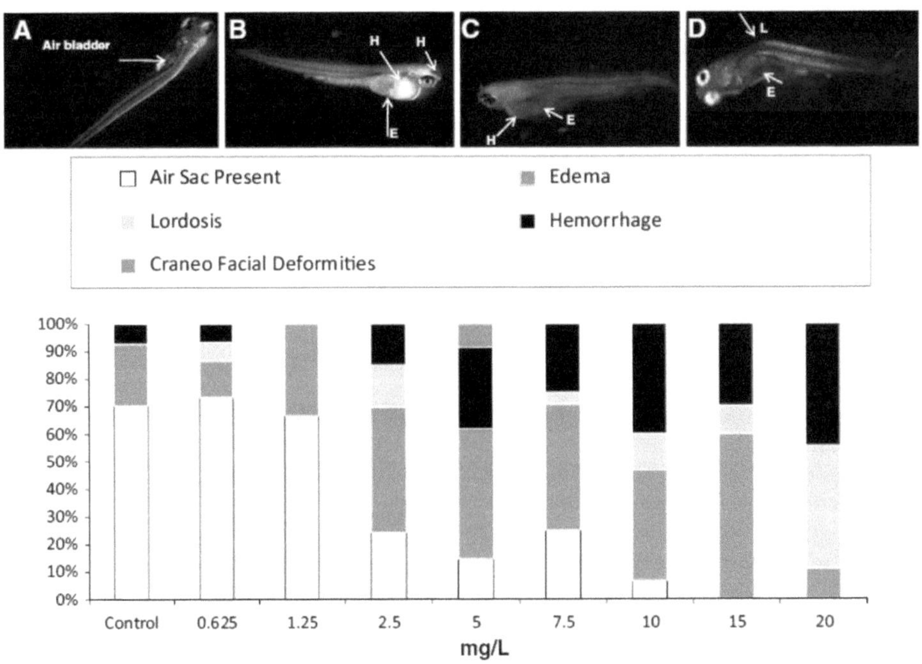

Figure 6. Captured images of fathead minnow larvae exposed to silver nanoparticles (Ag NPs) (a control; b and c exposed to 10 mg/L of NanoAmor and Sigma Ag NPs, respectively; and d exposed to 2.5 mg/L of NanoAmor Ag NPs). Note the absence of air bladder in treatment larvae. Bar graph summarizes the types and frequency of deformities per treatment group with N = 30 observed compared to controls (no Ag NP added).

Color version at the end of the book

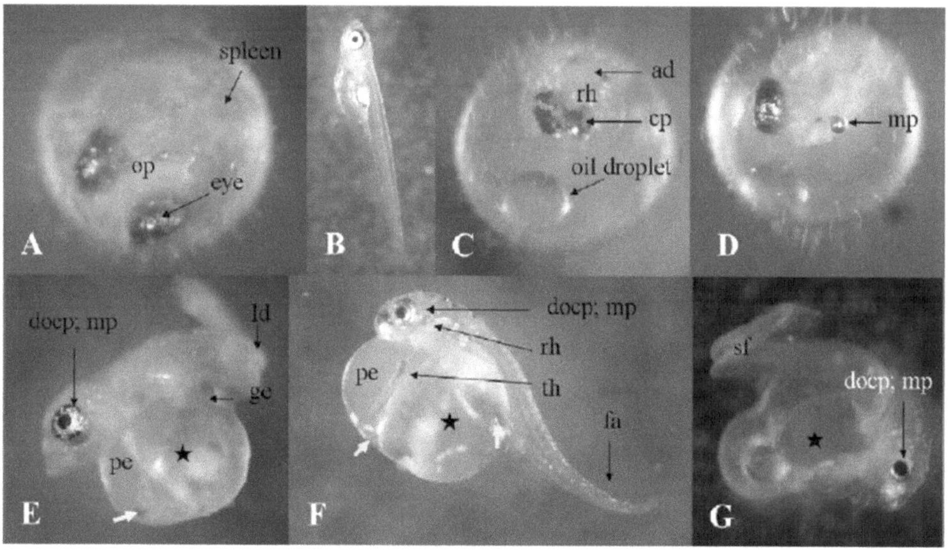

Figure 7. Representative morphological abnormalities of medaka larvae exposed to 62.5–1000 mg/L AgNPs during the embryonic stage. Embryos normally developed in the control (A and B). However, various abnormalities were observed in the AgNP-treated groups (C–G). ad, arrested development; cp, cyclopia; docp, decreased optic cup pigmentation; fa, finfold abnormality; ge, gallbladder oedema; Id, lordosis; mp, microphthalmia; op, optic tectum; pe, pericardial oedema; rh, reduced head; sf, skeletal flexture; and th, tubular heart. Stars mark the opaque and edematous yolk sac, whereas yellow arrowheads indicate hemostasis.

Color version at the end of the book

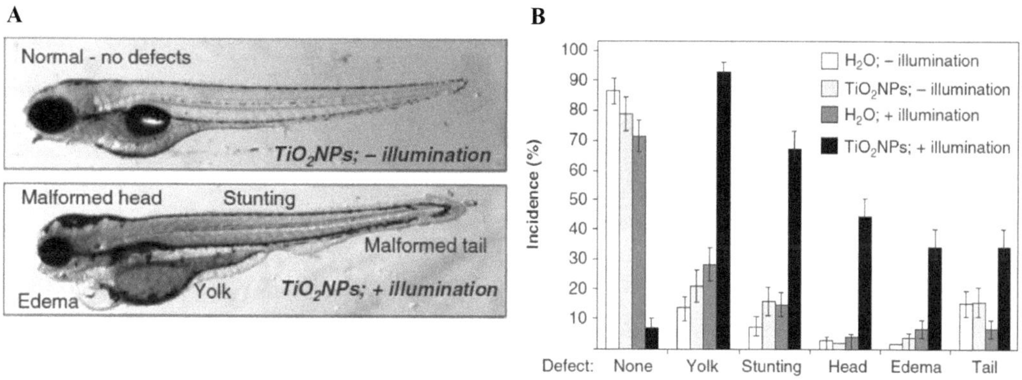

Figure 8. Malformations caused by TiO_2NPs exposure and illumination. (A) Representative micrographs of 120 hpf zebrafish embryo exposed to 100 mg/ml TiO_2NPs with and without illumination. (B) Incidence of morphological endpoints observed in embryos described above. Data are presented as mean ± SEM (n = 12 replicates, 6 fish/replicate point).

rutile/anatase at 36 percent (Clemente et al. 2014). However, when ultra-violet (UV) light was added, the mortality rate was increased (Clemente et al. 2014). This study concluded that the zebrafish larvae demonstrated equilibrium changes when exposed to both rutile/anatase TiO_2 nanoparticles and UV light (Clemente et al. 2014). This indicates that the structure of the nanomaterial and environmental conditions can play a significant role in increased ecotoxicity. Another study using TiO_2 nanoparticles (~ 25 nm) in a 3:1 anatase:rutile mixture were additionally found to cause illumination dependent toxicity, resulting in a variety of deformities in zebrafish embryonic development (Bar-Ilan et al. 2012). Zebrafish embryos exposed to TiO_2 nanoparticles and illuminated with a 250 W blue spectrum metal halide lamp displayed a malformed head, stunting, pericardial sac and yolk sac edema, and malformed tail (Fig. 8) (Bar-Ilan et al. 2012). Fathead minnow embryos exposed to TiO_2 nanoparticles

(~ 43 nm) did not suffer a significant change in mortality rate; however, these NMs did cause significant change in the innate immunity function demonstrated at the gene expression level and in cellular functioning (Jovanovic et al. 2011). Japanese medaka fish embryos exposed to TiO_2 nanoparticles having an average diameter of ~ 21 nm in a mixture of 80 percent/20 percentanatase/rutile, exhibited chorion deformity, early hatching from the chorion and pericardial sac edema (Paterson et al. 2011).

Zinc oxide (ZnO) nanoparticle toxicity has been studied in embryonic development of zebrafish. ZnO nanoparticles (~ 27 nm) and ZnO nanosticks with a width of ~32 nm and a length of ~ 81 nm were determined to cause toxicity, increased death and decreased hatching of embryos from the chorion in a concentration dependent manner (Hua et al. 2014). It was found that the ZnO nanosticks invoked more toxicity, having a higher death rate and lower hatching rate than the ZnO nanoparticles (Hua et al. 2014). This is interesting because it suggests that the shape of the nanomaterial plays a significant role in the rate of toxicity during embryonic development. A similar study using ZnO nanoparticles (~ 37 nm) found decreased hatching rates with increased concentrations;the NMs were able to cause toxicity and deformities in the form of shortened body length and tail malformations (Fig. 9) (Bai et al. 2010a).

Fullerene nanomaterial toxicity has been studied in embryonic development in different types of fish species. Fullerene nanospheres (C60 and C70) demonstrated toxicity and death in a dose dependent manner producing a wide variety of deformities (Fig. 10) (Usenko et al. 2007). Zebrafish embryos exhibit pericardial sac edema, yolk sac edema and finfold and tail malformations (Fig. 10) (Usenko et al. 2007). Another study using nanoscale buckminster fullerene (C60) showed increase in death, decrease in hatching rate and deformities in zebrafish embryo development (Zhu et al. 2007). Zebrafish embryos exposed to C60 had decreased heart rates as compared to the controls and demonstrated pericardial sac edema (Zhu et al. 2007). Further studies with fullerene single-walled carbon nanotubes (SWCNTs) having an average diameter of ~ 11 nm and length of 0.5 to 100 mm indicated there was a delay in hatching during zebrafish development, but these SWCNTs did not cause abnormal embryonic development (Cheng et al. 2007). This could be related to the inability of the SWCNTs to enter through the chorion surrounding the developing zebrafish embryo, thus providing protection during this most sensitive time. This is interesting because it suggests that the structure and size of the fullerene greatly influences the toxicity in the zebrafish embryonic development. Rare minnow (*Gobiocyprisrarus*) embryos were exposed to fullerene SWCNTs having an inner diameter of 0.8 to 1.6 nm and a length of 0 to 1300 nm; these SWCNTs did not cause toxicity to the developing embryos most likely due to the protection of the chorion. However, they were able to transport inside the rare minnow larvae post-hatching (Zhu et al. 2015). These SWCNTs were able to cause toxicity and deformities in the larvae in the form of decreased heart rate, shortened body length, altered swimming speed and increased cellular death (Zhu et al. 2015). The more exposure the nanomaterial has to the developing aquatic organism, the more likely there will be an increase in the severity of toxicity.

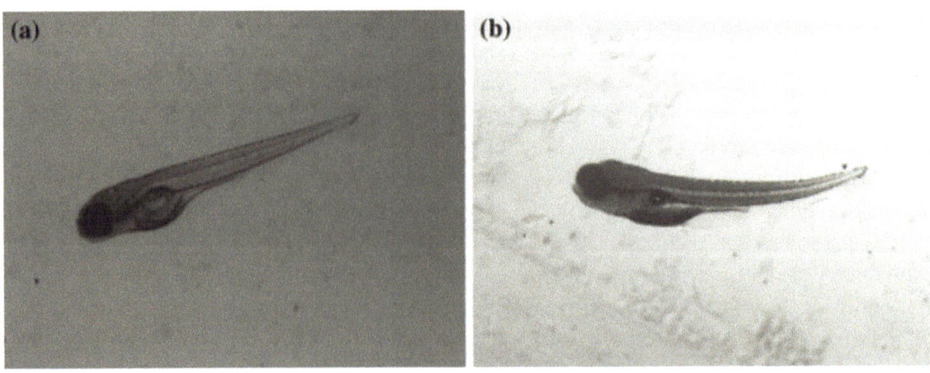

Figure 9. Representative images of the (a) normal and (b) malformed larvae (from ZnO NP treatment).

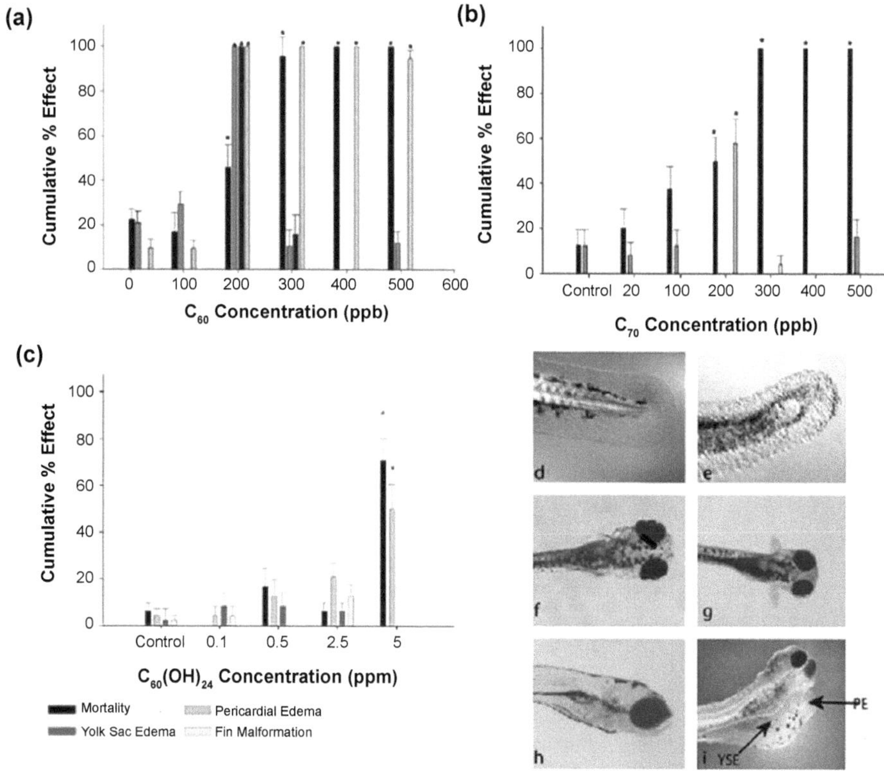

Figure 10. Concentration-responses observed for embryonic zebrafish exposed to (a) C_{60}, (b) C_{70}, and (c) $C_{60}(OH)_{24}$ from 24 to 96 hpf; evaluated daily until 6 dpf. Values represent the % showing the effect by day 6 (cumulative % effect): mortality, pericardial edema, yolk sac edema, and fin malformations. Representative images of the caudal fins for (d) control and (e) 200 ppb C_{60}-exposed animals are given. Representative images of the pectoral fin for (f) control and (g) 3500 ppb $C_{60}(OH)_{24}$-exposed animals. Representative images of (h) 1% DMSO control head at 6 dpf; arrows designate pericardial edema (PE) and yolk sac edema (YSE). Significance was determined using Fisher's Exact test (*$p < 0.05$) compared to 1% DMSO control (N = 24).

Fluorescent silica (SiO_2) nanoparticle ecotoxicity has also been studied in embryonic development of different types of fish species. SiO_2 nanoparticles (~ 60 nm), synthesized with fluorescent dye, Cy5.5, inside the core shell, were not able to transport through the chorion into developing zebrafish embryos and the Cy5.5-SiO_2 nanoparticles were only found to be on the chorion surface (Fig. 11) (Fent et al. 2010). Treatment with the Cy5.5-SiO_2 nanoparticles was determined to not cause toxicity in zebrafish embryo survival or development (Fent et al. 2010). The Cy5.5-SiO_2 inability to cross the chorion interface and come in contact with the developing embryo is why there was no toxicity observed. It is possible that the reason the SiO_2 nanoparticles were not able to enter the chorion is because they became clumped and aggregated on the chorion surface and thus, too large to penetrate the chorion and gain access to the developing embryo (Fent et al. 2010). Japanese medaka fish embryos exposed to SiO_2 nanoparticles having an average diameter of ~ 50 nm with rhodamine B (RhB) were found to attach to the chorion membrane surface and not able to transport through the chorion into developing medaka embryos (Fig. 12) (Lee et al. 2011). The RhB-SiO_2 nanoparticles were determined to not cause significant toxicity as compared to the controls related to medaka embryo hatching rate, mortality or developmental abnormalities (Lee et al. 2011). Similar to the Cy5.5-SiO_2 nanoparticles in zebrafish, this may be due to the inability of the RhB-SiO_2 nanoparticles to transport through the chorion and come in contact with the developing embryo. It was interesting, however, when embryos with RhB-SiO_2 nanoparticles were placed under sonication, there were significant increases in mortality and abnormalities, though hatching rate was not affected (Lee et al. 2011). Abnormalities in development observed consisted of delayed swimming post-hatching, ocular deformity, tube-shaped heart, blastula

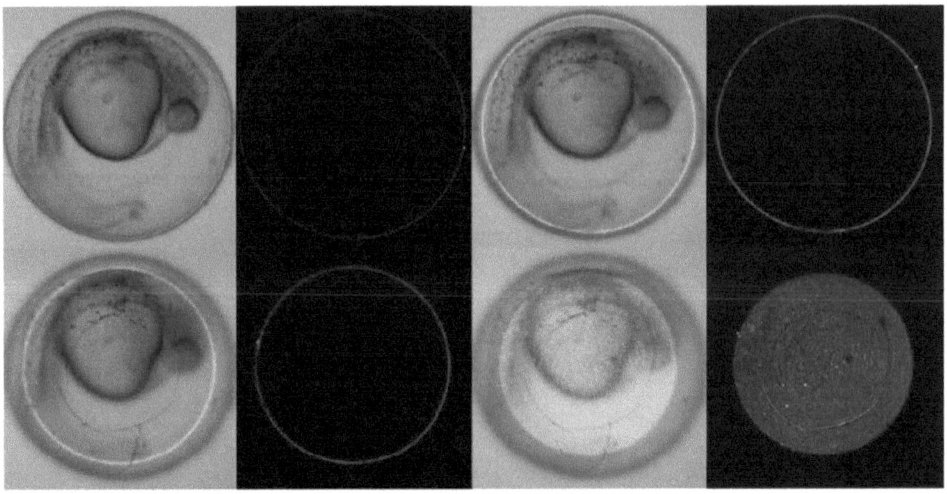

Figure 11. Sequence of images with increasing depth from the upper part to the lower part of the chorionated embryo from picture up left, down left to right up. Embryos were exposed for 24 hr to FSNP of 60 nm size. Left are transmission microscopic pictures, and on the right the same picture in the confocal microscope. Clearly visible is that green fluorescent FSNP adsorb to the chorion, but are not penetrating inside. The picture on the right side down stems from an egg that has been pushed tightly to the surface of the microscopic glass showing the chorion surface completely covered by FSNP.

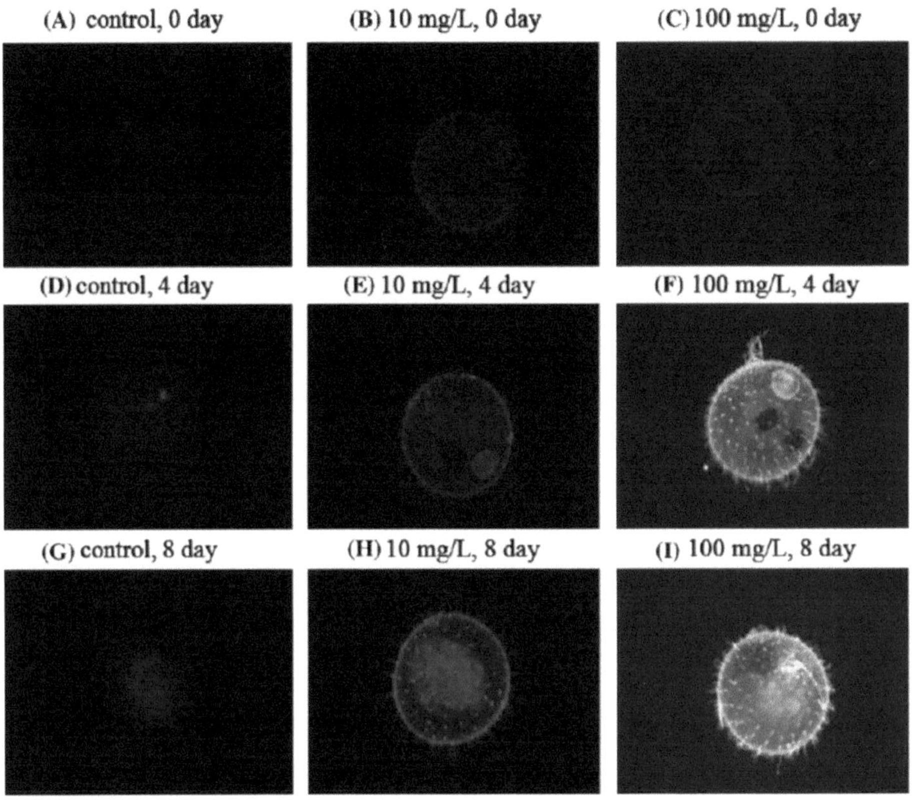

Figure 12. Fluorescent images of O. latipes embryos: (A) in the absence of Fluorescent nanoparticles (FNPs) (9–10 stages), (B) in the presence of 10 mg FNP (9–10 stage), (C) in the presence of 100 mg FNP (9–10 stage), (D) in the absence of FNPs after 4 d (31–33 stages), (E) in the presence of 10 mg FNP 4 d (31–33 stage), (F) in the presence of 100 mg FNP after 4 d (31–33 stages), (G) in the absence of NPs after 8 d (stage 38), (H) in the presence of 10 mg FNP after 8 d (stage 38), and (I) in the presence of 100 mg FNP after 8 d (stage 38).

stage edema and abnormal blood circulation in the tail (Lee et al. 2011). It was suggested that the toxicity could be due to the degradation of thc RhB-SiO2 nanoparticles and release of silicate ions, though future studies are required to determine this for certain (Lee et al. 2011). This study brings forth the realization that in a natural aquatic environment, the environmental parameters and interactions with nanomaterials can be unpredictable due to nanomaterial dissolution and can lead to a varied effect of ecotoxicity on the surrounding conditions.

Quantum dot (QD) nanoparticle toxicity has been studied in embryonic development of the zebrafish. Cadmium-telluride (CdTe) QDs, coated with a stabilizing agent, thioglycolic acid (TGA), of ultra-small size (~ 3.5 nm) were determined to cause toxicity in the form of deformities, increased death,and decreased hatching of embryos from the chorion in a concentration dependent manner (Zhang et al. 2012). TGA-CdTe QDs were able to cause toxicity and deformities in embryonic development presenting as eye malformation, abnormal pigment development, pericardial sac edema, abnormal tail and body flexure and decreased somite development (Fig. 13) (Zhang et al. 2012). This is noteworthy because it suggests that the type of the nanomaterial plays a significant role in the rate of toxicity during embryonic development. Another study explored zebrafish embryo nanotoxicity using a cadmium selenide (CdSe) core and zinc sulphide (ZnS) shell QDs, functionalized with poly-L-lysine (PLL) having a size of ~ 328 nm, methoxy-terminated polyethylene glycol (PEG)350-thiol (PEG350-OCH3) with a size of ~ 600 nm, and PEG5000-thiol terminated with carboxylate, methoxy, or amine functional groups (PEG5000-COO-, PEG5000-OCH3, PEG5000-NH3), with sizes of ~ 23, ~ 44, and 21 nm, respectively (King-Heiden et al. 2009). It was found that the smaller QDs, PEG5000-COO-, PEG5000-OCH3, PEG5000-NH3, caused toxicity in the form of deformities, ranging from abnormal axial curvature, submandibular edema, pericardial sac edema, yolk sac edema and malformation,

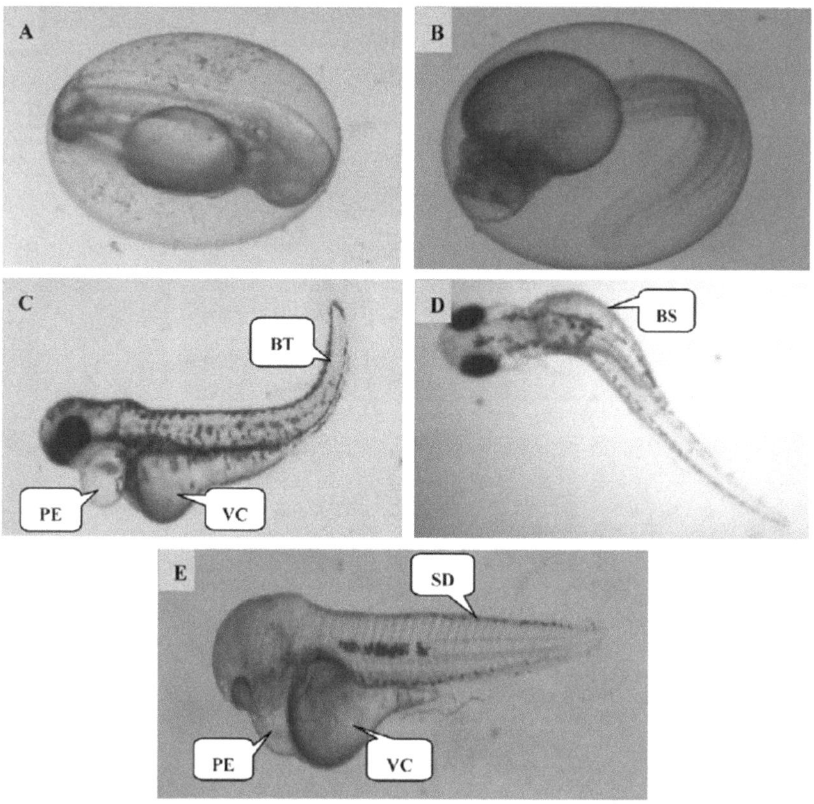

Figure 13. Malformation patterns of zebrafish after exposure to 200 nM TGA-CdTe: (A) eyespots developmental inhibition, (B) melanin developmental inhibition. (C) pericardial edema (PF), vitelline cyst (VC), bent tail (BT), (D) bent spine (BS), and (E) pericardial edema (PE), vitelline cyst (VC), somites decrease (SD).

ocular edema, and tail malformation (King-Heiden et al. 2009). The large particle QDs, PLL and PEG350-OCH3, were found to not cause deformities in embryonic development (King-Heiden et al. 2009). These results suggest that the size of the QD played a role in the toxicity. Similar to the SiO_2 nanoparticles, the large particle QD more than likely was not able to penetrate the zebrafish chorion during development, whereas the smaller QD nanoparticles were able to penetrate the chorion and developing embryo, causing specific developmental deformities. This is another example that the size of the nanoparticle can possibly contribute to the ecotoxicity of aquatic organisms.

Au nanoparticle transport and toxicity has been studied in zebrafish embryonic development. Au nanoparticles (~ 12 nm) were determined to be able to transport by simple diffusion into developing zebrafish (*Danio rerio*) embryos (Fig. 14) and cause some toxicity resulting in a small increase in death in a concentration dependent manner, producing only a few deformed zebrafish (Browning et al. 2009). The Au nanoparticles demonstrate less toxicity and death as compared to the similar sized Ag nanoparticle study mentioned earlier, supporting the idea that nanomaterials composed of different elements can have a significant difference on aquatic embryo developmental toxicity. Further studies done with larger sizes of Au nanoparticles (~ 86 nm) found that there was an even lesser amount of a toxicity effect on zebrafish embryo development (Browning et al. 2013b). It was determined that ~ 86 nm Au nanoparticles were able to enter embryos by simple diffusion (however, they caused less deformity and death in a concentration independent manner (Browning et al. 2013b). The toxicity with larger Au nanoparticles was less than the smaller Au nanoparticles suggesting that perhaps the smaller nanoparticles induce more toxicity related to their small size. An additional study using small Au nanoparticles of 1.3 nm in size with positive surface charge modifications demonstrated concentration dependent death and toxicity inducing eye defects, pericardial sac edema, yolk sac edema and abnormal tail flexure (Fig. 15) (Kim et al. 2013). This suggests that the surface modification

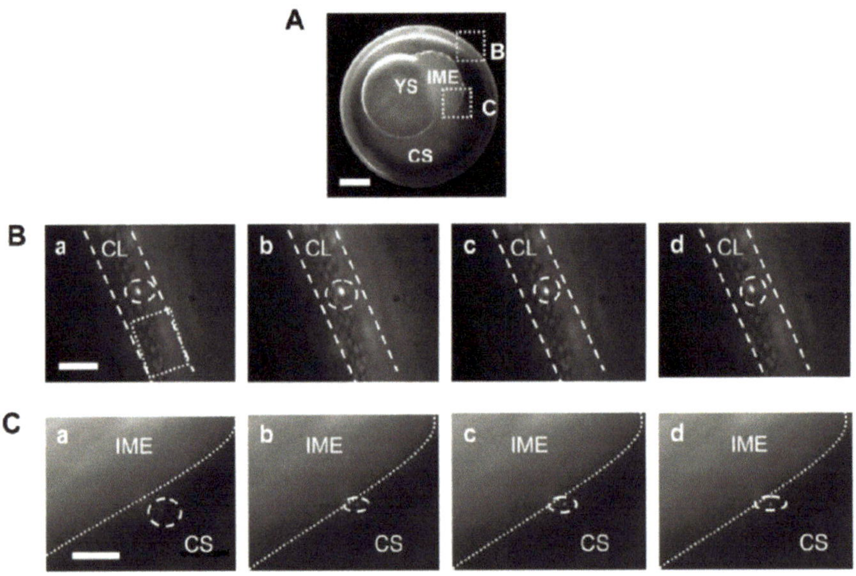

Figure 14. Real-time imaging of the diffusion and transport of single Au nanoparticles in a cleavage-stage zebrafish embryo. (A) Optical image of the cleavage-stage embryo shows chorion, chorionic space (CS), yolk sac (YS), and inner mass of embryo (IME), acquired by CCD camera. The transport of single Au nanoparticles at the interface of the chorion with egg water and at the interface of CS with the IME is illustrated in (B–C), respectively. Scale bar = 200 μm. (B) Sequential dark-field optical images of the chorionic layer (CL) illustrate the transport of single Au nanoparticles (circle), from the egg water into the CS via chorionic pore canals (square). The array of well-organized chronic pore canals is clearly visualized and determined as 0.5–0.7 μm in diameter, with each pore about 1.5–2.5 μm apart. The straight dashed lines outline the CL. The time interval between: (a) and (b) is 2.75 s; (b) and (c) is 3.92 s; and (c) and (d) is 5.10 s. Scale bar = 10 μm. (C) Sequential dark-field optical images of the interface of CS with the IME illustrate the transport of single Au nanoparticles (circle), from the CS into the IME. The dotted lines outline the interface of the CS and the IME. The time interval between: (a) and (b) is 5.88 s: (d) and (c) is 5.89 s; and (c) and (d) is 3.14 s. Scale bar = 20 μm.

of even a less toxic type of nanomaterial has the ability to cause ecotoxicity; thus, it is important to consider this fact for all manufactured nanomaterials before releasing them into the environment.

Magnetic nanoparticles in the form of iron oxide (Fe_3O_4) and cobalt ferrite ($CoFe_2O_4$) have been studied for toxicity during embryonic development of zebrafish. Fe_3O_4 nanoparticles (~ 30 nm) were used to treat developing zebrafish embryos in a dose dependent manner; however, when the Fe_3O_4 nanoparticles were placed in the zebrafish egg water media, they would aggregate to a larger particle size of ~ 1 mm (Zhu et al. 2012). The larger aggregated Fe_3O_4 particles were found to cause toxicity in the form of deformities, increased death and delayed hatching of zebrafish embryos from the chorion in a concentration dependent manner (Zhu et al. 2012). Aggregated Fe_3O_4 particles were able to cause deformities in the form of pericardial sac edema, tissue ulceration and abnormal body flexure (Fig. 16) (Zhu et al. 2012). This is noteworthy because it suggests that the stability of the nanoparticle is questionable when considering the environment of living aquatic organisms, can play a significant role in the rate of aggregation of particles and can have an outcome on the toxicity during embryonic development. Another study using $CoFe_2O_4$ nanoparticles (~ 40 nm) determined that $CoFe_2O_4$ nanoparticles caused delayed hatching rates, apoptosis, unstable heart rate and decreased metabolic rate with increased concentrations (Ahmad et al. 2015). $CoFe_2O_4$ nanoparticles were able to cause toxicity and deformities the form of pericardial sac edema, yolk sac edema and abnormal tail flexure (Ahmad et al. 2015).

Copper (Cu) and copper-based (CuO) nanoparticles have been studied for toxicity during embryonic development in zebrafish. Cu nanoparticles (~ 70 nm) were used to treat developing

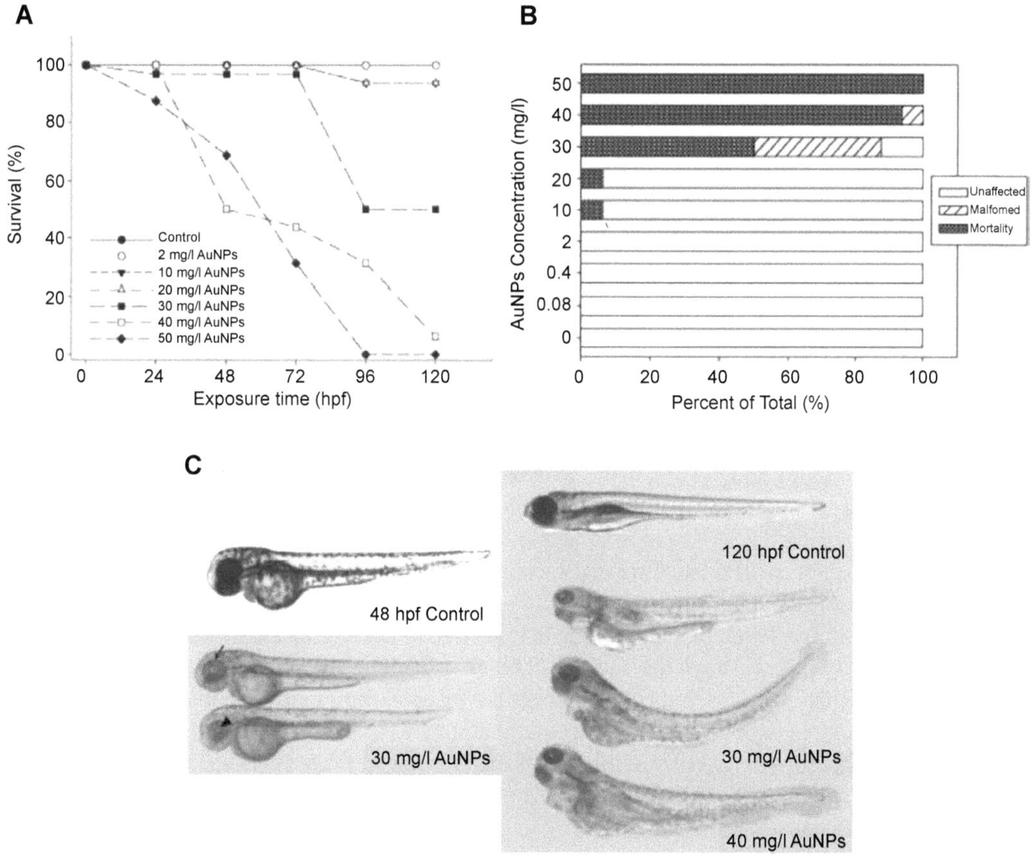

Figure 15. (A) Concentration-response relationship (n = 32). (B) The percentage of unaffected, malformed, and mortality at 120 hpf (n = 32). (C) Examples of malformations at 30 and 40 mg/l at 48 and 120 hpf. Eye defects were observed from 48 hpf, and other malformations such as pericardial edema, yolk sac edema, and bent spine were observed at 120 hpf. Black arrow and black arrowhead indicate the boundary of neural retina and the lens, respectively.

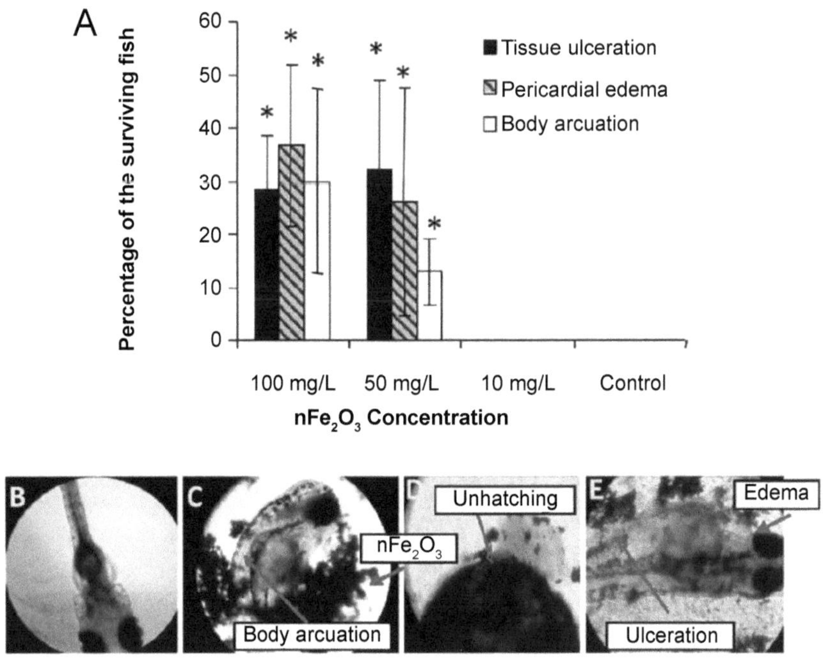

Figure 16. Malformations (e.g., pericardial edema, tissue ulceration, and body arcuation) induced by nFe_2O_3 at 168 hpf. (A) Malformation percentage in the surviving fish; (B) control fish; (C) hatching fish with body arcuation, treated with 50 mg/L of nFe_2O_3 aggregates; (D) unhatching embryos, treated with 50 mg/L of n Fe_2O_3 aggregate, dead at 168 hpf (E) hatching fish with pericardial edema, treated with 100 mg/L of nFe_2O_3 aggregates. Error bars represent 6 one standard deviation from the mean of three replicates. Significance indicated by: *p, 0.05.

Color version at the end of the book

zebrafish embryos, however, when the Cu nanoparticles were placed in the zebrafish media, they would aggregate to a larger particle size of ~ 107 nm and ~ 162 nm at the beginning of exposure (Bai et al. 2010b). At 24 hr the Cu nanoparticles had aggregated to even larger particles with sizes of ~ 178 nm and ~ 355 nm, and these particles were found to cause toxicity in the form of deformities, increased death and delayed hatching of zebrafish embryos from the chorion in a concentration dependent manner (Bai et al. 2010b). The aggregated Cu particles caused malformations in the form of yolk sac edema, notochord deformity and abnormal body flexure (Fig. 17) (Bai et al. 2010b). This is important because it suggests that the stability of the nanoparticle is questionable when considering the environment of living aquatic organisms and can play a significant role in the rate of aggregation of particles as well as have an outcome on the toxicity during embryonic development. Another study using small CuO nanoparticles (~ 6 nm) that would aggregate to large CuO particles ~ 1000 nm in size were found to cause decreased hatching and increased death rate with increasing concentrations in zebrafish embryo development (Thit et al. 2017). CuO particles were not determined to cause specific deformities in zebrafish development and this may be due to the aggregation of CuO nanoparticles and their inability to enter through the chorion (Thit et al. 2017).

From this chapter it is clear that nanomaterials are being used more frequently in our daily lives and have great potential to contribute to the contamination of our aquatic environments. The release of nanomaterials via many different possible routes in our water sources will pose the risk of polluting the natural bodies of water and contaminating the aquatic environments. This will increase the chances of interactions with aquatic organisms in their natural environment and threat of their development being affected. It was demonstrated that there are many factors that can cause the ecotoxicity of nanomaterials to affect the development of fish embryos in an aquatic environment. The factors such as aggregation and the ability for nanomaterials to have varying sizes greatly contribute to the toxicity in the developing fish embryo. Another factor that was shown to differ the toxicity level is

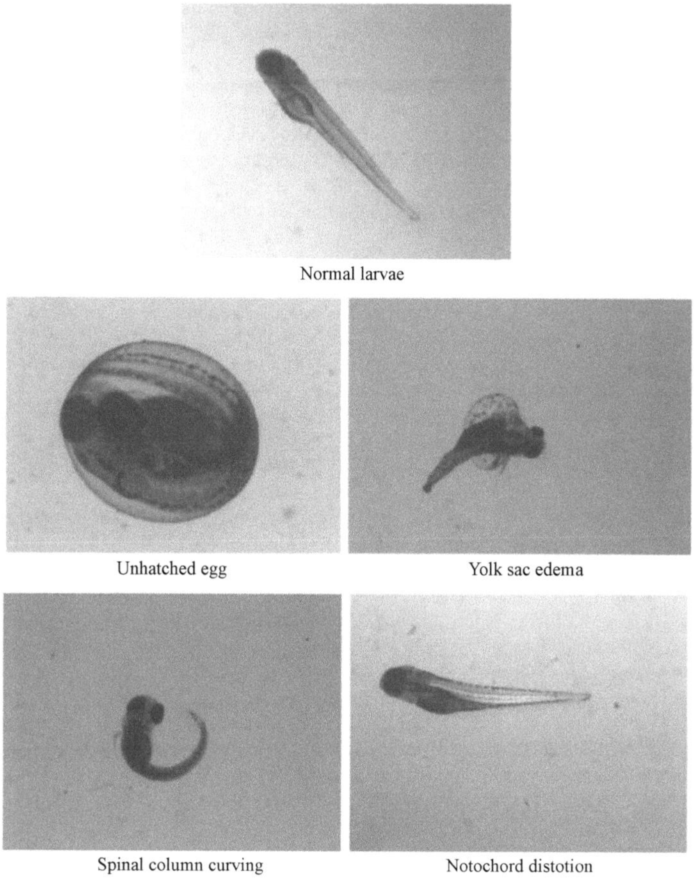

Figure 17. Representative images of the malformation of embryos and larvae (from Cu NP treatment).

the surface stabilizers and surface chemistry modifications on nanomaterials. It was also determined that the type and concentration of the nanomaterial used poses the greatest variation in risk for the severity in toxicity to aquatic organism embryo development.

References

Agrawal, A., T. Sathe and S. Nie. 2007. Nanoparticle Probes for Ultrasensitive Biological Detection and Imaging in New Frontiers in Ultrasensitive Bioanalysis: Advanced Analytical Chemistry Applications in Nanobiotechnology, Single Molecule Detection, and Single Cell Analysis. Wiley, New York.

Ahmad, F., X. Liu, Y. Zhou and H. Yao. 2015. An *in vivo* evaluation of acute toxicity of cobalt ferrite ($CoFe_2O_4$) nanoparticles in larval-embryo zebrafish (*Danio rerio*). Aquatic Toxicol. 166: 21–28.

Bai, W., Z. Zhang, W. Tian, X. He, Y. Ma, Y. Zhao et al. 2010a. Toxicity of zinc oxide nanoparticles to zebrafish embryos: a physicochemical study of toxicity mechanism. J. Nanopart. Res. 12: 1645–1654.

Bai, W., W. Tian, Z. Zhang, X. He, Y. Ma, N. Liu et al. 2010b. Effects of copper nanoparticles on the development of zebrafish embryos. J. Nanosci. Nanotechnol. 10: 1–7.

Bar-Ilan, O., K.M. Louis, S.P. Yang, J.A. Pederson, R.J. Hamers, Richard El Peterson et al. 2012. Titanium dioxide nanoparticles produce phototoxicity in the developing zebrafish. Nanotoxicol. 6: 670–679.

Boxall, A.B.A., K. Tiede and Q. Chaudhry. 2007. Engineered nanomaterials in soils and water: how do they behave and could they pose a risk to human health? Nanomedicine 2.6: 919–938.

Brigger, I., C. Dubernet and P. Couvreur. 2002. Nanoparticles in cancer therapy and diagnosis. Adv. Drug Deliv. Rev. 54: 631–651.

Brown, D.M., M.R. Wilson, W. MacNee, V. Stone and K. Donaldson. 2001. Size-dependent proinflammatory effects of ultrafine polystyrene paricles: a role for surface area and oxidative stress in the enhanced activity of ultrafines. Toxicol. Appl. Pharmacol. 175: 191–199.

Browning, L.M., K.J. Lee, T. Huang, P. Nallathamby and X. Xu. 2009. Random walk of single gold nanoparticles in zebrafish embryos leading to stochastic toxic effects on embryonic developments. Nanoscale. 1: 138–152.
Browning, L. M.,K. J. Lee, P. Nallathamby, X. Xu. 2013a. Silver nanoparticles incite size- and dose-dependent developmental phenotypes and nanotoxicity in zebrafish embryos. Chem. Res. Toxicol. 26: 1503–1513.
Browning, L.M.,T. Huang and X. Xu. 2013b. Real-time *in vivo* imaging of size-dependent transport and toxicity of gold nanoparticles in zebrafish embryos using single nanoparticle optical spectroscopy. Interface Focus. 3: 20120098.
Burns, A.A., J. Vider, H. Ow, E. Herz, O. Penate-Medina, M. Baumgart et al. 2008. Fluorescent silica nanoparticles with efficient uring excretion for nanomedicine. Nano. Lett. 9: 442–448.
Cattaruzza, E., G. Battaglin, F. Gonella, R. Polloni, B.F. Scremin, G. Mattei et al. 2007. Au-Cu nanoparticles in silica glass as composite material for photonic applications. Appl. Surf. Sci. 254: 1017–1021.
Chen, Y.-H., C.-Y. Tsai, P.-Y. Huang, M.-Y. Chang, P.-C. Cheng, C.-H. Chou et al. 2007. Methotrexate conjugated to gold nanoparticles inhibits tumor growth in a syngeneic lung tumor model. Mol. Pharm. 4: 713–722.
Cheng, J., E. Flahaut, and S. H. Cheng. 2007. Effect of carbon nanotubes on developing zebrafish (*Danio rerio*) embryos. Environ. Toxicol. and Chem. 26: 708–716.
Clemente, Z., V.L.S.S. Castro, M.A.M. Moura, C.M. Jonsson and L.F. Fraceto. 2014. Toxicity assessment of TiO_2 nanoparticles in zebrafish embryos under different exposure conditions. Aquatic Tox. 147: 129–139.
Cui, Z., P.R. Lockman, C.S. Atwood, C.-H. Hsu, A. Gupte, D.D. Allen et al. 2005. Novel D-penicillamine carrying nanoparticles for metal chelation therapy in Alzheimer's and other CNS diseases. Eur. J. Pharm. Biopharm. 59: 263–272.
Dahan, M., S. Levi, C. Luccardini, P. Rostaing, B. Riveau, A. Triller et al. 2003. Diffusion dynamics of glycine receptors revealed by single-quantum dot tracking. Science. 302: 442–445.
Dahl, A., A. Gharibi, E. Swietlicki, A. Gudmundsson, M. Bohgard, A. Ljungman et al. 2006. Traffic-generated emissions of ultrafine particles from pavement-tire interface. Atmospheric Environ. 40: 1314–1323.
Daniel, M.C. and D. Astruc. 2004. Gold nanoparticles: assembly, supramolecular chemistry, quantum size-related properties, and applications toward biology, catalysis, and nanotechnology. Chem. Rev. 104: 293–346.
Dobson, J. 2001. Nanoscale biogenic iron oxides and neurodegenerative disease. FEBS Lett. 496: 1–5.
Fent, K., C.J. Weisbrod, A. Wirth-Heller and U. Pieles. 2010. Assessment of uptake and toxicity of fluorescent silica nanoparticles in zebrafish (*Danio rerio*) early life stages. Aquatic Toxicol. 100: 218–228.
Gawande, M. B., A. Goswami, F-X. Felpin, T. Asefa, X. Huang, R. Silva et al. 2016. Cu and Cu-based nanoparticles: synthesis and applications in catalysis. Chem. Rev. 116: 3722–3811.
Guglielmo, C.D, D.R. Lopez, J. Lapuente De, J.M. LlobetMallafre, M.B. Suarez. 2010. Embryotoxicity of cobalt ferrite and gold nanoparticles: a first *in vitro* approach. Reprod. Toxicol. 30: 271–276.
Handley, D. A. 1989. Colloid gold: principles, methods and applications. Academic Press, New York.
Hoet, P.H.M., I. Bruske-Hohlfield, O.V. Salata. 2004. Nanoparticles—known and unknown health risks. J. Nanobiotechnol. 2: 12.
Hong, N.H., A.T. Raghavender, P.M.H. Ciftja, K.S.H. Stojak and Y.H. Zhang. 2013. Ferrite nanoparticles for future heart diagnostics. Appl. Phys. A. 112: 323–327.
Howard, C.V. 2004. Small particles—big problems. Int. Lab. News. 34: 28–29.
Hua, J., M.G. Vijver, M.K. Richardson, F. Ahmad and W.J.G.M. Peijnenburg. 2014. Particle-specific toxic effects of differently shaped zinc oxide nanoparticles to zebrafish embryos (*Danio rerio*). Environ. Toxicol. and Chem. 33: 2859–2868.
Jovanovic, B., L. Anastasova, E.W. Rowe, Y. Zhang, A.R. Clapp and D. Palic. 2011. Effects of nanosized titanium dioxide on innate immune system of fathead minnow (*Pimephalespromelas* Rafinesque, 1820). Ecotoxicol. Environ. Safety. 74: 675–683.
Khanna, P.K., S. Gaikwad, P.V. Adhyapak, N. Singh and R. Marimuthu. 2007. Synthesis of nano-particles of anatase-TiO_2 and preparation of its optically transpartent film in PVA. Mater. Lett. 61: 4725–4730.
Kim, K.-T., T. Zaikova, J.E. Hutchison and R.L. Tanguay. 2013. Gold nanoparticles disrupt zebrafish eye development and pigmentation. Toxicolog. Sci. 133: 275–288.
King-Heiden, T., P. Wiecinski, A. Mangham, K.M. Metz, D. Nesbit, J.A. Pederson et al. 2009. Quantum dot nanotoxicity assessment using the zebrafish embryo. Environ. Sci. Techol. 43: 1605–1611.
Laban, G., L.F. Nies, R.F. Turco, J.W. Bickham and M.S. Sepulveda. 2010. The effects of sliver nanoparticles on fathead minnow (*Pimephalespromelas*) embryos. Ecotoxicol. 19: 185–195.
Lee, K.J., P.D. Nallathamby, L.M. Browning, C.J. Osgood and X.H.N. Xu. 2007. *In vivo* imaging of transport and biocompatibility of single silver nanoparticles in early development of zebrafish embryos. ACS Nano. 1: 133–143.
Lee, K.J., L. Browning, P. Nallathamby, T. Desai, P. Cherukuri and X. Xu. 2012a. *In vivo* quantitative study of sized-dependent transport and toxicity of single silver nanoparticles using zebrafish embryos. Chem. Res. in Tox. 25: 1029–1046.
Lee, K.J., P. Nallathamby, L. Browning, T. Desai, P. Cherukuri and X. Xu. 2012b. Single nanoparticle spectroscopy for real-time *in vivo* quantitative analysis of transport and toxicity of single silver nanoparticles in single embryos. Analyst. 137: 2973–2986.
Lee, K.J., L. Browning, P. Nallathamby, C. Osgood and X. Xu. 2013a. Silver nanoparticles induce developmental stage-specific embryonic phenotypes in zebrafish. Nanoscale. 5: 11625–11636.
Lee, K.J., L. Browning, P. Nallathamby and X. Xu. 2013b. Study of charge-dependent transport and toxicity of peptide-functionalized silver nanoparticles using zebrafish embryos and single nanoparticle plasmonic spectroscopy. Chem. Res. in Tox. 26: 904–917.
Lee, W.-M., S.-W. Ha, C.-Y. Yang, J.-K. Lee and Y.-J. An. 2011. Effect of fluorescent silica nanoparticles in embryo and larva of *Oryzias latipes*: sonic effect in nanoparticle dispersion. Chemosphere. 82: 451–459.

Lin, C.J., S.L. Lo and Y.H. Liou. 2005. Degradation of aqueous carbon tetrachloride by nanoscale zerovalent copper on a cation resin. Chemosphere. 59: 1299–1307.

Moronne Jr., M.M.B., P. Gin, S. Weiss and A.P. Alivisatos. 1998. Semiconductor nanocrystals as fluorescent biological labels. Science 281: 2013–2016.

Murphy, C.J., A.M. Gole, J.W. Stone, P.N. Sisco, A.M. Alkilany, E.C. Goldsmith et al. 2008. Gold nanoparticles in biology: beyond toxicity to cellular imaging. Acc. Chem. Res. 41: 1721–1730.

Pankhurst, Q.A., J. Connolly, S.K. Jones and J. Dobson. 2003. Applications of magnetic nanoparticles in biomedicine. J. Phys. D. Appl. Phys. 36: 167–181.

Paterson, G., J.M. Ataria, M. Ehsanul Hoque, D.C. Burns and C.D. Metcalfe. 2011. The toxicity of titanium dioxide nanopowder to early life stages of the Japanese medaka (*Oryzias latipes*). Chemosphere 82: 1002–1009.

Perrault, S.D., C. Walkey, T. Jennings, H.C. Fischer and W. Chan. 2009. Mediating tumor targeting efficiency of nanoparticles through design. Nano Lett. 9: 1909–1915.

Sayes, C.M., J.D. Fortner, W. Guo, D. Lyon, A.M. Boyd, K.D. Ausman et al. 2004. The differential cytotoxicity of water-soluble fullerenes. Nano. Lett. 4: 1881–1887.

Sincai, M., D. Ganga, D. Bica and L. Vekas. 2001. The antitumor effect of locoregional magnetic cobalt ferrite in dog mammary adenocarcinoma. J. Magn. Magn. Mater. 225: 235–240.

Sun, R.W., R. Chen, N.P. Chung, C.M. Ho, C.L. Lin and C.M. Che. 2005. Silver nanoparticles fabricated in hepes buffer exhibit cytoprotective activities toward HIV-1 infected cells. Chem. Commun. 40: 5059–5061.

Thit, A., L.M. Skjolding, H. Selck and J. Sturve. 2017. Effects of copper oxide nanoparticles and copper ions to zebrafish (*Danio rerio*) cells, embryos, and fry. Toxicol. *In Vitro*. 45: 89–100.

Tiwari, S.B. and M.M. Amiji. 2006. A review of nanocarrier-based CNS delivery systems. Curr. Drug Deliv. 3: 219–232.

Tran, D.T. and R. Salmon. 2011. Potential photocarcinogenic effects of nanoparticle sunscreens. Australas. J. Dermatol. 52: 1–6.

Usenko, C.Y., S.L. Harper and R.L. Tanguay. 2007. *In vivo* evaluation of carbon fullerene toxicity using embryonic zebrafish. Carbon. 45: 1891–1898.

Wiesner, M., G.V. Lowry, P. Alvarez, D. Dionysiou and P. Biswas. 2006. Assessing the risks of manufactured nanomaterials. Environ. Sci. Tech. 4336–4345.

Wu, Y. and Q. Zhou. 2012. Dose- and time-related changes in aerobic metabolism, chorionic disruption, and oxidative stress in embryonic medaka (*Oryzias latipes*): Underlying mechanisms for silver nanoparticle developmental toxicity. Aquatic Toxicol. 124: 238–246.

Xu, X.-H.N., W.J. Brownlow, S.V. Kyriacou, Q. Wan and J.J. Viola. 2004. Real-time probing of membrane transport in living microbial cells using single nanoparticle optics and living cell imaging. Biochemistry. 43: 10400–10413.

Xu, X.-H.N. and R.P. Patel. 2005. Imaging and assembly of nanoparticles in biological systems. American Scientific Publishers, Stevenson Ranch, CA.

Yezhelyev, M.V., X. Gao, Y. Xing, A. Al Hajj, S.M. Nie and R.M. O'Regan. 2006. Emerging use of nanoparticles in diagnosis and treatment of breast cancer. Lancet Oncology. 7: 657–667.

Zhang, W., K. Lin, Y. Miao, Q. Dong, C. Huang, H. Wang et al. 2012. Toxicity assessment of zebrafish following exposure to CdTe QDs. J. Hazardous Mater. 213-214: 413–420.

Zhu, B., G.-L. Liu, F. Ling, L.-S. Song and G.-X. Wang. 2015. Development toxicity of functionalized single-walled carbon nanotubes on rare minnow embryos and larvae. Nanotoxicol. 9: 579–590.

Zhu, X., L. Zhu, Y. Li, Z. Duan, W. Chen and P.J.J. Alvarez. 2007. Developmental toxicity in zebrafish (*Danio rerio*) embryos after exposure to manufactured nanomaterials: buckminsterfullerene aggregates (nC60) and fullerol. Environ. Toxicol. and Chem. 26: 976–979.

Zhu, X., S. Tian and Z. Cai. 2012. Toxicity assessment of iron oxide nanoparticles in zebrafish (*Danio rerio*) early life stages. PLoS ONE 7: e46286.

Zhu, Z.-J., R. Carboni, M.J. Quercio Jr., B. Yan, O.R. Miranda, D.L. Anderton et al. 2010. Surface properties dictate uptake, distribution, excretion, and toxicity of nanoparticles in fish. Small. 6: 2261–2265.

6

Effects of Nanomaterials on the Body Systems of Fishes

An Overview from Target Organ Pathology

Richard D. Handy[1,*] and *Genan Al-Bairuty*[2]

Introduction

One of the central concepts in toxicology is that the internal dose of a substance informs on the disposition of the organism to adverse biological effects. This notion is also well-known in fish toxicology and the absorption, distribution, metabolism and excretion (ADME) of chemicals found in the environment has been studied for many years. The ADME approach to understanding toxicity has also been applied to engineered nanomaterials (ENMs) in fish (Handy et al. 2008). However, any biological effect of a foreign material or toxic substance would also depend on whereabouts in the body the test substance is accumulated and the damage it does to the internal organs at that location(s). For the pathologists, this aspect is the study of target organ pathology (Turton and Hooson 1998). The use of histopathology in fundamental ecotoxicological research is long-established and is beyond the scope here (see reviews by Handy et al. 2002a, Di Giulio and Hinton 2008). However, fish histopathology has been used as a biomarker of exposure and as a biomonitoring tool in ecosystems (Handy et al. 2002a). The application of histopathological biomarkers in field situations presents a number of practical advantages over other monitoring approaches. These include the ease of sample collection and storage, the ability to estimate effects on many body systems and cell types from the same fish, as well as the opportunity to examine very small fish that could be too small to dissect for biochemistry (Hinton and Laurén 1990). Several studies have shown that histopathologic biomarkers are very sensitive, and can elucidate toxicant aetiology consistent with the type of environmental pollution (e.g., Norrgren et al. 2000, Handy et al. 2002b).

For nanoscience, there are currently no nationally agreed monitoring programmes for ENMs, and certainly none that use fish histopathology. However, controlled exposures to ENMs in the laboratory are beginning to provide some information on the organ pathologies in different species of fish. Much of this fundamental research reports incidental aspects of histology from a few internal organs to help explain the bigger picture of toxicity of an ENM, and is not necessarily intended for

[1] School of Biological and Marine Sciences, University of Plymouth, Plymouth, UK.
[2] Department of Biology, College of Education for Pure Science-Ibn AlHaitham, University of Bagdad, Iraq.
* Corresponding author: rhandy@plymouth.ac.uk

the expert pathologist. Nonetheless, some fundamental research has been specifically directed at organ pathology from ENMs (e.g., Cu ENMs, Al-Bairuty et al. 2013) and the target organ concept can be applied to the physiological processes and body systems of fish exposed to ENMs (Handy et al. 2011). The aim of the present chapter is to take the classical approach of target organ pathology, and to identify the types of pathology observed so far from ENM exposures on the body systems of fish *in vivo*. There are some ethical justifications needed for using vertebrate animals for *in vivo* research, especially where toxicity is concerned. All of the new data presented here was collected under ethical approval in the United Kingdom under the Animals (Scientific Procedures) Act 1986, and its amendments by the European Directive 2010/63/EU on the protection of animals used for scientific purposes. There is much cross-talk between the body systems of fish, and this is inevitably needed to enable physiological integration so that the whole organism functions. Fishes and other animals also have a number of defences against chemical insult and homeostatic mechanisms that tend to prioritise the functions that preserve life (e.g., cardiac function, oxygenation of the brain). This may involve compromising a less important tissue so that the functions of the vital organs are preserved. Thus, for the pathologist, target organ effects and the aetiology of morphological change are best revealed *in vivo* where this pathophysiological integration occurs. The focus here is on reporting altered morphological aspects of organs and tissues *in vivo* that are relevant to ENMs, although a context with fish health and survival is also indicated where the organ pathologies might define an adverse outcome for the animal.

External Barriers: The Skin, Gills and Gut Epithelium

The gills are the major organs for respiratory gas exchange and osmoregulation in fish. Similar to traditional chemicals, the concern is that waterborne exposure to ENMs may compromise the gill epithelium such that the animals suffer hypoxia and/or life-threatening osmotic disturbances to the blood. The freshwater-adapted fish gill is regarded as one of the tightest epithelia in animal biology. Nonetheless, acute respiratory distress has been documented. Smith et al. (2007) showed that waterborne exposure to 1 mg L^{-1} of single-walled carbon nanotubes (SWCNTs, length 5–30 μm) in rainbow trout (*Oncorhynchus mykiss*) causes an increase in the ventilation rate and coagulation of the ENMs in the gill mucus. Overt gill pathology has also been observed in rainbow trout with high mg L^{-1} concentrations of TiO_2 ENMs (TiO_2 particle diameter, 21 nm, Federici et al. 2007). Perhaps a more interesting concern is whether or not gill injury is repairable, and whether or not there are nano-specific pathologies in the gill that are not seen with other chemicals.

In epithelial tissues, and soft parenchyma such as the liver, cell turnover is the usual way of refreshing the cells to maintain a healthy organ. It is expected that some organs can proliferate more (healthy) cells in response to an insult (i.e., a reactive hyperplasia). This phenomenon has been observed in fish gills following exposure to ENMs in the water column. For example, Griffitt et al. (2007) noted proliferation of the epithelial cells and oedema of both the primary and secondary filaments in zebrafish (*Danio rerio*) exposed to 1.5 mg L^{-1} of Cu ENMs (80 nm diameter) for 48 h. Interestingly, the location of reactive hyperplasia in the gill from a ENM exposure is sometimes different to other chemicals. For example, waterborne exposure to SWCNTs in rainbow trout caused unusual hyperplasia of the epithelial cells on the leading edge of the primary filament and at the base of the secondary lamellae on the primary filament (Smith et al. 2007). Normally one would expect hyperplasia to be on the secondary lamellae proper. This uncommon form of hyperplasia was also observed in the gills of *O. mykiss* after 14 d of waterborne exposure to TiO_2 ENMs (Federici et al. 2007).

Other aspects of gill injury are entirely consistent with traditional chemicals. For example, Griffitt et al. (2009) found that Cu ENMs (27 nm diameter) caused increased filament width in zebrafish gills by three to four fold between 24 and 48 h of exposure. This is consistent with the idea of Cu being an osmoregulatory toxicant that disturbs water balance in the gill, with subsequent cell swelling and/or oedema (Handy 2002). However, silver is also known to be a osmoregulatory toxicant to freshwater fish, but Ag ENMs (27 nm diameter) did not alter the gill filament width in zebrafish (Griffitt et al. 2009). Clearly, not every aspect of dissolved metal pathology is also observed with an equivalent nanomaterial. Regardless, the reports of oedema, lamellar fusion and hyperplasia in the secondary lamellae of fish gills (e.g., Hao et al. 2009, Li et al. 2009) show that high mg L^{-1} concentrations of ENMs can damage fish gills, such that respiratory functions and/or osmoregulation might be compromised.

The gut is a leakier epithelium than the gill, but still remains a formidable mucous barrier to chemicals. Exposure of the gut is a concern for food contaminated with ENMs, but fish can also drink the surrounding water. Stress-induced drinking of ENM-contaminated water in rainbow trout has been observed during waterborne exposure to TiO_2 ENMs (Federici et al. 2007), SWCNT (Smith et al. 2007), and Cu ENMs (particle diameter, 87 nm, Shaw et al. 2012). Federici et al. (2007) suggested that stress-induced drinking by rainbow trout exposed to 1 mg L^{-1} suspension of TiO_2 ENMs for 14 d caused erosion and a fusion of the intestinal villi, as well as the appearance of vacuoles in the mucosa. More severe erosion and inflammation of the gut were observed with rainbow trout ingesting SWCNT in suspension for 10 d of exposure (Smith et al. 2007), whereas for the same SWCNTs added in the food, overt gut pathology was not observed (Fraser et al. 2010). The level of gut pathology from drinking ENMs also appears to be material-specific. Waterborne exposure to $CuSO_4$ or Cu ENMs (87 nm diameter) in rainbow trout showed very mild injuries to the gut (Al-Bairuty et al. 2013) compared to the TiO_2 or SWCNT reports above. However, there remain many knowledge gaps. The matrix of the food in the gut lumen is likely to affect both the bioavailability and toxicity of ENMs to the gut, as it does for soluble chemicals (Handy et al. 2005); but data demonstrating this bioavailability is scarce. Details of alterations in the regional anatomy of different parts of the gut are also not reported for most ENMs or species of fish. There appear to be no reports of skin pathology from ENMs at this time.

Heart, Cardiovascular System and Red Blood Cells

Information on the organ pathology in fish heart from ENMs and the integrity of the major blood vessels including the arteries and veins is very limited indeed. Smith et al. (2007) reported rupturing of the cerebral vasculature in rainbow trout exposed to SWCNTs for 10 d that was consistent with a stroke-like event in the cardiovascular system. Notably, the injury seem to be specific to the type of vascular bed, with arterial injury absent in other blood vessels around the brain. The absence of data on cardiac injury is a knowledge gap of concern. Clearly, the heart is a vital organ and it is very sensitive to oxidative stress (Patel et al. 2004). Some ENMs are designed with the ability to generate reactive oxygen species in mind, such as anatase TiO_2 and zerovalent iron. However, the cardiopathologies of oxidising ENMs have not been investigated in fish hearts.

Several studies have reported haematology in rainbow trout from ENM exposures (TiO_2, Federici et al. 2007, SWCNTs, Smith et al. 2007, Cu ENMs, Shaw et al. 2012, Ag ENMs, Clark et al. 2018). In most cases, and despite mg L^{-1} exposures via the water, there are only transient and mild changes in the haematology that would not compromise the oxygen carrying or acid-base balance functions of the red blood cells. So far, there is no evidence of overt haemolysis or haemolytic pathologies of the red blood cells in fish.

Haematopoietic System

There have been some concerns that ENMs are particulate materials of a similar size to virus particles, and might therefore be seen as 'antigen' by the immune system (Dobrovolskaia and McNeil 2007). *In vitro* studies with mammalian cells have shown the release of inflammatory cytokines and altered expression of immune-related genes (Kinaret et al. 2017), and similar *in vitro* reports are emerging from fish (Jovanović et al. 2011). However, these observations have not been translated into overt organ pathology or inflammation in the haematopoietic system *in vivo* in fish, at least so far (Handy et al. 2011, Jovanović and Palić 2012). Unlike mammals which have a thymus, lymph nodes, etc., the haematopoietic tissues of fish are rather diffuse and present as clusters of cells (e.g., immune-like cells in the head kidney). The primary discrete organ of the haematopoietic system in fishes is the spleen. The spleen functions in the housekeeping of the blood cells and has a role in maintaining normal healthy blood cells in the circulation (Handy et al. 2011). It contains red and white pulp where the erythrocytes and immune cells are stored, respectively. A decrease in red pulp might indicate the spleen is needing to release more red cells into the circulation to maintain the blood supply. Similarly, changes in the white pulp might indicate an immune challenge.

For example, in rainbow trout, waterborne exposure to 1 mg L^{-1} of either TiO_2 ENMs (24 nm diameter) or bulk TiO_2 (134 nm diameter) for 14 d caused a decline in the proportion of red pulp in the spleen with a concomitant elevation in sinusoid space when compared to the unexposed control without TiO_2 (Boyle et al. 2013a). Interestingly, there was no material-type effect in the spleen structure with both bulk and nano forms causing similar changes in the spleen. The number of melanomacrophages was also doubled compared to the controls. Boyle et al. (2013a) suggested that the changes in the melanomacrophage deposits reflect the normal activity of the spleen in processing damaged blood cells; which indicated that the spleen was working harder to maintain the normal counts of circulating blood cells, especially in the ENM-exposed fish. In a similar study, Boyle et al. (2013b) injected 50 µg of each of TiO_2 ENMs and bulk TiO_2 into the caudal vein of rainbow trout. After 96 h, there was a decrease in the proportion of red pulp in the ENM-injected fish. Whereas, the proportion of white pulp was increased. However, the sinusoid space proportion remaining around 5–8 percent in the spleens of fish from all treatments. The authors suggested that these changes were not pathological and were within the normal working range of trout spleen.

Al-Bairuty et al. (2013) investigated the organ pathologies of juvenile rainbow trout exposed to 20 or 100 µg L^{-1} of either dissolved Cu as $CuSO_4$ or Cu ENMs (87 nm diameter) for up to 10 d. The pathologies in gill, gut, liver, kidney and muscle are presented in Al-Bairuty et al. (2013). However, in the same study, the spleen was also collected (Al-Bairuty and Handy, unpublished observations). An increase in the proportion of white pulp and a decrease in the proportion of red pulp was observed in all Cu treatments compared to the control group, but there was no material-type effect between the metal salt and the nano forms of Cu (Fig. 1). There was also some evidence of enhanced melanomacrophage

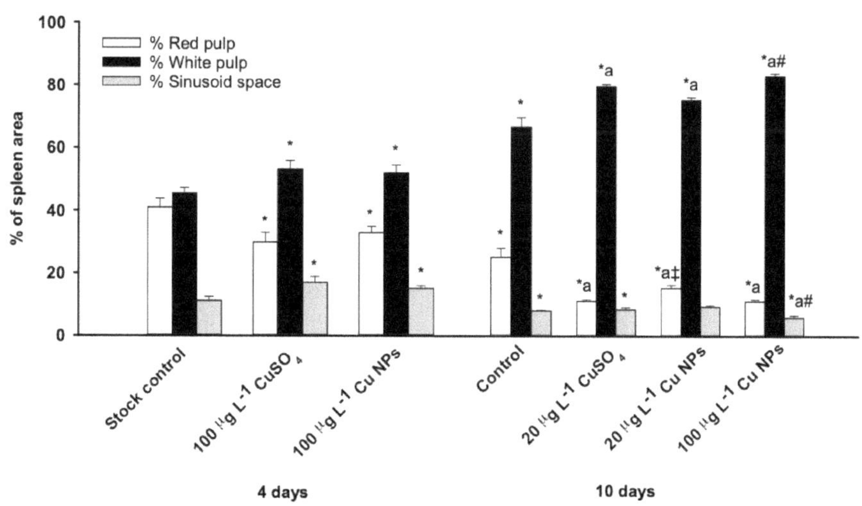

Figure 1. Changes in the proportions of red and white pulp as well as sinusoids space in the spleens of rainbow trout exposed to dissolved Cu as $CuSO_4$ (20 µg L^{-1}) or Cu ENMs (20 or 100 µg L^{-1}) for 0, 4, and 10 d. Data are means of proportional areas ± S.E.M. $n = 6$ fish/treatment. (*), significant difference from initial fish (stock fish at time zero, ANOVA, $P < 0.05$). (a), significant difference from control within treatment (ANOVA, $P < 0.05$).(#), significant difference from the previous Cu-ENMs concentration (dose effect, ANOVA, $P < 0.05$). (‡), significant difference between low concentration of Cu and Cu-ENMs (nano-effects, *t*-test, $P < 0.05$). Note the termination of the 100 µg L^{-1} $CuSO_4$ treatment after 4 d for ethical reasons. Data are from Al-Bairuty and Handy (unpublished).

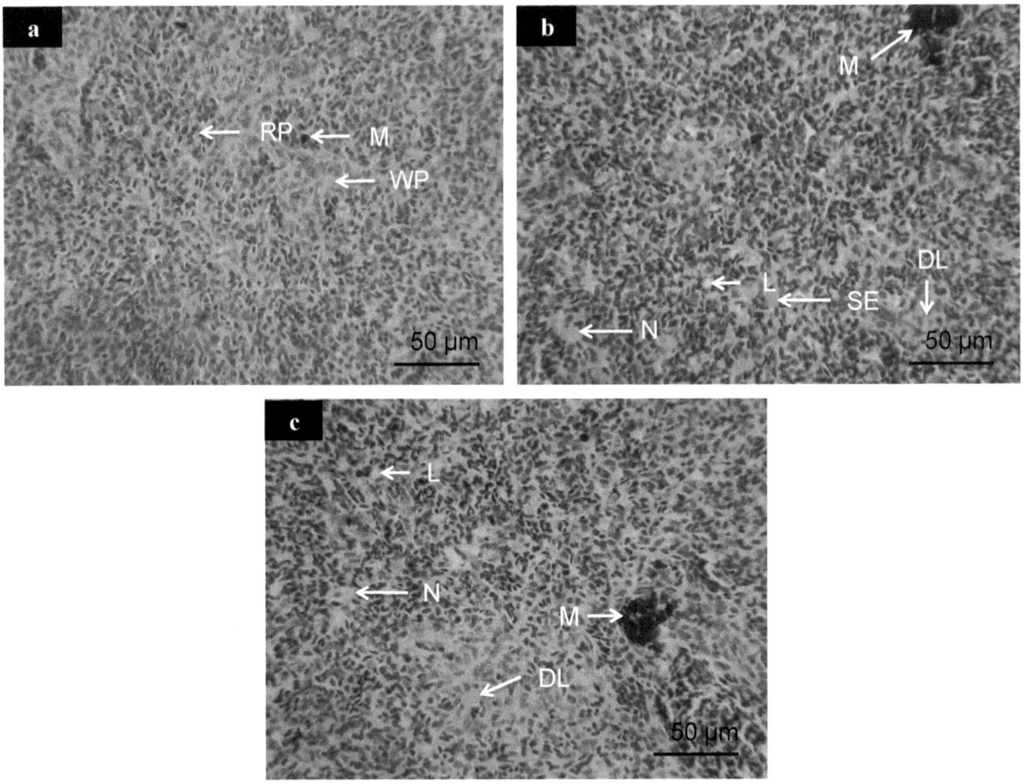

Figure 2. Spleen morphology in rainbow trout following exposure to $CuSO_4$ or Cu ENMs for 4 d. Panels include (a) control, (b) 100 μg L^{-1} $CuSO_4$, (c) 100 μg L^{-1} Cu ENMs. Spleen from control fish showed normal histology, with defined red (RP) and white (WP) pulp. All treatments showed similar types of lesions. These lesions include necrosis (N), depletion of lymphoid tissues (DL), melanomacrophage deposits (M), Lipidosis (L), and swollen erythrocyte (SE). Scale bar indicates magnification; sections were 7 μm thick and stained with H&E. Note the termination of the 100 μg L^{-1} $CuSO_4$ treatment after 4 d for ethical reasons (Al-Bairuty and Handy, unpublished).

Color version at the end of the book

activity, but without overt pathology of the surrounding tissue (Fig. 2). Similar to the study by Boyle et al. (2013a), the data is interpreted as non-pathological, but with physiological change in the activity of the spleen that attempts to maintain normal circulation in the animal.

Liver and Biliary System

The liver is an important excretory organ in fish and it is likely that ENMs in the circulation are excreted via the liver into the bile, rather than via the kidney (see discussion in Handy et al. 2008). The tissue of fish liver is organised into functional units that are somewhat different to that of mammals (Hampton et al. 2008). For traditional chemicals, the pathologies of the liver are reasonably well-known. For example, pesticides (Glover et al. 2007) or trace metals (Cu et al. 2006) may cause fatty change in the liver. The latter, with lipidosis also present. However, these alterations in liver histology are difficult to associate with any loss of liver function *in vivo* because bile production and composition is rarely measured in such studies. Early research with ENMs in rainbow trout also showed that fish could also suffer fatty change in the liver (TiO_2, Federici et al. 2007, SWCNTs, Smith et al. 2007). In the livers of juvenile carp (*Cyprinus carpio*), Hao et al. (2009) reported necrotic cells in exposures up to 200 mg L^{-1} TiO_2 ENMs (50 nm diameter) for 20 d. Interestingly, there have been several studies on nano Cu and its effects on fish liver. The liver is a central compartment for Cu metabolism in fish and a target

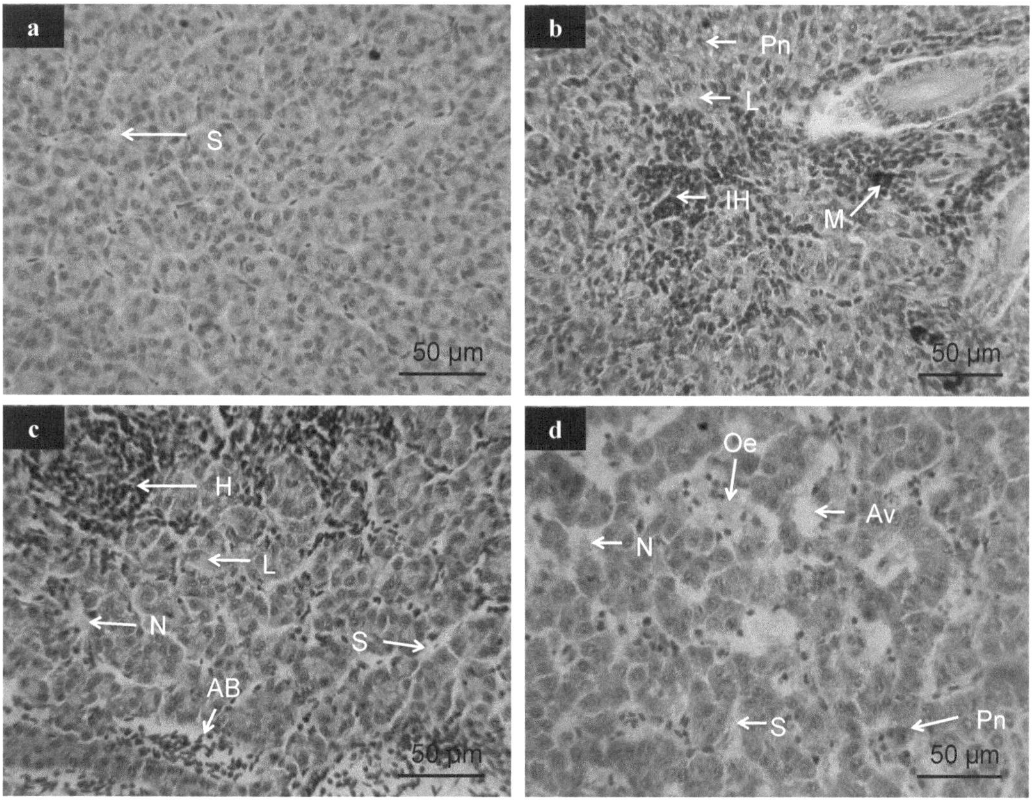

Figure 3. Liver morphology in rainbow trout following waterborne exposure to $CuSO_4$ or Cu ENMs for 10 d. Panels include (a) control, (b) 20 µg L^{-1} $CuSO_4$, (c) 20 µg L^{-1} Cu ENMs and (d) 100 µg L^{-1} Cu ENMs. The livers of control fish showed normal histology with sinusoid space (S). Both materials caused similar types of lesions, although these were severe in the equivalent Cu ENM treatment. These lesions include cells with pyknotic nuclei (Pn), foci of hepatitis-like injury (H), foci of melanomacrophages (M), lipidosis (L), vacuole formation (V), necrosis (N), oedema in the tissue (Oe), and aggregation of blood cell (AB). Scale bar indicates magnification; sections were 7 µm thick and stained with haematoxylin and eosin. Note the termination of the 100 µg L^{-1} $CuSO_4$ treatment after 4 d for ethical reasons (modified from Al-Bairuty et al. 2013).

Color version at the end of the book

organ for Cu accumulation (Grosell et al. 1996, Handy et al. 1999). Al-Bairuty et al. (2013) showed that exposure to 100 µg L^{-1} of Cu as $CuSO_4$ or Cu ENMs for 4 d caused similar types of lesion in the liver. These included the occasional cell with pyknotic nuclei, cytoplasmic vacuoles indicative of the early stages of necrosis in a few hepatic cells, and small foci of hepatitis-like cell injury. After 10 d, these lesions persisted (Fig. 3), with the some additional minor injury of the endothelium of the blood vessels in the liver. At a much higher concentrations of 1.5 mg L^{-1} of Cu ENMs for 48 h, Griffitt et al. (2007) found no strong evidence of hepatocellular necrosis in the liver. It should be noted that rainbow trout liver cells *in vivo* can synthesis endogenous Cu-containing granules (Lanno et al. 1987). This ability to make metal storage granules may also be protective of liver pathology. At least, so far, these studies on the liver suggest the organ is robust with respect to pathology from ENM exposure. However, there are many data gaps on the different forms and compositions of ENMs, and also on exposure routes. For example, there is very little information on liver pathology from dietary exposures to ENMs. A single incidence of hepatitis was observed in two rainbow trout exposed through the food to C60 (500 mg SWCNT kg^{-1} food for six weeks, Fraser et al. 2010). However, such low incidence rates are hard to associate with any treatment effect and might easily be attributed to a random spontaneous pathology that can occur from time to time in any animal.

Kidney and the Urinary System

The vertebrate kidney functions by filtering solutes from the blood into the pre-urine; and then the renal tubules selectively reabsorb essential solutes such as sodium, calcium, glucose and amino acids. The main concerns for pathology include damage to the glomerulus, such that the molecular cut off of the renal filter (approximately 60 kDa, see Handy et al. 2008) is not maintained, and also for the integrity of the epithelial cells in the renal tubules so that processes such as selective reabsorption and secretion can occur. There are only a few reports on renal histopathology with ENMs in fish, and all of these are on freshwater-adapted animals. There appears to be no information on ENM pathologies in the kidney of stenohaline marine species. Scown et al. (2009) infused rainbow trout with TiO_2 ENMs (34 nm diameter), and reported no effect on creatinine clearance; suggesting that the glomerular filtration rate (GFR) of the animals was preserved. Indeed, Scown et al. (2009) showed histological sections of the renal tubules and reported no tubular pathologies, but the study did report some particulate matter in electron micrographs of the kidney. These latter deposits were not verified as TiO_2 particles and were possibly melanomacrophage deposits in the haematopoietic cells of the kidney. Al-Bairuty et al. (2013) also studied renal histology in trout exposed via the water to either $CuSO_4$ or Cu ENMs for 10 d. The pathologies observed with Cu ENMs were broadly of the same type as $CuSO_4$ (Fig. 4) including: degeneration of some parts of the renal tubules, enlargement of the Bowman's space, increases in the sinusoid space and foci of necrosis in the glomerular cells. Interestingly, the number of melanomacrophage deposits increased throughout the kidney sections (Fig. 4) and a few necrotic cells were found in the haematopoietic tissue, suggesting the immune functions of the kidney may have also been affected (Al-Bairuty et al. 2013).

The vertebrate kidney also functions in acid-base balance to manage the long-term ingested load from food (i.e., the net acid or alkali load from ingested prey items). However, there is only one study of ENM exposure at different pH values that also collected information on the kidney (Al-Bairuty et al. 2016). In this study, trout were exposed to 20 $\mu g\ L^{-1}$ of either $CuSO_4$ or Cu ENMs at pH 7 and pH 5. Similar to a previous study (Al-Bairuty et al. 2013), the kidney showed comparable types of lesions for both substances. The lesions included: occasional necrotic cells in the haematopoietic tissue, vacuoles in the renal epithelial tubules and haematopoietic area, some renal tubule degeneration and enlargement of the Bowman's space. However, the effects were generally worse at pH 5 compared to pH 7 (Fig. 5).

The Musculo-Skeletal System

The use of video tracking methods has shown that fish behaviour is very sensitive to pollutant exposure via the water or food (e.g., Cu, Campbell et al. 2002). Predatory fish such as trout need to spend time swimming in order to find and capture prey items. However, when decreases in locomotion and other swimming behaviours occur, there may be several explanations. For example, if the animal has chosen not to swim, or whether the animal lacks the capacity for exercise because of pathology. Attention is usually focussed on the gills and respiratory functions in this regard, but an alternative explanation is that the skeletal muscle of the fish is injured to the detriment of swimming behaviours. Altered locomotor behaviours have been implicated with ENMs (see below), but there are only a few studies that document the histology of the musculo-skeletal system in fish. Al-Bairuty et al. (2013) showed that rainbow trout exposure to 100 $\mu g\ L^{-1}$ of Cu as $CuSO_4$ or Cu ENMs for 4 d showed similar types of lesions in the muscle tissues;including a decrease in the relative proportion of muscle fibre area compared to controls (Al-Bairuty et al. 2013). A material–type effect was also confirmed by quantitative histological analysis of the muscle fibre dimensions, which showed a smaller proportion of muscle fibre area in the $CuSO_4$ compared to the Cu ENM treatment (Al-Bairuty et al. 2013).

This mild degeneration of the muscle fibres with an increased dorso-ventral thickness of the fibre bundles has also been observed in rainbow trout during waterborne exposure to 1 mg L^{-1} of either bulk or TiO_2 ENMs for 14 d (Fig. 6); and was potentially associated with changes in animal

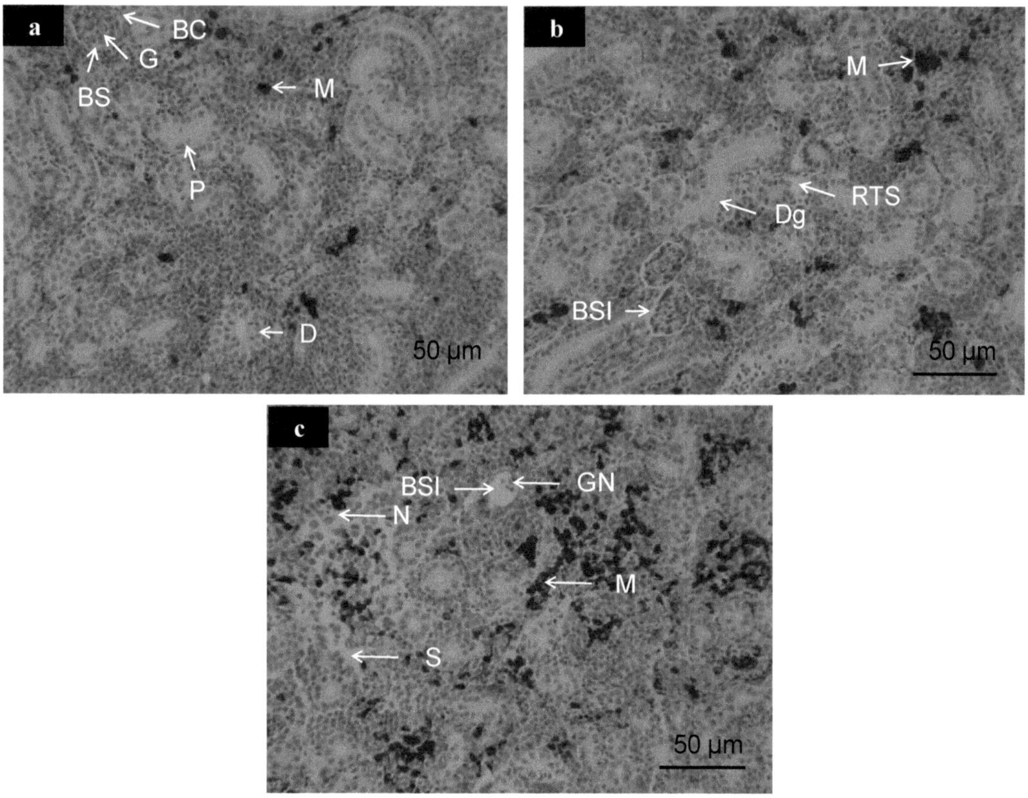

Figure 4. Kidney morphology in rainbow trout following waterborne exposure to (a) control (b) 20 μg L^{-1} of Cu as $CuSO_4$, and (c) 20 μg L^{-1} of Cu as Cu ENMs for 10 d. Kidney of control fish showed normal histology structure with parietal epithelium of Bowman's capsule (BC), glomerulus (G), Bowman's space (BS), proximal tubules (P), distal tubules (D) and melanomacrophages (M). The types of pathologies were similar for $CuSO_4$ and Cu ENMs, but with more deposit of melanomacrophage in the latter. These lesions include degeneration of renal tubule (Dg), increased Bowman's space (BSI), melanomacrophage aggregate (M), sinusoids were enlarged (S), renal tubular separation (RTS), necrosis of haematopoietic tissue (N) and glomerular necrosis (GN). Scale bar indicates magnification, section were 7 μm thick and stained with haematoxylin and eosin (modified from Al-Bairuty et al. 2013).

Color version at the end of the book

behaviour (Boyle et al. 2013a). A subtle decline in time spent swimming at high speed in juvenile trout exposed to 1 mg L^{-1} TiO_2 ENMs for 14 d compared to controls may have been partly explained by muscle injury (Boyle et al. 2013a). Interestingly, the direct injection of TiO_2 ENMs into muscle tissue can result in localised muscle pathology. Boyle et al. (2013b) conducted an intramuscular injection of trout (near the anterior insertion of the dorsal fin) with 50 μg of TiO_2 ENMs or bulk TiO_2 in saline. After 96 h, the injection site showed particles deposits along the margins of adjacent blood capillaries in the muscle tissue and with evidence of necrotic muscle fibres. This at least demonstrates that muscle pathology from direct contact with ENMs is possible.

Brain and the Peripheral Nervous System

To date, there has been no unequivocal demonstration of ENMs inside the primary brain tissue of fishes arising from an *in vivo* exposure (i.e., uptake of intact ENMs into the grey or white matter while the animal was still alive). For some of the organic ENMs, such as C60 that might diffuse through the lipid of the blood brain barrier, early reports of biochemical injury to fish brain turned out to be attributed to the organic solvent used, not the nanomaterial (Henry et al. 2007, Shinohara et al. 2009).

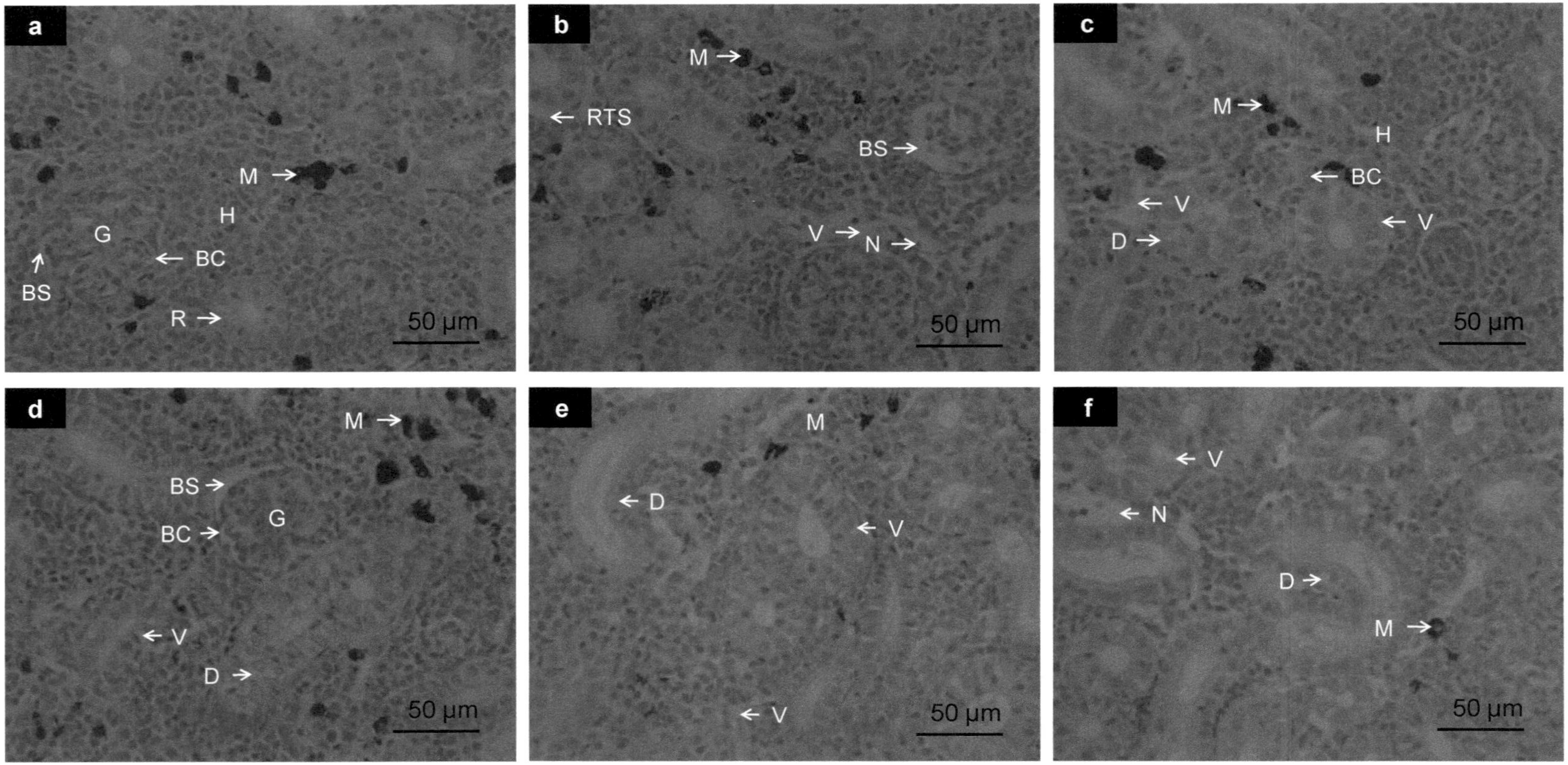

Figure 5. Kidney morphology in rainbow trout after 7 d of waterborne exposure to (a) control (b) $CuSO_4$, (c) Cu ENMs at pH 7 and (d) control (e) $CuSO_4$ (f) Cu ENMs at pH 5. Kidney of control fish showed normal histology with parietal epithelium of Bowman's capsule (BC), glomerulus (G), Bowman's space (BS), renal tubules (R), haematopoietic tissues (H) and melanomacrophages deposits (M). Similar pathologies were observed with $CuSO_4$ and Cu ENMs at pH 7 and 5, with worse effects at pH 5 than 7. These pathologies include necrosis (N), vacuolisation (V), and degeneration of renal tubule (D). Scale bar indicates magnification, section were 7 μm thick and stained with haematoxylin and eosin (Al-Bairuty and Handy, unpublished observations from the exposure reported in Al-Bairuty et al. 2016).

Color version at the end of the book

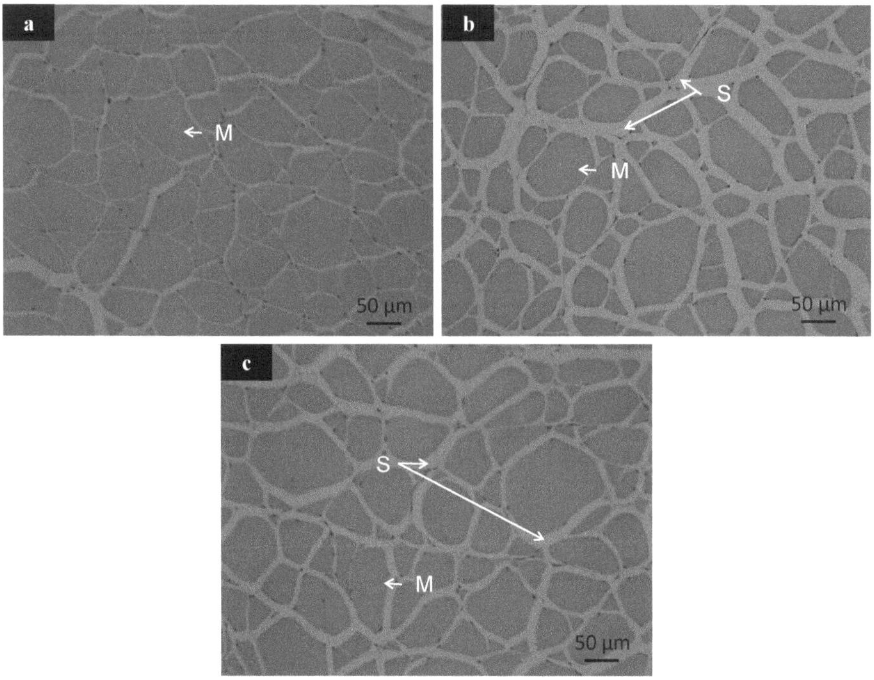

Figure 6. Muscle morphology in trout after 14 d of exposure to (a) control (b) 1 mg L^{-1} bulk TiO_2 (c) 1 mg L^{-1} TiO_2 ENMs. Muscles of control groups showed normal structure of muscle fibres (M). Both treatments showed increased space among muscles fibres (S). Scale bar indicates magnification. Sections were 7 µm thick and stained with haematoxylin and eosin. (Al-Bairuty, Boyle and Handy, unpublished observations).

Color version at the end of the book

Despite a lack of evidence of direct *in vivo* exposure of the brain to ENMs in fish, the possibility of indirect toxicity to the central nervous system remains. For example, the brain is intolerant of systemic hypoxia. Rainbow trout that showed respiratory distress during SWNCT exposure (Smith et al. 2007), also showed vascular injury to the blood vessels in the brain, and necrotic cell bodies and foci of vacuoles in several regions of the brain tissue itself. This latter effect was not present in the solvent controls. Al-Bairuty et al. (2013) revealed some effects on the brain tissue of rainbow trout following exposure to $CuSO_4$ or Cu ENMs. The effects were mostly very mild in the cerebellum with some blood vessel abnormalities on the ventral surface, and some necrosis between the molecular and granular layers of the tissue. However, in the mesencephalon, there were alteration in the thickness of mesencephalon tissue layers of the brain, with the stratum periventriculare adjacent to the ventricular fluid showing considerable injury. This latter effect may have been mediated by an osmotic disturbance (e.g., altered hydrostatic pressure) in the cerebral spinal fluid, causing mechanical injury of the surrounding tissue. There are many data gaps on brain histology with ENMs, with only one or two materials studied so far in rainbow trout. It is also worth recognising that synaptic plasticity and the complex neural networks in the vertebrate brain do allow the organ to function in the face of injury. The mild pathologies reported so far would seem unlikely to compromise the neurological functions of the central nervous system (CNS).

The peripheral nervous system is inevitable regarded as more accessible or vulnerable to chemicals, and comprises the sensory pathways leading to the CNS and the motor neurons coming from it. There is no direct evidence from electrical recordings of sensory nerves or histological observations of such nerves to indicate that sensory input pathways are damaged by ENM exposures. However, there is evidence that sense organs can be affected. Sovová et al. (2014) investigated the alarm response of trout during exposures to Cu ENMs (20 nm diameter) or $CuSO_4$ and found that the response was attenuated. In the $CuSO_4$-exposed animals, the olfactory rosettes showed the characteristic erosion associated with the metal pathology of this ciliated organ. However, animals from the Cu ENM treatment did not show any pathology of the olfactory rosettes, despite changes in animal behaviour. One possible explanation was that the sense organ and neural pathway was intact, but the canals leading to the sense organ were blocked with particles such that the fish were no longer able to smell the water. McNeil et al. (2014) investigated the effects of Cu ENMs (20 nm diameter) and Ag ENMs (58 nm diameter) on the lateral line of zebrafish embryos and found that exposure to $CuSO_4$, and to a lesser extent Cu ENMs, decreased the number of visible neuromast cells. These changes were associated with the animals being unable to maintain a positive rheotaxis (i.e., inability to orientate the body correctly). Clearly, there are some concerns for the sense organs that are in direct contact with the external water. Evidence that the motor pathways from the CNS are directly affected by ENM exposure is sparse. Application of unrealistic mg L^{-1} concentrations of ENMs directly onto the compound nerves of shore crabs (*Carcinus maenas*) had no effect on the ability of the nerve to generate single action potentials or action potential trains (Windeatt and Handy 2013). Given that such a high, non-physiological exposure had no effects on action potentials, it seems unlikely that short-term effects on motor nerves will be found *in vivo*.

Reproductive System and Endocrinology

There is a considerable body of evidence showing that ENMs are toxic to the early life stages of fish (Handy et al. 2011, Shaw et al. 2016). However, direct *in vivo* evidence of pathology in the testis or ovary of adult fish due to ENM exposure is mostly lacking. Indeed, exposure of adult zebrafish to 1 mg L^{-1} TiO_2 ENMs (24 nm diameter) had no effect on the ability of the animals to breed, although there were some concerns about the cumulative survival of the unexposed off-spring (Ramsden et al. 2013). In a dietary exposure study, Blickley et al. (2014) fed killifish, *Fundulus heteroclitus*, CdSe/ZnS quantum dots for 85 d and reported a reduction in the fecundity of the adults. However, so far, such studies on adult fish have not reported the histopathology of the gonads. Similarly, whether or not changes in fecundity or survival of unexposed *F1* embryos might be attributed to defects in the endocrine control of reproduction in the parents is unclear. So far, there are no reports of EMN mediated pathology in the hypothalamic-pituitary axis of fish.

Conclusions and Data Gaps

Overall, information on the organ pathologies of fish associated with ENM exposures is very limited and there is not enough data to reach a consensus opinion on whether or not there are nano-specific pathologies. So far, only a few unusual pathologies have been observed with ENMs, with most pathologies from ENMs being similar to their metal salt or other nearest substance controls. However, there are subtle differences in the aetiology of the pathology with some ENMs; with the pathology taking longer to appear, and sometimes eventually being worse than the equivalent metal salt control. Despite the limited data sets, pathology has been observed in all the major body systems of the animals and the target organ approach to pathology is valid. However, the data gaps on species of fish and types of ENMs are considerable. Almost all of the histopathology has been conducted on fish that are large enough to easily dissect for the internal organs (i.e., rainbow trout). While there is also some information on zebrafish, most other species of fish are not yet represented. There is no histopathology of ENMs reported for marine species. Another data gap is concerning the range of ENMs that have been studied. The work has mostly been limited to a few pristine or first generation ENMs such as TiO_2, Cu ENMs and unmodified CNTs. Nanocomposites, nano-hybrids and coated ENMs have not been documented for organ pathologies. There is also a need for more quantitative histology in the literature, where the dimensions of organs are measured and the pathologies counted. It is both the incidence rate (i.e., how many animals per treatment) and the severity of the pathology (numbers and types of injury per animal) that needs to be reported in order to adequately understand the hazard. Nonetheless, histopathology is proving a useful tool to understand the types and mechanisms of injury in fish due to ENM exposure.

References

Al-Bairuty, G.A., D. Boyle, T.B. Henry and R.D. Handy. 2016. Sublethal effects of copper sulphate compared to copper nanoparticles in rainbow trout (*Oncorhynchus mykiss*) at low pH: Physiology and metal accumulation. Aquat. Toxicol. 174: 188–198.

Al-Bairuty, G.A., B.J. Shaw, R.D. Handy and T.B. Henry. 2013. Histopathological effects of waterborne copper nanoparticles and copper sulphate on the organs of rainbow trout (*Oncorhynchus mykiss*). Aquat. Toxicol. 126: 104–115.

Blickley, T.M., C.W. Matson, W.N. Vreeland, D. Rittschof, R.T. Di Giulio and P.D. McClellan-Green. 2014. Dietary CdSe/ZnS quantum dot exposure in estuarine fish: bioavailability, oxidative stress responses, reproduction, and maternal transfer. Aquat. Toxicol. 148: 27–39.

Boyle, D., G.A. Al-Bairuty, C.S. Ramsden, K.A. Sloman, T.B. Henry and R.D. Handy. 2013a. Subtle alterations in swimming speed distributions of rainbow trout exposed to titanium dioxide nanoparticles are associated with gill rather than brain injury. Aquat. Toxicol. 126: 116–127.

Boyle, D., G.A. Al-Bairuty, T.B. Henry and R.D. Handy. 2013b. Critical comparison of intravenous injection of TiO_2 nanoparticles with waterborne and dietary exposures concludes minimal environmentally-relevant toxicity in juvenile rainbow trout *Oncorhynchus mykiss*. Environ. Pollut. 182: 70–79.

Campbell, H.A., R.D. Handy and D.W. Sims. 2002. Increased metabolic cost of swimming and consequent alterations to circadian activity in rainbow trout (*Oncorhynchus mykiss*) exposed to dietary copper. Can. J. Fish. Aquat. Sci. 59: 768–777.

Clark, N.J., B.J. Shaw and R.D. Handy. 2018. Low hazard of silver nanoparticles and silver nitrate to the haematopoietic system of rainbow trout. Ecotoxicol. Environ. Saf. 152: 121–131.

Di Giulio, R.T. and D.E. Hinton. 2008. The Toxicology of Fishes. Taylor and Francis Group, CRC Press. Boca Raton, USA.

Dobrovolskaia, M.A. and S.E. McNeil. 2007. Immunological properties of manufactured nanomaterials. Nat. Nanotechnol. 2: 469–478.

Glover, C.N., D. Petri, K.-E. Tollefsen, N. Jørum, R.D. Handy and M.H.G. Berntssen. 2007. Assessing the sensitivity of Atlantic salmon (*Salmo salar*) to dietary endosulfan exposure using tissue biochemistry and histology. Aquat. Toxicol. 84: 346–355.

Grosell, M., I. Boetius, H.J.M. Hansen and P. Rosenkilde. 1996. Influence of preexposure to sublethal levels of copper on ^{64}Cu uptake and distribution among tissues of the european eel (*Anguilla anguilla*). Comp. Biochem. Physiol. 114C: 229–235.

Federici, G., B.J. Shaw and R.D. Handy. 2007. Toxicity of titanium dioxide nanoparticles to rainbow trout, (*Oncorhynchus mykiss*): Gill injury, oxidative stress, and other physiological effects. Aquat. Toxicol. 84: 415–430.

Fraser, T.W.K., H.C. Reinardy, B.J. Shaw, T.B. Henry and R.D. Handy. 2011. Dietary toxicity of single-walled carbon nanotubes and fullerenes (C60) in rainbow trout (*Oncorhynchus mykiss*). Nanotoxicology. 5: 98–108.

Griffitt, R.J., R. Weil, K.A. Hyndman, N.D. Denslow, K. Powers, D. Taylor and D.S. Barber. 2007. Exposure to copper nanoparticles caused gill injury and acute lethality in zebrafish (*Danio rerio*). Environ. Sci. Technol. 41: 8178–8186.

Griffitt, R.J., K. Hyndman, N.D. Denslow and D.S. Barber. 2009. Comparison of molecular and histological changes in zebrafish gills exposed to metallic nanoparticles. Toxicol. Sci. 107: 404–415.

Hampton, J.A., R.C. Lantz, P.J. Goldblatt, D.J. Lauren and D.E. Hinton. 1988. Functional units in rainbow trout (*Salmo gairdneri*, Richardson) liver: II. The biliary system. The Anat. Rec. 221: 619–634.

Hao, L., Z. Wang and B. Xing. 2009. Effect of sub-acute exposure to TiO_2 nanoparticles on oxidative stress and histopathological changes in juvenile carp (*Cyprinus carpio*). J. Environ. Sci. 21: 1459–1466.

Hinton, D.E. and D.J. Laurén. 1990. Integrative histopathological approches to detecting effects of environmental stressors on fish. pp. 51–66. *In*: S.M. Adams (ed.). Biological Indicators of Stress In Fish. American Fisheries Society, Bethesda Maryland, USA.

Handy, R.D., T. Runnalls and P.M. Russell. 2002b. Histopathologic biomarkers in in three spined sticklebacks, *Gasterosteus aculeatus*, from several rivers in southern England that meet the Freshwater Fisheries Directive. Ecotoxicology 11: 467–479.

Handy, R.D., A.N. Jha and M.H. Depledge. 2002a. Biomarker approaches for ecotoxicological biomonitoring at different levels of biological organisation. pp. 9.1–9.32. *In*: F. Burden, I. McKelvie, U. Förstner and A. Guenther (eds.). Handbook of Environmental Monitoring. McGraw Hill, New York.

Handy, R.D., G. Al-Bairuty, A. Al-Jubory, C.S. Ramsden, D. Boyle, B.J. Shaw and T.B. Henry. 2011. Effects of manufactured nanomaterials on fishes: A target organ and body systems physiology approach. J. Fish Biol. 79: 821–853.

Handy, R.D., T.B. Henry, T.M. Scown, B.D. Johnston and C.R. Tyler. 2008. Manufactured nanoparticles: their uptake and effects on fish—a mechanistic analysis. Ecotoxicology 17: 396–409.

Handy, R.D., D.W. Sims, A. Giles, H.A. Campbell and M.M. Musonda. 1999. Metabolic trade-off between locomotion and detoxification for maintenance of blood chemistry and growth parameters by rainbow trout (*Oncorhynchus mykiss*) during chronic dietary exposure to copper. Aquat. Toxicol. 47: 23–41.

Handy, R.D., J.C. McGeer, H.E. Allen, P.E. Drevnick, J.W. Gorsuch, A.S. Green et al. 2005. Toxic effects of dietborne metals: Laboratory studies. pp. 59–112. *In*: J.S. Meyer, W.J. Adams, K.V. Brix, S.N. Luoma, D.R. Mount, W.A. Stubblefield et al. (eds.). Toxicity of Dietborne Metals to Aquatic Organisms. SETAC Press, Pensacola, USA.

Henry, T.B., F.M. Menn, J.T. Fleming, J. Wilgus, R.N. Compton and G.S. Sayler. 2007. Attributing effects of aqueous C-60 nano-aggregates to tetrahydrofuran decomposition products in larval zebrafish by assessment of gene expression. Environ. Health Perspect. 115: 1059–1065.

Jovanović, B., L. Anastasova, E.W. Rowe, Y. Zhang, A.R. Clapp and D. Palić. 2011. Effects of nanosized titanium dioxide on innate immune system of fathead minnow (*Pimephales promelas* Rafinesque, 1820). Ecotoxicol. Environ. Saf. 74: 675–683.

Jovanović, B. and D. Palić. 2012. Immunotoxicology of non-functionalized engineered nanoparticles in aquatic organisms with special emphasis on fish—Review of current knowledge, gap identification, and call for further research. Aquat. Toxicol. 118: 141–151.

Kinaret, P., M. Ilves, V. Fortino, E. Rydman, P. Karisola, A. Lähde et al. 2017. Inhalation and oropharyngeal aspiration exposure to rod-like carbon nanotubes induce similar airway inflammation and biological responses in mouse lungs. ACS Nano. 11: 291–303.

Li, H., Q. Zhou, Y. Wu, J. Fu, T. Wang and G. Jiang. 2009. Effect of waterborne nano iron on medaka (*Oryzias latipes*): Antioxidant enzymatic activity, lipid peroxidation and histopathology. Ecotoxicol. Environ. Saf. 72: 684–692.

Lanno, R.P., B. Hicks and J.W. Hilton. 1987. Histological observations on intrahepatocytic copper-containing granules in rainbow trout reared on diets containing elevated levels of copper. Aquat. Toxicol. 10: 251–263.

McNeil, P.L., D. Boyle, T.B. Henry, R.D. Handy and K.A. Sloman. 2014. Effects of metal nanoparticles on the lateral line system and behaviour in early life stages of zebrafish (*Danio rerio*). Aquat. Toxicol. 152: 318–323.

Norrgren, L., U. Pettersson, S. Örn and P.-A. Bergqvist. 2000. Environmental monitoring of Kafue river, located in the Copperbelt, Zambia. Arch. Environ. Contam. Toxicol. 38: 334–41.

Patel, A.P., A.J. Moody, J.R. Sneyd and R.D. Handy. 2004. Carbon monoxide exposure in rat heart: evidence for two modes of toxicity. Biochem. Biophys. Res. Commun. 321: 241–246.

Ramsden, C.S., T.B. Henry and R.D. Handy. 2013. Sub-lethal effects of titanium dioxide nanoparticles on the physiology and reproduction of zebrafish. Aquat. Toxicol. 126: 404–413.

Shaw, B.J., G.A. Al-Bairuty and R.D. Handy. 2012. Effects of waterborne copper nanoparticles and copper sulphate on rainbow trout (*Oncorhynchus mykiss*): physiology and accumulation. Aquat. Toxicol. 116–117: 90–101.

Scown, T.M., R. van Aerle, B.D. Johnston, S. Cumberland, J.R. Lead, R. Owen et al. 2009. High doses of intravenously administered titanium dioxide nanoparticles accumulate in the kidneys of rainbow trout but with no observable impairment of renal function. Toxicol. Sci. 109: 372–380.

Shaw, B.J. and R.D. Handy. 2006. Dietary copper exposure and recovery in Nile tilapia, *Oreochromis niloticus*. Aquat. Toxicol. 76: 111–121.

Shaw, B.J., C.C. Liddle, K.M. Windeatt and R.D. Handy. 2016. A critical evaluation of the fish early life stage toxicity test for engineered nanomaterials: experimental modifications and recommendations. Arch. Toxicol. 90: 2077–2107.

Shinohara, N., T. Matsumoto, M. Gamo, A. Miyauchi, S. Endo, Y. Yonezawa and J. Nakanishi. 2009. Is lipid peroxidation induced by the aqueous suspension of fullerene C60 nanoparticles in the brains of *Cyprinus carpio*? Environ. Sci. Technol. 43: 948–953.

Smith, C.J., B.J. Shaw and R.D. Handy. 2007. Toxicity of single walled carbon nanotubes on rainbow trout, (*Oncorhynchus mykiss*): Respiratory toxicity, organ pathologies, and other physiological effects. Aquat. Toxicol. 82: 94-109.

Sovová, T., D. Boyle, K.A. Sloman, C. Vanegas Pérez and R.D. Handy. 2014. Impaired behavioural response to alarm substance in rainbow trout exposed to copper nanoparticles. Aquat. Toxicol. 152: 195–204.

Turton, J. and J. Hooson. 2008. Target Organ Pathology. Taylor and Frances Ltd, London.

Windeatt, K.M. and R.D. Handy. 2013. Effect of nanomaterials on the compound action potential of the shore crab, *Carcinus maenas*. Nanotoxicology. 7: 378–388.

7

Nanoparticles Under the Spotlight

Intracellular Fate and Toxic Effects on Cells of Aquatic Organisms as Revealed by Microscopy

Alba Jimeno-Romero,[1,2,*] *Yvonne Kohl,*[2] *Ionan Marigómez*[1] and *Manu Soto*[1,*]

Microscopic Visualization of Engineered NPs in Target Cell and Tissue Compartments

During the last decades, the use of engineered nanoparticles (NPs) has become massive, and so has increased the number of products featuring nanotechnology. Although no exact production quantities are currently available, the most produced NPs are TiO_2, SiO_2, ZnO, carbon nanomaterials (fullerenes; nanotubes, CNTs), FeOx, Ag, Au and different Cd-based quantum dots (Krysanov et al. 2010, Piccinno et al. 2012, Skjolding et al. 2016). Nanotechnology is in the cutting edge of the development of new materials used in targeted cancer therapy, superconductors design, cosmetics or renewable energy capture. The performance of this technology is defined by physical properties, which include size, size distribution, shape and surface interactions (Su 2017). Thus, in the first steps of engineered NP production, accurate characterization is crucial and, for that, the capabilities provided by electron microscopy are highly useful.

Qualitative microscopy visualization techniques, including Scanning Electron Microscopy (SEM), Transmission Electron Microscopy (TEM), Fluorescence Microscopy and Confocal and Multiphoton Laser Scanning Microscopy, offer opportunities of high sensitivity and high spatial resolution analysis of NP barrier penetration (Zou et al. 2017). Transmission Electron Microscopy

[1] CBET Research Group, Dept. Zoology and Animal Cell Biology, Research Centre for Experimental Marine Biology and Biotechnology (PiE-UPV/EHU) and Faculty of Science and Technology (ZTF/FCT), University of the Basque Country (UPV/EHU), Basque Country, Spain.
[2] Fraunhofer Institute for Biomedical Engineering, IBMT, Sulzbach/Saar, Germany.
* Corresponding authors: manu.soto@ehu.eus; alba.jimeno.romero@ibmt.fraunhofer.de

(TEM) and Scanning Electron Microscopy (SEM) are the most commonly used approaches for NP visualization in cells. SEM provides information on the composition of the surface of the sample whereas TEM gives information about internal details (Buhr et al. 2009, Smith 2015, Williams 2015) or internal accumulation of nanoparticles in special cell organelles. SEM maps interactions between electrons and the sample based on the detection of secondary electrons emitted from the surface (Behzadi et al. 2017) and thus could be used to analyze the cellular binding of nanoparticles to the cell membrane. In contrast, TEM relies on electron transmission across the sample, therefore samples must be sectioned (ultramicrotomy), and hence it is crucial that the section thickness is carefully selected according to the size of the NP to visualize. The transmitted 2D images produced at TEM require further interpretation but they provide the observer with much greater resolution than SEM 3D images. The spatial resolution given by TEM is 0.05 nm at energies above 100 keV (Su 2017) whereas moderate electron energies of 10 keV typically found in SEM provide 1 nm resolution, which nevertheless can be sufficient to visualize nanosize scale particles (Buhr et al. 2009). Interestingly, operating at low voltage in SEM is less restrictive regarding vacuum conditions, which has rendered SEM at near-environmental conditions to become a reality. Recently developed Scanning Helium Ion Microscopy (SHIM), which uses helium ions instead of electrons, and since He ions can be focused to a smaller size than electrons or X-rays, also provides sub-nanometer resolution (~ 0.25 nm) at low voltage without damaging the specimen (Chen et al. 2011). Moreover, SHIM enhances surface contrast without the need of metal coating (as the sample does not charge) and increases depth of field (Joens et al. 2013, Leppänen et al. 2017). All these advantages could make possible very detailed studies of cell membrane and nanoparticle interactions, as well as ENP localization inside the cell, by measuring the energy-loss for He ions as they go through the sample (Leppänen et al. 2017).

Other electron microscopy techniques include Environmental SEM (ESEM) and Liquid Cell TEM (LCTEM). ESEM is a scanning electron microscope that allows for the option of collecting electron micrographs of specimens that are "wet", uncoated or both. LCTEM is especially interesting since allows for imaging chemical and biological processes in a completely liquid state, with the usual expected resolution. This technique can be used to image living cells, although cell death due to radiation from the electron beam can be expected to occur over minutes. In any case, it is an invaluable tool to acquire a very high resolution snapshot of molecular localizations, allowing the study of NP internalisation processes and mechanics in depth. Peckys and de Jonge (2011) have already used a similar technique to visualize the uptake of Au NPs in COS-7 cell line enclosed in a SiN chamber within a microfluidic microchip. Nowadays, holders and chips especially designed for this application can be added to any TEM. One of the most disheartening flaws of LCTEM is that, in preserving the sample as little processed as possible, the staining of intracellular structures is impossible, and if at all, it would be limited to electrondense stainings.

Recently, the combination of a variety of microscopical techniques is gaining interest. This approach is known as correlative microscopy and is based on the integration of two or more microscopy techniques (Light Microscopy, Electron Microscopy, Atomic Force Microscopy and Super-Resolution Microscopy) to analyze the same sample (Behzadi et al. 2017). The aim of this recently introduced new approach (more than new methodology) is to obtain more in-depth information of the sample using the advantages of the different classes of microscopy overcoming their limitations. Up-to-now it has been mostly used in nanomedicine to decipher intracellular trafficking of NPs.

In addition, electron microscopy can be used along with spectroscopy techniques such as X-Ray Energy Dispersive Spectroscopy (EDS), Electron Energy-Loss Spectroscopy (EELS) and diffraction techniques such as X-Ray Diffraction or XRD (Su 2017). Both detectors measure the behaviour of electrons after interacting with the sample, and are able to provide information on elemental composition, elemental bonds, valence... etc., and are best suited for "lighter" and "heavier" elements, respectively. Spectroscopy is essential to confirm the elemental composition of NPs. For instance, the size and size distribution of NPs can be calculated after XRD (assuming the particle is crystalline in nature) and after electron microscopy imaging whereas their composition is normally determined using EDS or EELS (Buhr et al. 2009, Smith 2015, Su 2017).

At the light microscope, confocal laser microscopy (CLSM) has been often applied to visualize NPs (Behrens et al. 2012). CLSM excludes out-of-focus objects through point scanning and, due to its narrow field depth, it provides us with optical sections suitable for subcellular localization of NPs (Figs. 1–2). CLSMs are highly valuable to clinical research; for instance, Zou et al. (2017) used CLSM to study the skin penetration of NPs.Its usefulness for environmental studies on NPs is clearly limited because in order to be visualized the NPs must be fluorescent or functionalized with antibodies (Behrens et al. 2012). Recently, observation of cells in their native state has been shown by Liu et al. (2018) who, using noninvasive lattice light sheet microscopy combined with aberration-correcting adaptive optics, have been able to study subcellular events *in vivo*. This technique offers incredibly high resolution, and could be used to monitor intracellular fate of nanoparticles very precisely. Multiphoton laser scanning microscopy enable researchers to obtain 3D images of NP distribution at micrometer resolution by way of optical sectioning (Labouta and Schneider 2013). Optical detection of nanoparticles has been achieved with the use of surface-enhanced Raman scattering. Individual silver colloidal nanoparticles were screened from a large heterogeneous population for special size-dependent properties and were then used to amplify the spectroscopic signatures of adsorbed molecules (Nie and Emory 1997). For single rhodamine 6G molecules adsorbed on the selected nanoparticles, the intrinsic Raman enhancement factors were on the order of 10^{14} to 10^{15}, much larger than the ensemble-averaged values derived from conventional measurements (Nie and Emory 1997). This enormous enhancement leads to vibrational Raman signals that are more intense and more stable than single-molecule fluorescence (Nie and Emory 1997). A summary of the main methods for imaging with emphasis in the interactions with cellular systems is illustrated in Table 1.

Being increasingly produced and used worldwide, NPs input into environmental systems is considerable and persistent, thus posing a growing threat for the health of the environment. Particularly, the aquatic environment, which receives runoff and wastewater from domestic and industrial sources, is considered as a main sink for NPs (Scown et al. 2010, Canesi et al. 2012). As a result, exposure to NPs and their subsequent effects on the aquatic biota are issues of major concern, and great research efforts are being addressed and aimed at assessing the ecotoxicological risk of NPs in aquatic systems. A few target species are optimal for this evaluation. Outstandingly, suspension-feeding mollusks are

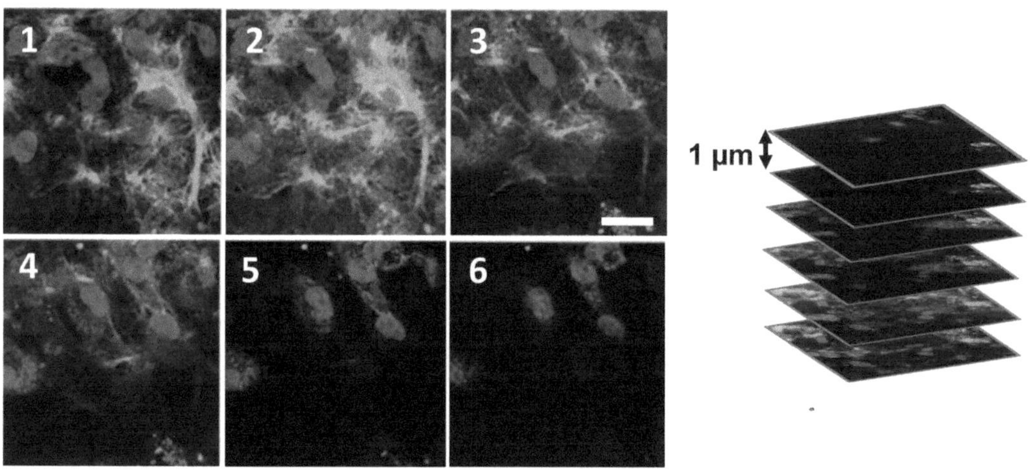

Figure 1. CLSM 2D image of osteogen differentiated human stem cells under chronic exposure to polystyrene NPs over 21 days (1 µg/ml). Nano- and microplastics are a present topic in the field of ecotoxicology, because the number of microplastics in the environment, especially in the sea, increases dramatically due to plastic waste. Image 1–6 show the individual stacks of a 1 µm-z-stack as overlay of the fluorescence images. Green: Hydroxylapatite, Blue: DAPI staining of nucleus, Red: Polystyrene NPs labeled with red fluorecence. Scale bar 25 µm. Image from Patrizia Komo (Fraunhofer Institute for Biomedical Engineering, Sulzbach, Germany), "Effect of nanoparticles on the differentiation of mesenchymal stem cells in 2D and 3D *in vitro* model".

Color version at the end of the book

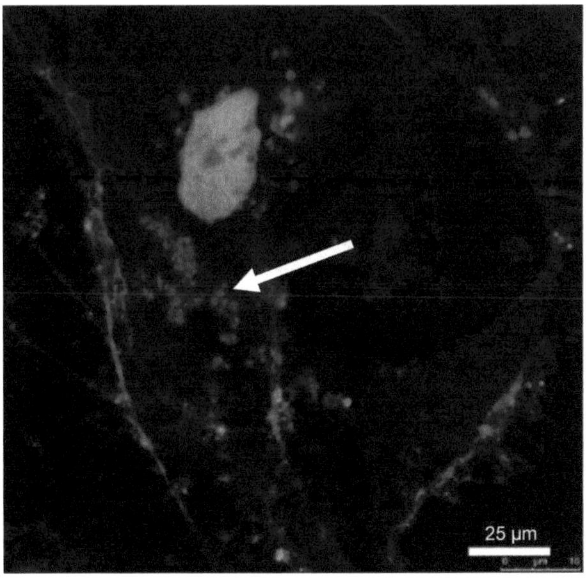

Figure 2. CLSM 2D image of adipogenc differentiated human stem cells under chronic exposure to polystyrene NPs over 21 days (1 µg/ml). Image shows the overlay of the fluorescence images. Blue: DAPI staining of nucleus, Green: Concanavalin A 488 staining of the cell structure. Red: Polystyrene NPs labeled with red fluorescence. Scale bar 25 µm. Image from Patrizia Komo (Fraunhofer Institute for Biomedical Engineering, Sulzbach, Germany), "Effect of nanoparticles on the differentiation of mesenchymal stem cells in 2D and 3D *in vitro* model".

Color version at the end of the book

considered unique model animals for nanotoxicology studies; among other reasons, this is because their feeding mechanism implies specialized processes for internalization of nano- and micro-scale sized (food) particles (Moore 2006). Indeed, the effects produced by NPs (mortality, developmental effects, growth inhibition, immune dysfunction and reproduction impairment) have been studied not only in mollusks but in a vast variety of other marine organisms, including for example diatoms, cyanobacteria, macroalgae, copepods, rag worms and several fish species (Moore 2006, Krysanov et al. 2010, Handy et al. 2011, Hull et al. 2011, Burchardt et al. 2012, Canesi et al. 2012, Maurer-Jones et al. 2013, Leclerc and Wilkinson 2014, Sohn et al. 2015, Skjolding et al. 2017, Ostaszewska et al. 2018).

Once introduced into aquatic ecosystems, the fate and transport of NPs is poorly known, although it is well-known that the fate of NPs will be determined by their physical, chemical and nano-scale properties (Krysanov et al. 2010). Thus, engineered NP properties (e.g., shape, surface area, surface coating, surface charge and reactivity) and water chemistry (e.g., salinity, dissolved organic carbon, ionic strength, pH and temperature) will determine whether the NPs will remain as single particles in suspension, partitioned to dissolved organic carbon in the water column, aggregated or adsorbed to suspended matter (USEPA 2012). Regardless, these new materials will be incorporated into the biota through different pathways; for example, either actively via ingestion (e.g., NPs in colloidal suspended matter or adhered to small organisms) or passively (e.g., by contact with body barriers such as the skin -integument- or the gill epithelium). As a result, cellular uptake, *in vivo* reactivity and distribution across tissues will vary depending on the internalization pathway (Walters et al. 2016). Moreover, since very little is known regarding NP fate and properties in the environment, identifying the mechanisms underneath bioavailability and toxicity of environmental NPs is certainly a difficult endeavour (Skjolding et al. 2016). To add complexity to the problem, NPs can be fully or partially dissolved and exert their toxic effects through released ions rather than through their NP properties (i.e., Ag, ZnO, CuO; Jimeno-Romero et al. 2017a). Likewise, different substances such as stabilizing agents (i.e., citrate), or coatings (polyethylene glycol, PEG; polyvinylpyrrolidone, PVP) are added in their formulation, and these additives can modify the environmental fate, the internalization pathway,

Table 1. (Microscopical) Methods used to determine different endpoints and interactions between NPs and cellular systems (modified from Lowry and Wiesner 2013).

Endpoint	Microscopy Method	Possibilities	Problems/Challenges
Particle Size & Shape, Size Distribution	Dynamic Light Scattering (DLS)	Measures particle size distribution (PSD) in situ. Applicable to a size range of 0.6 nm to 6 μm.	Disperse samples make data interpretation difficult. Cannot distinguish particle types and provides no information on shape. No chemical information.
	Atomic Force Microscopy (AFM)	Determines particle size and shape. Yields topographical and mechanical information at high resolution. Indirect image of shape.	No chemical information. Particles must be adhered to a surface.
	Transmission Electron Microscopy (TEM)	Direct imaging of particle size and shape. Can provide chemical information with proper detectors.	Requires dried samples which aggregates samples and introduces other sample preparation artifacts.
	Multiphoton laser scanning microscopy	3D image of NP distribution at micrometer resolution by optical sectioning.	It can produce phototoxicity and damage living cells.
Elemental Composition	TEM with Energy Dispersive X-Ray Detection (EDS)	Provides chemical information at specific locations on a particle.	Only semi-quantitative in most cases.
	Scanning Electron Microscopy (SEM)	Provides elemental information for labeling experiments with appropriate detectors.	Only surface information unless complicate processing is carried out. Limitation of artifacts due to sample handling.
	X-ray photoelectron spectroscopy (XPS)	Provides surface chemical composition and oxidation state.	Vacuum technique and usually semi quantitative. No shape or size information. Absolute quantification requires the use of standard samples.
	X-ray Powder Diffraction (XRD)	Determines crystalline phases present. Can be an indirect measure of crystal size in a particle.	Low sensitivity. Overlapping peaks in a complex sample make positive identification difficult.
	X-ray fluorescence (XRF)	Non-destructive analytical technique used to determine the elemental composition	It is necessary to ensure that the sample is sufficiently thick to absorb the entire primary beam.
	Synchrotron-based XRF microscope	Improves spatial resolution to allow the localization of diffusely distributed NPs. Discriminate between nanogold or soluble gold.	Cannot discriminate between nano or soluble forms.
Interactions	Flow Cytometry*	Improves objectivity and statistical rigor through high-throughput single-cell analysis.	*It is not a microscopic technique.
	Scanning Helium Ion Microscopy (SHIM)	Minimizes sample damage, enhances surface contrast without metal coating and increases depth of field.	The high resolution afforded needs the development of new sample preparation protocols to mitigate the possibility of fixation and drying artifacts (i.e., cryo-immobilization).
	Transmission Electron Microscopy (TEM)	High resolution imaging.	Long prep times. Low contrast. Need of sectioning.
	Confocal laser scanning microscopy (CLSM)	Excludes out-of-focus objects through point scanning. Provides optical sections suitable for subcellular localization of NPs.	Limited usefulness for environmental studies: NPs must be fluorescent or functionalized with antibodies.
	Atomic Force Microscopy (AFM)	Can be performed in the aqueous state maintaining native hydration of the sample.	Poor discrimination.
	Scanning Electron Microscopy (SEM)	Provides lateral resolution (1.0 nm). Biological tissues can also be imaged in serial sections, providing precise tomographic information.	Only surface information.

Table 1 contd. ...

... *Table 1 contd.*

Endpoint	Microscopy Method	Possibilities	Problems/Challenges
	Dynamic Light Scattering Microscopy (elastic/inelastic)	Provides information regarding local chemical environment in biological samples (i.e., orientation).	Unable to precisely resolve polydisperse samples.
	Dark-field microscopy	Enables imaging using scattered light. Useful in imaging translucent objects with refractive indices similar to the surrounding. Tracks plasmonic NPs** in live cells.	Only highly scattering objects are visible on a black background.
	Photoacoustic (PA) microscopy	Uses laser-induced thermoelastic expansion to map cells and tissues based on variations in subsequent ultrasound emission.	Emerging method. Few applications available.
Intracellular trafficking	Correlative microscopy***	Super-resolution microscopy in a correlative microscopy can overcome the conventional limitation of light microscopy, which is poor resolution below 300 nm.	Emerging approach.
	Raman-microscopy	Tracking of individual NPs in real time (live-cell). Allows elemental composition determination. Raman signals that are more intense and more stable than single-molecule fluorescence.	Sample heating through the intense laser radiation can destroy the sample or mask the Raman spectrum. Large background signals from fluorescence from impurities.
	Confocal laser scanning microscopy (CLSM)	Super resolution microscopy that allows intracellular trafficking routes determination (endosomes).	NPs must be labelled with fluorochromes to be tracked.
	liquid cell TEM (LCTEM)	Allows for imaging chemical and biological processes in a completely liquid state.	Can be used to image living cells, although cell death due to radiation from the electron beam has to be expected over minutes.

* It is not a microscopic technique but has recently emerged as an indispensable tool to combine with microscopy in order to shed light on the heterogeneity of cell (monoclonal)-NP interactions.

** Particles whose electron density can couple with electromagnetic radiation of wavelengths that are far larger than the particle due to the nature of the dielectric-metal interface between the medium and the particles.

*** Integration of two or more microscopy techniques (light microscopy, electron microscopy, atomic force microscopy and super-resolution microscopy) to analyze the same sample.

the uptake mechanism and the toxicity of the engineered NPs (Jimeno-Romero et al. 2016, 2017b). Finally, it is worth noting that engineered NPs that are highly degradable in the aquatic environment (e.g., lipid-based or metallic ones that dissolve relatively fast) cannot be studied by microscopic techniques.

This review deals with how the microscopic visualization of engineered NPs in target cell and tissue compartments can enhance our understanding of the NP-cell interactions in aquatic organisms and thus provide support for the environmental risk assessment of NPs in aquatic systems. We presume that the reader has basic knowledge about electron microscopy and therefore the principles behind the variants of electron microscopy will not be detailed. Thus, this chapter will focus on how advances in microscopy contribute to getting a deeper knowledge of the environmental and systemic fate of engineered NPs and their mechanisms of intracellular uptake and toxicity in aquatic organisms.

Imaging Nanoparticles in the Environment

Nanoparticles enter aquatic environments from diverse sources including wastewater treatment plants effluents, direct use and air deposition (Vale et al. 2016). The natural physicochemical properties of the aquatic environments (pH, salinity, ionic strength, salt composition and suspended organic matter) are continuously changing; thus affecting the chemistry, physicochemical properties, behaviour and fate of the NPs (Lowry and Wiesmer 2007, Klaine et al. 2009, Ward and Kach 2009, Canesi et al. 2012, Brunelli et al. 2013). As a result, dissolution, aggregation and sedimentation of NPs will be altered depending on the physicochemical properties of both the NPs and the environment where these are released (Fig. 3, Vale et al. 2016).

Literature dealing with NPs in freshwater ecosystems is more extensive than literature regarding coastal ecosystems but great research efforts are being addressed to estuaries and littoral areas in the recent years. Existing data clearly reveal that both freshwater and marine systems are quite different regarding relevant issues for NP fate and toxicity. Indeed, marine waters are deeper, present a variety of colloids[1] natural or artificial, and are generally more alkaline, with a greater ionic strength than freshwater ecosystems (Klaine et al. 2008). Likewise, major differences exist in the fate of NPs in seawater compared to freshwater due the higher diversity of algae and invertebrates that can interact with the NPs. Noteworthy, the majority of industrial discharges occurs in the coastal zone and, furthermore, marine systems are considered the ultimate fate for NPs released to freshwater systems (Klaine et al. 2008).

In aquatic systems, colloids can strongly interact with NPs and produce aggregates, most remarkably in seawater, thus conditioning NP bioavailability when these aggregates get in contact with biological barriers. Particle size, shape, surface charge and hydrophobicity are crucial to go across these barriers, and these properties can be largely modified in aggregates. For instance, the optimal NP size for cellular uptake in several cells has been reported to be in the range of 50 nm, or maybe lower according to some theoretical models (Chithrani et al. 2006). Aggregates largely exceed this

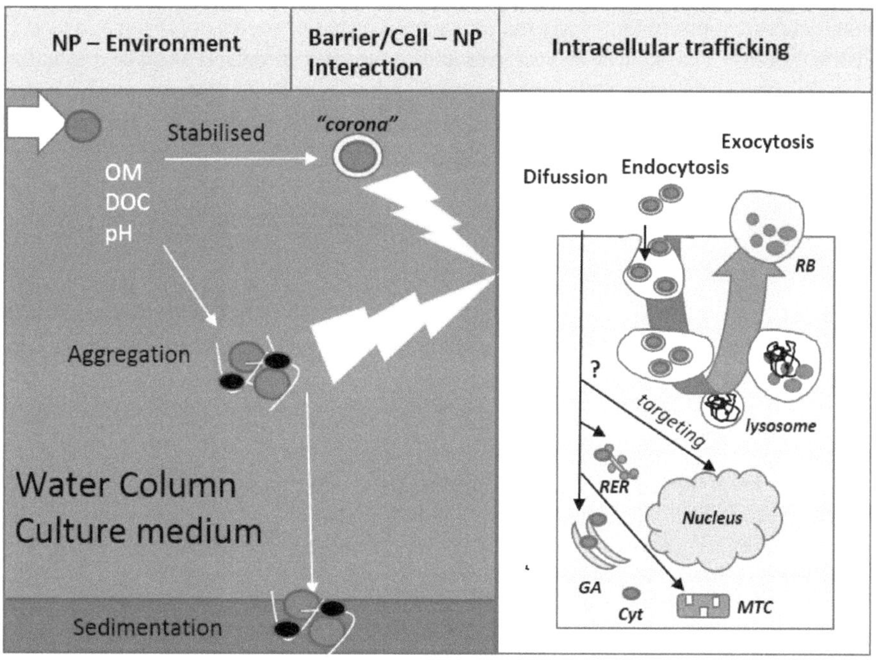

Figure 3. Representation of the transformations when nanoparticles (red dots) enter the aquatic environment and culture media and interact with organic matter (OM), dissolved organic carbon (DOC), pH and other physicochemical entities such as mineral surfaces, humic substances, colloids, biopolymers and so on (not shown). In the interaction between NPs and cell barriers, the role of the corona is crucial to produce the internalization of NPs. This corona can be lost during the internalization through diffusion or endocytosis to cross the cell membrane. The internalized NP undergoes different processes following the endocytic and exocytic pathways meanwhile affecting different subcellular compartments (mitochondria–MTC-), endoplasmic reticulum (rough, RER), Golgi apparatus (GA). RB, residual bodies.

[1] In aquatic systems, colloid is the generic term applied to particles between 1 nm and 1 μm size range (Klaine et al. 2009). It includes organic and inorganic matter originated from both natural and anthropic sources (Vale et al. 2016).

size that otherwise can be common for NPs, thus preventing internalization of NPs. Moreover, for a small aggregate less than 50 nm the internalization may be also hampered by, for example, its shape, surface charge and hydrophobicity (Behzadi et al. 2017). The presence of additives in engineered NPs can be relevant in this respect, as it may prevent the formation of aggregates in seawater (Stakenborg et al. 2008). For instance, the functionalization of NPs with, for example, citrate-capping reduces the formation of aggregates in seawater, as shown in the case of Au NPs (Su and Kanjanawarut 2009, Hull et al. 2011, Jimeno-Romero et al. 2017b).

The degree of NP aggregation in water solutions can be determined by measuring time-dependent changes in size and surface charge on the basis of TEM observations and dynamic light scattering (DLS; Oh and Park 2014) or field flow fractionation (von der Kammer et al. 2011). However, to our knowledge, this approach has been applied in culture media for *in vitro* assays but not in seawater. Electron microscopy is currently commonly used to characterize engineered NPs by manufacturers, and the same approach is being used to a limited extent to identify engineered NPs in the aquatic environment. Alas, most of the techniques employed for NP characterization require the sample to be dried for examination under high vacuum, which causes major alterations such as additional aggregation (Lowry and Wisner 2007).

Imaging Nanoparticles Interaction with Biological Barriers

The main question raised herein deals with how microscopy can contribute to identifying persistent NPs in aquatic environments and their internalization and accumulation in living organisms. The accumulation of NPs by aquatic organisms is the result of the balance between internalization and elimination (detoxification) of the NPs in the organism. The major routes of NP entry are via ingestion (digestive tract) or direct passage across the gills and/or the integument. NPs can be internalized alone or aggregated,for example, with suspended organic matter or colloids (Levard et al. 2012, Behra et al. 2013, Vale et al. 2016). For instance, mollusks and fish are likely to incorporate NPs, both alone or aggregated, because they possess well developed mechanisms for the internalization of nano- and microparticles into cells via endocytosis or phagocytosis (Moore 2006, Canesi et al. 2012). In contrast, mucus secretion and other polysaccharidic layers often appear covering the gills and the integument in mollusks and fish, and these layers can act as protective barriers preventing NP internalization.

In mollusks, gills are an important biological barrier as they comprise a large respiratory surface located in the interface between the surrounding milieu and the circulatory system. This barrier is comprised of a respiratory epithelium and the underlying blood cells, known as hemocytes. Hemocytes are endowed with well-developed endo-lysosomal system and highly active in endocytosis and phagocytosis, and are active in the systemic transport of foreign particles and substances, including also NPs (Canesi et al. 2012, Katsumiti et al. 2015a,b, Fig. 3). In marine filter-feeding bivalves, together with the respiratory function, gills are also involved in feeding, particularly in the sorting and ingestion of food particles. Food particles are trapped by the gill sieve are moved towards the labial palps and the mouth, thus entering the gut, and reaching the digestive gland, where intracellular digestion occurs in digestive cells (Vale et al. 2016). Digestive cells have an extremely developed lysosomal system for intracellular digestion (Marigómez et al. 2002).

TEM and X-ray microanalysis in combination have been useful to demonstrate that gills are a major site for waterborne NPs' entry into the mollusks. Using conventional TEM, Au NPs were observed in the gills of clams, *Scrobicularia plana*, after *in vivo* exposure but the resolution was limited (Koehler et al. 2008, Joubert et al. 2013, McCarthy et al. 2013). Later on, García-Negrete et al. (2015) applied the STEM-in-SEM approach to investigate the uptake of citrate stabilized Au NPs (average particle size: 23.5 ± 4.0 nm) into gills of the clam, *Ruditapes philippinarum*, after *in vitro* exposure experiments with gill explants (Fig. 4). STEM-in-SEM was optimized for achieving optimum resolution under SEM low voltage operating conditions (20–30 kV). Resolution below 10 nm was properly achieved with 200–300 nm thick specimens at magnifications over 100k×. These relatively thick sections of the gills appear to be stable under the beam and prevent NP dislocation

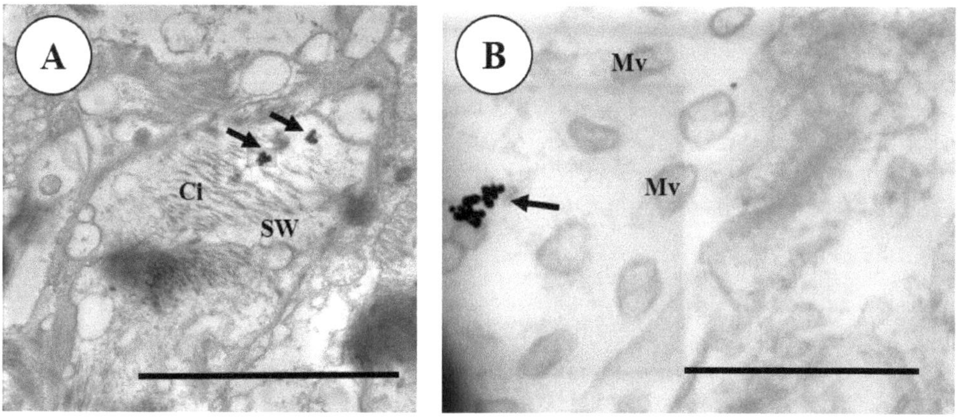

Figure 4. STEM-in-SEM high resolution imaging of gold nanoparticles (22 nm) in gill explants of clams (*Ruditapes decussatus*) (exposure: 60 mg Au/L; 1 h). (C) Overview of the apex of the gill epithelium with NPs (arrow) associated to cilia (Ai) and microvilli (SW, seawater). (B) Detail of the apex with NPs between microvilli (mv). Scale bars: A, 5 µm; B, 500 nm. C-D courtesy of M.C. Jiménez de Haro, C. García and A. Fernández (ICMS-CSIC, Sevilla, Spain).

during cutting (Garcia-Negrete et al. 2015). Overall, this enhanced resolution approach allowed both localizing internalized NPs and identifying ultrastructural alterations in gill cells, using thicker sections than those normally employed for standard TEM at higher voltages. In addition, the resolution improves as magnification increases (Fig. 4), so that at magnifications > 100 k×, resolution can go beyond < 10 nm for 200–300 nm sections (Garcia-Negrete et al. 2015). Using a synchrotron-based XRF microscope, Hull et al. (2011) demonstrated that gold is internalized in the digestive gland of the freshwater clam, *Corbicula fluminea*, after exposure to Au NPs. This technique is more sensitive than TEM-EDX and provides a better spatial resolution (beam size: 3×3 µm^2) through which these authors localized diffusely distributed gold spots in the digestive gland and other zones of the digestive tract but could not discriminate between nanogold or soluble gold. Scanning acoustic microscopy (SAM) is another technique which could be used for analyzing localization of nanoparticles inside the cell. Here, you distinguish high resolution ultrasound imaging and photoacoustic imaging (Strohm et al. 2016).

In fish, gills, liver and kidney are organs involved in NP accumulation while the importance of the spleen, blood and brain is residual in many fish species (Scown et al. 2009, Krysanov et al. 2010). The gastrointestinal tract represents the main route of NP entry in fish (Fröhlich and Fröhlich 2016). The liver is a major target organ and the hepatocytes the main target cell type (Lee and An 2015, Asztemborska et al. 2014, Maes et al. 2014, Dedeh et al. 2015, Ostaszewska et al. 2018). Electron-dense granules can be identified as NPs with the aid of TEM observation coupled with EDX analysis, which is a highly recommended approach, especially in the case of metallic NPs. Conventional TEM examination is still sensitive enough to demonstrate that certain NPs can penetrate into gill filaments from water (Moger et al. 2008), say: Ti (from TiO_2) in carps (*Cyprinus carpio*; Sun et al. 2007) and rainbow trout (Federici et al. 2007, Johnston et al. 2010); and Sn (from SnO_2 NPs) in the guppy (*Poecilia reticulata*; Krysanov et al. 2010).

Modifications of NPs Prior to Internalisation by Living Systems in Natural and Culture Media: *Knockin' on Cells Doors with a Brilliant Disguise*

The main question raised herein deals with how microscopy can contribute to visualizing and deciphering interactions between NPs and living entities.

In vivo, a crucial aspect is that the surface of the NPs can be dramatically modified along the internalization pathway in the organisms. For instance, internalized NPs acquire newly adsorbed molecules (including proteins, colloids and so on) that form the so-called "**protein corona**", thus leading to changes in their chemical composition and surface charge (Klaine et al. 2009, Behzadi et al. 2017). That is how NPs disguise themselves to cross the door. In simple words, what cells "see" when interacting with NPs is corona-coated NPs rather than pure NPs with pristine surface; this can be a crucial issue to be considered as regards NP fate, toxicity and risk (Domingos and Pinheiro 2014). Proteomic approaches have been most commonly used to identify the components of NPs' protein corona (Walkey and Chan 2012). Recently, an outstanding work by Kokkinopoulou et al. (2017), successfully applied conventional TEM combined with tomography to give insights into the thickness ("soft" and "hard") and 3D conformation of protein corona. Authors investigated the corona formed around engineered polystyrene NPs with surface charge modifications (mediated by carboxil- and amino-coating). These modified NPs were stabilized in a trehalose film with negative staining properties and the images obtained were compared with those obtained after stabilisation of the NPs by plunge-freezing for cryo-TEM. This method shows great ingenuity; not only does it use electron microscopy equipment that is readily available in many environmental toxicology labs, it also has the great advantage of allowing the imaging of both the hard corona and the elusive soft protein corona (Fig. 5). It is precisely this soft corona that makes up the bulk of what cells "see" when facing ENPs that have undergone environmental modifications.

In vitro assays, in which cells are directly exposed to NPs in the culture media, are commonly used in clinical, toxicological and ecotoxicological research. Great efforts have been addressed to characterize NP behaviour in these culture media, which are rich in proteins, antibiotics, amino acids, nutrients and other compounds that are essential for the maintenance of cells but often may promote particle aggregation (Michen et al. 2015). Whereas *in vitro* toxicity assays using vertebrate cell lines as models in nanotoxicology are extensive, *in vitro* studies in aquatic animal cells are scarce. For instance, Katsumiti et al. (2015a, 2015b, 2016) used mussel hemocytes and gill cells to assess the toxic effects of Au, ZnO, TiO_2, Ag and SiO_2 NPs with different sizes and shapes. A thorough review on this issue has been carried out in the present book (see Chapter by Katsumiti and Cajaraville).

Interactions between NPs and cells in culture have been investigated by conventional TEM; however, although the resolution of TEM can reach up to 0.2 nm, this approach is less useful in the case of relatively small size NPs (< 10 nm) due to the complexity of specimens made of cells in culture. SEM has also been successfully used to visualize the internalization of Au NPs in human stem

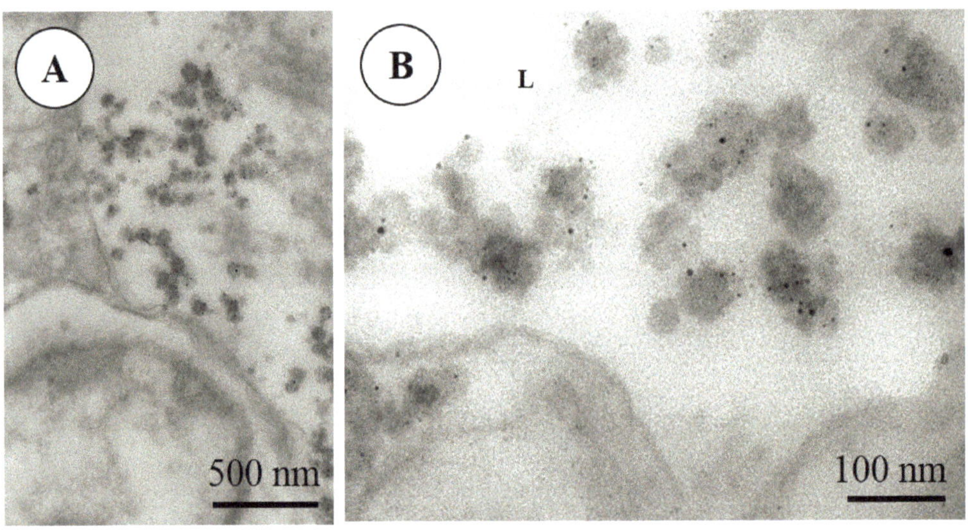

Figure 5. TEM (operating at 120 kev) micrographs (unstained sections) of the digestive gland of mussels *Mytilus galloprovincialis* exposed to 180 nm size TiO_2 NPs (1 mg Ti/L, 3 d). (a) NPs in the lumen of digestive epithelium related to the (probably) proteic material which forms the corona. (b) Detail of NPs-corona interaction close to the cell membrane. L: lumen. Scale bars: A, 500 nm; B, 100 nm.

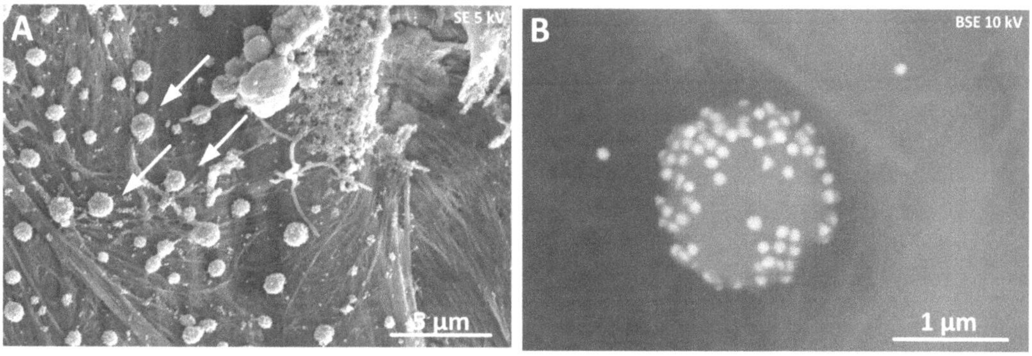

Figure 6. SEM image of the internalization of gold NPs in human stem cells. Human mesenchymal stem cells (hMSCs) were differentiated to adipocytes whilel exposed to 22,8 µg/ml gold nanoparticles (80 nm) for 21 days. Afterwards, cells were analyzed via SEM, (A) Lipid droplets on differentiated cell. (B) Lipid droplet with gold nanoparticles.

cells. Kohl et al. (2011) studied the internalization of gold NPs in adipogenic differentiated human mesenchymal stem cells during the differentiation process as well as their impact on the differentiation efficacy (Fig. 6). Alternatively, CLSM has been applied to visualize endosome-entrapped Au NPs (4–6 nm), to determine their subcellular distribution and to quantify NP uptake into HeLa and MCF 7 cells (Kim et al. 2015).

Regarding aquatic animals, a pioneering study was carried out by Ringwood et al. (2005). These authors evaluated the accumulation and potential toxicity of Cd/Se QDs under different salinity regimes using primary cell preparations of digestive cells of oysters, *Crassostrea virginica*. In this work, QDs accumulation into digestive cells was demonstrated by CLSM and TEM and QD degradation was shown by TEM. In a later study, the same approach was applied to assess the effects produced by fullerenes in oyster digestive cells *in vitro* (Ringwood et al. 2009).

Interactions between NPs and Cell Membranes for Internalization: Endocytosis and Phagocytosis ... *Journey to the Centre of the Cell*

When NPs approach the membrane of a cell, they interact with its components or with the extracellular matrix produced by the cell itself (Voigt et al. 2014). The eukaryotic alga *Chlorella pyrenoidosa* hinders CuO NP internalization by secreting a thick layer of extracellular polymeric substances (EPS) that retain NPs outside the cell as observed by SEM (Zhao et al. 2016). The authors concluded that the secretion of EPS can act as a protective mechanism to reduce the toxic effects of NPs. Behra et al. (2013) used dark field STEM found Ag NPs in the thylakoid membranes of the algae, *Chlamydomonas reinhardtii,* and concluded that NPs had passed through the cell wall, cell membrane and the chloroplast envelope; which has been reported for other NPs such as Ag, CuO, and TiO_2 NPs in several algae species (von Moos and Slaveykova 2014). However, how NPs passed across the cell wall is not fully understood.

If aggregates are too large they cannot be directly transported across the cell membrane and the NPs enter the cell through endocytosis[2] or phagocytosis, depending on the size of the aggregates. Endocytosis is a universal mechanism present in all eukaryotic cells that serves a variety of housekeeping and specialized cellular functions and can be mediated by distinct molecular pathways (Ivanov 2008). Among them, internalization via clathrin-coated pits, lipid raft/caveolae-mediated endocytosis and macropinocytosis/phagocytosis are the most extensively characterized. In almost all cells (with the exception of brain endothelial cells), internalized nanomaterials follow the endocytic pathway and are sequentially delivered to early endosomes ("sorting stations"), transported along with receptors to recycling endosomes and subsequently excreted. Others move (along microtubules)

[2] Endocytosis leads to the engulfment of NPs in membrane invaginations, followed by their budding and pinching off to form endocytic vesicles, which are then transported to specialized intracellular sorting/trafficking compartments.

and fuse with late endosomes, and these fuse with lysosomes, and some undergo exocytosis and are released out of the cell (Behzadi et al. 2017).

In filter-feeding bivalves, food particles are size sorted along the gills; those of the proper size (in the range of 4 to 25 μm) are further on carried by cilia and enter the gut whilst oversized ones are released as pseudofaeces before entering the body (Morton 1983). Thus, microscale-sized Au NP aggregates of the proper size can be internalized following the pathway of food particle intake whereas larger aggregates would be discarded as pseudofaeces and not internalized (García-Negrete et al. 2013, Hull et al. 2011, Pan et al. 2012, Jimeno-Romero et al. 2017a,b). Under the acidic conditions of the gut, agglomerates may undergo partial dissolution so that the NPs, which usually are loosely bonded by weak forces, would be released to the gut lumen, as particles of a suitable size to be internalized in cells by endocytosis (Oh and Park 2014). In contrast, more stable may occupy and obliterate the gut lumen, as reported in marine mussels, *M. galloprovincialis*, exposed to maltose-capped Ag NPs (Jimeno-Romero et al. 2017a). Nevertheless, slow but long-lasting release of these NPs seems to occur from these stable aggregates. This was demonstrated by the combination of conventional TEM and Light-Microscopy with autometallography in mussels exposed to Ag NPs and Au NPs (Fig. 7), in which, together with the presence of stacked aggregates in the gut lumen, NPs were incorporated into digestive cells via endocytosis (Jimeno-Romero et al. 2017a,b). In a similar way, glass wool NPs were shown after TEM to be internalized by endocytosis in the digestive cells of mussels, *Mytilus edulis* (Koehler et al. 2008).

In some cases, through the combination of TEM and X-ray microanalysis, Na-citrate capped Au NPs (most likely internalized by endocytosis due to their suitable size) were localized inside the digestive cells of marine mussels and clams; in contrast, the NPs could not be found in apical endocytosis vesicles (Joubert et al. 2013, Jimeno-Romero et al. 2017b). It is conceivable that the NPs entered the cells by very fast endocytosis and therefore they were not easily captured while in the endocytic vesicles. These Au NPs can be of optimal size for cell uptake through what is so-called the "wrapping effect", which refers to how the plasma membrane encloses NPs. It seems that 27–30 nm NPs would have the fastest wrapping time and thus the fastest receptor-mediated endocytosis (Gao et al. 2005). Smaller particles would dissociate from the cell membrane and larger particles would reduce the membrane wrapping necessary for endocytosis to occur (Jain and Stylianopoulos 2010). The trade-off between Au NP size and the optimal size for endocytosis in mussels could thus explain, at least partially, the subcellular distribution of these NPs observed by Jimeno-Romero et al. (2017b).

Although endocytosis seems to be the most common pathway for the internalization of individual NPs in cells, in phagocytic (specialized) cells, large aggregates can be internalized by phagocytosis, thus incorporating materials with an average caliper, larger than 0.5 μm (Oh and Park 2014). In marine mussels, hemocytes are major phagocytic cells that accumulate NPs (Canesi et al. 2010, 2015, Gomes et al. 2013a, Katsumiti et al. 2015b); however, although digestive cells are epithelial cells, they also exhibit phagocytic capacity. For instance, they can phagocytose exogenous cells such as, for example, epithelial apoptotic cells (Chabikovsky et al. 2004). Thus, Jimeno-Romero et al. (2017b), on the basis of combined TEM observation and EDS microanalysis, concluded that in mussel digestive cells, some Au NP aggregates might be incorporated by either phagocytosis (mid-sized aggregates) or endocytosis (small-sized aggregates) depending on their size, whilst the largest aggregates would have remained stacked in the gut lumen.

The lysosomal compartment is the main intracellular target as regards sequestration and degradation of endocytosed and phagocytosed NPs (Alkilany and Murphy 2010, Stern et al. 2012). For example, TiO_2, Ag and Au NPs and CdS quantum dots (Jimeno-Romero et al. 2016, 2017a,b, 2019) (Fig. 8) have been localized, using TEM combined with electron probe X-ray microanalysis, in different compartments of the endo-lysosomal system in digestive cells of mussels *Mytilus galloprovincialis in vivo*. Comparable results with the same techniques were obtained in clams *Ruditapes philippinarum* (Garcia-Negrete et al. 2013) and in the freshwater bivalve, *Corbicula fluminea* (Renault et al. 2008, Hull et al. 2011). In summary, the presence of NPs in the lysosomes seems to be a general principle, as it has been demonstrated also in annelid gut cells and in mammalian macrophages and Kupffer cells (Cho et al. 2009, Sadauskas et al. 2009, de Jong et al. 2010, Unrine et al. 2010). Thus, the sequestration of NPs in lysosomes and other membrane-bound vesicles of the endo-lysosomal system would hamper their direct interactions with the nuclear components and with other cellular organelles and thus they would not induce toxic events (Hauck et al. 2008).

One of the main problems with conventional TEM is that samples undergo a fixation procedure that is undoubtedly invasive, and there is high probability to disturb, or even remove, NPs from the sample. Even if NPs remain relatively intact in the sample, it is difficult to locate them. However, there are techniques that can be applied in order to ease the visualization of NPs. Silver enhancement, for instance, is commonly used to enhance antibody-bound nanogold in immunolabeled TEM

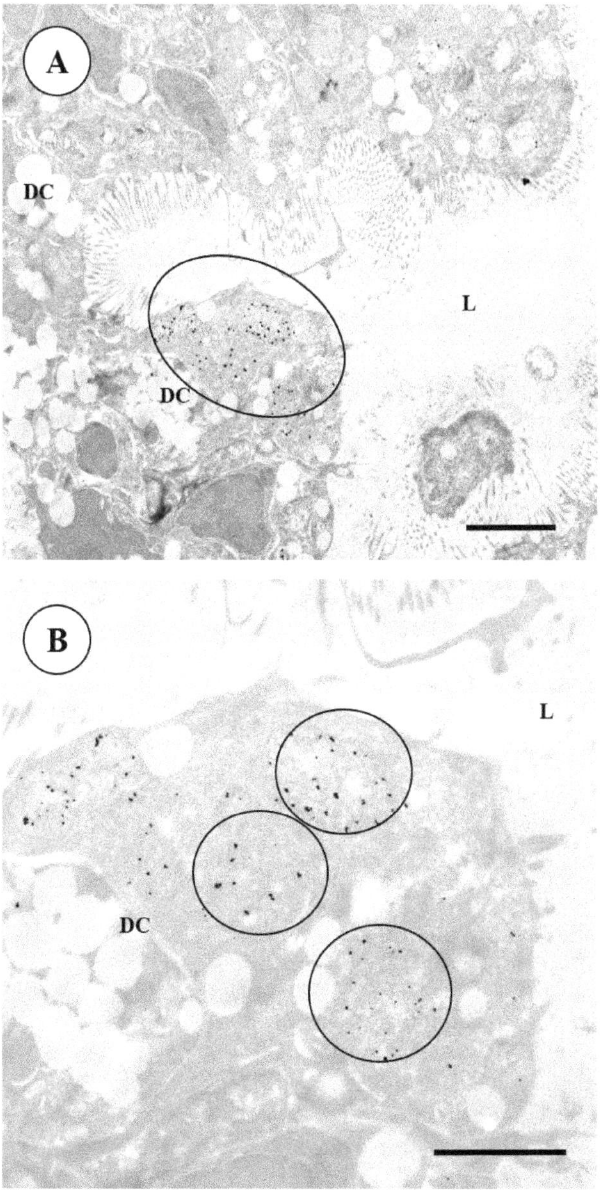

Figure 7. TEM micrographs (unstained sections) of digestive gland of mussels *Mytilus galloprovincialis* exposed to 5 nm size Au NPs (75 μg Au/L, 3 d). (a) Au NPs are present in the endosomes located in the apex of the digestive cells (circle) of the digestive epithelium (b) Detail of endosomes bearing Au NPs (circles). L: lumen; DC: digestive cell. Scale bars: A 4 μm; B 2 μm.

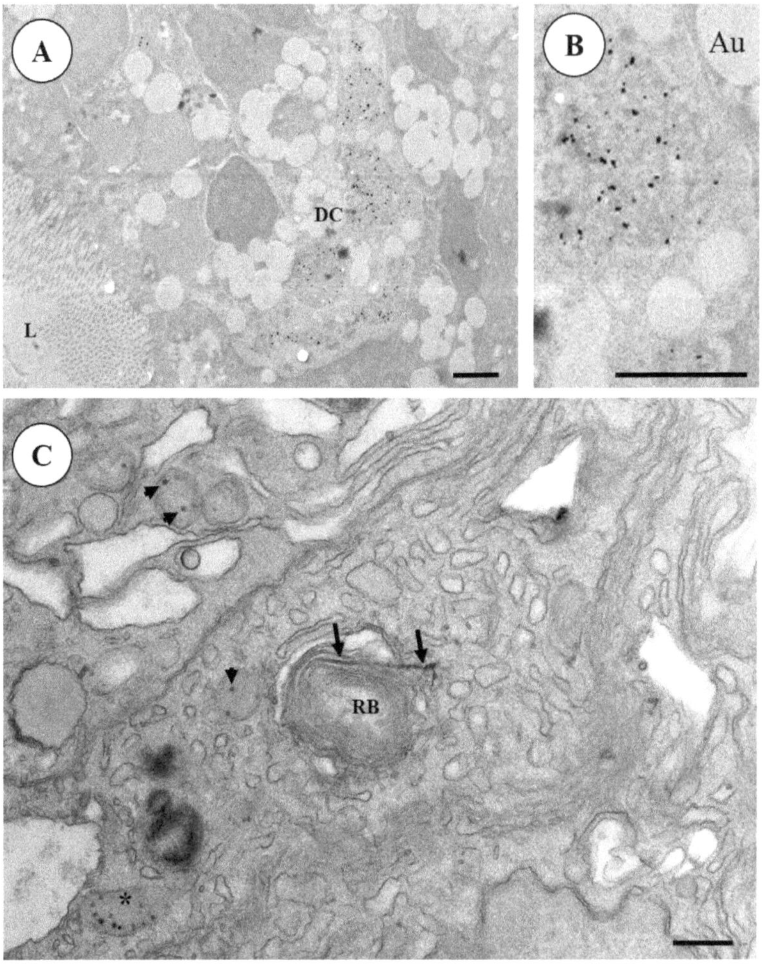

Figure 8. TEM micrographs (unstained sections) of digestive gland of mussels *Mytilus galloprovincialis* exposed to (A, B) 5 nm size Au NPs (75 μg Au/L, 3 d) and (C) 5 nm size CdS QDs (5 mg Cd/L, 1 d). (a) Overall view of the digestive epithelium with the lumen (L) on the left hand side. (b) Detail of Au NPs in the digestive cell lysosomes. (c) NPs in residual bodies (RB), mitochondria (arrowheads) and lysosomes (asterisk) L: lumen; DC: digestive cell. Scale bars: A, 4 μm; B, 2 μm; C, 500 nm.

samples for quantification purposes (Lui et al. 2014) and can be successfully applied to enhance the detection of several metal-bearing NPs such as TiO_2, ZnO, CuO and Au in environmental nanotoxicology studies (Jimeno-Romero et al. 2016, 2017a,b). However, this silver enhancement approach introduces a confounding factor for X-ray scattering detection, and thus, that no overlap with the element of interest occurs has to be checked. In addition, the size of the silver conjugates that appear as "background" should, ideally, not be larger than the NP of interest, so it is not advised to use this technique to enhance the detection of very small NPs such as quantum dots. When working with resin embedded samples, it should be taken into consideration that silver can only nucleate to exposed surfaces, and thus, NPs that remain completely embedded in the resin and whose surface have not been exposed when cutting the sample will not react to this technique, and will appear as less electrondense bodies in the sample. TEM is also suitable for studying internalization of NPs in specific cell compartment which are formed during the differentiation processes in the organism. Using the example of adipogenic differentiation of human stem cells, Kohl et al. (2011) verified the internalization of gold NPs in vesicles under chronic exposure of the cells during the 21 days of their adipogenic differentiation phase (Fig. 9).

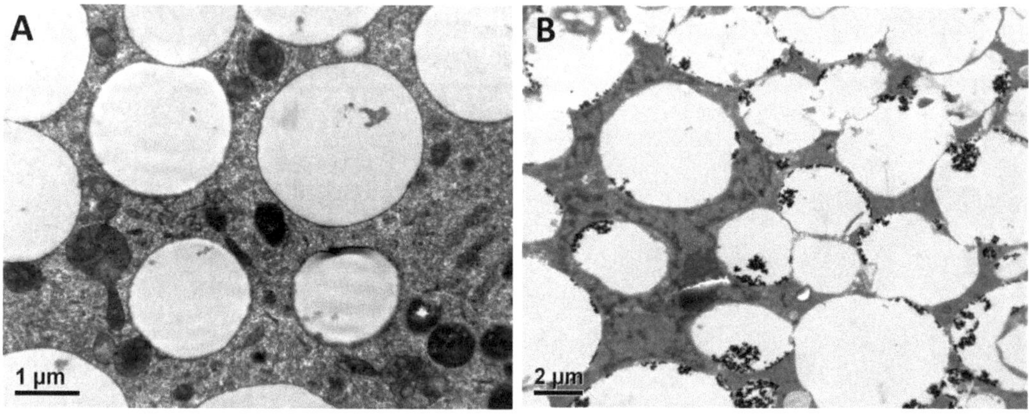

Figure 9. TEM images of adipogenic differentiated Human mesenchymal stem cells (hMSCs) treated with 22.8 μg/mL Gold NPs (80 nm) and untreated adipogenic differentiated hMSCs. The internalization of AuNPs (visible as black dots in image B) in hMSCs was investigated by TEM after 21 days of chronically treatment with 22.8 μg/mL Gold NPs. (A) After a period of 21 days, hMSCs were differentiated to adipocytes shown by incorporated lipid vesicles. Each vesicle is surrounded by a black protein membrane. If treated chronically with Gold NPs during the differentiation process, hMSCs internalized the AuNPs. The AuNPs are located in vesicles which differ in morphology (not round) and color (bright) from the lipid vesicles in the untreated control cells.

Scanning TEM techniques are particularly useful when dealing with NPs that could be degraded (even partially) in, for example, acidic environments inside the cell. It is even possible to have both bright field and/or high-angle annular dark-field imaging (HAADF) for STEM, the latter being especially useful to quickly localize ENPs with high atomic numbers (Fig. 10).

After the action of hydrolytic enzymes, the lysosomes become residual bodies that contain the undigested material (including primary or secondary NPs) and are expelled from the cell via **exocytosis.**[3] A great variety of NPs (Au, Ag, Ti) have been identified in residual bodies and cell debris excreted to the lumen of digestive alveoli in mussels (Jimeno-Romero et al. 2016, 2017a,b) and clams (Gu et al. 2009, Garcia-Negrete et al. 2013). Further on, the NPs are released to the external millieu as a part of the faeces where they can be microscopically identified as electron-dense particles resembling the original size of the source NPs (Hull et al. 2011).

In the field of applied biomedical nanotechnology, it has been reported that NPs can break away from the endo-lysosomal route and emerge in the cytoplasm and even enter the nucleus, mitochondria, endoplasmic reticulum and Golgi apparatus (Behzadi et al. 2017). Novel challenging applications in DNA labeling or DNA cleavage are based on the interactions of small sized engineered NPs with,for example, nucleic acids. These small NPs can go across the nuclear envelope pores in order to get the nucleotides tagged with small NPs (nanoprobes and nanodiamonds)[4] that act as labeling agents for bioimaging applications and therapeutic purposes (Dubertret et al. 2002; Klaine et al. 2006). Despite this intentional targeting, the majority of studies reviewed did not find Au NPs in the nucleus (Renault et al. 2008, Hull et al. 2011, Garcia-Negrete et al. 2013, Jimeno-Romero et al. 2016, 2017a,b); the

[3] Exocytosis: the active transport in which the cell transports molecules out of the cell.

[4] Engineered nanoprobes and nanodiamonds are quantum dots and fluorescent NPs especially designed for applications in biomedicine (Larson et al. 2003, Klaine et al. 2008). They are intended to be injected and then surpass the most effective body barriers by default, even though they still have to face immune clearance in the liver and spleen, permeation across the endothelium and endocytosis in target cells, among others (Barua and Mitragotri 2014). They can be equipped with a "homing device" that could guide the engineered NPs to the intended target and specifically recognize the target site. The intracellular fate of ENPs is critical to their success, considering that these carriers are intended to deliver specific molecules (i.e., genes, drugs and contrast agents) to components of the cytosol, the nucleus or to other specific intracellular sites (Behzadi et al. 2017). They can be localized by different varieties of microscopes depending on the targeting molecules they carry on their surface.

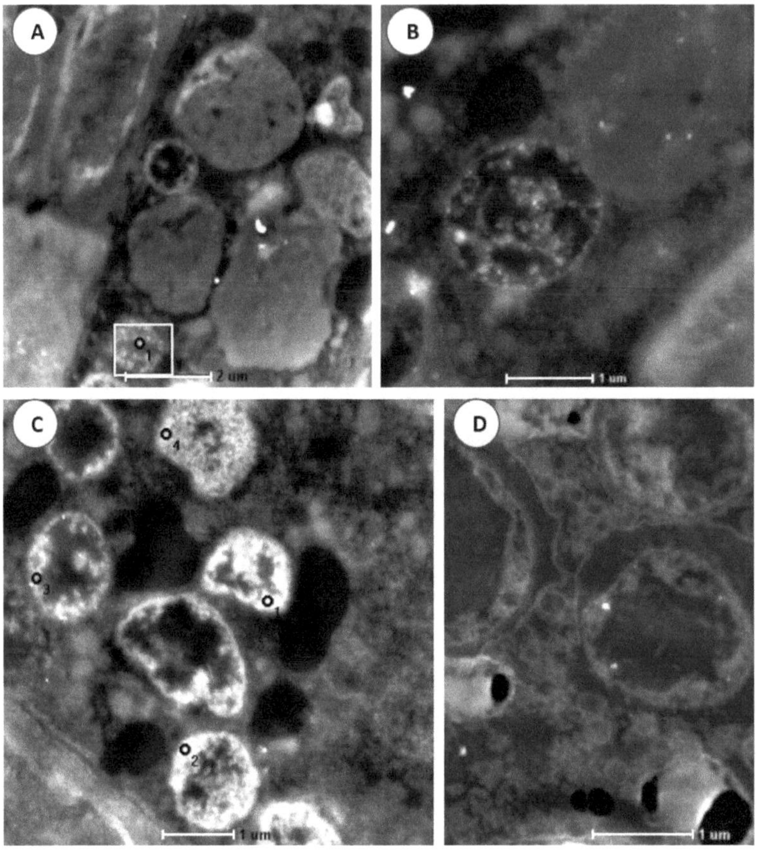

Figure 10. TEM micrographs (high-angle annular dark-field imaging –HAADF- imaging of unstained sections) of digestive gland of mussels *Mytilus galloprovincialis* exposed to 180 nm size TiO_2 NPs (1 mg Ti/L, 3 d). (a–d) Images show different compartments of the endolysosomal system of digestive cells. Bright signals inside lysosomes correspond to high-energy atoms. Confirmation of elemental composition by X Ray microanalysis is still necessary (a) (ROI 1) and (c) (ROI 1–4). Scale bars: A, 2 µm; B–D, 1 µm.

only exception was the work by Joubert et al. (2013), who reported Au NPs in the nucleus of cells of quahogs (*Scrobicularia plana*), although this finding was not fully conclusive.

Conclusions and Future Perspectives

Despite a high amount of literature available on the synthesis and physicochemical properties of NPs, the research on the interactions of NPs with biological systems is hampered by the lack of analytical techniques to accurately and easily detect and quantify NPs in complex matrices. From our point of view, there is a great challenge to understand the interactions of NPs at the single-cell level, regarding their uptake and trafficking. Vanhecke et al. (2014) pointed out that a precise knowledge about the possible uptake into any cell is necessary to shed some light on the subsequent intracellular particle distribution. With such a purpose, promising methods should include highly resolved imaging techniques combined with image analysis tools. However, in order to overcome the limitations of all the techniques, the use of more than one (correlative microscopy, Behzadi et al. 2017) is recommended (Light Microscopy, Electron Microscopy, Atomic Force Microscopy, Super-Resolution Microscopy, CLSM). However, microscopical approaches should be also combined with others such as Cell Culturing, Flow Cytometry and others to allow the visualization of NPs in living systems (cells and tissues, see also Table 1 for a review of the challenges of different microscopical techniques). In summary, the main challenge for the near future is to localize and quantify intracellular NPs to understand the possible correlation between physicochemical characteristics, their topology and subsequent toxicity.

References

Ahmad, A., S. Senapati, M.I. Khan, R. Kumar, R. Ramani, V. Srinivas and M. Sastry. 2003. Intracellular synthesis of gold nanoparticles by a novel alkalotolerant actinomycete, *Rhodococcus* species. Nanotechnology 14: 824–828.

Alkilany, A.M. and C.J. Murphy. 2010. Toxicity and cellular uptake of gold nanoparticles: What we have learned so far? J. Nanopart. Res. 12: 2313–2333.

Asztemborska, M., M. Jakubiak, M. Książyk, R. Stęborowski, H. Polkowska-Motrenko and G. Bystrzejewska-Piotrowska. 2014. Silver nanoparticle accumulation by aquatic organisms—neutron activation as a tool for the environmental fate of nanoparticles tracing. Nukleonica 59(4): 169–173.

Barua, S. and S. Mitragotri. 2014. Challenges associated with penetration of nanoparticles across cell and tissue barriers: A review of current status and future prospects. Nano Today 1; 9(2): 223–243.

Behra, R., L. Sigg, M.J.D. Clift, F. Herzog, M. Minghetti and B. Johnston. 2013. Bioavailability of silver nanoparticles and ions: From a chemical and biochemical perspective. J. R. Soc. Interface. 10: 20130396.

Behrens, I., A.I. Vila Pena, M.J. Alonso and T. Kissel. 2002. Comparative uptake studies of bioadhesive and non-bioadhesive nanoparticles in human intestinal cell lines and rats: The effects of mucus on particle adsorption and transport. Pharm. Res. 19(8): 1185–1193.

Behzadi, S., V. Serpooshan, W. Tao, M.A. Hamaly, M.Y. Alkawareek and E.C. Dreaden. 2017. Cellular Uptake of Nanoparticles: Journey Inside the Cell. Chem. Soc. Rev. 46: 4218.

Brunelli, A., G. Pojana, S. Callegaro and A. Marcomini. 2013. Agglomeration and sedimentation of titanium dioxide nanoparticles (n-TiO2) in synthetic and real waters. J. Nanopart. Res. 15: 1684.

Buhr, E., N. Senftleben, T. Klein, D. Bergmann, D. Gnieser, C.G. Frase et al. 2009. Characterization of nanoparticles by scanning electron microscopy in transmission mode. Meas. Sci. Technol. 20: 084025: 9 pp.

Burchardt, A.D., R.N. Carvalho, A. Valente, P. Nativo, D. Gilliland, C.P. Garcia et al. 2012. Effects of silver nanoparticles in diatom *Thalassiosira pseudonana* and *Cyanobacterium synechococcus* sp. Environ. Sci. Technol. 46: 11336–11344.

Canesi, L., C. Ciacci, R. Fabbri, A. Marcomini, G. Pojana and G. Gallo. 2012. Bivalve molluscs as a unique target group for nanoparticle toxicity. Mar. Environ. Res. 76: 16–21.

Canesi, L., C. Ciacci, D. Vallotto, G. Gallo, A. Marcomini and G. Pojana. 2010. *In vitro* effects of suspensions of selected nanoparticles (C60 fullerene, TiO_2, SiO_2) on Mytilus hemocytes. Aquat. Toxicol. 96(2): 151–158.

Canesi, L., C. Ciacci, E. Bergami, M.P. Monopoli, K.A. Dawson, S. Papa et al. 2015. Evidence for immunomodulation and apoptotic processes induced by cationic polystyrene nanoparticles in the hemocytes of the marine bivalve Mytilus. Mar. Environ. Res. 111: 34–40.

Chabicovsky, M., W. Klepal and R. Dallinger. 2004. Mechanisms of cadmium toxicity in terrestrial pulmonates: Programmed cell death and metallothionein overload. Environ. Toxicol. Chem. 23(3): 648–655

Chen, X., C.N.B. Udalagama, C. Chen, A.A. Bettiol, D.S. Pickard and T. Venkatesan. 2011. Whole-cell imaging at nanometer resolutions using fast and slow focused helium ions. Biophys. J. 101(7): 1788–1793.

Chithrani, B.D., A. Ghazani and W.C. Chan. 2006. Determining the size and shape dependence of gold nanoparticle uptake into mammalian cells. Nano Lett.6: 662–668.

Cho, W.S., S. Kim, B.S. Han, W.C. Son and J. Jeong. 2009b. Comparison of gene expression profiles in mice liver following intravenous injection of 4 and 100 nm-sized PEG-coated gold nanoparticles. Toxicol. Lett. 191: 96–102.

De Jong, W.H., Burger, M.A. Verheijen and R.E. Geertsma. 2010. Detection of the presence of gold nanoparticles in organs by transmission electron microscopy. Materials 3: 4681–4694.

Dedeh, A., A. Ciutat, M. Treguer-Delapierre and J.P. Bourdineaud. 2015. Impact of gold nanoparticles on zebrafish exposed to a spiked sediment. Nanotoxicology 9(1): 71–80.

Domingos, R.F. and J.P. Pinheiro. 2014. Implications of the use of Nanomaterials for Environmental Protection, in Nanomaterials for Environmental Protection. *In*: B.I. Kharisov, O.V. Kharissova and H.V.R. Dias (eds.). John Wiley & Sons, Inc, Hoboken, NJ. doi:10.1002/9781118845530.ch32.

Dubertret, B., P. Skourides, D.J. Norris, V. Noireaux, A.H. Brivanlou and A. Libchaber. 2002. *In vivo* imaging of quantum dots encapsulated in phospholipid micelles. Science 298: 1759–1762.

Federici, G., B.J. Shaw and R.D. Handy. 2007. Toxicity of titanium dioxide nanoparticles to rainbow trout (*Oncorhynchus mykiss*): Gill injury, oxidative stress, and other physiological effects. Aquat. Toxicol. 84: 415–430.

Fröhlich, E.E. and E. Fröhlich. 2016. Cytotoxicity of nanoparticles contained in food on intestinal cells and the gut microbiota. Int. J. Mol. Sci. 17(4): 509. doi:10.3390/ijms17040509.

García-Negrete, C.A., J. Blasco, M. Volland, T.C. Rojas, M. Hampel, A. Lapresta-Fernández et al. 2013. Behaviour of Au-citrate nanoparticles in seawater and accumulation in bivalves at environmentally relevant concentrations. Environ. Poll. 174: 134–141.

García-Negrete, C.A., M.C. Jiménez de Haro, J. Blasco, M. Soto and A. Fernández. 2015. STEM-in-SEM high resolution imaging of gold nanoparticles and bivalve tissues in bioaccumulation experiments. Analyst. 140: 3082–3089.

Gao, H., W. Shi and L.B. Freund. 2005. Mechanics of receptor-mediated endocytosis. PNAS. 102(27): 9469–9474.

Gomes, T., O. Araújo, R. Pereira, A.C. Almeida, A. Cravo and M.J. Bebianno. 2013a. Genotoxicity of copper oxide and silver nanoparticles in the mussel *Mytilus galloprovincialis*. Mar. Environ. Res. 84: 51–59.

Gomes, T., C.G. Pereira, C. Cardoso and M.J. Bebianno. 2013b. Differential protein expression in mussels *Mytilus galloprovincialis* exposed to nano and ionic Ag. Aquat Toxicol. 136-137: 79–90.

Handy, R.D., G. Al-Bairuty, A. Al-Jubory, C.S. Ramsden, D. Boyle, B.J. Shaw et al. 2011. Effects of manufactured nanomaterials on fishes: A target organ and body systems physiology approach. J. Fish Biol. 79: 821–853.

Hauck, T.S., A.A. Ghazani and W.C. Chan. 2008. Assessing the effect of surface chemistry on gold nanorod uptake, toxicity, and gene expression in mammalian cells. Small 4: 153–159.

Hull, M.S., P. Chaurand, J. Rose, M. Auffan, J.-Y. Bottero, J.C. Jones et al. 2011. Filter-feeding bivalves store and biodeposit colloidally stable gold nanoparticles. Environ. Sci. Technol. 45: 6592–6599.

Ivanov, A.I. 2008. Pharmacological inhibition of endocytic pathways: Is it specific enough to be useful?.Methods Mol. Biol. 440: 15–33.

Jain, R.K. and T. Stylianopoulos. 2010. Delivering nanomedicine to solid tumors. Nat. Rev. Clin. Oncol. 7(11): 653–664.

Jimeno-Romero, A., M. Oron, M.P. Cajaraville, M. Soto and I. Marigómez. 2016. Nanoparticle size and combined toxicity of TiO_2 and DSLS (surfactant) contribute to lysosomal responses in digestive cells of mussels exposed to TiO_2 nanoparticles. Nanotoxicology 10(8): 1168–1176.

Jimeno-Romero, A., E. Bilbao, U. Izagirre, M.P. Cajaraville, I. Marigómez and M. Soto. 2017a. Digestive cell lysosomes as main targets for Ag accumulation and toxicity in marine mussels, *Mytilus galloprovincialis,* exposed to maltose-stabilised Ag nanoparticles of different sizes. Nanotoxicology 11: 168–183.

Jimeno-Romero, A., U. Izagirre, D. Gilliland, A. Warley, M.P. Cajaraville, I. Marigómez et al. 2017b. Lysosomal responses to different gold forms (nanoparticles, aqueous, bulk) in mussel digestive cells: A trade-off between the toxicity of the capping agent and form, size and exposure concentration. Nanotoxicology 11: 658–670.

Jimeno-Romero, A., E. Bilbao, E. Valsami-Jones, M.P. Cajaraville, M. Soto and I. Marigómez. 2019. Bioaccumulation, tissue and cell distribution, biomarkers and toxicopathic effects of CdS quantum dots in mussels, *Mytilus galloprovincialis*. Ecotox. Environ. Safety 167: 288–300.

Joens, M.S., C. Huynh, J.M. Kasuboski, D. Ferranti, T.J. Sigal, F. Zeitvogel et al. 2013. Helium Ion Microscopy (HIM) for the imaging of biological samples at sub-nanometer resolution. Sci. Rep. 3: 3514.

Johnston, B.D., T.M. Scown, J. Moger, S.A. Cumberland, M. Baalousha, K. Linge et al. 2010. Bioavailability of nanoscale metal oxides TiO_2, CeO_2, and ZnO to fish. Environ. Sci. Technol. 44: 1144–1151.

Joubert, Y., J.F. Pan, P.E. Buffet, P. Pilet, D. Gilliland, E. Valsami-Jones et al. 2013. Subcellular localization of gold nanoparticles in the estuarine bivalve Scrobicularia plana after exposure through the water. Gold Bull. 46: 47–56.

Katsumiti, A., D. Berhanu, K.T. Howard, I. Arostegui, M. Oron, P. Reip et al. 2015a. Cytotoxicity of TiO_2 nanoparticles to mussel hemocytes and gill cells *in vitro*: Influence of synthesis method, crystalline structure, size and additive. Nanotoxicology 9: 543–553.

Katsumiti, A., D. Gilliland, I. Arostegui and M.P. Cajaraville. 2015b. Mechanisms of toxicity of Ag nanoparticles in comparison to bulk and ionic Ag on mussel hemocytes and gill cells. PLoS ONE 10: e0129039. https://doi.org/10.1371/journal.pone.0129039.

Katsumiti, A., I. Arostegui, M. Oron, D. Gilliland, E. Valsami-Jones and M.P. Cajaraville. 2016. Cytotoxicity of Au, ZnO and SiO_2 NPs using *in vitro* assays with mussel hemocytes and gill cells: Relevance of size, shape and additives. Nanotoxicology 10(2): 185–193.

Kim, C.S., X. Li, Y. Jiang, B. Yan, G.Y. Tonga, M. Ray et al. 2015. Cellular imaging of endosome entrapped small gold nanoparticles. MethodsX 2: 306–315.

Koehler, A., U. Marx, K. Broeg, S. Bahns and J. Bressling. 2008. Effects of nanoparticles in *Mytilus edulis* gills and hepatopancreas. A new threat to marine life. Mar. Environ. Res. 66: 12–14.

Kohl, Y., E. Gorjup, A. Katsen-Globa, C. Büchel, H. von Briesen and H. Thielecke. 2011. Effect of gold nanoparticles on adipogenic differentiation of human mesenchymal stem cells. J. Nanopart. Res. 13: 6789–6803.

Klaine, S.J., P.J.J. Alvarez, G.E. Batley, T.F. Fernandes, R.D. Handy, D.Y. Luyon et al. 2008. Nanomaterials in the environment: behavior, fate, bioavailability and effects. Environ. Toxicol. Chem. 27(9): 1825–1851.

Kokkinopoulou, M., J. Simon, K. Landfester, V. Mailänder and I. Lieberwirth. 2017. Visualization of the protein corona: Towards a biomolecular understanding of nanoparticle cell-interactions. Nanoscale 9: 8858.

Krysanov, E.Y.S., D.S. Pavlov, T.B. Demidova and Y.Y. Dgebuadze. 2010. Effect of nanoparticles on aquatic organisms. Biol. Bull. 37(4): 406–412.

Labouta, H.I. and M. Scheneider. 2013. Interaction of inorganic nanoparticles with the skin barrier: Current status and critical review. Nanomedicine: NBM 9: 39–54.

Larson, D.R., W.R. Zipfel, R.M. Williams, S.W. Clark, M.P. Bruchez, F.W. Wise. 2003. Water-soluble quantum dots for multiphoton fluorescence imaging *in vivo*. Science 300: 1434–6

Leclerc, S. and K.J. Wilkinson. 2014. Bioaccumulation of Nanosilver by *Chlamydomonas reinhardtii*. Nanoparticle or the Free Ion?. Environ. Sci. Technol. 48: 358−364.

Lee, W.M. and Y.J. An. 2015. Evidence of three-level trophic transfer of quantum dots in aquatic food chain by using bioimaging. Nanotoxicology 9: 407–412.

Leppänen, M., L.R. Sundberg, E. Laanto, G.M. de Freitas Almeida, P. Papponen and I.J. Maasilta. 2017. Imaging bacterial colonies and phage-bacterium interaction at sub-nanometer resolution using helium-ion microscopy. Adv. Biosys. 1, 1700070.

Levard, C., E.M. Hotze, G.V. Lowry and G.E. Brown. 2012. Environmental transformations of silver nanoparticles: impact on stability and toxicity. Environ. Sci. Technol. 46: 6900–6914. doi:10.1021/es2037405.

Liu, T.L., S. Upadhyayula, D.E. Milkie, V. Singh, K. Wang, I.A. Swinburne et al. 2018. Observing the cell in its native state: Imaging subcellular dynamics in multicellular organisms. Science 360: 6386.

Lowry, G.V. and M.R. Wiesner. 2007. Environmental *Considerations occurrences*, fate, and characterization of nanoparticles in the environment. pp. 369–390. *In*: N.A. Monteiro-Riviere and L.C. Tran (eds.).Nanotoxicology. Characterization, Dosing and Health Effects. Taylor. and Francis. Boca Raton FL.

Lui, R., Y. Zhang, S. Zhang, W. Qiu and Y. Gao. 2014. Silver enhancement of gold nanoparticles for biosensing: From qualitative to quantitative. Appl. Spectrosc. Rev. 49(2): 121–138.

Maes, H.M., F. Stibany, S. Giefers, B. Daniels, B. Deutschmann, W. Baumgartner et al. 2014. Accumulation and distribution of multiwalled carbon nanotubes in zebrafish (*Danio rerio*). Environ. Sci Technol. 48(20): 12256–64.

Marigómez, I., M. Soto, M. Cajaraville, E. Angulo and L. Giamberini. 2002 Cellular and subcellular distribution of metals in mollusc. Microsc. Res. Tech. 56: 358–392.

Maurer-Jones, M.A., I.L. Gunsolus, C.J. Murphy and C.L. Haynes. 2013. Toxicity of engineered nanoparticles in the environment. Anal. Chem. 85(6): 3036–3049.

McCarthy, M.P., D.L. Carroll and A.H. Ringwood. 2013. Tissue specific responses of oysters, *Crassostrea virginica* to silver nanoparticles. Aquat. Toxicol. 138-139: 123–128.

Michen, B., C. Geers, D. Vanhecke, C. Endes, B. Rothen-Rutishauser, S. Balog et al. 2015. Avoiding drying-artifacts in transmission electron microscopy: Characterizing the size and colloidal state of nanoparticles. Sci. Rep. volume 5, Article number: 9793.

Moger, J., B.D. Johnston and C.R. Tyler. 2008. Imaging metal oxide nanoparticles in biological structures with CARS microscopy. Opt. Express. 16(5): 3408–19.

Moore, M.N. 2006. Do nanoparticles present ecotoxicological risks for the health of the aquatic environment? Environ. Int. 32: 967–976.

Moore, M.N., J. Icarus Allen and A. McVeigh. 2006. Environmental prognostics: an integrated model supporting lysosomal stress responses as predictive biomarkers of animal health status. Part. Fiber Toxicol. 61: 278–304.

Morton, B.S. 1983. Feeding and digestion in Bivalvia. pp. 65–147. *In*: A.S.M. Saleuddin and K.M. Wilbur (eds.). The Mollusca, Vol. 5, Physiology, Part 2, Academic Press, New York.

Mühlfeld, C., B. Rothen-Rutishauser, D. Vanhecke, F. Blank, P. Gehr and M. Ochs. 2007. Visualization and quantitative analysis of nanoparticles in the respiratory tract by transmission electron microscopy. Particle and Fibre Toxicology 4: 11.

Nie, S. and S.R. Emory. 1997. Probing single molecules and single nanoparticles by surface-enhanced raman scattering. Science 275: 1102–1106.

Oh, N. and J.H. Park. 2014. Endocytosis and exocytosis of nanoparticles in mammalian cells. International Journal of Nanomedicine 9(supp. 1): 51–63.

Ostaszewska, T., J. Śliwiński, M. Kamaszewski, P. Sysa and M. Chojnacki. 2018. Cytotoxicity of silver and copper nanoparticles on rainbow trout (*Oncorhynchus mykiss*) hepatocytes. Environ. Sci. Pollut. Res. 25: 908–915.

Pan, J.F., P.E. Buffet, L. Poirier, C. Amiard-Triquet, D. Gilliland, Y. Joubert et al. 2012. Size dependent bioaccumulation and ecotoxicity of gold nanoparticles in an endobenthic invertebrate: The tellinid clam *Scrobicularia* plana. Environ. Poll. 168: 37–43.

Peckys and de Jonge. 2011.Visualizing gold nanoparticles uptake in live cells with liquid scanning transmission electron microscopy. Nano Lett 11(4): 1733–1738.

Piccinno, F., F. Gottschalk, S. Seeger and B. Nowack. 2012. Industrial production quantities and uses of ten engineered nanomaterials in europe and the world. J. Nanopart. Res. 14: 1109.

Renault, S., M. Baudrimont, N. Mesmer-Dudons, P. Gonzalez, S. Mornet and A. Brisson. 2008. Impacts of gold nanoparticle exposure on two freshwater species: a phytoplanktonic alga (*Scenedesmus subspicatus*) and a benthic bivalve (*Corbicula fluminea*). Gold Bull. 41: 116–126.

Ringwood, A.H., S. Khambhammettu, P. Santiago, E. Bealer, M. Stogner, J. Collins et al. 2005. Characterization, imaging and degradation studies of quantum dots in aquatic organisms. MRS Proceedings 895. doi:10.1557/PROC-0895-G04-06-S04-06.

Ringwood, A.H., N. Levi-Polyachenko and D.L. Carroll. 2009. Fullerene exposures with oysters: Embryonic, adult, and cellular responses. Environ. Sci. Technol. 43: 7136–7141.

Sadauskas, E., G. Danscher, M. Stoltenberg, U. Vogel, A. Larsen and H. Wallin. 2009. Protracted elimination of gold nanoparticles from mouse liver. Nanomed. Nanotechnol. Biol. Med. 5: 162–169.

Scown, T.M., E. Santos, B.D. Johnston, B. Gaiser, M. Baalousha, S. Mitov et al. 2010. Effects of aqueous exposure to silver nanoparticles of different sizes in rainbow trout. Toxicol. Sci. 115(2): 521–534.

Scown, T.M., R.M. Goodhead, B.D. Johnston, J. Moger, M. Baalousha, J.R. Lead et al. 2010. Assessment of cultured fish hepatocytes for studying cellular uptake and (eco)toxicity of nanoparticles. Environ. Chem. 7: 36–49.

Skjolding, L.M., S.N. Sørensen, N.B. Hartmann, R. Hjorth, S.F. Hansen and A. Baun. 2016. Aquatic ecotoxicity testing of nanoparticles—the quest to disclose nanoparticle effects. Angew. Chem. Int. Ed. 55: 15224–15239.

Skjolding, L.M., G. Ašmonaitė, R.I. Jølck, T.L. Andresen, H. Selck, A. Baun and J. Sturve. 2017. An assessment of the importance of exposure routes to the uptake and internal localisation of fluorescent nanoparticles in zebrafish (Danio rerio), using light sheet microscopy. Nanotoxicology 11(3): 351–359.

Sohn, E.K., S.A. Johari, T.G. Kim, J.K. Kim, E. Kim, J.H. Lee et al. 2015. Aquatic toxicology comparison of silver nanoparticles and silver nanowires. BioMed Research International. Volume 2015, Article ID 893049, 12 pages.

Stakenborg, T., S. Peeters, G. Reekmans, W. Laureyn, H. Jans, G. Borghs et al. 2008.Increasing the stability of DNA-functionalized gold nanoparticles using mercaptoalkanes. J. Nanopart. Res. 10: 143–15.

Stern, S., P. Adiseshaiah and R. Crist. 2012. Autophagy and lysosomal dysfunction as emerging mechanisms of nanomaterial toxicity. Part Fibre Toxicol. 9: 20.

Strohm, E.M., M.J. Moore and M.C. Kolios. 2016. High resolution ultrasound and photoacoustic imaging of single cells. Photoacoustics 4: 36–42.

Smith, D.J. 2015. Characterization of nanomaterials using transmission electron microscopy. *In*: Hierarchical Nanostructures for Energy Devices (37 ed., Vol. 2015-January, pp. 1–29). (RSC Nanoscience and Nanotechnology; Vol. 2015-January, No. 37). Royal Society of Chemistry. doi:10.1039/9781782621867-00001.

Su, D. 2017. Advanced electron microscopy characterization of nanomaterials for catalysis. Green Energy & Environment 2: 70–83.

Su, X. and R. Kanjanawarut. 2009. Control of metal nanoparticles aggregation and dispersion by PNA and PNA-DNA complexes, and its application for colorimetric DNA detection. ACS Nano. 3(9): 2751–9.

Sun, H., X. Zhang, Q. Niu, Y. Chen and J.C. Crittenden. 2007. Enhanced Accumulation of arsenate in carp in the presence of *Titanium dioxide* nanoparticles. Water. Air and Soil Pollution 178: 245–254.

Unrine, J.M., S. Hunyadi, O.V. Tsyusko and P. Bertsch. 2010. Evidence for bioavailability of Au nanoparticles from soil and biodistribution within earthworms (*Eisenia fetida*). Environ. Sci. Technol. 44: 8308–8313.

USEPA (United States Environmental Protection Agency). Nanomaterial Case Study: Nanoscale Silver in Disinfectant Spray. EPA/600/R-10/081F; 2012. 423.

Vale, G., K. Mehennaouic, S. Cambierc, G. Libralato, S. Jominie and R.F. Domingos. 2016. Manufactured nanoparticles in the aquatic environment-biochemical responses on freshwater organisms: A critical overview. Aquat. Toxicol. 170: 162–174.

Vanhecke, D., L. Rodriguez-Lorenzo, M.J. Clift, F. Blank, A. Petri-Fink and B. Rothen-Rutishauser. 2014. Quantification of nanoparticles at the single-cell level: An overview about state-of-the-art techniques and their limitations. Nanomedicine 9(12): 1885–900.

Voight, J., J. Christensen and V. Prasad Shastri. 2014. PNAS 111(8): 2942–2947.

von der Kammer, F., S. Legros, T. Hofmann, E.H. Larsen and K. Loeschner. 2011. Separation and characterization of nanoparticles in complex food and environmental samples by field-flow fractionation. Trends Anal. Chem. 30(3): 42–436.

von Moos, N. and V.I. Slaveykova. 2014. Oxidative stress induced by inorganic nanoparticles in bacteria and aquatic microalgae—state of the art and knowledge gaps. Nanotoxicology 8: 605–630.

Walters, C., E. Pool and V. Somerset. 2016. Nanotoxicity in Aquatic Invertebrates, In: Invertebrates—Experimental Models in Toxicity Screening, Dr. Marcelo Larramendy (ed.). InTech, doi:10.5772/61715. Available from:https://www.intechopen.com/books/invertebrates-experimental-models-in-toxicity-screening/nanotoxicity-in-aquatic-invertebrates.

Walkey, C.D. and W.C.W. Chan. 2012. Understanding and controlling the interaction of nanomaterials with proteins in a physiological environment. Chem. Soc. Rev. 41: 2780–2799.

Ward, J.E. and D.J. Kach. 2009. Marine aggregates facilitate ingestion of nanoparticles by suspension-feeding bivalves. Mar. Environ. Res. 68: 137–142.

Williams, D. 2015. Measuring & Characterizing Nanoparticle Size—TEM vs. SEM.https://www.azonano.com/article.aspx?ArticleID=4118.

Zhao, J., X. Cao, X. Liu, Z. Wang, C. Zhang, J.C. White et al. 2016. Interactions of CuO nanoparticles with the algae *Chlorella pyrenoidosa*: Adhesion, uptake, and toxicity. Nanotoxicology 10: 1297–1305.

Zou, D., W. Wang, D. Lei, Y. Yin, P. Ren, J. Chen et al. 2017. Penetration of blood–brain barrier and antitumor activity and nerve repair in glioma by doxorubicin-loaded monosialoganglioside micelles system. Int. J. Nanomed. 12: 4879–4889.

8

Insights from 'Omics on the Exposure and Effects of Engineered Nanomaterials on Aquatic Organisms

Helen C. Poynton

Introduction

Why Apply Genomics to Study Nanotoxicology?

Nanotechnology is an emerging area of both commercial interest and environmental concern. However, since engineered nanomaterials (ENMs) have been incorporated into over 3000 commercial products (http://nanodb.dk/en/) and have been detected in environmental samples, there is urgency to define the associated environmental risks. To date, the development of nanotechnologies and novel products containing ENMs has far out-paced research on their environmental and health effects (Klaine et al. 2012). Therefore, our current understanding of ENMs, their fate and transport, bioavailability and toxicity is not at a sufficient enough level to assess the increasing prevalence of these emerging contaminants (Selck et al. 2016).

Nanomaterials are defined as molecular or macromolecular structures with one or more dimensions between 1–100 nm (US EPA 2007). There are a diversity of particles, composed of different materials and with different chemical properties that are artificially grouped together based on this size definition. However, the grouping of these different materials as EMNs is relevant for several reasons. First, within the nano range, molecules gain high reactivity due to a high surface area to volume ratio and in some cases, materials take on unexpected catalytic properties. Secondly, since the size of ENMs is within the range of biomolecules,there is concern that they will interact with biomolecules, including proteins, membranes and genetic material (Klaine et al. 2012). Finally, because ENMs are colloids, they undergo physical and chemical transformations that affect their bioavailability in natural waters and soils. This creates similar challenges for their detection and characterization across the spectrum of ENMs.

School for the Environment, University of Massachusetts Boston, Boston, MA 02125, USA.
Email: Helen.poynton@umb.edu

There have been recent attempts to catalogue the commercial products that contain EMNs or use nanotechnology. The Project for Emerging Nanotechnologies has provided a "living inventory" of nanotechnology consumer products since 2006. In 2015, it reported 1827 consumer products (Vance et al. 2015). Recently, the Nanodatabase was established to provide a similar product list for European countries (Hansen et al. 2016). This database currently contains over 3000 consumer products (http://nanodb.dk/, accessed on January 20, 2018). Most analysts agree that the industry is still young and market research suggests a steady growth of the industry over the next few years (Mulvaney and Weiss 2016). In addition, new players to the EMNs industry, most prominently the agricultural market, are currently in the research and development stage, but also have the potential to grow exponentially. Agricultural products, such as bioactive compounds (i.e., pharmaceuticals, vitamins, probiotics) and nanoformulated agrochemicals are likely to pose new challenges for environmental risk assessments (Handford et al. 2014).

Environmental concentrations are thought to be low currently, but as the industry grows, these levels are expected to rise. The limited environmental monitoring for EMNs has detected Ag nanoparticles (Ag NPs), Au NPs, and TiO_2 NPs in waste water treatment plant (WWTP) influent and effluent as well as surface waters (Bäuerlein et al. 2017, Gondikas et al. 2014). C60 NPs have also been detected in WWTP effluent (Bäuerlein et al. 2017, Farré et al. 2010). For some of these NPs detected, it has not possible to attain accurate concentrations because it is difficult to analytically distinguish between metal NPs and dissolved metals. When possible, the levels of ENMs measured (ppt to ppb) are below concentrations shown to cause adverse effects in aquatic life; however, this does not exclude the possibility that other ENMs may be present in the environment at toxic levels. A number of modeling studies suggested relatively low risk for ENMs overall, but the most potent NPs (e.g., Ag NPs) exhibited a probable risk in some environmental matrices (Gottschalk et al. 2013). A more recent study, focused on the San Francisco Bay, suggested that metal oxide nanoparticles (i.e., CuO, TiO_2, ZnO) will accumulate over time, especially in freshwater and marine sediments, and are likely to reach toxic concentrations (Garner et al. 2017). The information available on environmental concentrations of ENMs suggests that levels are quickly approaching concentrations that will pose a risk for environmental health.

Accumulating evidence suggests that ENMs pose a unique risk to aquatic and marine ecosystems. Impacts on fish (Handy et al. 2008), invertebrates (Baun et al. 2008) and bivalves (Rocha et al. 2015) reveal varying levels of potency across nanomaterials and species (Kahru and Dubourguier 2010). However, on an organismal level, a consensus is emerging that many ENMs cause the production of reactive oxygen species (ROS), oxidative stress, immunotoxicity and genotoxicity across many diverse species (Ivask et al. 2014, Kahru and Dubourguier 2010). The solubilization of metal NPs is an important determinant of their toxicity (Kahru and Dubourguier 2010, Notter et al. 2014), but particle specific effects have also been documented (Fabrega et al. 2011, Ivask et al. 2014, Ma et al. 2013). This is likely related to exposure conditions, and in the field, will be dependent on physical and chemical transformations of the NPs (Selck et al. 2016). The potency of ENMs varies significantly, with LC50 values that span several orders of magnitude from low μg/L to high mg/L concentrations. Ag NPs show highest potency, and crustaceans and algae typically show the highest sensitivity (Kahru and Dubourguier 2010). Carbon-based ENMs are most toxic to bacteria and have the potential to disrupt bacterial communities (Kahru and Dubourguier 2010). Given the large span in toxicity values, understanding which taxa are most sensitive is a priority (Scown et al. 2010).

In addition to impacts on individual organisms, mesocosm studies demonstrate potential impacts on entire communities. Through mesocosm studies, bioaccumulation of NPs in bivalves and benthic species has been confirmed (Buffet et al. 2014, Ferry et al. 2009), which impacts immune responses, oxidative stress and genotoxicity (Buffet et al. 2014). Mecosoms studies investigating the effects of Ag NPs leached from biosoilds in terrestrial (Colman et al. 2013) and wetland mesocosms (Colman et al. 2014) revealed ecosystem level impacts on nitrogen cycling, bacterial composition and plant growth and biomass.

Environmental concentrations of ENM are increasing, and while current knowledge of their impacts is concerning, there are many outstanding questions especially related to how ENMs will behave and interact with organisms in natural environments. In addition, as ENM are developed for new sectors (e.g., agriculture, Handford et al. 2014) and take on more specialized functions (e.g., pharmaceuticals, sensors, robotics, US EPA 2007) new challenges and questions will undoubtedly arise. Multidisciplinary collaborations will be necessary to keep pace with these emerging technologies (Klaine et al. 2012) and provide an understanding of organismal responses to ENMs from the molecular to the ecosystem level (Holden et al. 2012). In this review, I focus on the role that 'omic technologies can play in advancing our understanding of the risks of ENMs and developing models to predict adverse effects. Used within an interdisciplinary assessment of ENMs, 'omic tools can assist in our understanding of exposures and mechanisms of toxicity (Poynton and Vulpe 2009). This chapter is not meant to provide a comprehensive review of 'omic studies applied to nanotoxicology (the reader is referred to Revel et al. 2017). Instead, my intention is to highlight how 'omics technologies have been applied to exposure and effects assessments of ENMs and provide examples to guide future studies.

The 'Omic Technologies and their Application in Ecotoxicology

Transcriptomics, proteomics and metabolomics, also referred to collectively as 'omic' technologies are methods to examine the fluctuating state of a cell, a tissue, or an organism by measuring gene expression, protein levels and metabolite fluxes (Poynton et al. 2008a). The overwhelming appeal of 'omic technologies is their ability to provide a global survey of cellular processes at the transcript, protein or metabolic level. Applications of these technologies has grown over the past fifteen years, with the publication of over a thousand studies using 'omic technologies in ecotoxicology (Martyniuk and Simmons 2016). In addition, the methods and technologies have seen significant advancements with the movement from lower throughput methods, such as 2D gel electrophoresis in proteomics and microarrays in transcriptomics, to more high-throughput methods. A detailed description of the most common technologies utilized in ecotoxicogenomics is provided in Table 1. Each technology provides a snapshot of cellular condition at the gene expression, protein expression or metabolite expression level, and provides certain advantages and limitations. Due to translational and post-translation regulation, transcriptomic and proteomic profiles do not always correspond directly with final enzymatic regulation, causing some to conclude that metabolomics is the ideal technology to directly link to phenotypic outcomes (Revel et al. 2017). However, gene and protein expression can tell us important information about receptor activation and cell signaling even if the final gene product is not activated. For example, when the cell experiences oxidative stress, transcription factors are activated that bind to the antioxidant response element (ARE) in the promoter region of genes involved in the oxidative stress response (e.g., glutathione S-transferase; GST, catalase, ferritin, thioredoxin). These genes are transcribed and will be detected by transcriptomic techniques. Even if the final proteins are not activated, the upregulation of these genes provides important information about the redox status of the cell (Nguyen et al. 2003).

In ecotoxicogenomics research, 'omic technologies have focused on two main areas of research, possibly arising from the early distinction between biomarkers of exposure and effect. First, to better characterize environmental exposures, 'omics have been used to identify chemical signatures and distinguish contaminants in chemical mixtures. Secondly, 'omics technologies have been promoted as a means to provide hypotheses regarding the mode of action (MOA) of pollutants, and more recently, incorporate this information into an Adverse Outcome Pathway (AOP) framework.

'Omics for Detecting Chemical Signatures

From the early studies that first detected chemical specific expression profiles in marine mussels (Shepard et al. 2000) and freshwater fish (Hook et al. 2006), ecotoxicogenomics has aimed to provide more specific biomarkers of exposure for environmental detection and monitoring (Ankley et al. 2006, Poynton et al. 2008a). Applications stemming from a Toxicant Identification Evaluation (TIE)

Table 1. 'Omic technology descriptions and common techniques for 'omic level assessments.

Technology	Description	Methods for quantification[1]
Transcriptomics	Measurement of the total mRNA in a cell, tissue or whole body of an organism. Transcriptomics provides a profile of the genes being actively transcribed in a cell. When two or more conditions are compared, genes which are up-regulated or down-regulated by a particular condition can be identified. While many genes are regulated at the transcriptional level, transcript levels do not always directly correspond with protein levels due to translational regulation and protein degradation.	microarray (Poynton and Vulpe 2009) RNA-seq (Mehinto et al. 2012)
Proteomics	Global measurement of the proteins currently expressed in a cell, tissue or whole body of an organism. Similar to transcriptomics, different conditions are compared to determine which proteins are being up-regulated or down-regulated by the treatment. In contrast to transcriptomics, proteomics does provide a real-time profile of the proteins present in the cell, but not all these proteins are necessarily active due to post-translational regulation.	2D gel electrophoresis (Gevaert and Vandekerckhove 2000) LC-MS/MS[2] (Gillet et al. 2016)
Metabolomics	Measurement of the total metabolites or small biomolecules in a cell, tissue or blood. The metabolic profile reflects products of differentially expressed proteins and genes taking into account all levels of regulation. Therefore, metabolic profiles are thought to be most closely linked to phenotypes compared with transcriptomics and proteomics (Bundy et al. 2009).	NMR[3] (Viant 2008) LC-MS GC-MS[4](García-Sevillano et al. 2015)
Emerging 'omic technologies for ecotoxicology	A few studies have begun to apply **Epigenomics** to ecotoxicological studies (Chen et al. 2016, Vandegehuchte and Janssen 2011). These technologies investigate the upstream processes responsible for gene regulation. Epigenomics refers to heritable changes that are not related to changes in DNA sequence and include DNA methylation, histone modification and microRNA expression. MicroRNAs regulate mRNA expression levels through interference, binding to and targeting particular mRNAs for degradation. **Metagenomics** is another emerging area that holds promise for monitoring global changes in community structure. While this technique has been applied widely to study bacterial community composition (Shahsavari et al. 2017), recent investigations have explored its utility for macroinvertebrate community analysis (Chariton et al. 2010).	Next generation sequencing technologies: Bisulfite sequencing, small RNA-seq, 16S-rRNA amplicon sequencing, whole genome sequencing (Tringe and Rubin 2005)

[1] Methods for quantification are continuously evolving as new techniques are developed and become available. An overview of these methods can be found in (Martyniuk and Simmons 2016, Poynton et al. 2008a, Sanchez et al. 2011) and see references for more specific details on techniques. [2]LC-MS/MS: liquid chromatography tandem mass spectrometry, [3]NMR: nuclear magnetic resonance, [4]GC-MS: gas chromatography mass spectrometry.

approach first showed that genomics could complement traditional TIEs by providing increased sensitivity, which facilitated the identification of toxicants in marine sediments (Biales et al. 2013). Others have suggested that gene expression signatures themselves can act as a molecular TIE, classifying chemicals based on gene expression profiles (Antczak et al. 2013, Osborn and Hook 2013) or metabolite signatures (Ekman et al. 2015), although some investigators have encountered challenges when applying this approach to complex mixtures (Garcia-Reyero et al. 2011c). Several field-based studies have successfully used 'omic technologies to identify chemicals responsible for toxicity in field samples. For example, gene expression profiles correctly predicted copper as the toxicant responsible for toxicity in mine site effluent containing a mixture of metals (Poynton et al. 2008b). In a proteomics study, protein signatures developed for perfluorooctanesulfonate (PFOS) in laboratory exposures were correlated with field exposures to develop specific biomarkers for PFOS exposure (Roland et al. 2014). Finally, a number of studies at wastewater treatment plants (WWTPs) have successfully identified estrogenic expression signatures correlated with measured chemicals (Garcia-Reyero et al. 2011b) and provided responses that reflected downstream pollution gradients using transcriptomics (Martinović-Weigelt et al. 2014) and metabolomics (Skelton et al. 2014). Given the current challenges associated with measuring ENMs in environmental samples, recent reviews have suggested that 'omics could assist in biomonitoring for ENM exposure by providing biomarkers or expression signatures specific for NP exposure (Kahru and Dubourguier 2010, Selck et al. 2016).

'Omics in the Development of Adverse Outcome Pathways

In ecotoxicology, impacts on chemicals are measured across different levels of biological organization recognizing that effects at the molecular or cellular level are more specific, but are limited in their ecological relevance. In contrast, impacts to populations and communities have high ecological relevance, but are difficult to attribute to particular pollutants. In order to define a mechanism of action (MOA) for a pollutant, studies must span these different levels of biological organization and phenotypically link molecular effects to higher orders of biological organization. Renewed interest in mechanistic studies was recently motivated by the National Research Council's report, *Toxicity Testing for the 21st Century* (NRC et al. 2007), which proposes a new paradigm in chemical testing moving away from whole organism toxicity tests and toward high throughput assays. These high throughput assays are designed to target molecular and cellular perturbations that predict adverse effects relevant to human and ecological health. However, before the *Toxicity Testing for the 21st Century* approach can be implemented, well-defined MOAs must be built so that molecular assays can predict higher order effects.

The Adverse Outcome Pathway (AOP) concept was developed to provide a framework to organize the mechanic information across levels of biological organization and highlights the importance of key events (KE) critical for translating effects from one level to the next (Ankley et al. 2010). As shown in Fig. 1, a chemical will interact with a target biomolecule leading to inhibition or altered function in a step called the Molecular Initiation Event (MIE). When concentrations of the chemical reach a critical level at the target site, the MIE will lead to impacts on the cell and tissue, possibly inhibiting the normal function of that tissue. These impacts will alter normal organ function, resulting in a disease state for that individual. In ecotoxicology, effects are not considered a concern until impacts on populations or communities of organisms are realized. Therefore, population models are needed to translate individual level impacts to population effects (Kramer et al. 2011). An important consideration in the AOP, is the threshold levels that (1) lead to initial inhibition at the target site (i.e., MIE) and (2) cause effects to reach the next level of biological organization (i.e., KE).

'Omic technologies have been recognized as important tools for AOP development. Initially, it was envisioned that 'omic technologies would help generate hypotheses for the creation of new

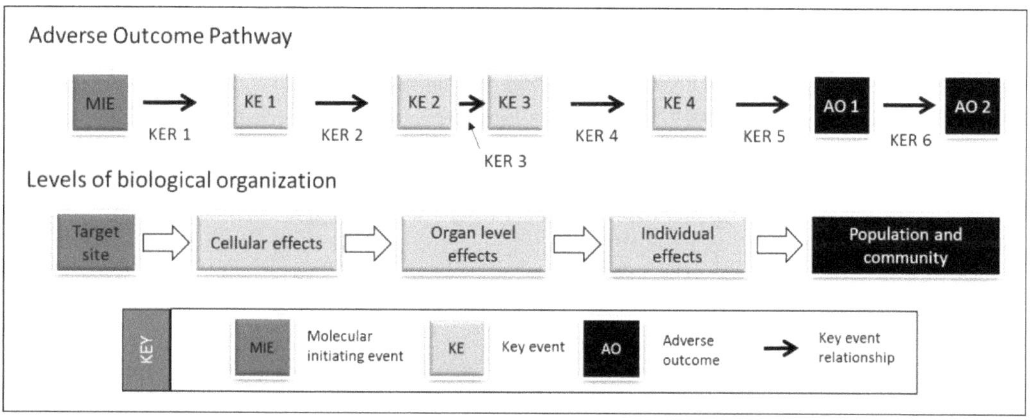

Figure 1. The Adverse Outcome Pathways framework and its relationship to effects at different levels of biological organization. A molecular initiating event (MIE) occurs when a chemical interacts with its target, triggering a set of key events (KEs) at the cellular level and organ level, ultimately leading to impacts on the individuals. As individuals experience increased mortality, decreased fitness or fecundity, population and community level adverse outcomes (AO) are seen. Key event relationships (KERs) between each KE and AO help to define the linkages between each level of biological organization and the threshold levels at each stage that will result in the impacts being translated to the next level.

AOPs and that the AOP framework would support linking genomic biomarkers to higher order effects (Ankley et al. 2010). Kramer et al. (2011) developed these ideas further, suggesting that 'omic tools can assist in identifying "system nodes" or essential physiological pathways that result in reduced growth or survival when perturbed. In this way, 'omics could help fill in knowledge gaps at the cellular and molecular levels and identify MIEs or KEs. There are many examples of 'omics studies that use this type of "discovery-based" approach to uncover molecular modes of action of chemicals (Garcia-Reyero and Perkins 2011), but Perkins et al. (2011) provided one of the first examples that aimed specifically to develop AOPs using transcriptomic data. In their study, a large gene expression data set from fathead minnow exposed to different endocrine disruption chemicals was used to construct gene regulatory networks. They investigated how one of these chemicals, flutamide, perturbed the gene networks allowing them to identify MIEs and perturbed pathways corresponding to an AO. Another important aspect of this study was that additional phenotypic data was available to link the molecular changes to adverse effects (Perkins et al. 2011). In the absence of the phenotypic evidence necessary to link 'omics data to an adverse outcome, only hypotheses or candidate AOPs can be generated and additional research will be needed to construct complete AOPs.

Adoption of the *Toxicity Testing for the 21st Century* approach provides a potential avenue for ecotoxicologists to keep pace with the development of new nanotechnologies. To accept this new direction in toxicological research, we must first develop AOPs that are relevant for ENMs. Due to the diversity of ENMs and the current uncertainty of the adverse effects that they pose to aquatic organisms, 'omic tools could become essential in the development of ENM-relevant AOPs.

'Omic Applications in Nanotoxicology

As 'omics technologies have expanded our understanding of environmental exposures and effects of chemical stressors, they should be considered important tools in understanding nanotoxicology. As reviewed below, they have already provided significant insight in defining exposure pathways and bioavailability, and helping us understand mechanisms of toxicity and develop AOPs.

Defining Exposure Pathways and Bioavailability

Nanoparticle Specific Expression Profiles

The utility of 'omic technologies in investigating ENM exposures requires that NPs produce chemical specific expression profiles. Griffitt et al. (2009) first demonstrated that copper (Cu NPs), silver (Ag NPs) and tianinium dioxide nanoparticles (TiO_2 NPs) produced reproducible and distinct gene expression profiles in zebrafish exposed to equitoxic (i.e., no observed effect concentration; NOEC) concentrations of each NP for 24 or 48 hours. In addition, the metal NP expression profiles could also be differentiated from metal ion exposures to either $Cu2^+$ or Ag^+ (Griffitt et al. 2009). Other studies have used a similar approach to determine if expression profiles can distinguish exposures to metal NPs and their corresponding metal ions. Two transcriptomics studies in *D. manga* (Poynton et al. 2012) and sheepshead minnow *Cyprinodon variegarus* (Griffitt et al. 2012) used hierarchical clustering analysis to illustrate the differences in gene expression profiles between Ag^+ and Ag NPs. Gomes et al. (2013) compared bioaccumulation in the gill and digestive gland of *Mytilus galloprovincialis* with proteomic responses, following a 15 day exposure to 10 mg/L Ag NPs or ionic silver. They found a similar level of accumulation in gills, but Ag NP exposure led to higher silver accumulation in the digestive gland compared to the Ag^+ exposure. However, in both tissues, the protein expression profiles differed between the Ag^+ and Ag NP exposures (Gomes et al. 2013). A metabolomics study in the earthworm *Eisenia fetida* also successfully distinguished Ag^+ and Ag NP exposures using metabolic profiles (Li et al. 2015).

In contrast, for metal-based NPs that readily dissolve under some experimental conditions, gene expression profiles have varied depending on the level of dissolution. This is best described by two recent transcriptomic studies of ZnO NPs in *D. magna.* In the first study, ZnO NP underwent only partial dissolution, and the gene expression profiles for ZnO NPs and Zn^{2+} were distinct suggesting that the NPs themselves were absorbed and responsible for toxicity (Poynton et al. 2011). The second study found that under their experimental conditions, the ZnO NPs completely dissolved to Zn^{2+} and the gene expression profiles between the Zn^{2+} and the ZnO NP exposures were very similar (Adam et al. 2015). Therefore, the gene expression response corresponded with the degree of physical transformation (i.e., dissolution) that the particles underwent. Taken together, these two studies confirm that gene expression profiles reflect what the organism "sees" providing insight into the physical transformations NPs may undergo before becoming bioavailable to an organism.

Differentiation of Exposures Using 'Omics

Several studies have used transcriptomics to determine whether coated or functionalized ENMs produce specific gene expression patterns. The results from a number of studies suggest that differences in expression profiles, when seen, are related to differences in bioavailability. In our laboratory, we exposed *D. magna* to Ag NPs coated with either citrate (cit-Ag NP) or 0.2% poly vinyl pyrrolidone (PVP-Ag NPs). There was a five-fold difference in the toxicity of these two particles, but when exposed to equitoxic concentrations of the NPs, the gene expression patterns were similar for the two particle types (Poynton et al. 2012). In a separate study, Scanlan et al. (2013) investigated the effects of different Ag nanowires (Ag NWs) to *D. magna.* Their study included Ag NWs of different sizes and surface coatings and they found that surface coatings affected the toxicity of the NWs. In contrast to the previous study, when *D. magna* were exposed to equitox concentrations of the Ag NWs, the gene expression profiles were different for each Ag NW type (Scanlan et al. 2013). The differentially expressed genes and impacted biological processes were also different when compared with Ag NP exposures (Poynton et al. 2012). The results suggest that Ag NP and Ag NWs cause toxicity through different mechanisms.

Additional studies have used specific biomarker genes to investigate how functionalization of NPs affects gene expression. Functionalization refers to the coating around the NPs, which alters the surface charge of the particles affecting their stabilization in water exposures and bioavailability to organisms. In two companion studies, several functionalized Au NPs with different surface charges were tested for toxicity to *D. magna.* Gene expression profiles differed between the each of the functionalized particles in both the gut (Dominguez et al. 2015) and whole organisms (Qiu et al. 2015). The authors suggest that the particle coatings caused some of the particles to aggregate resulting in differences in bioavailability that was reflected in the gene expression profiles. In the midge, *Chironomus riparius* unique gene expression patterns were identified across ten biomarker genes after exposure to uncoated Ag NPs or Ag NPs coated with citrate, gum arabic or PVP. The most prevalent changes in gene expression occurred with the cit-Ag NP, which also caused the greatest amount of DNA damage (Park et al. 2015). These studies in *D. magna* and *C. riparius* focused on particular biomarker genes related to reproduction and oxidative stress. In both cases, subtle, but significant differences were found between the functionalized NPs. These findings support a transcriptomic approach to differentiate NP exposures.

Because of the limited environmental concentrations of ENMs, there are currently no 'omic studies of ENMs in the field. However, a recent study brought environmental realism to their work by simulating the environmental impacts of ENMs present in biosolids resulting from the wastewater treatment process. These biosolids are often applied to agricultural fields as fertilizers, yet if ENMs are present in these biosoilds, there may be negative impacts on the crops farmers are trying fertilize. Chen et al. (2015) generated ENM-containing biosolids from a pilot WWTP by adding ENMs to the influent, and then aged the biosolids outdoors for six months. ENM-containing biosolids were mixed with soil and then used in exposures to the model legume *Medicago truncatula.* In comparison to biosolids containing dissolved metals, the ENM-containing biosolids caused higher metal accumulation in *M. truncatula* and reduced growth. Gene expression profiles from ENM-containing biosolid exposures were significantly different and separated from biosolids containing dissolved metals using hierarchical clustering analysis (Chen et al. 2015). Although not a true field study, Chen et al. (2015) clearly show the potential for 'omic technologies to distinguish between dissolved metal and ENM exposures in realistic exposure scenarios.

'Omics Tools to Explore Uptake Mechanisms

ENMs may interact with aquatic organisms through several different routes that will determine their bioavailability and ultimately toxicity. As described by Rocha et al. (2015), the initial exposure may consist of individual NPs, aggregations of ENMs, environmentally transformed NPs, ions released during dissolution or ENMs absorbed to food particles or algae (Rocha et al. 2015). Therefore, to begin to unravel mechanisms of toxicity, we must first understand what organisms are actually responding to and how ENMs are being taken up. In a study conducted in my laboratory, we used a transcriptomic approach to investigate the high sensitivity of the sediment dwelling crustacean *Hyalella azteca* to ZnO NPs. Dissolution experiments clearly showed that the concentration of Zn^{2+} in the ZnO NP exposure media was not sufficient to induce toxicity; however, the transcriptomic analysis showed similar gene expression responses to both ZnO NPs and Zn^{2+} when exposures were conducted at equitoxic concentrations. We interpreted these seemingly contradictory results that the ZnO NPs provided an enhanced mechanism for uptake (i.e., a "Trojan-horse" mechanism; Park et al. 2010), but following absorption, the NPs dissolved and caused cellular toxicity through a similar mode of action (Poynton et al. 2013). 'Omic tools may also help pin-point specific uptake mechanisms through the dysregulation of genes or proteins. For example, enrichment analysis of transcriptomic and proteomic data of multi-walled carbon nanotube (MWCNT) exposures to *Caenorhabditis elegans* revealed increased phagocytosis, suggesting a possible mechanism for uptake (Eom et al. 2015).

Use of 'Omics in Exposure Assessment

The studies summarized above demonstrate that 'omics techniques can often distinguish exposures from NPs of different composition, size and surface coatings and also differentiates between metal NPs and ions. In one study where 'omic profiles could not discriminate between ENMs and metal ions, the lack of differentiation was attributed to dissolution of the NPs (Adam et al. 2015). This implies that if ENMs undergo environmental transformations, 'omic tools may provide signatures related to these alterations and predict changes in bioavailability.

To date, it has been very difficult to characterize ENMs in environmental samples, including distinguishing between NP and ionic metal exposures (i.e., Bäuerlein et al. 2017) or determining whether ENMs have undergone environmental transformations. Figure 2 provides three examples, based on results from studies discussed above, of how 'omic tools can be applied in biomonitoring. Because 'omic profiles can distinguish between metal ions and metal-based NPs, these profiles could be used to identify whether an environmental sample contained metal NPs (Fig. 2A). Likewise, 'omic profiles may be able to distinguish between pristine NPs and those that have undergone environmental transformations (Fig. 2B). Finally, 'omic profiles have the potential to identify ENM exposures and characterize ENM bioavailability with unknown field samples (Fig. 2C). Overall, these examples, indicate that 'omic tools have strong potential to supplement chemical methods for ENM biomonitoring programs and these applications should be further explored and validated (Kahru and Dubourguier 2010, Selck et al. 2016).

Defining Mechanisms of Toxicity and Establishing AOPs

'Omics Tools to Uncover Mechanisms of Cellular Toxicity

'Omic studies have provided a wealth of information to date on the mechanisms of toxicity of ENMs. As recently reviewed by Revel et al. (2017), transcriptomic studies have identified several pathways

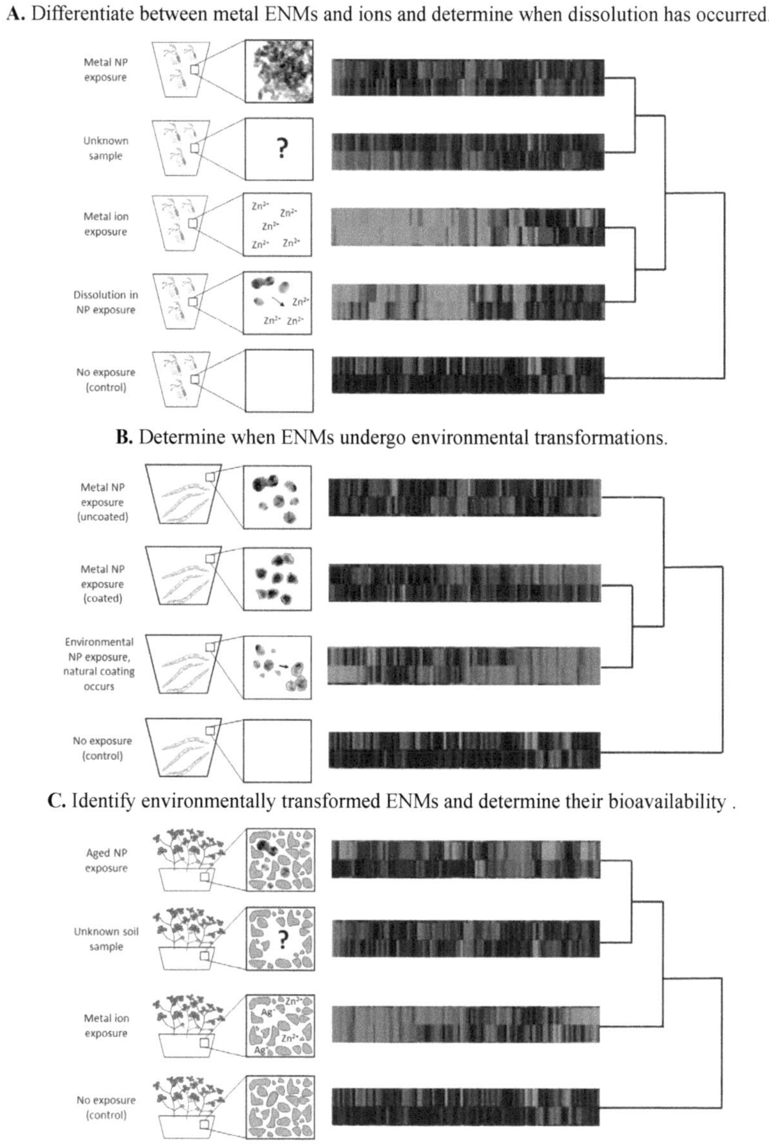

Figure 2. Potential applications of 'omic technologies for environmental monitoring of ENMs. (A) 'Omic expression profiles are able to distinguish between metal NP and metal ion exposures (Poynton et al. 2011). When dissolution of NPs occurs, gene expression profiles resemble ionic metal exposures (Adam et al. 2015). These results imply that 'omic expression profiles will differentiate between ionic metal and metal NP exposures in unknown environmental samples (although in cases where NPs act through a "Trojan-horse" like mechanism, this may not be possible, e.g., Poynton et al. 2013). In this example, the unknown environmental sample produces an 'omic profile similar to the metal NP exposure and is classified as a NP exposure. (B) 'Omic expression profiles are able to distinguish between pristine NPs and coated NPs (Park et al. 2015). This example suggests that if NPs undergo environmental transformations that result in organic coating of the NPs, 'omic profiles would differentiate the sample from pristine particles. In this example, the environmental sample contains NPs that were coated by natural organic matter and the 'omic profile is most similar to the coated NP exposure. (C) 'Omic profiles represent the bioavailability of ENMs in aged soils (Chen et al. 2015). In environmental samples, 'omic profiles should be able to distinguish between metal ion exposures and metal NP exposure even following environmental transformations. In this example, the unknown environmental samples produces an 'omic profile most similar to the aged ENMs exposure. For all panels, the different exposures are shown on the left in each panel, and heatmaps representative of different 'omic profiles are shown in the middle. The relationships between 'omic profiles are shown here using heatmaps and hierarchical clustering for simple visualization (Gehlenborg et al. 2010); however, more sophisticated classification models are usually employed for mTIE applications. For a discussion of different methods, see Kostich (2017).

Color version at the end of the book

and biological processes impacted by ENMs including oxidative response, immune system effects, DNA damage and repair, and effects to growth, development and reproduction, while proteomic studies have identified oxidative stress as a primary mechanism for many ENMs (Revel et al. 2017). Because oxidative stress is observed in many ENM studies, redox proteomics has emerged as a promising technique to investigate these mechanisms further. Redox proteomics focuses on protein modifications, including thiol oxidation and protein carbonylation, that provide an overview of the redox state of a cell while also identifying the proteins with these modifications. This may indicate damage and loss of function of particular proteins, or reveal signaling pathways as these modifications also play a role in signal transduction and gene regulation (Riebeling et al. 2016). A redox proteomics approach was taken in a series of papers by Tedesco et al. (2010) which explored the impacts of Au NPs on the blue mussel *Mytilus edulis*. By comparing levels of protein oxidation across different tissues, they identified the digestive gland as the target organ for oxidative damage (Tedesco et al. 2010a). Then, using activated thiol sepharose, they were able to isolate reduced-thiol containing proteins, or the proteome of the non-oxidized proteins. By comparing these protein sets in control and Au NP exposed animals, they uncovered approximately 50 thiol-containing proteins that were oxidized in the Au NP exposed animals (Tedesco et al. 2010b). Although they were not able to identify the proteins involved, their analysis, combined with other measures of lipid peroxidation and glutathione levels, enabled them to locate the target organ and propose oxidative stress as the primary mechanism of toxicity.

Two multi-omic studies in the green algae *Chlamydomonas reinhardtii* provided insight into the mechanisms of ENMs in this species. Pillai et al. (2014) used combined transcriptomics and proteomics to undercover several mechanisms of toxicity to Ag^{+}, recognizing that Ag^{+} is at least partially responsible for Ag NP toxicity in algae. They focused on proteins that were dysregulated at both the transcriptional and protein levels, identifying Cu^{2+} containing proteins involved in the photosynthetic electron transport chain, several genes involved in reactive oxygen species (ROS) scavenging (i.e., GSTs, glutathione peroxidases; GPXs, superoxide dismutases; SODs) and genes involved in repairing oxidized protein thiol groups (i.e., thioredoxin genes). The authors proposed a mechanism where Ag^{+} displaces Cu^{2+} in genes of the electron transport chain, shutting down Photosystem II while also producing ROS through photooxidative stress. At lower exposure concentrations, antioxidant responses can compensate for the increased ROS, thus avoiding adverse effects; however at higher concentrations, these responses are overwhelmed and lead to lipid peroxidation and reduced growth (Pillai et al. 2014). In another study using *C. reinhardtii* to examine the risk of Ceria NPs to phytoplankton, a combined transcriptomic and metabolomics approach found that effects of gene or metabolite expression were only seen at high concentrations that were not environmentally realistic. Although the study confirmed that ecological risk of Ce NPs was low, their analysis uncovered mechanisms of toxicity at the high exposure levels. Decreased expression of photosynthesis and carbon fixation genes resulted in a number of downstream impacts to carbohydrate, fatty acid and amino acid synthesis (Taylor et al. 2016). The power of using combined 'omic approaches is highlighted in these two papers that were able to concentrate on overlapping gene, protein and metabolic pathways to provide a much clearer and convincing picture of the toxic mechanisms involved.

Using 'Omics Technologies to Generate AOPs

While the progress in understanding the mechanisms of ENM toxicity is encouraging, placing this information within AOPs would provide vital information for risk assessments and identify molecular and cellular targets for high throughput assays. Indeed, there are currently several AOPs under development within the AOP wiki (aopwiki.org; Villeneuve et al. 2014b) for ENMs. In addition, there are now a few published studies that used 'omic data to generate AOPs for ENMs toxicity. These examples, which are illustrated in Fig. 3 and described below provide different approaches for using 'omics data for developing preliminary AOPs or refining AOPs by providing quantitative data to support KERs (Villeneuve et al. 2014a).

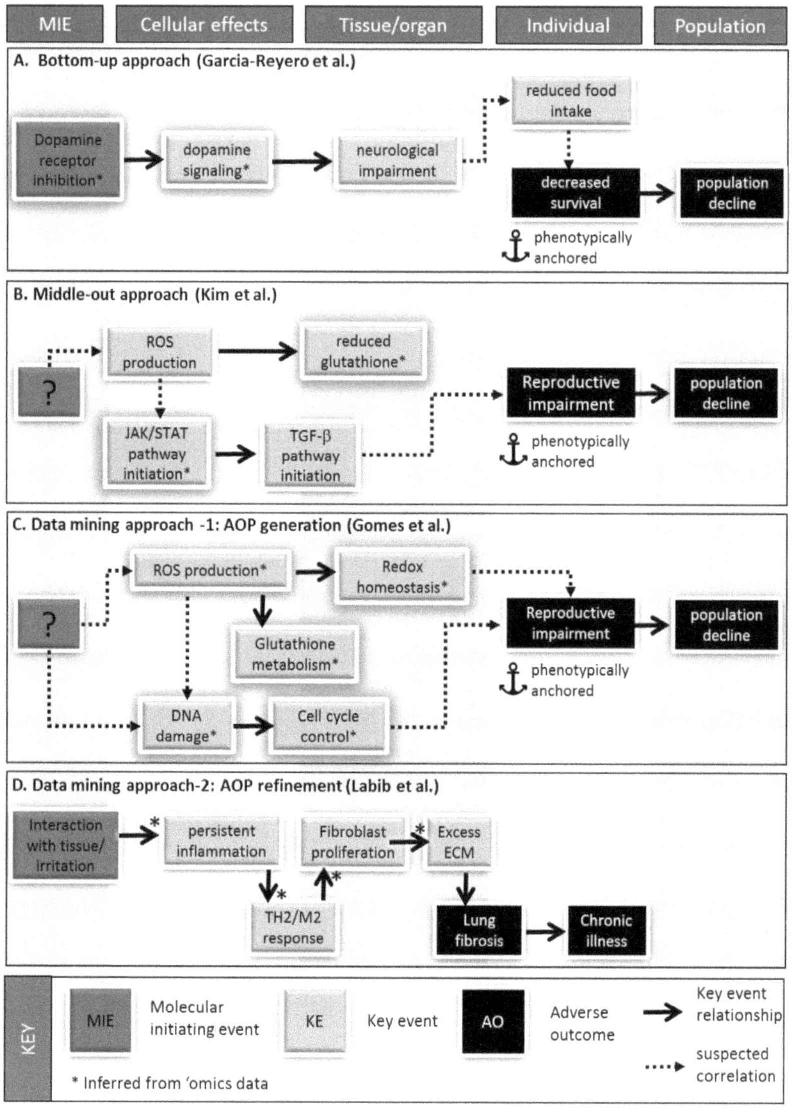

Figure 3. Approaches for applying 'omics data to AOP generation and refinement. Common AOP development strategies are reviewed in Villeneuve et al. (2014b). Four strategies are presented here due to their relevance to 'omics data and the availability of published studies that have applied these approaches. (A) The bottom-up approach uses 'omic data to infer early KEs or MIEs. In their study of Ag NPs, Garcia-Reyero et al. (2014), identified dopamine signaling as an impaired pathway using a transcriptomic approach, which implied that the dopamine receptor was inhibited by Ag NPs. This was then confirmed with an *in vitro* assay. Dopamine signaling has been shown to affect neurological systems which may cause reduced food intake and decreased survival of larval fish. The exposure concentrations used in this study allowed them to anchor the molecular responses to the adverse outcome of decreased survival. (B) The middle-out approach detects essential KEs with 'omics data and uses additional evidence or known relationships or to relate effects to an AO and develop hypotheses for an MIE. This approach was taken by Kim et al. (2017) who identified that the JAK/STAT pathway was initiated by TiO_2 NPs at concentrations leading to reproductive impairment. Additional studies revealed a connection to TGF-beta signaling which has been shown to affect growth and reproduction. An MIE was not inferred in this study, but was hypothesized to involve ROS production based on the increase of glutathione detected in their metabolomics study. (C) Data mining approaches involving large 'omics data sets can be used to develop preliminary AOPs. Gomes et al. (2017) took this approach, organizing GO data from an Ag NP transcriptomic experiment into KEs at different levels of biological organization. They used known or suspected relationships between KEs to link these effects to the reduction in reproduction seen at their exposure concentrations. (D) Data mining approaches may also be used to refine existing AOPs. In the case of Labib et al. (2015), the authors used existing transcriptomic data from MWCNT exposures to develop threshold doses (BMDs) for each of the known KEs enabling them to construct a quantitative AOP. Each example AOP provided is based on data from the corresponding papers and represent adaptations of the published AOPs. TH2: T-helper 2 cells, M2: macrophage type 2, ECM: extracellular matrix

In a study of Ag NPs, transcriptomics was used to identify impacts to the brain in the fathead minnow *Pimephales promelas* that maybe related to an AO. Their study revealed specific impacts of the NPs on a diversity of biological processes, including neurological disease functions. Through further investigation of upstream regulators, the authors identified dopamine signaling as a potential mechanism, which prompted them to perform *in vitro* receptor binding assays. Their results showed that PVP-Ag NPs inhibit dopamine binding to the dopamine receptor, providing a MIE for a proposed AOP. Given what is known of dopamine signaling and the adverse effects associated with dopamine receptor antagonism, they were able to use a "*bottom-up approach*" to link PVP-Ag NP exposure to behavioural effects in larval fish, such as decrease food intake that could lead to reduced survival (Fig. 3A, Garcia-Reyero et al. 2014).

Kim et al. (2017) used a combined transcriptomic and metabolomics investigation of UV activated TiO_2 NPs to determine the mechanisms responsible for reproductive impairment in *C. elegans*. They revealed that several genes in the Janus kinase/Signal transducers and activators of transcription (JAK/STAT) pathway were specifically affected by the TiO_2 NP + UV treatment. In addition, decreased glutathione levels seen through metabolomics analysis suggested a role for oxidative stress in initiating this signaling pathway. These holistic investigations motivated the investigators to study components of the JAK/STAT and transforming growth factor β (TGFβ) pathway, which is involved in regulation of growth and reproduction, through functional genetic assays. Loss of JAK/STAT function rescued the mutants from TiO_2 NP + UV toxicity, thus confirming an essential role of this pathway in the toxic response of these NPs. These authors used a "*middle-out approach*" for AOP generation (Villeneuve et al. 2014b) by identifying a KE (i.e., JAK/STAT pathway initiation) in their transcriptomic study and using additional functional genetics and pharmacological approaches to anchor the KE to reproductive effects (Fig. 3B, Kim et al. 2017). This AOP is currently under construction at aopwiki.org (AOP:208, accessed January 20, 2018).

In an investigation of the impacts of NPs on the soil oligochaete *Enchytraeus crypticus,* Gomes et al. (2017) used a transcriptomic approach to differentiate the effects of $AgNO_3$ from different Ag NPs and identify biological processes (i.e., gene ontology, or GO terms) that were deregulated at concentrations leading to reproductive failure (EC50). They organized the affected biological processes (e.g., apoptotic stimuli, DNA damage, glutathione metabolism, lipid oxidation) according to whether the impacts were best described at the molecular, cellular, or organ level and linked them to the AO of decreased reproduction. In this way they could organize several different KEs by the level of biological organization they impacted to help construct a preliminary AOP. Although many of the KERs and MIEs are not well established, the preliminary AOP organizes the data that is known, helps to identify knowledge gaps and provides a blueprint for future research needs (see Fig. 3C, Gomes et al. 2017).

A final human health study provides an interesting application for 'omics data in constructing a quantitative AOP (qAOP; Villeneuve et al. 2014a). Labib et al. (2015) first performed a literature review to construct a preliminary AOP for pulmonary injury leading to fibrosis (full description of the AOP is under development at AOPwiki.org, AOP:173, accessed January 30, 2018). They used data from three previously published *in vivo* transcriptomic studies of mouse lung tissue following exposure to multi-walled carbon nanotubes (MWCTs). For each differentially expressed gene identified in one of the studies, they derived a benchmark dose (BMD) by modeling the gene expression across a dose response. Genes were mapped to functional pathways and pathways containing at least five differentially expressed genes were considered impacted. The authors then calculated a BMD for each impacted pathway using the median BMD for all the differentially expressed genes in that pathway. Pathways were mapped to KEs in the AOP using expert judgement, which enabled them to generate BMDs for each KE, essentially constructing a qAOP based on the dysregulation of important genes and pathways (Fig. 3D). The transcriptomic-based BMDs correlated well to phenotypic endpoints indicative of each KE, which demonstrated the potential of their approach (Labib et al. 2015).

The studies reviewed above used 'omic technologies to hypothesize and generate AOPs. Kim et al. (2017) and Garcia-Reyero et al. (2014) followed up on transcriptomic "leads" through genetic experiments and *in vitro* assays to get better insight into important KEs within the AOP. Another important aspect of these studies is that they used concentrations and doses in the 'omic analysis that corresponded to reduced survival, reproductive impairment or lung fibrosis (an AO for human health) allowing their 'omics results to be anchored to AOs. Using the AOP constructed in each of these studies (Fig. 3), additional experiments can be proposed to test MIE hypotheses or confirm proposed KEs and ultimately fill in the KERs leading to an AO. Finally, the construction of AOPs through the approaches described here, will greatly facilitate the development of high-throughput screening methods. By targeting MIEs and KEs related to molecular or cellular perturbations, high-throughput assays can be developed to quickly screen new chemicals including ENMs for toxicity (Nel et al. 2013).

Future Directions

Emerging 'Omics Technologies

Transgenerational Effects and Epigenetics

Epigenetic responses to chemicals is arguably an understudied area of research, despite compelling examples of transgenerational effects being mediated by changes in the epigenome. Epigenetics refers to heritable phenotypic changes that occur independent of alterations to the genetic code through the processes of DNA methylation, histone tail modifications or small non-coding RNAs (ncRNAs) including microRNAs (miRNAs) (Vandegehuchte and Janssen 2011). Through these mechanisms, epigenetics induces long-term changes to gene transcription and protein translation, which can be passed down to daughter cells and in some cases, offspring (Vandegehuchte and Janssen 2011). For example, nutritional status and exposure to endocrine disrupting chemicals can induce transgenerational effects, which are mediated through inheritance of DNA methylation patterns (Vandegehuchte and Janssen 2011). For ENMs, there is preliminary evidence that they can cause transgenerational effects. Carbon-based ENMs and Au NPs have been shown to impact reproduction in subsequent generations in *D. magna* (Arndt et al. 2014) and *C. elegans* (Kim et al. 2013, Moon et al. 2017); although not yet investigated, it seems likely that epigenetic mechanisms are involved.

The advent of next generation sequencing technologies have provided opportunities to study epigenetic changes using an 'omics approach (Table 1). Despite this, investigations of epigenomic effects of ENMs are still limited and mostly concentrated in mammalian systems under a human health context. These studies have shown impacts on all three types of epigenetic changes: DNA methylation, histone tail modification and miRNA expression, associated with NP exposure (Yao and Costa 2013). A few studies with different ENMs have independently observed a repression of DNA methyltransferase I (DNMT1), the primary enzyme responsible for methylating DNA, suggesting that DNA methylation processes are critical components of NP toxicity. miRNA expression has been linked to NP toxicity in cell line studies after Ag NP (Eom et al. 2014) and Mn NP exposure (Grogg et al. 2016). In both studies, the authors identified specific miRNAs that played roles in NP-induced inflammation, DNA damage and apoptosis. In ecotoxicology, epigenomic approaches have been applied to study the transgenerational effects of metals in *D. magna* on DNA methylation patterns (Vandegehuchte et al. 2010) and miRNA expression (Chen et al. 2016). Because oxidative stress is a potential mechanism for these epigenetic changes (Vandegehuchte and Janssen 2011), it is possible that similar impacts could be induced by metal-based NPs. Indeed, Gomes et al. (2017) detected changes in genes involved in histone tail modification processes after exposure to Ag NPs, thus suggesting epigenetic effects (Gomes et al. 2017). Overall, these studies indicate that ENMs are likely affecting the epigenome of ecologically important aquatic species, that these epigenetic changes may be essential KEs in the toxicity of NPs, and that these effects may result in transgenerational effects.

Metagenomics

Microbial community analysis using whole genome shot-gun sequencing has allowed for the discovery and study of 99 percent of the microbes that are difficult or impossible to grow in culture. Whole microbial genomes can be assembled from the raw sequencing data, allowing for a detailed annotation and characterizing of all the genes they possess. This has allowed scientists to understand not only which microbes are present in an environmental sample, but also their ecological function (Tringe and Rubin 2005). Therefore, when the abundance of a particular microbe decreases due to changing environmental conditions or pollutants, investigators can also infer what ecosystem functions may be altered. In this way, the application of metagenomics to ENM assessment targets the highest levels of biological organization, that is, effects to community structure and ecosystem function.

Microbial community analysis has been investigated following ENM exposure to activated sludge in wastewater treatment processes and in natural soils and sediments. Many ENMs are predicted to enter the environment through waste water streams; therefore, their effects on activated sludge that helps breakdown organic waste in WWTPs, is an important area of research. For example, ZnO NP addition to activated sludge caused a decrease in diversity, which was related to specific microbes (i.e., operational taxonomic units; OTUs) (Meli et al. 2016). Zheng et al. (2011) investigated the impact of TiO_2 NPs on activated sludge. They specifically targeted OTUs associated with nitrification processes and found decreased abundance of ammonia oxidizing bacteria (AOB) and nitrite-oxidizing bacteria (NOB) that corresponded to a large drop in nitrogen removal from the wastewater (Zheng et al. 2011). To study the impacts of transformed ENMs to natural systems, Moore et al. (2016) constructed wetland mesocosms. Using 16S amplicon DNA sequencing, they observed an ecosystem shift from a well-oxygenated community able to support aquatic plants to an anoxic wetland with an increased abundance of methanogens (methane producing bacteria). Finally, to apply the metagenomics approach directly to risk assessment, Doolette et al. (2016) developed Ag NP toxicity thresholds by applying a dose-response model to the microbial abundance data. They emphasized abundance patterns for specific functional communities, including those related to nitrification (Doolette et al. 2016). Overall, these studies reveal that ENMs cause impacts to microbial community structure that translate to the loss of important ecosystem functions and highlight the potential of metagenomics for ENM assessment.

Cross Species Extrapolation

A major challenge in ecotoxicology is extrapolating effects seen in a small set of model organisms in the laboratory to ecosystems rich in biodiversity. Particularly for ENMs, sensitivity across species can range several orders of magnitude (Kahru and Dubourguier 2010). To begin to tackle this problem, we should first consider that many toxicological pathways are evolutionarily conserved across even highly divergent taxa. Therefore, a phylogenetic approach can assist in determining whether two species are likely to experience similar levels of toxicity (Reid and Whitehead 2016). This may include the comparison of homologous gene sequences to detect similar active sites or the analysis of gene expansions/losses in relevant gene families (Goldstone et al. 2006). In addition to comparative genomic approaches, 'omics tools may also assist in cross-species extrapolation. Because expression profiles reflect the toxicological response of an organism to the toxicant, conservation of 'omic profiles can infer similar mechanisms and levels of toxicity. For example, Garcia-Reyero (2011) compared transcriptomic responses to RDX, or cyclotrimethylenetrinitramine, across five divergent species. Pathway analysis revealed several conserved effects, but in general, the similarity in transcriptomic responses diverged with greater phylogenetic distance between the species (Garcia-Reyero et al. 2011a).

However, when attempting to make cross-species comparisons, we also must consider that ENMs undergo complex environmental transformations, concentrate in particular environmental compartments and maybe taken up through unconventional mechanisms. All of these factors impact the susceptibility of organisms to the ENMs; therefore, toxicity is likely governed by both the specific bioavailability of the EMN to a species as well as phylogenetic relationships (e.g., Poynton et al. 2013).

Incorporating 'Omics into Exposure and Effects Assessment of ENMs

In 1999, Nuwaysir et al. introduced the concept of toxicogenomics, setting the stage for the use of 'omics in toxicological and ecotoxicological research (Nuwaysir et al. 1999). They proposed that 'omic signatures could be used to identify mechanisms of action of novel chemicals and identify environmental exposures, predicting not only the chemicals organisms are exposed to, but also possible adverse effects. Over the past nearly two decades, toxicogenomics (Nuwaysir et al. 1999) and later ecotoxicogenomics (Snape et al. 2004) have devoted resources to prove the existence of 'omic signatures and test the feasibility of using these profiles for exposure and effects assessment.

Within exposure assessments, challenges still arise when applying 'omics to current monitoring programs (Bahamonde et al. 2016). However, environmental monitoring for ENMs presents its own unique hurdles that are not easily addressed by traditional chemical methods, thus creating a new space for biologically based approaches. Here, 'omics may be able to address difficulties in differentiating between metal NPs and metal ions or recognizing when NPs have undergone environmental transformations. These potential applications of 'omics to ENM exposure assessment are based on the observation that exposures to different ENMs produce specific 'omic signatures (Fig. 2). However, these signatures have not been found consistently across all species, treatments and sometimes, tissue type. More research is needed to understand why these discrepancies occur and to control for them in monitoring or exposure assessment programs.

In effects assessments, the emphasis has shifted from using 'omics signatures to classify chemicals and predict MOAs to developing AOPs. Although the AOP concept is not inherently different from defining a MOA, its implementation has allowed researchers to consider new avenues to apply 'omics data to effects assessment and link it to AOs of regulatory interest (Villeneuve et al. 2014b). For ENMs many of the MOAs are still unknown, especially for aquatic organisms of ecological relevance (Vale et al. 2016). However, using "*bottom-up*", "*middle-out*" or data mining approaches (Fig. 3), several investigators have applied 'omics to develop AOPs bringing us closer to implementing a *Toxicity Testing for the 21st Century* approach (NRC 2007) to ENM risk assessment. Although more research is needed to further develop and refine AOPs for ENMs, applying 'omics within an AOP framework provides a potential route to close the gap between nanotechnology development and ecotoxicology testing.

Acknowledgments

Financial support was provided by National Science Foundation grant CBET-1437409 to HCP. Special thanks to William Robinson and Albert Armstrong for their editing and thoughtful comments on this manuscript.

References

Adam, N., L. Vergauwen, R. Blust and D. Knapen. 2015. Gene transcription patterns and energy reserves in *Daphnia magna* show no nanoparticle specific toxicity when exposed to ZnO and CuO nanoparticles. Environ. Res. 138: 82–92.

Ankley, G.T., R.S. Bennett, R.J. Erickson, D.J. Hoff, M.W. Hornung, R.D. Johnson et al. 2010. Adverse outcome pathways: A conceptual framework to support ecotoxicology research and risk assessment. Environ Toxicol Chem. 29: 730–741.

Ankley, G.T., G.P. Daston, S.J. Degitz, N.D. Denslow, R.A. Hoke, S.W. Kennedy et al. 2006. Toxicogenomics in regulatory ecotoxicology. Environ. Sci. Technol. 40: 4055–4065.

Antczak, P., H.J. Jo, S. Woo, L. Scanlan, H. Poynton, A. Loguinov et al. 2013. Molecular toxicity identification evaluation (mTIE) approach predicts chemical exposure in *Daphnia magna*. Environ. Sci. Technol. 47: 11747–11756.

Arndt, D.A., J. Chen, M. Moua and R.D. Klaper. 2014. Multigeneration impacts on *Daphnia magna* of carbon nanomaterials with differing core structures and functionalizations. Environ. Toxicol. Chem. 33: 541–547.

Bahamonde, P.A., A. Feswick, M.A. Isaacs, K.R. Munkittrick and C.J. Martyniuk. 2016. Defining the role of omics in assessing ecosystem health: Perspectives from the Canadian environmental monitoring program. Environ. Toxicol. Chem. 35: 20-35.

Bäuerlein, P.S., E. Emke, P. Tromp, J.A. Hofman, A. Carboni, F. Schooneman et al. 2017. Is there evidence for man-made nanoparticles in the dutch environment? Sci. Total Environ. 576: 273–283.

Baun, A., N.B. Hartmann, K. Grieger and K.O. Kusk. 2008. Ecotoxicity of engineered nanoparticles to aquatic invertebrates: A brief review and recommendations for future toxicity testing. Ecotoxicology 17: 387–395.

Biales, A.D., M. Kostich, R.M. Burgess, K.T. Ho, D.C. Bencic, R.L. Flick et al. 2013. Linkage of genomic biomarkers to whole organism end points in a toxicity identification evaluation (TIE). Envion. Sci. Technol. 47: 1306–1312.

Buffet, P.-E., A. Zalouk-Vergnoux, A. Châtel, B. Berthet, I. Métais, H. Perrein-Ettajani et al. 2014. A marine mesocosm study on the environmental fate of silver nanoparticles and toxicity effects on two endobenthic species: The ragworm *Hediste diversicolor* and the bivalve mollusc *Scrobicularia plana*. Sci. Total Environ. 470: 1151–1159.

Bundy, J.G., M.P. Davey and M.R. Viant. 2009. Environmental metabolomics: A critical review and future perspectives. Metabolomics 5: 3.

Chariton, A.A., L.N. Court, D.M. Hartley, M.J. Colloff and C.M. Hardy. 2010. Ecological assessment of estuarine sediments by pyrosequencing eukaryotic ribosomal DNA. Frontiers in Ecology and the Environment. 8: 233–238.

Chen, C., J.M. Unrine, J.D. Judy, R.W. Lewis, J. Guo, D.H. McNear et al. 2015. Toxicogenomic responses of the model legume medicago truncatula to aged biosolids containing a mixture of nanomaterials (TiO_2, Ag, and ZnO) from a pilot wastewater treatment plant. Envrion. Sci. Technol. 49: 8759–8768.

Chen, S., K.M. Nichols, H.C. Poynton and M.S. Sepulveda. 2016. MicroRNAs are involved in cadmium tolerance in *Daphnia pulex*. Aquat Toxicol. 175: 241–248.

Colman, B.P., C.L. Arnaout, S. Anciaux, C.K. Gunsch, M.F. Hochella Jr., B. Kim et al. 2013. Low concentrations of silver nanoparticles in biosolids cause adverse ecosystem responses under realistic field scenario. PloS One 8: e57189.

Colman, B.P., B. Espinasse, C.J. Richardson, C.W. Matson, G.V. Lowry, D.E. Hunt et al. 2014. Emerging contaminant or an old toxin in disguise? Silver nanoparticle impacts on ecosystems. Envrion. Sci. Technol. 48: 5229–5236.

Dominguez, G.A., S.E. Lohse, M.D. Torelli, C.J. Murphy, R.J. Hamers, G. Orr et al. 2015. Effects of charge and surface ligand properties of nanoparticles on oxidative stress and gene expression within the gut of *Daphnia magna*. Aquat. Toxicol. 162: 1–9.

Doolette, C.L., V.V. Gupta, Y. Lu, J.L. Payne, D.J. Batstone, J.K. Kirby et al. 2016. Quantifying the sensitivity of soil microbial communities to silver sulfide nanoparticles using metagenome sequencing. PloS one 11: e0161979.

Ekman, D., D. Skelton, J. Davis, D. Villeneuve, J. Cavallin, A. Schroeder et al. 2015. Metabolite profiling of fish skin mucus: A novel approach for minimally-invasive environmental exposure monitoring and surveillance. Envrion. Sci. Technol. 49: 3091–3100.

Eom, H.-J., N. Chatterjee, J. Lee and J. Choi. 2014. Integrated mRNA and microRNA profiling reveals epigenetic mechanism of differential sensitivity of Jurkat T cells to AgNPs and Ag ions. Toxicology Letters 229: 311–318.

Eom, H.-J., C.P. Roca, J.-Y. Roh, N. Chatterjee, J.-S. Jeong, I. Shim et al. 2015. A systems toxicology approach on the mechanism of uptake and toxicity of MWCNT in *Caenorhabditis elegans*. Chemico-Biological Interactions 239: 153–163.

Fabrega, J., S.N. Luoma, C.R. Tyler, T.S. Galloway and J.R. Lead. 2011. Silver nanoparticles: Behaviour and effects in the aquatic environment. Environ. International. 37: 517–531.

Farré, M., S. Pérez, K. Gajda-Schrantz, V. Osorio, L. Kantiani et al. 2010. First determination of C60 and C70 fullerenes and n-methylfulleropyrrolidine C60 on the suspended material of wastewater effluents by liquid chromatography hybrid quadrupole linear ion trap tandem mass spectrometry. J. of Hydrology 383: 44–51.

Ferry, J.L., P. Craig, C. Hexel, P. Sisco, R. Frey, P.L. Pennington et al. 2009. Transfer of gold nanoparticles from the water column to the estuarine food web. Nat. Nanotechnol. 4: 441–444.

Garcia-Reyero, N., T. Habib, M. Pirooznia, K.A. Gust, P. Gong, C. Warner et al. 2011a. Conserved toxic responses across divergent phylogenetic lineages: A meta-analysis of the neurotoxic effects of RDX among multiple species using toxicogenomics. Ecotoxicology 20: 580.

Garcia-Reyero, N., C.M. Lavelle, B.L. Escalon, D. Martinović, K.J. Kroll, P.W. Sorensen et al. 2011b. Behavioral and genomic impacts of a wastewater effluent on the fathead minnow. Aquat. Toxicol. 101: 38–48.

Garcia-Reyero, N. and E.J. Perkins. 2011. Systems biology: Leading the revolution in ecotoxicology. Environ. Toxicol. Chem. 30: 265–273.

Garcia-Reyero, N.I., B.L. Escalon, P.R. Loh, J.G. Laird, A.J. Kennedy, B. Berger et al. 2011c. Assessment of chemical mixtures and groundwater effects on *Daphnia magna* transcriptomics. Envrion. Sci. Technol. 46: 42–50.

Garcia-Reyero, N.I., A.J. Kennedy, B.L. Escalon, T. Habib, J.G. Laird, A. Rawat et al. 2014. Differential effects and potential adverse outcomes of ionic silver and silver nanoparticles *in vivo* and *in vitro*. Envrion. Sci. Technol. 48: 4546–4555.

García-Sevillano, M.Á., T. García-Barrera and J.L. Gómez-Ariza. 2015. Environmental metabolomics: Biological markers for metal toxicity. Electrophoresis 36: 2348–2365.

Garner, K.L., S. Suh and A.A. Keller. 2017. Assessing the risk of engineered nanomaterials in the environment: Development and application of the nanofate model. Envrion. Sci. Technol. 51: 5541–5551.

Gehlenborg, N., S.I. O'donoghue, N.S. Baliga, A. Goesmann, M.A. Hibbs, H. Kitano et al. 2010. Visualization of omics data for systems biology. Nature Methods 7: S56.

Gevaert, K. and J. Vandekerckhove. 2000. Protein identification methods in proteomics. Electrophoresis. 21: 1145-1154.

Gillet, L.C., A. Leitner and R. Aebersold. 2016. Mass spectrometry applied to bottom-up proteomics: Entering the high-throughput era for hypothesis testing. Annual Review of Analytical Chemistry 9: 449–472.
Goldstone, J., A. Hamdoun, B. Cole, M. Howard-Ashby, D. Nebert, M. Scally et al. 2006. The chemical defensome: Environmental sensing and response genes in the *Strongylocentrotus purpuratus* genome. Developmental Biology 300: 366–384.
Gomes, S.I., C.P. Roca, J.J. Scott-Fordsmand and M.J. Amorim. 2017. High-throughput transcriptomics reveals uniquely affected pathways: AgNPs, PVP-coated AgNPs, and Ag NM300k case studies. Environ. Sci: Nano. 4: 929–937.
Gomes, T., C.G. Pereira, C. Cardoso and M.J. Bebianno. 2013. Differential protein expression in mussels *Mytilus galloprovincialis* exposed to nano and ionic Ag. Aquat. Toxicol. 136: 79–90.
Gondikas, A.P., F. von der Kammer, R.B. Reed, S. Wagner, J.F. Ranville and T. Hofmann. 2014. Release of TiO_2 nanoparticles from sunscreens into surface waters: A one-year survey at the Old Danube recreational lake. Environ. Sci. Technol. 48: 5415–5422.
Gottschalk, F., T. Sun and B. Nowack. 2013. Environmental concentrations of engineered nanomaterials: Review of modeling and analytical studies. Environ. Pollut. 181: 287–300.
Griffitt, R.J., K. Hyndman, N.D. Denslow and D.S. Barber. 2009. Comparison of molecular and histological changes in zebrafish gills exposed to metallic nanoparticles. Toxicol. Sci. 107: 404–415.
Griffitt, R.J., N.J. Brown-Peterson, D.A. Savin, C.S. Manning, I. Boube, R. Ryan et al. 2012. Effects of chronic nanoparticulate silver exposure to adult and juvenile sheepshead minnows (*Cyprinodon variegatus*). Environ. Toxicol. Chem. 31: 160–167.
Grogg, M.W., L.K. Braydich-Stolle, E.I. Maurer-Gardner, N.T. Hill, S. Sakaram, M.P. Kadakia et al. 2016. Modulation of miRNA-155 alters manganese nanoparticle-induced inflammatory response. Toxicol. Res. 5: 1733–1743.
Handford, C.E., M. Dean, M. Henchion, M. Spence, C.T. Elliott and K. Campbell. 2014. Implications of nanotechnology for the agri-food industry: Opportunities, benefits and risks. Trends in Food Science & Technology 40: 226–241.
Handy, R.D., T.B. Henry, T.M. Scown, B.D. Johnston and C.R. Tyler. 2008. Manufactured nanoparticles: Their uptake and effects on fish—a mechanistic analysis. Ecotoxicology 17: 396–409.
Hansen, S.F., L.R. Heggelund, P.R. Besora, A. Mackevica, A. Boldrin and A. Baun. 2016. Nanoproducts–what is actually available to european consumers? Environ. Sci: Nano. 3: 169–180.
Holden, P.A., R.M. Nisbet, H.S. Lenihan, R.J. Miller, G.N. Cherr, J.P. Schimel et al. 2012. Ecological nanotoxicology: Integrating nanomaterial hazard considerations across the subcellular, population, community, and ecosystems levels. Accounts of Chemical Research 46: 813–822.
Hook, S.E., A.D. Skillman, J.A. Small and I.R. Schultz. 2006. Gene expression patterns in rainbow trout, *Oncorhynchus mykiss*, exposed to a suite of model toxicants. Aquat. Toxicol. 77: 372–385.
Ivask, A., K. Juganson, O. Bondarenko, M. Mortimer, V. Aruoja, K. Kasemets et al. 2014. Mechanisms of toxic action of Ag, AnO, and CuO nanoparticles to selected ecotoxicological test organisms and mammalian cells *in vitro*: A comparative review. Nanotoxicology 8: 57–71.
Kahru, A. and H.-C. Dubourguier. 2010. From ecotoxicology to nanoecotoxicology. Toxicology 269: 105–119.
Kim, H., J. Jeong, N. Chatterjee, C.P. Roca, D. Yoon, S. Kim et al. 2017. JAK/STAT and TGF-ß activation as potential adverse outcome pathway of TiO_2 NPs phototoxicity in *Caenorhabditis elegans*. Scientific Reports 7: 17833.
Kim, S.W., J.I. Kwak and Y.-J. An. 2013. Multigenerational study of gold nanoparticles in *Caenorhabditis elegans*: Transgenerational effect of maternal exposure. Envrion. Sci. Technol. 47: 5393–5399.
Klaine, S.J., A.A. Koelmans, N. Horne, S. Carley, R.D. Handy, L. Kapustka et al. 2012. Paradigms to assess the environmental impact of manufactured nanomaterials. Environ. Toxicol. Chem. 31: 3–14.
Kostich, M.S. 2017. A statistical framework for applying RNA profiling to chemical hazard detection. Chemosphere 188: 49–59.
Kramer, V.J., M.A. Etterson, M. Hecker, C.A. Murphy, G. Roesijadi, D.J. Spade et al. 2011. Adverse outcome pathways and ecological risk assessment: Bridging to population-level effects. Environ. Toxicol. Chem. 30: 64–76.
Labib, S., A. Williams, C.L. Yauk, J.K. Nikota, H. Wallin, U. Vogel et al. 2015. Nano-risk science: Application of toxicogenomics in an adverse outcome pathway framework for risk assessment of multi-walled carbon nanotubes. Particle and Fibre Toxicology 13: 15.
Li, L., H. Wu, W.J. Peijnenburg and C.A. van Gestel. 2015. Both released silver ions and particulate Ag contribute to the toxicity of AgNPs to earthworm *Eisenia fetida*. Nanotoxicology 9: 792–801.
Ma, H., P.L. Williams and S.A. Diamond. 2013. Ecotoxicity of manufactured ZnO nanoparticles—a review. Environ. Pollut. 172: 76–85.
Martinović-Weigelt, D., A.C. Mehinto, G.T. Ankley, N.D. Denslow, L.B. Barber, K.E. Lee et al. 2014. Transcriptomic effects-based monitoring for endocrine active chemicals: Assessing relative contribution of treated wastewater to downstream pollution. Envrion. Sci. Technol. 48: 2385–2394.
Martyniuk, C.J. and D.B. Simmons. 2016. Spotlight on environmental omics and toxicology: A long way in a short time. Comp. Biochem. Physiol. Part D Genomics Proteomics. 19: 97–101.
Mehinto, A.C., C.J. Martyniuk, D.J. Spade and N.D. Denslow. 2012. Applications for next-generation sequencing in fish ecotoxicogenomics. Frontiers in Genetics. 3(62): 1–10.
Meli, K., I. Kamika, J. Keshri and M. Momba. 2016. The impact of zinc oxide nanoparticles on the bacterial microbiome of activated sludge systems. Scientific Reports 6: 39176.
Moon, J., J.I. Kwak, S.W. Kim and Y.-J. An. 2017. Multigenerational effects of gold nanoparticles in *Caenorhabditis elegans*: Continuous versus intermittent exposures. Environ. Pollut. 220: 46–52.

Moore, J.D., J.P. Stegemeier, K. Bibby, S.M. Marinakos, G.V. Lowry and K.B. Gregory. 2016. Impacts of pristine and transformed Ag and Cu engineered nanomaterials on surficial sediment microbial communities appear short-lived. Envrion. Sci. Technol. 50: 2641–2651.

Mulvaney, P. and P.S. Weiss. 2016. Have nanoscience and nanotechnology delivered? ACS Nano. 10: 7225–7226.

Nel, A., T. Xia, H. Meng, X. Wang, S. Lin, Z. Ji et al. 2013. Nanomaterial toxicity testing in the 21st century: Use of a predictive toxicological approach and high-throughput screening. Acc. Chem. Res. 46: 607–621.

Nguyen, T., P.J. Sherratt and C.B. Pickett. 2003. Regulatory mechanisms controlling gene expression mediated by the antioxidant response element. Annual Review of Pharmacology and Toxicology 43: 233–260.

Notter, D.A., D.M. Mitrano and B. Nowack. 2014. Are nanosized or dissolved metals more toxic in the environment? A meta-analysis. Environ Toxicol Chem. 33: 2733–2739.

NRC (National Research Council). 2007. Toxicity testing for the 21st century: A vision and strategy. National Academies Press, Washington D.C.

Nuwaysir, E.F., M. Bittner, J. Trent, J.C. Barrett, C.A. Afshari. 1999. Microarrays and toxicology: The advent of toxicogenomics. Mol Carcinog. 24: 153–159.

Osborn, H.L. and S.E. Hook. 2013. Using transcriptomic profiles in the diatom *Phaeodactylum tricornutum* to identify and prioritize stressors. Aquat. Toxicol. 138: 12–25.

Park, E.J., J. Yi, Y. Kim, K. Choi and K. Park. 2010. Silver nanoparticles induce cytotoxicity by a trojan-horse type mechanism. Toxicol. *In Vitro* 24: 872–878.

Park, S.Y., J. Chung, B.P. Colman, C.W. Matson, Y. Kim, B.C. Lee et al. 2015. Ecotoxicity of bare and coated silver nanoparticles in the aquatic midge, *Chironomus riparius*. Environ. Toxicol. Chem. 34: 2023–2032.

Perkins, E.J., J.K. Chipman, S. Edwards, T. Habib, F. Falciani, R. Taylor et al. 2011. Reverse engineering adverse outcome pathways. Environ. Toxicol. Chem. 30: 22–38.

Pillai, S., R. Behra, H. Nestler, M.J.-F. Suter, L. Sigg and K. Schirmer. 2014. Linking toxicity and adaptive responses across the transcriptome, proteome, and phenotype of *Chlamydomonas reinhardtii* exposed to silver. Proceedings of the National Academy of Sciences 111: 3490–3495.

Poynton, H.C., J.M. Lazorchak, C.A. Impellitteri, B. Blalock, M.E. Smith, K. Struewing et al. 2013. Toxicity and transcriptomic analysis in *Hyalella azteca* suggests increased exposure and susceptibility of epibenthic organisms to zinc oxide nanoparticles. Environ. Sci. Technol. 47: 9453–9460.

Poynton, H.C., J.M. Lazorchak, C.A. Impellitteri, B.J. Blalock, K. Rogers, H.J. Allen et al. 2012. Toxicogenomic responses of nanotoxicity in *Daphnia magna* exposed to silver nitrate and coated silver nanoparticles. Environ. Sci. Technol. 46: 6288–6296.

Poynton, H.C., J.M. Lazorchak, C.A. Impellitteri, M.E. Smith, K. Rogers, M. Patra et al. 2011. Differential gene expression in *Daphnia magna* suggests distinct modes of action and bioavailability for ZnO nanoparticles and Zn ions. Environ. Sci. Technol. 45: 762–768.

Poynton, H.C. and C.D. Vulpe. 2009. Ecotoxicogenomics: Emerging technologies for emerging contaminants. J. American Water Resources Association 45: 83–96.

Poynton, H.C., H. Wintz and C.D. Vulpe. 2008a. Progress in ecotoxicogenomics for environmental monitoring, mode of action, and toxicant identification. pp. 21–74. *In*: C. Hogstrand and P. Kille (eds.). Comparative Toxicogenomics. Elsevier Science, Amsterdam, The Netherlands.

Poynton, H.C., R. Zuzow, A.V. Loguinov, E.J. Perkins and C.D. Vulpe. 2008b. Gene expression profiling in *Daphnia magna*, Part II: Validation of a copper specific gene expression signature with effluent from two copper mines in California. Environ. Sci. Technol. 42: 6257–6263.

Qiu, T., J. Bozich, S. Lohse, A. Vartanian, L. Jacob, B. Meyer et al. 2015. Gene expression as an indicator of the molecular response and toxicity in the bacterium *Shewanella oneidensis* and the water flea *Daphnia magna* exposed to functionalized gold nanoparticles. Environ. Sci: Nano. 2: 615–629.

Reid, N.M. and A. Whitehead. 2016. Functional genomics to assess biological responses to marine pollution at physiological and evolutionary timescales: Toward a vision of predictive ecotoxicology. Briefings in Functional Genomics 15: 358–364.

Revel, M., A. Châtel and C. Mouneyrac. 2017. Omics tools: New challenges in aquatic nanotoxicology? Aquat. Toxicol.

Riebeling, C., M. Wiemann, J. Schnekenburger, T.A. Kuhlbusch, W. Wohlleben, A. Luch et al. 2016. A redox proteomics approach to investigate the mode of action of nanomaterials. Toxicology and Applied Pharmacology 299: 24–29.

Rocha, T.L., T. Gomes, V.S. Sousa, N.C. Mestre and M.J. Bebianno. 2015. Ecotoxicological impact of engineered nanomaterials in bivalve molluscs: An overview. Marine Environ. Res. 111: 74–88.

Roland, K., P. Kestemont, R. Loos, S. Tavazzi, B. Paracchini, C. Belpaire et al. 2014. Looking for protein expression signatures in European eel peripheral blood mononuclear cells after *in vivo* exposure to Perfluorooctane sulfonate and a real world field study. Sci. Total Environ. 468: 958–967.

Sanchez, B.C., K. Ralston-Hooper and M.S. Sepúlveda. 2011. Review of recent proteomic applications in aquatic toxicology. Environ. Toxicol. Chem. 30: 274–282.

Scanlan, L.D., R.B. Reed, A.V. Loguinov, P. Antczak, A. Tagmount, S. Aloni et al. 2013. Silver nanowire exposure results in internalization and toxicity to *Daphnia magna*. Acs. Nano. 7: 10681–10694.

Scown, T.M., R. van Aerle and C.R. Tyler. 2010. Review: Do engineered nanoparticles pose a significant threat to the aquatic environment? Crit. Rev. Toxicol. 40: 653–670.

Selck, H., R.D. Handy, T.F. Fernandes, S.J. Klaine and E.J. Petersen. 2016. Nanomaterials in the aquatic environment: A european union–united states perspective on the status of ecotoxicity testing, research priorities, and challenges ahead. Environ. Toxicol. Chem. 35: 1055–1067.

Shahsavari, E., A. Aburto-Medina, L.S. Khudur, M. Taha and A.S. Ball. 2017. From microbial ecology to microbial ecotoxicology. pp. 17–38. *In*: Microbial Ecotoxicology. Springer,

Shepard, J., B. Olsson, M. Tedengren and B. Bradley. 2000. Protein expression signatures identified in *Mytilus edulis* exposed to PCBs, copper and salinity stress. Marine Environ. Res. 50: 337–340.

Skelton, D., D. Ekman, D. Martinovic-Weigelt, G. Ankley, D. Villeneuve, Q. Teng et al. 2014. Metabolomics for *in situ* environmental monitoring of surface waters impacted by contaminants from both point and nonpoint sources. Environ. Sci. Technol. 48: 2395–2403.

Snape, J.R., S.J. Maund, D.B. Pickford and T.H. Hutchinson. 2004. Ecotoxicogenomics: The challenge of integrating genomics into aquatic and terrestrial ecotoxicology. Aquat Toxicol. 67: 143–154.

Taylor, N.S., R. Merrifield, T.D. Williams, J.K. Chipman, J.R. Lead and M.R. Viant. 2016. Molecular toxicity of cerium oxide nanoparticles to the freshwater alga *Chlamydomonas reinhardtii* is associated with supra-environmental exposure concentrations. Nanotoxicology 10: 32–41.

Tedesco, S., H. Doyle, J. Blasco, G. Redmond and D. Sheehan. 2010a. Exposure of the blue mussel, *Mytilus edulis*, to gold nanoparticles and the pro-oxidant menadione. Comparative Biochemistry and Physiology Part C: Toxicology & Pharmacology 151: 167–174.

Tedesco, S., H. Doyle, J. Blasco, G. Redmond and D. Sheehan. 2010b. Oxidative stress and toxicity of gold nanoparticles in *Mytilus edulis*. Aquat. Toxicol. 100: 178–186.

Tringe, S.G. and E.M. Rubin. 2005. Metagenomics: DNA sequencing of environmental samples. Nature Rev. Genetics 6: 805.

US EPA. 2007. Nanotechnology White Paper. U.S. EPA, Washington D.C.

Vale, G., K. Mehennaoui, S. Cambier, G. Libralato, S. Jomini and R.F. Domingos. 2016. Manufactured nanoparticles in the aquatic environment-biochemical responses on freshwater organisms: A critical overview. Aquat. Toxicol. 170: 162–174.

Vance, M.E., T. Kuiken, E.P. Vejerano, S.P. McGinnis, M.F. Hochella Jr., D. Rejeski et al. 2015. Nanotechnology in the real world: Redeveloping the nanomaterial consumer products inventory. Beilstein Journal of Nanotechnology 6: 1769–1780.

Vandegehuchte, M.B., D. De Coninck, T. Vandenbrouck, W.M. De Coen and C.R. Janssen. 2010. Gene transcription profiles, global DNA methylation and potential transgenerational epigenetic effects related to Zn exposure history in *Daphnia magna*. Environ. Pollut. 158: 3323–3329.

Vandegehuchte, M.B. and C.R. Janssen. 2011. Epigenetics and its implications for ecotoxicology. Ecotoxicology 20: 607–624.

Viant, M.R. 2008. Recent developments in environmental metabolomics. Molecular Biosystems 4: 980–986.

Villeneuve, D.L., D. Crump, N. Garcia-Reyero, M. Hecker, T.H. Hutchinson, C.A. LaLone et al. 2014a. Adverse outcome pathway development II: Best practices. Toxicol. Sci. 142: 321–330.

Villeneuve, D.L., D. Crump, N. Garcia-Reyero, M. Hecker, T.H. Hutchinson, C.A. LaLone et al. 2014b. Adverse outcome pathway (AOP) development I: Strategies and principles. Toxicol. Sci. 142: 312–320.

Yao, Y. and M. Costa. 2013. Genetic and epigenetic effects of nanoparticles. J. Mol. Genet. Med. 7: 1747-0862.100008.

Zheng, X., Y. Chen and R. Wu. 2011. Long-term effects of titanium dioxide nanoparticles on nitrogen and phosphorus removal from wastewater and bacterial community shift in activated sludge. Envrion. Sci. Technol. 45: 7284–7290.

9

Nanomaterials in Aquatic Sediments

Simon Little and *Teresa F. Fernandes**

Introduction

A broad definition of sediments includes all consolidated and unconsolidated materials at the bottom of aquatic systems. These materials commonly vary in physical and chemical composition and can include but are not limited to sand, clay, silt, gravel and rocks (Palmer et al. 2000). Within this matrix, organic matter exists which confers the sediment matrix specific properties affecting interaction with environmental chemicals and organisms living therein (Shaw 2018). Both freshwater and marine sediments provide a habitat for a large range of animal and plant species. Although difficult to quantify, estimates for sediment biodiversity stand at approximately 100,000 species in freshwater (Dudgeon et al. 2006) while approximately 100,000 marine benthic species have been described, it is estimated that a further 100,000,000 remain undescribed (Snelgrove 1998).

Sediment-associated biota are integral to ecological functions including the decomposition of organic materials and nutrient cycling. Sediment reworking and bioturbation by invertebrate species via feeding and foraging have profound ecological impacts for sediments and overlying waters (McCall and Tevezs 1982), while suspension feeding can transport sediments and a variety of other material, such as seston across the sediment-water interface. The ecological functions described not only benefit immediate ecosystems but also hold global importance in relation to carbon, nitrogen and sulphur cycling (Snelgrove 1998).

Through the release of industrial and municipal wastes to the environment, sediments have experienced contamination for several decades. Sources of sediment contamination can generally be divided into two categories: point and non-point (diffuse). Point sources (i.e., wastewater effluents) are consistent in relation to flow, easy to identify and are not influenced by meteorological factors (Vink et al. 2001). Diffuse sources (surface runoff, sewer overflows), conversely, are more dynamic and challenging to predict owing to their dependence on meteorological factors. When introduced to water systems, contaminants undergo partitioning between the aqueous (pore and overlying water) and solid (sediment, suspended material and biota) phases. The solid/liquid partition coefficient (K_d) is determined by the properties of the contaminant itself and the abiotic factors of the system it is introduced to. Common sediment pollutants include nutrients (predominately phosphorus and

Institute of Life and Earth Sciences, Heriot-Watt University, Edinburgh EH14 4AS, UK.
* Corresponding author: T.Fernandes@hw.ac.uk

nitrogen compounds), bulk organics (oil and grease), halogenated hydrocarbons (DDT, PCBs, PAHs) and metals (lead, cadmium, magnesium, iron and mercury). Although each of these contaminants behave differently, they all exhibit a strong affinity for sediments and are anticipated to persist for long periods (years or decades). Due to their close association with sediments, benthic organisms (and by extension, the functioning of aquatic ecosystems) are potentially jeopardised in cases of contamination.

As with 'traditional' contaminants, the fate and behaviour of NMs are heavily influenced by the abiotic factors, such as sediment grain size and organic matter, of the media in which they are received. However, unlike other contaminants, NMs exhibit unique and specific properties such as surface charge, surface coating, size, composition and shape which further complicate the understanding of their interactions in the environment. Due to the very large diversity of potential NM properties (i.e., no two NMs are identical) and the unique ways in which they behave in different abiotic conditions, which themselves are transient, it is impossible to apply a hard and fast rule towards NM fate and behaviour. For this reason, the characterisation of NMs, particularly under environmentally relevant conditions, has been a key area of research within nano-ecotoxicology. This allows the understanding of how organisms might be exposed to NMs and the derivation of cause-effect relationships.

Aggregation and Sedimentation of Nanomaterials in Aquatic Systems

Aggregation (irreversible) and agglomeration (reversible) are key processes in understanding how NM reach freshwater and marine sediments. It is extremely difficult to distinguish between these two processes using available techniques such as dynamic light scattering (DLS), nanoparticle tracking analysis (NTA), scanning electron microscopy (SEM) or transmission electron microscopy (TEM), as such techniques only generally confirm the presence of large particle clusters (Sokolov et al. 2015). Therefore, the term aggregation will be used in this text to represent both. The balance between attractive *van der Walls* and repulsive electrical double layer (EDL) forces are the basis of the *Derjaguin-Landau-Verwey-Overbeek* (DVLO) theory which dictates NM aggregation states, and in turn, NM fate and toxicity. In natural waters where conditions are optimal, NMs will aggregate with one another (homoaggregation) predominantly via Brownian motion—the random movement of particles due to the bombardment from molecules of the surrounding media. Increased aggregation reduces the potential for NMs to be transported long distances and increases the rate at which they experience sedimentation.

Numerous studies have investigated the effects of environmental conditions on NM stability and aggregation. It is generally understood that when NMs experience a pH close to their point of zero charge (PZC) whereby electrostatic forces are reduced, NMs will undergo aggregation. Further away from the PZC, NMs will exhibit greater repulsion and experience less aggregation. Similarly, in strongly ionic media, the electrical double layer (EDL) of NMs (which determines their repulsive force), becomes more compressed, limiting repulsion and promoting aggregation.

Surface coating is also known to influence the aggregation kinetics of NMs in aquatic environments and hence the rate at which they experience sedimentation. Using bare (non-coated) and citrate-coated cerium oxide nanoparticles (CeO_2NP), Tella et al. (2015) observed reduced stability and rapid homoaggregation of bare CeO_2NP within a mesocosm study, whilst citrate-coated CeO_2NPs experienced greater stability—experiencing both homo- and heteroaggregation (aggregation with other particles) only after several days following the degradation of their citrate coating. NM concentration will also determine aggregation and sedimentation rates, whereby at higher concentrations, there is greater potential for particles to interact with one another and undergo homoaggregation.

Natural colloids, comprised of inorganic solids, organic compounds and biopolymers are ubiquitous in natural surface waters and have demonstrated their ability to influence NM sedimentation. For example, Ferry et al. (2009) observed that gold NMs within an estuarine mesocosm predominantly partitioned to biofilms, while Kiser et al. (2010) demonstrated the affinity of silica NMs towards wastewater biomass. The presence of natural organic matter (NOM), typically humic and fulvic acids, is also known to influence NM stability and thereby aggregation/sedimentation. This has been demonstrated by Delay et al. (2011) who found the stability of silver NMs to improve in typically unstable conditions (high ionic strength) via the presence of NOM. Increased stability was attributed to the adsorption of NOM molecules to the surface of NMs, effectively 'coating' them and increasing steric repulsion due to their negative charge.

Fate, Behaviour and Detection of NMs in Sediment

Once associated with sediments, NMs will undergo further physical, biological and chemical transformations (dependent on sediment characteristics), which will ultimately shape their bioavailability and toxicity. The fate and behaviour of NMs in sediments are considerably less well understood in comparison to the water column and are often extrapolated from results obtained from soil studies. A comprehensive overview of the potential fate and transformations of NMs in sediments is offered by Cross et al. (2015). A balance between the aforementioned attractive (van der Waals) and repulsive (electrostatic) forces are cited as the determining factors in NM deposition to sediment grains. Where the charge of particles and sediments and abiotic conditions are optimal (i.e., ionic strength and pH), NMs may become strongly or weakly associated with sediment grains. In instances where pore sizes are too small for large aggregates (~ 1000 nm) to pass, NM aggregates can essentially become trapped between sediment grains in a process described as 'straining' which has been observed with large TiO_2NP aggregates (Chowdhury et al. 2011). Strained aggregates can either 'collect' from or 'release' particles to the pore water, in processes termed ripening and shearing, respectively. Therefore, while straining may initially reduce the transport of aggregated NMs, the shearing of individual particles over time may serve as a longer term NM source (Chowdhury et al. 2011). Where high NM concentrations prevail, sediment surfaces may become saturated and due to the charge of NMs, block further deposition via electrostatic repulsion.

The influence of sediment NOM on NM fate and behaviour is complex, with certain components (humic and fulvic acids) reported to increase steric repulsion between AgNP and thus, mobility (Cornelis et al. 2013) while NOM rich soils have reported to limit the mobility of AgNPs in comparison to mineral rich soils (Coutris et al. 2011). Conflicting studies also exist concerning the influence of NOM on NM dissolution—the chemical process whereby NMs release potentially highly toxic ions. Liu and Hurt (2010) reported a decrease in AgNP dissolution in increasing presence of humic or fulvic acids; however, greater dissolution of zinc oxide NPs (ZnO NPs) has been observed in soils containing higher organic matter (Waalewijn-Kool et al. 2014). As the dissolution of NMs is often related to their toxicity—and aids in distinguishing between ion and nano-specific effects, understanding the processes which influence dissolution in complex media is of upmost importance.

Due to the transportation and transformation of NMs in the environment, detecting or predicting their concentration and quantifying their characteristics in natural media, particularly sediment, is extremely difficult. A host of techniques are available to measure the mass concentration of NMs in sediments; for example, inductively coupled mass spectrometry (ICP-MS), atomic adsorption spectrophotometer (AAS), liquid and gas chromatography-mass spectrometry (LC-MS and GC-MS) and inductively coupled atomic emission spectroscopy (ICP-AES). AAS and ICP-MS were employed by Lowry et al. (2012) to quantify the migration of over 60 percent of AgNPs introduced to a mesocosm via water columns to sediments after 18 months. While such techniques are vital within ecotoxicology, they are not able to elucidate NM-specific properties such as particle size distribution or particle size concentration. As NMs are anticipated to potentially exhibit nano-specific forms of toxicity (i.e., related to size), this information is crucial in terms of their ecotoxicological effects on sediments and decision-making in the formation of risk assessments.

Furthermore, size measurement and distribution techniques such as electron microscopy and DLS are not without limitations, particularly in complex matrices such as sediments. Sample preparation is often complicated for electron microscopy and can also introduce external artefacts, potentially compromising measurements, while measurements of hydrodynamic diameter are not possible. DLS offers ease and efficiency; however, it provides a narrow measurement of size distribution (2–500 nm) and struggles at lower, more environmentally relevant concentrations and is prone to skewing results in the presence of larger aggregates (Tomaszewska et al. 2013).

By separating NMs according to particle size prior to analytical measurement, typically using ICP-MS, field flow fractionation (FFF) has been advocated in the measurement of NMs in complex media. Similarly, single-particle ICPS-MS is able to detect and measure individual NPs in complex samples. Both techniques have been used in combination for the detection of NMs in complex matrices at low concentrations as demonstrated by El Hadri and Hackley (2017) who measured AgNPs in spiked soils at a concentration of 10–5000 μg/kg. Moreover, Poda et al. (2011) were able to use FFF-ICP-MS to measure and characterise AgNPs ingested by the freshwater sediment worm, *Lumbriculus variegatus* following a two week exposure to 100 μg/kg spiked sediments.

Nanoparticle tracking analysis (NTA)—a laser-based technique which combines the properties of light scattering and Brownian motion to calculate particle size distribution is typically applied to aquatic samples; however, its application for NM analysis in sediments has recently been investigated. Luo et al. (2017) were able to measure the particle size distribution and particle concentration of gold and magnetite NMs spiked to natural sediments (Derwent River, York, UK). Within the same study, Luo et al. (2017) also underline the importance played by NM coatings in determining their affinity with sediments. NMs with positively-charged coatings interacted with sediments to a greater degree than those with a negative or neutral charge.

When combined, the aforementioned techniques offer insight into the concentrations and characteristics of NMs in sediments; however, they are not yet fully optimised. Taking this into consideration and in the absence of routine NM monitoring in sediments for regulatory purposes, it is necessary to rely on models to generate predicted environmental concentrations (PECs) for NMs in sediments. Using available data related to worldwide annual production, potential release pathways, NM characterisation and sedimentation, Sun et al. (2014) modelled PECs for a variety of NMs (titanium dioxide, zinc oxide, silver, carbon nanotubes and fullerenes) in different environmental compartments (surface water, soil, sediment and air). Results concluded that the highest concentration of NMs were anticipated to be deposited in sediments with concentrations typically within the low mg/kg or in most cases, the μg/kg range. Similar models ran by Garner et al. (2017) also anticipate the highest concentrations of NMs (CeO_2, CuO, TiO_2, and ZnO) to be highest in freshwater and marine sediments, along with agricultural land receiving biosolids,and display a potential to increase over time. While the output from such models are by no means comprehensive, owing to the fact that there still remain large data gaps for NM characterisation, ecotoxicity and environmental transformation in freshwater and marine sediments, the results from Garner et al. (2017) nevertheless highlight the potential risk posed to sediments and are a useful tool on which to base risk assessments.

Ecotoxicity of NMs in Sediments and the Applicability of Current Protocols

As with their fate and behaviour in the environment, the mode through which NMs exhibit toxicity has been linked to their unique size and physicochemical attributes. Sediment biota, or those at the sediment-water interface, are at risk of accumulating potentially high concentration of sediment-associated NMs, either dermally (e.g., via bioturbation) or through ingestion, especially in the case of non-selective or filter-feeders. The predation of such species (predominantly), by fish, enables the potential trophic transfer of NMs, whilst overall changes in sediment species diversity or density could incur wider ecological implications. Taking this into consideration, understanding the ecotoxicological effects of NMs on sediment species is of upmost importance, especially as the latter are anticipated to be major recipients.

Although the toxicological effects of NMs towards humans have been researched for almost 30 years, comparatively, ecotoxicological studies are still in their infancy. Furthermore, within the ecotoxicological testing of NMs, there is often a bias towards acute, aquatic tests using pelagic organisms. This bias has recently been brought to light by Selck et al. (2016) who estimated there to be approximately 31 times less sediment studies in the available literature in comparison with water column studies. A tendency towards the testing of benthic species in water only exposures is also evident. As many benthic species inhabit the sediment-water interface, these studies still provide valuable data; however, the use of sediments increases environmental relevance and hence the comprehension of the behaviour of NMs in the environment.

The ecotoxicity of particular substances in their bulk form are reasonably well understood; however, it cannot be assumed that the same element in the nano-form will exhibit the same level of ecotoxicity, nor the same mode of action. To this end, it is also important to consider the applicability of current sediment testing protocols. Several sediment toxicity and protocols currently exist (OECD 315, OECD 225, OECD 218, OECD 219, ASTM E1611, ASTM E1706-05 and ASTM E1367-03) and while there is a consensus that they are generally applicable for use with NMs, it is important that where possible, they can be optimised or improved for use with NMs (Handy et al. 2012a,b).

More recently, protocol applicability was assessed by Hund-Rinke et al. (2016) who used a range of protocols set by the Organisation for Economic Cooperation and Development (OECD) using two common NMs, silver and titanium dioxide. OECD test guideline 225 (sediment-water *Lumbriculus variegatus* toxicity test using spiked sediment) was investigated and while some practical alterations were suggested (i.e., method for the spiking of sediments), others, such as the lowering of organic matter content in formulated sediments (to represent more environmentally representative conditions) could have greater implications given the reactive nature of NMs. By standardising and adapting regulatory sediment toxicity tests for use with NMs, greater confidence can be applied to hazard assessments and in theory, assisting in the identification of potential environmental risks.

Ecotoxicity of NMs Towards Sediment Prokaryotes

Determining the potential ecotoxicological effects of NMs on prokaryotic sediment communities is a key area of research as they comprise the bulk of sediment biomass. Of particular interest is the effects of silver NMs given their widespread production (400 tonnes per annum—PEN (2016)) primarily for their use in medical and commercially available products due to their antibacterial properties. Despite this, Bradford et al. (2009) found that the bacterial assemblage of an estuarine sediment exposed to AgNPs for 30 d experienced no or little changes in diversity (in sediment) or abundance (in water) up to a concentration of 1000 µg/l of the overlying water. Mühling et al. (2009) also found AgNPs to have no influence on the antibiotic resistance of the naturally occurring sediment bacteria, despite well-documented incidents of such following mesocosm experiments with heavy metals. These authors suggested that the effects of AgNPs towards sediment bacteria are dependent on abiotic factors that may 'deactivate' their antibacterial properties.

Colman et al. (2012) also found minimal effects of AgNPs on sediment bacteria—with only a 14 percent decrease in sulphate concentration and no significant impact on biomass, respiration or community composition. These results contrast with the same experiments conducted with Ag in its dissolved form ($AgNO_3$) whereby respiration (34%) and biomass (55%) were significantly reduced, while shifts in community composition and changes in enzyme activity were detected. Colman et al. (2012) attributed the lack of observed toxicity to limited Ag dissolution from AgNPs in sediments and increased aggregation, both of which are influenced by NM-specific properties and abiotic conditions.

Conversely, Beddow et al. (2017) observed significant effects towards ammonia-oxidising bacteria (AOB) following the exposure of estuarine sediments to AgNPs. Using sediments collected from sites along the Colne estuary, United Kingdom, mesocosm experiments revealed AgNPs (PEG-coated, 50 mg/l) significantly decreased AOB ammonia monooxygenasegene abundance and the nitrification potential rates (NPR) of bacteria in a low-salinity and mesohalite sediment after 7 and 14 d. These results are particularly pertinent as ammonia-oxidation is the first and rate-limiting stage in nitrification and thus, their inhibition could have major implications on nitrogen cycling in aquatic systems.

Although Miao et al. (2017) observed no effect on the oxygen consumption of benthic microbes in mesocosms containing freshwater sediment (Taihu Lake, China) following exposure to PVP-coated AgNPs at 1 mg/l, significant inhibition was recorded following exposure to uncoated AgNPs at 10 mg/l.

The disparity between the AgNP studies presented underline the uncertainty surrounding the ecotoxicological effects of NMs towards sediment bacteria. While minimal effects may be assumed from the studies of Bradford et al. (2009), Mühling et al. (2009) and Colman et al. (2012), Beddow et al. (2017) and Miao et al. (2017) clearly demonstrate the potential for AgNPs to significantly affect sediment bacteria. While Beddow et al. (2017) used concentrations far greater than Bradford et al.(2009) and Mühling et al. (2009) (50 mg/l compared to 1 mg/l), the concentrations were comparable to those used by Colman et al. (2012) (75 mg/l) where no adverse effects were observed.

Aside from AgNPs, adverse effects on microbial sediment communities have also been recorded for titanium dioxide (TiO_2) and cerium dioxide (CeO_2) NPs. Using mesocosm experiments containing freshwater sediment from Taihu Lake, China and 5 mg/l of each NP, Miao et al. (2018) recorded rapid sedimentation, with up to 85 percent accumulated on the superficial sediment after 5d. Furthermore, significant inhibition of the oxygen flux at the sediment-water interface was recorded, with the oxygen consumption of sediment microbial communities also inhibited. The observed inhibition of oxygen consumption was linked to a reduction in the aerobic bacteria *Methylotenera* and *Cytophagceae.* Conversely, the diversity of anaerobic bacteria increased, altering the overall microbial composition of sediments. These results are of ecological significance as the biogeochemical cycling of elements commonly occurs at the sediment-water interface and heavily relies on benthic microbial communities.

There is, therefore, some controversy surrounding the degree to which NMs will affect benthic bacteria or whether they will exert any influence at all. There is also evidence presented by Moore et al. (2016) to suggest that while NMs (silver and copper) may exert short-term effects on microbial communities, long-term impacts (monitored after 300 d) may be limited. Similarly, Ozaki et al. (2016) found that a one-time application of TiO_2NPs (P25, 1 mg/l) initiated a decline in bacterial abundance and increases in respiration and denitrification enzyme activity after 1 day but found levels to return to that of controls after 3 weeks with no significant changes in community composition. The habitat of bacteria within the sediment must also be considered when conducting ecotoxicological tests. As highlighted by Miao et al. (2018), it is anticipated that when NMs undergo aggregation and sedimentation in aquatic environments, surficial sediments will be the major recipient. Due to the potential size of NM aggregates, their reworking into deeper sediment layers might be limited, and thus deeper-lying bacteria have less potential for exposure. Surficial sediment bacteria may, therefore offer a more sensitive and environmentally relevant gauge of NM toxicity in sediments. Conflicting results highlight the complexity of NM and abiotic variables that must be considered and reinforce the need for further research.

Ecotoxicity of NMs Towards Benthic Invertebrates

Higher taxa organisms at the sediment-water interface play important roles in aquatic trophic food webs and are integral in sediment bioturbation and thus, the oxygenation of subsurface pore waters. Feeding on particulate matter from surficial sediments and resuspended sediments, they are prone to sediment-associated contamination and are commonly used within standardised testing. Several studies have investigated the effects of NMs on freshwater and marine amphipods, polychaetes, oligochaetes, molluscs and copepods; however, a number of these are based on exposures in water only. By comparison, whole sediment or microcosm/mesocosm studies are limited for NMs partly due to their cost and complexity.

Ten-day sediment exposures have demonstrated the ability of copper and zinc oxide NMs to bioaccumulate and exert toxicity in the estuarine amphipod, *Leptocheirus plumulosus* (Hanna et al. 2013). Although toxicity was incurred at relatively high concentrations (LC_{50} values of 868 and 763 µg/g for CuO and ZnO, respectively), the linear increase in metal bioaccumulation with NM sediment concentration indicates the potential transfer of NMs through trophic levels given that amphipods are prey for numerous invertebrate, fish and mammal species.

In addition to traditional measures of toxicity such as mortality, the potential effects on the behaviour of sediment-dwelling species, commonly burrowing and feeding behaviour are often employed within toxicity testing. In the absence of mortality, changes to key organism behaviours can be used as a more sensitive early-stage warning towards NM toxicity. Limited or delayed burrowing may lead to increased predation and limit sediment bioturbation, while changes in feeding behaviours may also have wider implications upon nutrient cycling.

Sediment exposures of NMs have led to significant effects on the feeding behaviours of the polychaete *Arenicola marina* (TiO_2, 23.2 nm, 1 g/kg) (Galloway et al. 2010) and the freshwater oligochaete, *Lumbriculus variegatus* (Rajala et al. 2016) using AgNP (30–50 nm, 1098 mg/kg). Burrowing has been seen to be inhibited following AgNP (20 and 80 nm) sediment exposure in *Nereis diversicolor* (Cong et al. 2014) as well as CuONP (100 nm, 150 μg/g) sediment exposuresas reported by Thit et al. (2015). Although Buffet et al. (2012) also observed inhibitions in the burrowing behaviour of the mollusc *Scrobicularia plana* and polychaete *Hediste diversicolor* exposed to sediments spiked with ZnONP (21–34 nm, 3 mg/kg), results could not be attributed to NPs alone due to the observed influence of the stabiliser, diethylene glycol (DEG).

As with behavioural endpoints, subcellular organism effects offer a more sensitive endpoint within sediment ecotoxicology and help elucidate the mode through which NMs exhibit toxicity and whether said modes are nano-specific. NM toxicity is commonly associated with the onset of oxidative stress. Consequently, the activity of defence parameters (predominantly antioxidant enzymes) and damage associated with oxidative stress (e.g., lipid, protein, lysosome and DNA damage) are commonly employed as toxicity biomarkers.

In relation to antioxidant enzymes, Buffet et al. (2012) witnessed increased activity of glutathione-S_tranferase (GST) in *H. diversicolor* and catalase (CAT) in *S. plana* following ZnONP (21–24 nm) exposure, while no changes in superoxide dismutase (SOD) or lipid peroxidation were detected. Short-term (< 14 day) sediment exposure of the freshwater gastropod *Bellamya aeruginosa* to CuONP (40 nm, 180 μg/g) resulted in significantly increased activities of SOD, CAT and GST while long-term (28 day) exposure suggested oxidative damage in the form of lipid peroxidation (Ma et al. 2017).

A concentration-dependent increase in lysosomal membrane permeability and DNA damage were reported in the coelomocytes of *H. diversicolor* exposed to AgNP-spiked sediments (20 and 80 nm, PVP coated, 0–100 μg/g sediment) by Galloway et al. (2010). Despite comparable Ag bioaccumulation between AgNP and soluble Ag-exposed *N. diversicolor*, significantly greater toxicity in relation to lysosomal stability and DNA damage were incurred following AgNP exposure, suggesting a nano-specific toxicity. Detrimental effects on lysosomal stability and DNA damage have also been recorded in the coelomocytes of *A. marina* exposed to TiO_2NP-spiked sediments (23.2 nm, 1 g/kg) (Galloway et al. 2010).

Microcosm and Mesocosm Sediment Studies

Sediment-spiked exposures allow a more standardised approach to NM toxicity testing, collecting data with greater environmental relevancy than exposures conducted with benthic species in water alone. As previously indicated, such studies are integral in determining the bioaccumulation and toxicity of NMs in sediments; however, the use of microcosm and mesocosm experiments are considered means to further improve environmental relevance, given the use of multiple species, and a potentially greater likeness of environmental abiotic conditions.

Using a multi-species, outdoor mesocosm to investigate the effects of AgNPs and $AgNO_3$ (both at 10 μg/l), Buffet et al. (2014) recorded a range of sub-lethal effects towards the marine polychaete *H. diversicolor* and the mollusc *S. plana*. After 21 d, significant Ag bioaccumulation in both species was measured alongside changes in defences against oxidative stress, apoptosis, detoxification, genotoxicity and immuno-modulation. Whilst these changes were attributed at least in part to the dissolution of AgNPs and the release of soluble Ag, DNA damage incurred in the digestive gland of *S. plana* and changes in phenoloxidase and lysosome activities in both species were deemed to be nano-specific. The results from this study are of interest due to their increased environmental relevance in terms of the abiotic conditions, the multi-species nature of the exposure and the low NP concentration selected, while the comparison with a dissolved Ag form provides an opportunity to identify nano-specific effects. Silver NPs have also been incorporated into microcosm experiments by Jiang et al. (2017) who cited surficial sediments as the main sink for AgNPs (20 nm, PVP) and demonstrated the ability of benthic species (*Radix* spp.) to accumulate Ag and thus play an important role in their potential trophic transfer.

Ferry et al. (2009) have also demonstrated the ability to monitor the trophic transfer of NMs using multi-species mesocosms. Gold NPs (65 nm) applied to the mesocosm were measured in the water, sediment, biofilm and within several species included in the mesocosm. Amongst the species studied, the benthic mollusc, *Mercenaria mercenaria* accumulated approximately 5 percent of the AuNPs added despite only accounting for 0.01 percent of the total mass of the mesocosm. These results highlight benthic feeders as potential sinks for NMs and given the commercial importance of shellfish for human consumption, offer a potential route for NMs into the human food chain.

Mesocosms have also been employed by Tella et al. (2014 and 2015) to study the effects of NM surface coating on the bioavailability of CeO_2NPs towards the benthic gastropod, *Planorbarius corneus*. As *P. corneus* are benthic grazers, their exposure to, and likelihood of ingesting NMs in aquatic environments is dependent on their settling kinetics and resultant concentration at the sediment-water interface. Tella et al. (2014 and 2015) observed a slower sedimentation rate of coated particles and whilst coating had no effect on ultimate sediment or *P. corneus* NM concentrations, greater sub-lethal toxicity (lipid peroxidation) was observed in *P. corneus* exposure to bare NMs, suggesting the presence of defensive capabilities to NM coatings. These studies highlight the multifaceted analysis capable via mesocosms in more environmentally relevant scenarios.

The negative buoyancy of a number of planktonic and pelagic larvae result in them spending a significant amount of their life-cycle at the sediment-water interface where they are prone to sediment-associated contamination (Picone et al. 2016). Developmental-stage organisms are potentially more sensitive targets of contaminants; however, preference is commonly given to adult amphipod, oligochaete and polychaete species in sediment toxicity testing. Larval development experiments are commonly employed using *Danio rerio*, in aqueous exposures; however, sediment larval development assays are seldom investigated within NM ecotoxicological research. The use of sediment-based larval development toxicity tests would provide valuable chronic data relating to the sub-lethal effects of NM exposure at the sediment-water interface and potentially offer greater relevancy compared to traditional lethality-based tests.

Summary

The relationship between NMs and sediments is highly complex given the number of variables to be considered. It is clear that NM transformations, initiated both by their unique characteristics and the abiotic conditions in which they are received will determine the rate at which they experience sedimentation, how they behave and are transported within sediments and ultimately, their toxic potential towards sediment biota. The continued development and optimisation of analytical tools for the characterisation of NMs in sediment will enhance understanding of their transformations in complex matrices and in turn, their fate and bioavailability. Despite a unison between fate and behaviour models which overwhelmingly consider sediments to be the main recipient of nano-wastes throughout their life-cycle, there is a disparity in the amount of environmentally relevant sediment toxicity data available for NMs. Of the available studies, there is evidence to suggest that benthic species are susceptible to NM accumulation and resultant toxic effects in the form of mortality, behavioural inhibition, the onset of oxidative stress and associated damage towards proteins, lipids and DNA. Toxicity towards benthic species and the potential NM trophic transfer could significantly alter the composition of aquatic ecosystems and consequently impair their functioning. Continuing research on the effects of NMs on sediments will ultimately aid the development of risk assessments to protect the environment and enable the sustainable development of nanotechnologies.

References

ASTM. 1995. Standard test methods for measuring the toxicity of sediment-associated contaminants with fresh water invertebrates. Philadelphia: ASTM.

ASTM. 2003. Standard test method for measuring the toxicity of sediment-associated contaminants with estuarine and marine invertebrates. Annual Book of ASTM Standards, 444–505.

ASTM. 2007. Standard guidelines for conducting sediment toxicity tests with polychaetous annelids. ASTM International, West Conshohocken.

Beddow, J., B. Stolpe, P.A. Cole, J.R. Lead, M. Sapp, B.P. Lyons et al. 2017. Nanosilver inhibits nitrification and reduces ammonia-oxidising bacterial but not archaeal amoA gene abundance in estuarine sediments. Environ. Microbiol. 19(2): 500–510.

Buffet, P.E., C. Amiard-Triquet, A. Dybowska, C. Risso-de Faverney, M. Guibbolini, E. Valsami-Jones et al. 2012. Fate of isotopically labeled zinc oxide nanoparticles in sediment and effects on two endobenthic species, the clam *Scrobicularia plana* and the ragworm *Hediste diversicolor*. Ecotoxicol. Environ. Saf. 84: 191–198.

Buffet, P.E., A. Zalouk-Vergnoux, A. Châtel, B. Berthet, I. Métais, H. Perrein-Ettajani et al. 2014. A marine mesocosm study on the environmental fate of silver nanoparticles and toxicity effects on two endobenthic species: The ragworm *Hediste diversicolor* and the bivalve mollusc *Scrobicularia plana*. Sci. Total Environ. 470: 1151–1159.

Bradford, A., R.D. Handy, J.W. Readman, A. Atfield and M. Mühling. 2009. Impact of silver nanoparticle contamination on the genetic diversity of natural bacterial assemblages in estuarine sediments. Environ. Sci. Technol. 43(12): 4530–4536.

Chowdhury, I., Y. Hong, R.J. Honda and S.L. Walker. 2011. Mechanisms of TiO_2 nanoparticle transport in porous media: role of solution chemistry, nanoparticle concentration, and flowrate. J. Colloid Interface Sci. 360: 548. doi:10.1016/J.JCIS.2011.04.111.

Colman, B.P., S.Y. Wang, M. Auffan, M.R. Wiesner and E.S. Bernhardt. 2012. Antimicrobial effects of commercial silver nanoparticles are attenuated in natural streamwater and sediment. Ecotoxicology 21(7): 1867–1877.

Cong, Y., G.T. Banta, H. Selck, D. Berhanu, E. Valsami-Jones and V.E. Forbes. 2014. Toxicity and bioaccumulation of sediment-associated silver nanoparticles in the estuarine polychaete, *Nereis (Hediste) diversicolor*. Aquat. Toxicol. 156: 106–115.

Cornelis, G., L. Pang, C. Doolette, J.K. Kirby and M.J. McLaughlin. 2013. Transport of silver nanoparticles in saturated columns of natural soils. Sci. Total Environ. 463: 120–130.

Coutris, C., T. Hertel-Aas, E. Lapeid, E.J. Joner and D.H. Oughton. 2011. Bioavailability ofcobalt and silver nanoparticles to the earthworm *Eiseniafetida*. Nanotoxicology 6(2): 186–195.

Cross, R.K., C. Tyler and T.S. Galloway. 2015. Transformations that affect fate, form and bioavailability of inorganic nanoparticles in aquatic sediments. Environ. Chem. 12(6): 627–642.

Delay, M., T. Dolt, A. Woellhaf, R. Sembritzki and F.H. Frimmel. 2011. Interactions and stability of silver nanoparticles in the aqueous phase: Influence of natural organic matter (NOM) and ionic strength. J. Chromotogr. A 1218(27): 4206–4212.

Dudgeon, D., A.H. Arthington, M.O. Gessner, Z.I. Kawabata, D.J. Knowler, C. Lévêque et al. 2006. Freshwater biodiversity: importance, threats, status and conservation challenges. Biol. Rev. 81(2): 163–182.

El Hadri, H. and V.A. Hackley. 2017. Investigation of cloud point extraction for the analysis of metallic nanoparticles in a soil matrix. Environ. Sci. Nano 4(1): 105–116.

Ferry, J.L., P. Craig, C. Hexel, P. Sisco, R. Frey, P.L. Pennington et al.. 2009. Transfer of gold nanoparticles from the water column to the estuarine food web. Nat. Nanotechnol. 4(7): 441.

Galloway, T., C. Lewis, I. Dolciotti, B.D. Johnston, J. Moger and F. Regoli. 2010. Sublethal toxicity of nano-titanium dioxide and carbon nanotubes in a sediment dwelling marine polychaete. Environ. Pollut. 158(5): 1748–1755.

Garner, K.L., S. Suh and A.A. Keller. 2017. Assessing the risk of engineered nanomaterials in the environment: Development and application of the nanoFate model. Environ. Sci. Technol. 51(10): 5541–5551.

Handy, R.D., G. Cornelis, T.F. Fernandes, O. Tsyusko, A. Decho, T. Sabo-Attwood et al. 2012. Ecotoxicity test methods for engineered nanomaterials: Practical experiences and recommendations from the bench. Environ. Toxicol. Chem. 31(1): 15–31.

Handy, R.D., N. van den Brink, M. Chappell, M. Muhling, R. Behra, M. Dusinska et al. 2012. Practical considerations for conducting ecotoxicity test methods with manufactured nanomaterials: What have we learned so far? Ecotoxicology 21(4): 933–72.

Hanna, S.K., R.J. Miller, D. Zhou, A.A. Keller and H.S. Lenihan. 2013. Accumulation and toxicity of metal oxide nanoparticles in a soft-sediment estuarine amphipod. Aquat. Toxicol. 142: 441–446.

Hund-Rinke, K., A. Baun, D. Cupi, T.F. Fernandes, R. Handy, J.H. Kinross et al. 2016. Regulatory ecotoxicity testing of nanomaterials–proposed modifications of OECD test guidelines based on laboratory experience with silver and titanium dioxide nanoparticles. Nanotoxicology 10(10): 1442–1447.

Jiang, H.S., L. Yin, N.N. Ren, L. Xian, S. Zhao, W. Li et al. 2017. The effect of chronic silver nanoparticles on aquatic system in microcosms. Environ. Pollut. 223: 395–402.

Kiser, M.A., H. Ryu, H. Jang, K. Hristovski and P. Westerhoff. 2010. Biosorption of nanoparticles to heterotrophic wastewater biomass. Water Res. 44(14): 4105–4114.

Lead, J.R., G.E. Batley, P.J.J. Alvarez, M.-C. Croteau, R.D. Handy, M.J. McLaughlin et al. 2018. Nanomaterials in the environment: Behavior, fate, bioavailability, and effects—an updated review. Environ. Toxicol. Chem. 37: 2029–2063.

Liu, J. and R.H. Hurt. 2010. Ion release kinetics and particle persistence in aqueous nano-silver colloids. Environ. Sci. Technol. 44(6): 2169–2175.

Lowry, G.V., B.P. Espinasse, A.R. Badireddy, C. Richardson, B. Reinsch, L. Bryant et al. 2012. Long-term transformation and fate of manufactured Ag nanoparticles in asimulated large scale freshwater emergent wetland. Environ. Sci. Technol. 46(13): 7027–7036.

Luo, P., A. Roca, K. Tiede, K. Privett, J. Jiang, J. Pinkstone et al. 2017. Application of nanoparticle tracking analysis for characterising the fate of engineered nanoparticles in sediment-water systems. J. Environ. Sci. 64: 62–71.

Ma, T., S. Gong and B. Tian. 2017. Effects of sediment-associated CuO nanoparticles on Cu bioaccumulation and oxidative stress responses in freshwater snail *Bellamya aeruginosa*. Sci. Total Environ. 580: 797–804.

McCall, P.L. and M.J. Tevesz. 1982. The effects of benthos on physical properties of freshwater sediments. Animal-Sediment Relations 105–176.

Miao, L., C. Wang, J. Hou, P. Wang, Y. Ao, Y. Li et al. 2017. Influence of silver nanoparticles on benthic oxygen consumption of microbial communities in freshwater sediments determined by microelectrodes. Environ. Pollut. 224: 771–778.

Miao, L., P. Wang, C. Wang, J. Hou, Y. Yao, J. Liu and Z. Liu. 2018. Effect of TiO_2 and CeO_2 nanoparticles on the metabolic activity of surficial sediment microbial communities based on oxygen microelectrodes and high-throughput sequencing. Water Res. 129: 287–296.

Moore, J.D., J.P. Stegemeier, K. Bibby, S.M. Marinakos, G.V. Lowry and K.B. Gregory. 2016. Impacts of pristine and transformed Ag and Cu engineered nanomaterials on surficial sediment microbial communities appear short-lived. Environ. Sci. Technol. 50(5): 2641–2651.

Mühling, M., A. Bradford, J.W. Readman, P.J. Somerfield and R.D. Handy. 2009. An investigation into the effects of silver nanoparticles on antibiotic resistance of naturally occurring bacteria in an estuarine sediment. Mar. Environ. Res. 68(5): 278–283.

Organisation for Economic Co-operation and Development. 2004. Test No. 218: Sediment-Water Chironomid Toxicity Using Spiked Sediment. OECD Publishing.

Organisation for Economic Co-operation and Development). 2004. Test No. 219: Sediment-Water Chironomid Toxicity Using Spiked Water, Section 2.

Organisation for Economic Co-operation and Development. 2007. Test No. 225: Sediment-Water Lumbriculus Toxicity Test Using Spiked Sediment. OECD Publishing.

Organisation for Economic Co-operation and Development. 2008. Test No. 315: Bioaccumulation in Sediment-dwelling Benthic Oligochaetes. OECD Publishing.

Ozaki, A., E. Adams, C.T.T. Binh, T. Tong, J.F. Gaillard, K.A. Gray et al. 2016. One-time addition of nano-TiO_2triggers short-term responses in benthic bacterial communities in artificial streams. Microb. Ecol. 71(2): 266e275.

Palmer, M., A. Covich, B. Finlay, J. Gilbert, K. Hyde, R. Johnson et al. 1997. Biodiversity and ecosystem processes in freshwater sediments. Ambio 571–577.

PEN. 2016. The Woodrow Wilson International Center for Scholars, Project on Emerging Nanotechnologies website.URL www.nanotechproject.org/.

Picone, M., M. Bergamin, C. Losso, E. Delaney, A. Arizzi Novelli and A. Volpi Ghirardini. 2016. Assessment of sediment toxicity in the Lagoon of Venice (Italy) using a multi-species set of bioassays. Ecotox. Environ. Saf. 123: 32–44.

Poda, A.R., A.J. Bednar, A.J. Kennedy, A. Harmon, M. Hull, D.M. Mitrano et al. 2011. Characterization of silver nanoparticles using flow-field flow fractionation interfaced to inductively coupled plasma mass spectrometry. J. Chromotogr. A 1218(27): 4219–4225.

Rajala, J.E., K. Mäenpää, E.R. Vehniäinen, A. Väisänen, J.J. Scott-Fordsmand, J. Akkanen et al. 2016. Toxicity testing of silver nanoparticles in artificial and natural sediments using the benthic organism *Lumbriculus variegatus*. Arch. Environ. Contam. 71(3): 405–414.

Selck, H., R.D. Handy, T.F. Fernandes, S.J. Klaine and E.J. Petersen. 2016. Nanomaterials in the aquatic environment: A European Union–United States perspective on the status of ecotoxicity testing, research priorities, and challenges ahead. Environ. Toxicol. Chem. 35(5): 1055–1067.

Shaw, E. 2018. How does sediment affect the ecosystem? Sciencing, https://sciencing.com/sediment-affect-ecosystem-6772.html.

Snelgrove, P.V. 1998. The biodiversity of macrofaunal organisms in marine sediments. Biodiverse. Conserv. 7(9): 1123–1132.

Sokolov, S.V., Kristina Tschulik, Christopher Batchelor-McAuley, Kerstin Jurkschat and G. Richard. 2015. Compton Reversible or not? Distinguishing agglomeration and aggregation at the nanoscale. Anal. Chem. 87(19): 10033–10039.

Sun, T.Y., D.M. Mitrano, N.A. Bornhöft, M. Scheringer, K. Hungerbühler and B. Nowack. 2017. Envisioning nano release dynamics in a changing world: using dynamic probabilistic modeling to assess future environmental emissions of engineered nanomaterials. Environ. Sci. Technol. 51(5): 2854–2863.

Tella, M., M. Auffan, L. Brousset, J. Issartel, I. Kieffer, C. Pailles et al. 2014. Transfer,transformation, and impacts of ceria nanomaterials in aquatic mesocosms simulating a pond ecosystem. Environ. Sci. Technol. 48: 9004–9013.

Tella, M., M. Auffan, L. Brousset, E. Morel, O. Proux, C. Chanéac et al. 2015. Chronic dosing of a simulated pond ecosystem in indoor aquatic mesocosms: fate and transport of CeO_2 nanoparticles. Environ. Sci, Nano 2(6): 653–663.

Thit, A., G.T. Banta and H. Selck. 2015. Bioaccumulation, subcellular distribution and toxicity of sediment-associated copper in the ragworm *Nereis diversicolor*: The relative importance of aqueous copper, copper oxide nanoparticles and microparticles. Environ. Pollut. 202: 50–57.

Tomaszewska, E., K. Soliwoda, K. Kadziola, B. Tkacz-Szczesna, G. Celichowski, M. Cichomski et al. 2013. Detection limits of DLS and UV-Vis spectroscopy in characterization of polydisperse nanoparticles colloids. J. Nanomater. 2013: 60.

Vink, R., H. Behrendt and W. Salomons. 2001. Present and future quality of sediments in the Rhine Catchment Area–heavy metals. Dredged Material in the Port of Rotterdam–Interface between Rhine Catchment Area and North Sea. GKSS Research Centre, Geesthacht, Germany 23–85.

Waalewijn-Kool, P.L., S. Rupp, S. Lofts, C. Svendsen and C.A. van Gestel. 2014. Effect of soil organic matter content and pH on the toxicity of ZnO nanoparticles to *Folsomia candida*. Ecotox. Environ, Safe. 108: 9–15.

10

The Role of Ecotoxicology in the Eco-Design of Nanomaterials for Water Remediation

*Ilaria Corsi** and *Giacomo Grassi*

Introduction

The increasing and rapid deterioration and degradation of the water quality is one of the most challenging issue to be faced in the 21st century. It can be ascribed to a series of factors such as the population growth, the effects of climate change on the hydrologic cycle and the increasing pollution. Although there is a growing interest in nanotechnological solutions for water pollution remediation, with significant economic investments worldwide, environmental and human risk assessment associated with the use of engineered nanomaterials or nanoparticles (ENM/Ps) is still a matter of debate and nanoremediation is even now considered as an emerging technology (Karn et al. 2009).

Nanotechnology has emerged as a robust and efficient technology that overcomes the limits of existing processes, due to the tunable properties and outstanding features of ENMs (e.g., high reactivity and surface area), which make them able to quickly and efficiently remove a wide spectra of hazardous water pollutants (Otto et al. 2008, Grieger et al. 2015). However, the ecological risks associated with nanoremediation have only just begun to be assessed and more emphasis need to be addressed on environmentally relevant testing strategies and conditions (Grieger et al. 2010, Sánchez et al. 2011, Trujillo-Reyes et al. 2014, Patil et al. 2016). In this context, the pursuit of efficient, ecologically safe and sustainable nanotechnologies for water remediation represents a major breakthrough for different stakeholders, that tackle cleanup/remediation tasks, either from private and public sectors (Holland et al. 2011, Grieger et al. 2012). Therefore, as the potential and efficacy of nanotechnology for wastewater treatment is well established, several drawbacks related to the full-scale application should be overcome. In this regard, more efforts should be placed at developing innovative, green and sustainable nanosolutions for water pollution remediation, which their own ecosafe features that protect against hazards to humans and wildlife (Corsi et al. 2014, 2018a). Further efforts in developing ad hocrisk assessment strategies, which are able to assess ENMs remediation effectiveness together with environmental safety and economic sustainability within the context of existing environmental regulations, are therefore strongly needed.

Department of Physical, Earth and Environmental Sciences, University of Siena, via Mattioli, 4, Siena, Italy.
* Corresponding author: ilaria.corsi@unisi.it

The aim of the present chapter is to describe how ecotoxicology can be used for environmental risk assessment of ENMs designed for water nanoremediation by providing support in the design of more ecologically sound nanomaterials as well as in developing strategies for ecosafe and sustainable nanoremediation.

Nanoremediation: Advantages and Limitations

To date, extensive research has been performed to address urgent environmental issues, such as the extensive pollution of water bodies by both organic and inorganic contaminants. Nanotechnology has significantly contributed to remarkable industrial and societal changes over the last decades and among the dazzling variety of fields of its application, considerable efforts have been devoted in exploiting the potential of ENMs for environmental remediation, commonly referred to as nanoremediation (Vaseashta et al. 2007).

Compared to conventional *in situ* remediation techniques, known to be expensive, partially effective and time-consuming, nanoremediation has emerged as a new clean up technology that is less costly, more effective and environmentally and socially sustainable (Otto et al. 2008, USEPA 2013, EEA 2014). In fact, nanotechnologies allow for the treatment of treat contaminated media *in situ* and the minimisation of the use of additional chemicals to aid the clean up process (Holland 2011).

Nanoremediation relies on the peculiar properties of ENMs, which have two primary advantages over conventional bulk materials: (1) very small size and huge specific surface area, which make them able to treat large volumes of water and to remove a wide spectrum of hazardous pollutants, including organoalogenated compounds (OA), hydrocarbons and heavy metals and (2) tunable chemical, physical, optical, electronic and mechanical properties, which can be rationally adjusted by controlling their size, surface morphology and shape (Karn et al. 2009, Müller and Nowack 2010). Such properties, obtained by tailoring the synthetic process, allow researchers to meet case-specific needs and overcome applicative limitations stemming from the complexity of the environmental matrices to be treated.

Additionally, the far-reaching mobility of ENMs in aquatic media significantly enhances their potential for treating not only large volumes of contaminated waters but also various typologies such as ground water, river and salt waters (Sánchez et al. 2011). This is particularly advantageous for large-scale *in situ* remediation of contaminated waters and have certainly boosted the efficiency of nanotechnology-based decontamination strategies, compared to "conventional" approaches (Grieger et al. 2015).

Based on various shapes and morphologies, ENMs include particles, tubes, wires, fibres, etc., which function as adsorbents and/or catalysts for the removal of legacy and emerging contaminants (such as pharmaceuticals) as well as disinfection from viruses, bacteria and parasites (De Luca and Ferrer 2017) (Fig. 1).

Currently, there are three main water nanoremediation strategies, which exploit diverse approaches, encompassing different ENMs types. These are: (i) nano-adsorbents, either made of carbon- or metal-based, with high adsorption efficiency towards organic and metal pollutants, due to extremely high specific surface area, more accessible sorption sites and lower intraparticle diffusion (Lofrano et al. 2016); (ii) membrane systems based on nanofibers or nanocomposites, which offer great opportunities to improve membrane permeability, fouling resistance, mechanical and thermal stability and provide new functions for contaminant degradation (Liu et al. 2015) and (iii) nano catalysts as photocatalyst such as TiO_2 (Carotenuto et al. 2014, Lofrano et al. 2016). Such applications for the wastewater treatment allows fast and efficient removal of inorganic and organic pollutants but also pharmaceuticals and personal care products as well as pathogens (Shao et al. 2013, Bethi et al. 2016).

According to the "Project of Environmental Nanotechnology" website and the Unites States Environmental Protection Agency (USEPA), overthe last ten years, almost 70 sites worldwide have been successfully treated using nanoremediation techniques, which in comparison with conventional

Figure 1. Schematic description of main ENMs used for water remediation (fibers, tubes, particles, sponges, wires).

methods have significantly reduced time frame (days vs months) and operational costs (up to 80%) (USEPA 2009, PEN 2015). Nevertheless, despite such promising results, nanoremediation has been seldom applied in Europe (JRC 2007) probably as a consequence of the emerging societal worries on nanotechnologies and the current lack of regulatory and proper legislative supports (Nature Nanotechnology 2007, Grieger et al. 2012).

It has been estimated that there are more than 2.5 million potentially polluted sites in Europe, which need to be remediated, and that 350,000 sites may pose a potential risk to humans or the environment (EEA 2014). A remediation technology must attend to cost-benefit approaches considering both practical immediate issues and long-term prospects. Costs and benefits are not always easy to handle, especially for emerging materials where the number of pros and cons are almost the same, at least at the beginning, when unexplored aspects are still present, and contradictory results exist considering both human health and environmental impacts (Lofrano et al. 2017).

Certainly, a major concern related to the use of ENMs in contaminated water bodies is the limited knowledge of their fate once released in the field: once dispersed in a contaminated site, their mobility could facilitate their contact with the organisms and consequently their uptake by plants or animals at the site or further away. They can be also transformed by interaction with abiotic and biotic factors and acquire new properties which could adversely affect them. Therefore, their use and behaviour pose many questions regarding environmental fate and impact, as their benefits could be overcome by environmental costs. Hence, the current debate relies on the balance between known benefits of nanoremediation and potential risks associated with the use of ENMs in natural environments, mainly due to uncertainties in terms of their mobility, transformations and potential ecotoxicity (Corsi et al. 2018a).

In the context of monitoring their fate and behaviour, the prevailing technical limitations on ENMs detection in environmental matrices, as well as on proper risk assessment strategies, provide new challenges and significantly limit their validation, both at the laboratory and field scale (Nowack et al. 2015). Such scientific gaps calls for a thorough evaluation of potential "side effects" of *in situ* nanoremediation; therefore, a case-by-case analysis must be undertaken to assess their actual applicability (Garner and Keller 2014).

All peculiar and desirable properties of ENMs can in fact rebound on the safety of their application in natural waters (Patil et al. 2016). Their small size, mobility and overall reactivity can dramatically improve their transport in water compartments, as they could potentially reach undesired targets and led to hazardous effects, beyond the envisaged benefits in terms of contaminants removal and/or degradation. For instance, ENMs bioaccumulation due to ingestion, dermal contact and inhalation is still controversial as well as their potential role to act as a Trojan horse and increase the uptake of adsorbed pre-existing contaminants (Trujillo-Reyes et al. 2014).

Controversial results have been published concerning such behaviour of ENMs, which is strongly affected by their stability in natural aquatic media and consequently, affects their interactions with contaminants to be remediated. Synergistic as well as antagonistic effects, in terms of impact towards biota, have been reported, and more research is therefore strongly needed in order to avoid any potentially unexpected outcomes occurring during the nanoremediation process (Banni et al. 2016, Canesi et al. 2014, Nigro et al. 2015, Rocco et al. 2015, Vannuccini et al. 2015). Indeed, one promising solution is the possibility to retrieve and/or remove ENMs after they have exerted their action. This is to some extent possible with magnetic ENMs, like magnetite NPs, which can be recovered with the application of weak electromagnetic fields once the remediation process is over (Yavuz et al. 2006). Nevertheless, major challenges have been experienced with the vast majority of non-magnetic ENMs, as their effective removal from remediated environmental media is often challenging or completely impractical. To this end, efforts to assess and model the fate of different ENMs in a wide range of environmental matrices, and to track relevant physico-chemical transformations, are much needed to anticipate natural scenarios and potential endpoints (Garner and Keller 2014, Nowack and Bucheli 2007, Nowack et al. 2012). Additionally, the complexity of natural environmental matrices requires novel and tailored detection and characterization technologies and strategies (Petersen et al. 2016).

Currently a certain level of uncertainty is related to ENMs instability in water media,for instance the tendency to form aggregates with different physico-chemical properties, with respect to the bare particles/materials (Klaine et al. 2008, Lowry et al. 2012, Corsi et al. 2014). Moreover, the possibility that such behaviour may be influenced by ENMs' interactions with biomolecules such as proteins both in purely biological and abiotic environments, forming new entities named *eco-/bio-coronas,* clearly underlines that information solely on their intrinsic properties is insufficient, and a deeper understanding of their environmental transformations is therefore urgently required (Canesi and Corsi 2016) (Fig. 2). NPs' extrinsic properties related to the formation of an *eco-corona* or *bio* (protein)

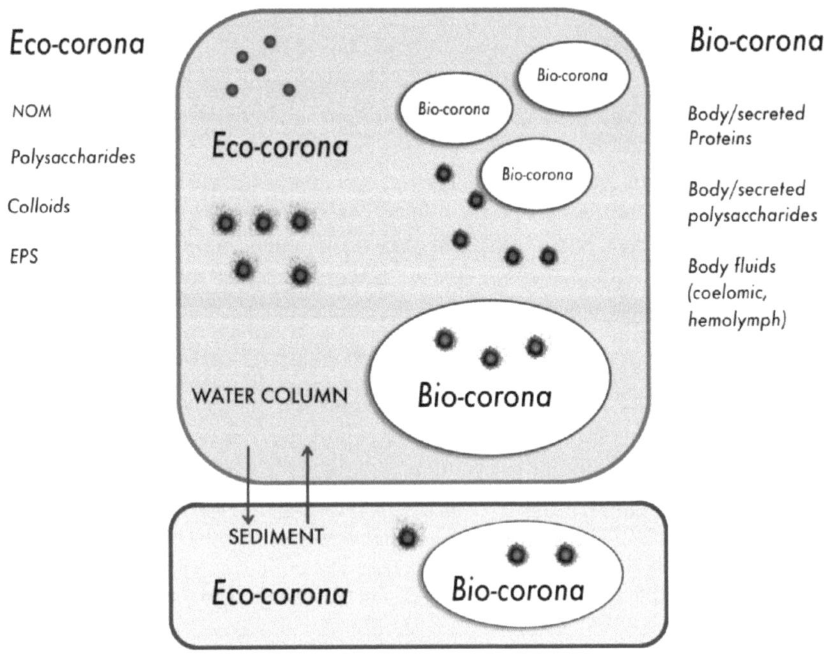

Figure 2. Schematic representation of ENMs fate and transformations in natural environmental media: the eco-bio-nanointeractions (from Canesi and Corsi 2016).

corona will consistently affect their ultimate fate in the water column, their biological activity and consequently, toxicity. For instance, dispersion and aggregation or agglomeration in aqueous solutions are driven by size and surface charges of the ENMs, but also by several parameters of the receiving media such as pH, ionic strength (osmolarity), dissolved oxygen, temperature and the presence of natural organic matter (NOM) (Klaine et al. 2008, Keller et al. 2010). Natural environmental exposure scenarios have been barely investigated in nano-ecotoxicological studies so far, causing a substantial lack of information on the eco-bio-interactions occurring, for instance, in natural waters (Elliot et al. 2016, Holden et al. 2016, Selck et al. 2016). The high ionic strength of marine waters, which also affects pH, as well as organic matter content, has been reported to play a significant role on ENMs aggregation and is able to affect their fate and toxicity with regard to aquatic organisms. Our records show how the behaviour and related toxicological outcome of titanium dioxide NPs, a known photocatalyst used for water treatment, are very much related to the strong aggregation occurring once dispersed in salt waters and enhanced in natural marine waters (Della Torre et al. 2015a,b). Heteroaggregation of n-TiO_2 with microalgae can play an important role in their toxicity, especially in high ionic strength conditions (Sendra et al. 2017). One of the main components of NOM are Exo-Polymeric Substances (EPS) excreted by bacteria and phytoplankton and mainly composed of high molecular weight polysaccharides and proteins, with variable chemistry depending on species and environmental conditions (Verdugo et al. 2004). The initial attempt to characterize EPS- n-TiO_2 complexes highlighted the propensity of NPs to bind proteins rather than carbohydrates, suggesting the formation of an eco-corona around n-TiO_2 in natural marine waters (Morelli et al. 2018).

Ecotoxicity of ENMs Currently Used for Remediation and Future Recommendations

In order to optimize nanoremediation processes, limiting potential environmental and human risk associated with the use of ENMs, fate scenarios should be predicted, carefully including any transformation occurring in environmental media, starting from the ENMs introduction into a polluted site until their removal or degradation, after the remediation of the target pollutants (Nowack et al. 2012).

Besides the current limitations due to lack of methods for *in-situ* assessment of ENMs speciation, ageing and agglomeration/aggregation state, including the eco-corona formation (Peijnenburg et al. 2016), predictive fate and transport models for ENMs are strongly recommended in the design and selection of a nanoremediation strategy for a specific contaminated area. Environmental transformation of ENMs should be therefore deeply investigated in order to fit data into models that could be used for risk assessment strategies. Transformations are not only affecting ENMs' behavior and fate in environmental media but more importantly are essential for recognizing exposure scenarios as well as pattern of uptake and toxicity, which still represent a major challenge for aquatic nano-ecotoxicologists and a fundamental hindrance for the use of ENMs in water remediation.

It is hence wise to foresee possible scenarios of ENMs interactions with natural ecosystems and to screen for their potential ecotoxicity towards different levels of biological organization (Bour et al. 2015, Garner et al. 2015, Gottschalk et al. 2013, Scott-Fordsmand et al. 2017).

A recent body of literature has been produced over the past few years concerning the hazard posed by ENMs and different nanoformulations currently employed for nanoremediation purposes, as reported in Table 1. Indeed, for the most of the applied nanoscale materials in nanoremediation, several adverse effects in both terrestrial and aquatic organisms have been reported, thus certainly increasing governmental as well as public concerns related to their *in situ* application (Hjorth et al. 2017, Nguyen et al. 2018) (see Table 1). For instance, among the tested ENMs, nZVI and iron oxides-based formulations received much attention, compared to other ENMs types, due to their consistent usage in ground- and surface water remediation (Müller et al. 2010). Notably, such studies reported toxicity for ENMs concentrations close to environmentally realistic usage scenarios and in many

Table 1. Documented ecotoxicity of selected ENMs for environmental remediation.

Nanoparticles Type	Remediation Mechanism	Remediated Contaminants	Ecotoxicity	Test Organisms	References
nZVI	Adsorption; oxidation; reduction	metals; chlorinated pollutants	Algal growth inhibition; ROS generation; oxidative stress; disruption of membrane integrity; genotoxicity; morphological alterations of roots; oxygen consumption	bacteria; freshwater microalga; freshwater crustaceans; earthworm; plant	Hjort et al. 2017, Nguyen et al. 2018, Schiwy et al., 2016, Ghosh et al. 2017
Iron-based ENMs	Adsorption; oxidation; reduction	metals; microbiological contaminants	Algal growth inhibition; ROS generation; oxidative stress; disruption of membrane integrity; genotoxicity; mutagenicity; reproduction impairment	bacteria; freshwater microalga; freshwater crustaceans; earthworm; plant; fish	Hjort et al. 2017, Nguyen et al. 2018, Qualhato et al. 2017
TiO_2	Photodegradation	organic contaminants	ROS generation; oxidative stress; genotoxicity; membrane damage; cell viability reduction; mutagenicity; reproduction impairment; tissues alterations and gill histopathology; neurotoxicty	bacteria; crustacean; plant; fish	Mathur et al. 2017, Nogueira et al. 2015, Callaghan et al. 2017, Yang et al. 2013
ZnO	Photocatalysis; photodegradation; adsorption	organic contaminants; heavy metals	Algal growth inhibition; ROS generation; gill damage; embriotoxicity; metal stress via dissolution and ion release; membrane damages	bacteria; freshwater microalga; crutacean; fish	Mathur et al. 2017, Nogueira et al. 2015, NaveedUlHaq et al. 2017, Chang et al. 2012
CNT-based ENMs	Catalytic facilitation; adsorption	organic contaminants; heavy metals	toxicity enhancement of contaminants; carrying of pollutants; ROS generation; growth rate inhibition; membrane damage	bacteria; microalga; crustaceans; molluscs	Callaghan et al. 2017, Mottier et al. 2017, Boncel et al. 2015, Hanna et al. 2014.

cases below the concentration of 100 mg/L, which is, according to EU regulations, the baseline concentration for environmental hazard labelling of chemicals (Hjorth et al. 2017). Moreover, such effects have been described over different level of biological organizations, ranging from plants (Zhang et al. 2018) and algae (Nguyen et al. 2018) to aquatic invertebrate and vertebrate species (Qualhato et al. 2017), identifying diverse toxicological endpoints. Such evidences highlight the necessity to move towards different nanoformulations and usage strategies when applying ENMs and ENMs-based products to natural waters.

Therefore, to implement the effective application of nanotechnology, a thorough ecologically safe (ecosafe) predictive assessment approach should be performed addressing the following key aspects:

a) realistically estimate the behaviour of ENMs in the media to be remediated, with particular focus on the physico/chemical modifications induced by environmental factors including biological ones, which might affect their reactivity and fate, also impairing their remediation potential;
b) consider the nature of the pollutants and the characteristics of the polluted media/area and its surroundings;
c) identify possible toxicological targets of ENMs and provide a mechanism-based evaluation of ecotoxicity in different species and more importantly, at the ecosystem level.

Ecotoxicology can provide suitable useful tools to select ecofriendly and sustainable ENMs for environmental remediation (Corsi et al. 2014, 2018a). Therefore, there is an urgent need to develop a comprehensive guidance on how to perform ecotoxicological testing of ENMs (Petersen et al. 2015) in order to address current limitations and difficulties and support regulatory measures and environmental policies. Regulators are expected to take decisions on the permitted level of ENMs released in the environment, as strongly required by stakeholders and industries. While standardized ad hoc ecotoxicity bioassays can be used as screening tools for selecting the best eco-design of ENMs used for remediation, any risk associated with their fate, behaviour and interaction with biological components of the media under remediation should be carefully investigated by using a more ecosystem-based approach.

Relevant environmental exposure scenarios which will include micro- and mesocosm studies and multi-trophic level approach are thus particularly needed in order to address the ENMs hazard at ecosystem level (Corsi et al. 2014). Trojan horse mechanism in cellular uptake of ENMs enhancing bioavailability and accumulation of contaminant to be remediated, as well as its trophic transfer up to the food chain, should be carefully considered and addressed by ecotoxicologists using an ecosystem-based approach. A more ecologically oriented hazard assessment of ENMs entering the natural environment has already been proposed and can take several advantages from the application in nanoremediation where size, properties and quantities of ENMs are known, as well as their potential biological effects from organism to population up to ecosystem level.

The Need for New Approaches and Safer Solutions

A multitude of studies have failed to reveal the risk of materials in the nano-dimension per se, as it is hard to differentiate ENMs' effects to those of the bulk material (Laux et al. 2017). Nevertheless, due to this uncertainty, national and international regulations often adopt a conservative approach, banning the *in situ* application of ENMs. This suggests the necessity to design new solutions, capable to take into account such critical aspects. In this context, a valuable alternative strategy to overcome the ecotoxicology and legislative issues related to the use of ENMs for environmental remediation is based on the simple yet radical concept of moving from discrete nano-sized materials to nano-structured or nano-embedded and immobilized devices, transferring the advantages of nanotechnology to macro-dimensioned systems and thus reducing dispersability and mobility of ENMs.

In the framework of the NANOBOND project (Nanomaterials for Remediation of Environmental Matrices associated to Dewatering), hydrogels obtained from oxidized cellulose nanofibers (TOCNF) branched with polyethyleneimine (bPEI) (Melone et al. 2015), have been proposed for sludge and

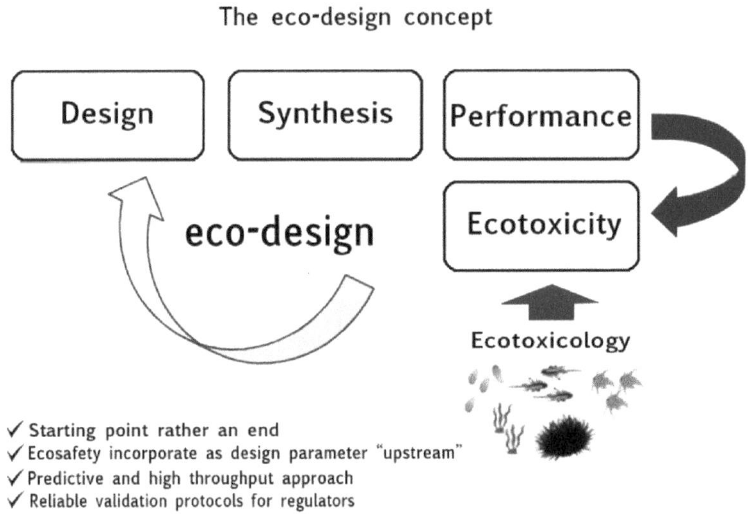

Figure 3. Schematic description of eco-design concept: from ecotoxicity to eco-design during product-development (from Corsi et al. 2018b).

dredged sediment remediation. Funded in the framework POR CReOFESRTuscany 2014–2020, the NANOBOND project strives to develop an innovative system, by coupling the use of nanostructured eco-friendly and sustainable materials with the classical geotextile dewatering tubes (Corsi et al. 2018b). This new solution has the goal to reduce contaminated sludge and sediments, in terms of volumes and costs of transport, but also to convert the resulting solid and liquid wastes into renewable clean resource to be smoothly used, for instance, in riverbanks settlements and similar applications. By developing nanoremediation techniques associated with dewatering, NANOBOND intends to explore new solutions for dredging and sludge management linked to hydrogeological disruption and maintenance of harbour areas, emerging issues which are tremendously increasingly worldwide. This innovative solution aims to become an efficient strategy to significantly reduce sludge and sediment contamination through nanoremediation since it is also easy to scale up for large-scale *in situ* applications with competitive costs.

Further examples include the INTERREG EUROPE project TANIA (TreAting contamination through NanoremedIAtion) with the aim to improve EU regional policies on treating contamination through nanoremediation in European countries and to implement regional development policies in the field of the environmental prevention and protection by pollutants. TANIA specifically addresses innovative and low cost technological solutions for the (nano)remediation of contaminated soil and water (https://www.interregeurope.eu/tania/).

Conclusions and Perspectives

Nanoremediation is able to provide enormous benefits and with appropriate strategies and ecosafe nanosolutions, allowing researchers to reduce uncertainties and environmental and human risks, satisfy regulatory requirements, boost circular economy and support a fully effective deployment of the sector. Main recommendation is that ecosafety obtained by an eco-design approach should be recognized as a priority feature (Corsi et al. 2018a,b). Ecotoxicity testing should be ecologically sound and include more realistic environmental scenarios. Research and innovation should focus on greener, sustainable and smart nanosolutions with the aim of provide a more ecofriendly nanoremediation.

To further promote the application of nanoremediation, regional policy makers must work together and with main stakeholders in order to: (i) support research and innovation for identification of ecosafe and sustainable nanosolutions; (ii) define a standardized methodology to evaluate ENMs effectiveness, ecosafety and economic sustainability within the context of existing environmental regulations at national and international level; (iii) support patenting and pilot applications of new ENMs developed on the basis of eco-design approach; (iv) develop a policy framework to provide incentives for *in situ* use of ENMs for treatment of contaminated water; (v) raise awareness about the process of nanoremediation, its benefits and means of application. In this context, ecotoxicology, as well as predictive models, can be extremely helpful in risk assessment for regulatory needs. Greener and sustainable solutions such as *ecofriendly* ENMs will alsobe mandatory for supporting industrial competitiveness, innovation and sustainability of the sector. A specific legislation is necessary to regulate their emissions and field application. Overall, the generation of ENMs that meet the highest standards of environmental safety will therefore support the effective deployment of nanoremediation.

Conflict of interest: The authors declare no conflict of interest.

Acknowledgements

This work was made within the framework of the project NANOBOND (Nanomaterials for Remediation of Environmental Matrices associated with Dewatering, Nanomateriali per la Bonifica associata a Dewatering di matrici ambientali) POR CReO FESR Toscana 2014-2020 - 30/07/2014-LA 1.1.5 CUP 3389.30072014.067000007.

References

Banni, M., S. Sforzini, T. Balbi, I. Corsi, A. Viarengo and L. Canesi. 2016. Combined effectsof n-TiO_2 and 2,3,7,8-TCDD in *Mytilus galloprovincialis* digestive gland: a transcriptomic and immunohistochemical study. Environ. Res. 145: 135–144.

Bethi, B., S.H. Sonawane, B.A. Bhanvase and S.P. Gumfekar. 2016. Nanomaterials-based advanced oxidation processes for wastewater treatment: A review. Chem. Engineer. Proc. 109: 178–189.

Boncel, S., J. Kyzioł-Komosińska, I. Krzyżewska and J. Czupioł. 2015. Interactions of carbon nanotubes with aqueous/aquatic media containing organic/inorganic contaminants and selected organisms of aquatic ecosystems—A review. Chemosphere 136: 211–221.

Bour, A., F. Mouchet, J. Silvestre, L. Gauthier and E. Pinelli. 2015. Environmentally relevant approaches to assess nanoparticles ecotoxicity: A review. J. Hazard. Mater. 283: 764–777.

Callaghan, N.I. and T.J. MacCormack. 2017. Ecophysiological perspectives on engineered nanomaterial toxicity in fish and crustaceans. Comp. Biochem. Physiol. Part C. Toxicol. Pharmacol. 193: 30–41.

Canesi, L., G. Frenzilli, T. Balbi, M. Bernardeschi, C. Ciacci, S. Corsolini et al. 2014. Interactive effects on n-TiO_2 and 2,3,7,8-TCDD on the marine bivalve *Mytilus galloprovincialis*. Aquat. Toxicol. 153: 53–65.

Canesi, L. and I. Corsi. 2016. Effects of nanomaterials in marine invertebrates. Sci. Total Environ. 565: 393–340.

Carotenuto M., G. Lofrano, A. Siciliano, F. Aliberti and M. Guida 2014. TiO2 photocatalytic degradation of caffeine and ecotoxicological assessment of oxidation by-products. Global Nest. J. 16: 463–473.

Chang, Y.-N., M. Zhang, L. Xia, J. Zhang and G. Xing. 2012. The toxic effects and mechanisms of CuO and ZnO nanoparticles. Materials 5: 2850–2871.

Corsi, I., G.N. Cherr, H.S. Lenihan, J. Labille, M. Hasselov, L. Canesi et al. 2014. Common strategies and technologies for the ecosafety assessment and design of nanomaterials entering the marine environment. ACS Nano 8: 9694–9709.

Corsi, I., M. Winther-Nielsen, R. Sethi, C. Punta, C. Della Torre, G. Libralato et al. 2018a. *Ecofriendly* nanotechnologies and nanomaterials for environmental applications: key issue and consensus recommendations for sustainable and ecosafe nanoremediation. Ecotox. Environ. Safe. 154: 237–244.

Corsi, I., A. Fiorati, G. Grassi, I. Bartolozzi, I. Daddi, L. Melone et al. 2018b. Environmentally sustainable and ecosafe polysaccharide-based materials for water nano-treatment: An ecodesign study. Materials 11(7): 1228.

De Luca, A. and B.B. Ferrer. 2017. Nanomaterials for water remediation: Synthesis, applications and environmental fate. *In*: G. Lofrano, G. Libralato and J. Brown (eds.). Nanotechnologies for Environmental Remediation. Applications and Implications. eBook Springer.

Della Torre, C., T. Balbi, G. Grassi, G. Frenzilli, M. Bernardeschi, A. Smerilli et al. 2015a. Titanium dioxide nanoparticles modulate the toxicological response to cadmium in the gills of *Mytilus galloprovincialis*. J. Hazard. Mater. 297: 92–100.

Della Torre, C., F. Buonocore, G. Frenzilli, S. Corsolini, A. Brunelli, P. Guidi et al. 2015b. Influence of titanium dioxide nanoparticles on 2,3,7,8-tetrachlorodibenzo-*p*-dioxin bioconcentration and toxicity in the marine fish European sea bass (*Dicentrarchus labrax*). Environ. Pollut. 196: 185–193.

EEA-European Environment Agency. 2014. Progress in management of contaminated sites. Report CSI 015. Copenhagen, Denmark. Available: http://www.eea.europa.eu/data-and-maps/indicators/progress-in-management-of-contaminated-sites-3/assessment.

Elliott, J. 2016. Toward achieving harmonization in a nano-cytotoxicity assay measurement through an interlaboratory comparison study. ALTEX-Altern. Anim. Ex. 34(2): 201–2018.

Garner, K.L. and A.A. Keller. 2014. Emerging patterns for engineered nanomaterials in the environment: A review of fate and toxicity studies. J. Nanopart. Res. 16: 2503.
Ghosh, I., A. Mukherjee and A. Mukherjee. 2017. In planta genotoxicity of nZVI: Influence of colloidal stability on uptake, DNA damage, oxidative stress and cell death. Mutagenesis 32: 371–387.
Gottschalk, F., T. Sun and B. Nowack. 2013. Environmental concentrations of engineered nanomaterials: Review of modeling and analytical studies. Environ. Pollut. 181: 287–300.
Grieger, K., F. Wickson, H.B. Andersen and O. Renn. 2012. Improving risk governance of emerging technologies through public engagement: the neglected case of nano-remediation? Int. J. Emerging Techn. Soc. 10: 61–78.
Grieger, K.D., A. Fjordbøge, N.B. Hartmann, E. Eriksson, P.L. Bjerg and A. Baun. 2010. Environmental benefits and risks of zero-valent iron nanoparticles (nZVI) for *in situ* remediation: risk mitigation or trade-off? J. Cont. Hydrol. 118: 165–83.
Grieger, K.D., R. Hjorth, J. Rice, N. Kumar and J. Bang. 2015. Nano-remediation: tiny particles cleaning up big environmental problems. IUCN. Accessed Jul 09, 2015, from http://cmsdata.iucn.org/downloads/nanoremediation.pdf.
Hanna, S.K., R.J. Miller and H.S. Lenihan. 2014. Deposition of carbon nanotubes by a marine suspension feeder revealed by chemical and isotopic tracers. J. Hazard. Mater. 279: 32–37.
Hjorth, R., C. Coutris, N.H.A. Nguyen, A. Sevcu, J.A. Gallego-Urrea, A. Baun et al. 2017. Ecotoxicity testing and environmental risk assessment of iron nanomaterials for sub-surface remediation—Recommendations from the FP7 project NanoRem. Chemosphere 182: 525–531.
Hjorth, R., L. van Hove and F. Wickson. 2017. What can nanosafety learn from drug development? The feasibility of "safety by design". Nanotoxicology 11: 305–312.
Holden, P.A., J.L. Gardea-Torresdey, F. Klaessig, R.F. Turco, M. Mortimer, K. Hund-Rinke et al. 2016. Considerations of environmentally relevant test conditions for improved evaluation of ecological hazards of engineered nanomaterials. Environ. Sci. Technol. 50: 6124–6145.
Holland, K.S. 2011. A framework for sustainable remediation. Environ. Sci. Technol. 45: 7116–7117.
Joint Research Centre (JRC). 2007. Report from the Workshop on Nanotechnologies for Environmental Remediation. JRC Ispra 16–17 April 2007. David Rickerby and Mark Morrison. www.nanowerk.com/nanotechnology/reports/reportpdf/report101.pdf.
Karn, B., T. Kuiken and M. Otto. 2009. Nanotechnology and in situ remediation: A review of the benefits and potential risks. Environ. Health Perspec. 117: 1823–1831.
Keller, A.A., H. Wang, D. Zhou, H.S. Lenihan, G. Cherr, B.J. Cardinale et al. 2010. Stability and aggregation of metal oxide nanoparticles in natural aqueous matrices. Environ. Sci. Technol. 44(6): 1962–1967.
Klaine, S.J., P.J.J. Alvarez, G.E. Batley, T.F. Fernandes, R.D. Handy, D.Y. Lyon et al. 2008. Nanomaterials in the environment: behavior, fate, bioavailability, and effects. Environ. Toxicol. Chem. 27: 1825–1851.
Liu, Q., Y. Zheng, L. Zhong and X. Cheng. 2015. Removal of tetracycline from aqueous solution by a Fe3O4 incorporated PAN electrospun nanofiber material. J. Environ. Sci. 28: 29–36.
Lofrano, G., M. Carotenuto, G. Libralato, R.F. Domingos, A. Markus, L. Dini et al. 2016. Polymer functionalized nanocomposites for metals removal from water and wastewater: An overview. Water Res. 92: 22–37.
Lofrano, G., G. Libralato and J. Brown. 2017. Nanotechnologies for Environmental Remediation—Applications and Implications. Springer.
Lowry, G.V., K.B. Gregory, S.C. Apte and J.R. Lead. 2012. Transformations of nanomaterials in the environment. Environ. Sci. Technol. 46: 6893–9.
Mathur, A., A. Parashar, N. Chandrasekaran and A. Mukherjee. 2017. Nano-TiO_2 enhances biofilm formation in a bacterial isolate from activated sludge of a waste water treatment plant. Int. Biodeterior. Biodegrad. 116: 17–25.
Matranga, V. and I. Corsi. 2012. Toxic effects of engineered nanoparticles in the marine environment: model organisms and molecular approaches. Mar. Environ. Res. 76: 32–40.
Melone, L., B. Rossi, N. Pastori, W. Panzeri, A. Mele and C. Punta. 2015. TEMPO-oxidized cellulose cross-linked with branched polyethyleneimine: Nanostructured adsorbent sponges for water remediation. Chem. Plus. Chem. 80: 1408–1415.
Morelli, E., E. Gabellieri, A. Bonomini, D. Tognotti, G. Grassi and I. Corsi. 2018. TiO_2 nanoparticles in seawater: Aggregation and interactions with the green alga *Dunaliella tertiolecta*. Ecotox. Environ. Safe. 148: 184–193.
Mottier, A., F. Mouchet, É. Pinelli, L. Gauthier and E. Flahaut. 2017. Environmental impact of engineered carbon nanoparticles: from releases to effects on the aquatic biota. Curr. Opin. Biotechnol. 46: 1–6.
Müller, N.C. and B. Nowack. 2010. Nano Zero Valent Iron—the solution for water and soil remediation? Observatory NANO Focus Report.
Nature Nanotechnology. 2007. A little Knowledge. Editorial. 12: 731.
Naveed UlHaq, A., A. Nadhman, I. Ullah, G. Mustafa, M. Yasinzai and I. Khan. 2017. Synthesis approaches of Zinc Oxide nanoparticles: the dilemma of ecotoxicity. J. Nanomater.
Nguyen, N.H.A., N.R. Von Moos, V.I. Slaveykova, K. Mackenzie, R.U. Meckenstock, S. Thűmmler et al. 2018. Biological effects of four iron-containing nanoremediation materials on the green alga *Chlamydomonas* sp. Ecotoxicol. Environ. Saf. 154: 36–44.
Nigro, M., M. Berbardeschi, D. Costagliola, C. Della Torre, G. Frenzilli, P. Guidi et al. 2015. Co-exposure to titanium dioxide nanoparticles and cadmium: genomic, DNA and chromosomal damage evaluation in the marine fish European sea bass *Dicentrarchus labrax*. Aquat. Toxicol. 168: 72–77.
Nogueira, V., I. Lopes, T.A.P. Rocha-Santos, M.G. Rasteiro, N. Abrantes, F. Gonçalves et al. 2015. Assessing the ecotoxicity of metal nano-oxides with potential for wastewater treatment. Environ. Sci. Pollut. Res. 22: 13212–13224.

Nowack, B. and T.D. Bucheli. 2007. Occurrence, behavior and effects of nanoparticles in the environment. Environ. Pollut. 150: 5–22.
Nowack, B., J.F. Ranville, S. Diamond, J.A. Gallego-Urrea, C. Metcalfe, J. Rose et al. 2012. Potential scenarios for nanomaterial release and subsequent alteration in the environment. Environ. Tox. Chem. 31: 50–59.
Nowack, B., M. Baalousha, N. Bornhöft, Q. Chaudhry, G. Cornelis, J. Cotterill et al. 2015. Progress towards the validation of modeled environmental concentrations of engineered nanomaterials by analytical measurements. Env. Sci Nano 2: 421–428.
Otto, M., M. Floyd and S. Bajpai. 2008. Nanotechnology for site remediation. Remediation 19: 99–108.
Patil, S.S., U.U. Shedbalkar, A. Truskewycz, B.A. Chopade and A. Ball. 2016. Nanoparticles for environmental clean up: a review of potential risks and emerging solutions. Env. Technol. Innov. 5: 10–21.
Peijnenburg, W., A. Praetorius, J. Scott-Fordsmand and G. Cornelis. 2016. Fate assessment of engineered nanoparticles in solids dominated media—Current insights and the way forward. Environ. Pollut. 218: 1365–1369.
PEN, The Project on Emerging Nanotechnologies. 2015. Nanoremediation Map. Available: http://www.nanotechproject.org/inventories/remediation_map/
Petersen, E.J., S.A. Diamond, A.J. Kennedy, G.G. Goss, K. Ho, J. Lead et al. 2015. Adapting OECD aquatic toxicity tests for use with manufactured nanomaterials: key issues and consensus recommendations. Environ. Sci. Technol. 49: 9532–9547.
Petersen, E.J., D.X. Flores-Cervantes, T.D. Bucheli, L.C.C. Elliott, J.A. Fagan, A. Gogos et al. 2016. Quantification of carbon nanotubes in environmental matrices: current capabilities, case studies, and future prospects. Environ. Sci. Technol. 50: 4587–4605.
Qualhato, G., T.L. Rocha, E.C. de Oliveira Lima, D.M. Silva, J.R. Cardoso, C. Koppe Grisolia et al. 2017. Genotoxic and mutagenic assessment of iron oxide (maghemite-γ-Fe_2O_3) nanoparticle in the guppy *Poecilia reticulata*. Chemosphere 183: 305–314.
Rocco, L., M. Santonastasio, M. Nigro, F. Mottola, D. Costagliola, M. Bernardeschi et al. 2015. Genomic and chromosomal damage in the marine mussel *Mytilus galloprovincialis*: effects of the combined exposure titanium dioxide nanoparticles and cadmium chloride. Mar. Environ. Res. 111: 144–148.
Sánchez, A., S. Recillas, X. Font, E. Casals, E. González and V. Puntes. 2011. Ecotoxicity of, and remediation with, engineered inorganic nanoparticles in the environment. TrAC-Trend. Anal. Chem. 30: 507–516.
Schiwy, A., H.M. Maes, D. Koske, M. Flecken, K.R. Schmidt, H. Schell et al. 2016. The ecotoxic potential of a new zero-valent iron nanomaterial, designed for the elimination of halogenated pollutants, and its effect on reductive dechlorinating microbial communities. Environ. Pollut. 216: 419–427.
Scott-Fordsmand, J., W. Peijnenburg, E. Semenzin, B. Nowack, N. Hunt, D. Hristozov et al. 2017. Environmental risk assessment strategy for nanomaterials. Int. J. Environ. Res. Public. Health 14: 1251.
Selck, H., R.D. Handy, T.F. Fernandes, S.J. Klaine and E.J. Petersen. 2016. Nanomaterials in the aquatic environment: A European Union–United States perspective on the status of ecotoxicity testing, research priorities, and challenges ahead. Environ. Toxicol. Chem. 35: 1055–1067.
Sendra, M., M.P. Yeste, J.M. Gatica, I. Moreno-Garrido and J. Blasco. 2017. Homoagglomeration and heteroagglomeration of TiO_2, in nanoparticle and bulk form, onto freshwater and marine microalgae. Sci. Total Environ. 592: 403–411.
Shao, T., P. Zhang, Z. Li and L. Jin. 2013. Photocatalytic decomposition of perfluorooctanoic acid in pure water and wastewater by needle like nanostructured gallium oxide. Chin. J. Catal. 34: 1551–1559.
Trujillo-Reyes, J., J.R. Peralta-Videa and J.L. Gardea-Torresdey. 2014. Supported and unsupported nanomaterials for water and soil remediation: Are they a useful solution for worldwide pollution? J. Hazard. Mat. 280: 487–503.
USEPA, Superfund: National Priorities List (NPL) Available online: https://www.epa.gov/superfund/superfund-national-priorities-list-npl (accessed on May 23, 2018).
Vannuccini, M.L., G. Grassi, M.J. Leaver and I. Corsi. 2015. Combination effects of nano-TiO_2 and 2,3,7,8-tetrachlorodibenzo-*p*-dioxin (TCDD) on biotransformation genes expression in the liver of European sea bass *Dicentrarchus labrax*. Comp. Biochem. Phys. C 176-177: 71–78.
Vaseashta, A., M. Vaclavikova, S. Vaseashta, G. Gallios, P. Roy and O. Pummakarnchana. 2007. Nanostructures in environmental pollution detection, monitoring, and remediation. Sci. Technol. Adv. Mater. 8: 47–59.
Verdugo, P., A.L. Alldredge, F. Azam, D.L. Kirchman, U. Passow and P.H. Santschi. 2004. The oceanic gel phase: a bridge in the DOM-POM continuum. Mar. Chem. 92: 67–85.
Yang, Y., C. Zhang and Z. Hu. 2013. Impact of metallic and metal oxide nanoparticles on wastewater treatment and anaerobic digestion. Env. Sci-Proc. Imp. 15: 39–48.
Yavuz, C.T., J.T. Mayo, W.W. Yu, A. Prakash, J.C. Falkner, S. Yean et al. 2006. Low-field magnetic separation of monodisperse Fe_3O_4 nanocrystals. Science 314: 964–967.
Zhang, W., I.M.C. Lo, L. Hu, C.P. Voon, B.L. Lim and W.K. Versaw. 2018. Environmental risks of nano zerovalent iron for arsenate remediation: impacts on cytosolic levels of inorganic phosphate and $MgATP^{2-}$ in *Arabidopsis thaliana*. Environ. Sci. Technol. 52: 4385–4392.

11

Analytical Tools Able to Detect ENP/NM/MNs in both Artificial and Natural Environmental Water Media

Yolanda Picó and *Vicente Andreu*

Introduction

Engineered nanoparticles (ENPs), nanomaterials (NMs) and/or manufactured nanomaterials (MNs) are now regularly used as fillers, catalysts, transporters, additives, semiconductors, cosmetics, microelectronics and biomedical devices. In 2012, the European Union defined nanomaterial (NM) as "... a natural, incidental or manufactured material containing particles, in an unbound state or as an aggregate or as an agglomerate and where, for 50% or more of the particles in the number size distribution, one or more external dimensions is in the size range 1–100 nm" (Recommendation of 18 October 2011). Following their use and disposal, NMs are released into the environment and subsequently transported, aggregated, transformed and even degraded. The increasing NMs production rates, their potential release to the environment and resultant effects on ecosystems health are becoming of increasing concern and need to be addressed. Various studies have demonstrated that NMs have undesirable effects on many species by interacting with DNA or producing cellular stress, and can be bioaccumulated in aquatic organisms (Hu et al. 2018, Naasz et al. 2018). Considering the toxicity and the large production volumes of ENP/NM/MNs, there is an increasing concern about their occurrence in the environment, and a consensus on the use of nanotechnology that must be introduced in a responsible and environmentally acceptable way to avoid the mistakes of the past with other contaminants (Bundschuh et al. 2018). The analysis of NMs' in artificial and natural environmental water media has become a hot topic regarding to the environment preservation. This is justified by the growing and widespread use of nanomaterials and also the increase of their release to the environment either as a waste product or as a by product. Nanoparticles or nanoscale materials are increasingly detected in almost any type of water.

Research Center on Desertification CIDE (CSIC-Universitat de València-GV), Carretera Moncada-Náquera, Km 4.5,46113-Moncada. Valencia, Spain.

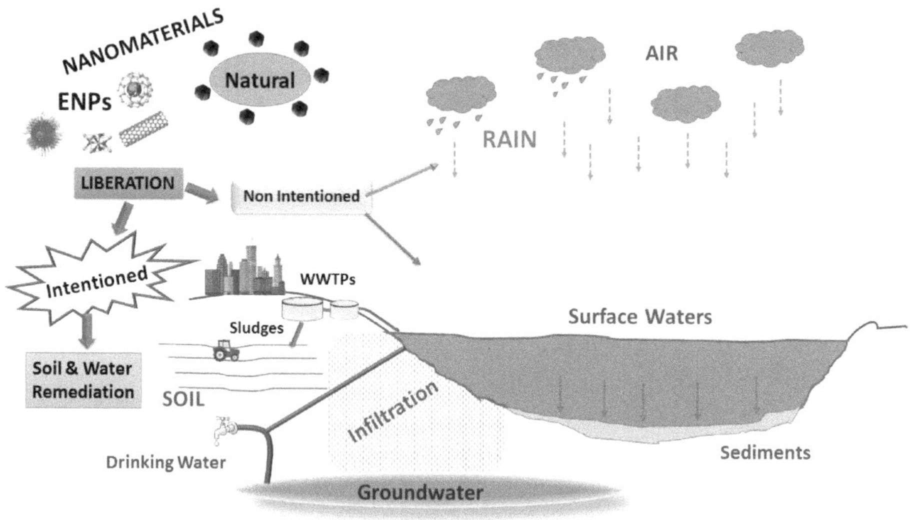

Figure 1. Sources and fate of NMs in the environment.

Over the past decade, our understanding of sources, fate and effects of ENMs in the environment has made significant progress as shown in the previous chapter. As a summary, Fig. 1 shows the different sources of NMs in the environment. Pathways for the release of nanoparticles (NPs) into surface waters are, among others, wastewater from households and atmospheric deposition (Baun et al. 2017, Bundschuh et al. 2018, Laux et al. 2018, Scott-Fordsmand et al. 2017). Due to the unique characteristics of ENMs (compared to other environmental contaminants), limited information concerning their transport, fate, exposure, dose and potential effects is available (Laux et al. 2018, Shevlin et al. 2018). Interestingly, Giese et al. (2018) investigated the current and predicted future environmental concentrations for frequently used ENPs, CeO_2, SiO_2 and Ag. Those predictions for freshwater CeO_2-ENPs range from 1 pg L^{-1} (2017) to a few hundred ng L^{-1} (2050), whereas predictions for SiO_2-ENPs are approximately 1,000 times higher, and for Ag-ENPs 10 times lower (Giese et al. 2018).

The analysis of ENP/NM/MNs is not an easy task, because there are in fact many types of them, depending on the origin, composition, dimensions and structural configurations. The Project on Emerging Nanotechnologies (PEN) established that the most commonly found ENPs in consumer products are based on silver (383 products). Titanium (179), which includes titanium dioxide, has surpassed Carbon (87), which includes fullerenes, followed by silica (52), zinc (including zinc oxide) (36) and gold (19) (PEN 2018). It is extremely difficult to simultaneously isolate mixtures of NMs according to their properties (De Marchi et al. 2018). Up to date, several excellent review articles have focused on the analysis of NMs in the environment. Leopold et al. (2016) reviewed the advances in the analytical methodologies for the determination of metal-containing NMs in the environment. Sample treatment and instrumental detection strategies are critically discussed, including analytical parameters related to quality assurance and quality control. Baun et al. (2017) summarized the analysis, occurrence and biodegradation of NMs. Picó et al. (2017) covered current analytical techniques, instruments and methodologies used to determine emerging contaminants and ENMs from irrigation water and/or soil amendments in plants. Nevertheless, there are very few reviews on the application of analytical techniques to determine NMs in artificial and natural environmental waters (Chadha et al. 2016, Furtado et al. 2016, Schaumann et al. 2015, Séby et al. 2015).

Therefore, the main purpose of this chapter is to present an updated overview of the most recent advances in the analysis of ENMs in water, published since 2014. The following aspects are presented and discussed:

- The application of different sample preparation techniques, especially those that involve use of NMs to determine other NMs. This has been coined as the "Third Way", the use of nanotools (mainly sorbents or sensors) to determine and/or characterize nanoparticles in water.
- The comparison of advanced separation, identification and quantification systems; for example, asymmetric field flow fractionation (AF^4) or hydrodynamic chromatography (HC) combined with inductively coupled plasma mass spectrometry (ICP-MS) or other detectors, keeping in mind that it is not only important to determine ENP/NM/Ms concentrations, but also other characteristics such as shape, size distribution and chemical composition.
- The benefits of effectively combining the information of different techniques in the study of sources, transport, transformation and fate of ENP/NM/MNs in aqueous systems.

Analytical Methodology

Many analytical techniques have been reported as complementary methods to characterize ENPs/NM/MNs, such as flow-cytometry, elemental analysis (EA), nuclear magnetic resonance (NMR), X-ray crystallography (XRD), infrared spectroscopy, scanning mobility particle sizer, atomic force microscopy, light scattering, X-ray diffraction, capillary electrophoresis, optical spectroscopy, etc. However, most of them are rarely used to quantify NMs in environmental samples, probably because they need milligrams of analyte in a pristine solution to operate effectively.

Apart from sensitivity, there are a number of considerations in choosing the most appropriate analytical tool to detect ENP/NM/MNs in both artificial and natural environmental water media. For example, according to the purpose of the study, toxicological studies might focus on the particle size distribution and effects, while environmental ones are mostly focused on detection and quantitation of ENPs. However, composition, structure, particle shape and size, surface area and charge and concentration, along with agglomeration, need to be measured to fully describe a NM, including also origin, dimensions and their structural configuration. The most used classification now divides NMs into metal-based, carbon based, dendrimers and composites. Analytical tools are much more developed for determining metal-based NMs than for the others in environmental matrices. As could be observed in Table 1, which covers studies since 2016, there are few methods available to determine carbon-based NMs (with the exception of fullerenes) and dendrimers, and no method has been reported for composites yet. On dendrimers in water, some pioneering studies were published in 2013 (Uclés et al. 2013a, Uclés et al. 2013b, Ulaszewska et al. 2013, Ulaszewska et al. 2012) but they did not find continuity. Table 1 outlines the techniques reported in the last three years to study NMs in in both artificial and natural environmental water media. The analytical tools reported can be categorized into:

- Electron microscopy techniques coupled or not to Energy-dispersive X-ray spectroscopy (EDX)
- Inductively coupled plasma mass spectrometry (ICP-MS)-based methods
- Liquid chromatography-mass spectrometry approaches
- A hotchpotch of very specific or recently applied analytical tools

The outline of the analytical tools able to detect ENP/NM/MNs will follow this schematization. Each section is intended to serve as a practical guideline that remark advantages, limitations, gaps, advances and prospects within each of the applied techniques.

Electron Microscopy and Electron Microscopy Coupled to Energy-Dispersive X-ray Spectroscopy (EDX)

The most well-known example of routine method successful to identify NPs in the liquid phase is electron microscopy, such as Scanning Electron Microscopy (SEM), Transmission Electron Microscopy (TEM) and Scanning Transmission Electron Microscopy (STEM). These techniques have higher resolving power than light microscopes, and can reveal the structure of ENP/NM/MNs. TEM

Table 1. Selected applications to determine NMs in the environment 2013–2017.

Matrix	Analytes	Extraction	Determination	Recovery (%)	LOD Particle Size (nm)	LOD [] ng L^{-1}	Ref.
Artificial seawater	Ag-NP	Direct determination	NP characterization: AF4-UV/Vis sp-ICP-MS NP agglomeration: UV/Vis	NR	NR	NR	António et al. 2015
Surface waters	Ag, CeO_2 TiO_2	Sonication for 10 min	sp-ICP-MS	NR	14 (Ag) 10 (Ce) 100 (Ti)	0.1 (Ag) 0.05 (Ce) 10 (Ti)	Peters et al. 2018
Seawater	Ag-NP	Direct determination	UV/Vis	NR	NR	NR	Sikder et al. 2018
Wastewater	Fe0-NP	Direct determination	Sizing and quantification sp-ICP-MS/MS	NR	36	0.709	Vidmar et al. 2018
Surface waters	TiO_2-NP, Au-NP, ZnO-NP, CeO-NP, AgO-NP, Zr-NP. La-NP	Ultrafiltration (SEM) Acidification (ICP)	FEG-SEM-EDX HR-ICP-MS	NR	NR	NR	Markus et al. 2018
Environmental water	Au-NPs (5 -30 nm) and Au+	MAE HNO_3+HCl (total Au) Other direct determination	RP-HPLC-ICP-MS DLS TEM ICP-MS (total Au)	93-99	5–30	28-36	Yang et al. 2018
River water	AgO	Stabilization of the suspension with gelatin	sp-ICP-MS TEM	NR	15	97	Loula et al. 2018
Surface water	Metal based-ENPs Al, As, Ag, Mn, Fe, Ti, and Zn	Not reported for any of the techniques	TEM-EDX ICP-MS DLS	NR	NR	NR	Baranidharan and Kumar 2018
Recreational water	TiO2-NP	Direct analysis	Detection and sizing sp-ICP-MS	NR	91	NR	Venkatesan et al. 2018
Drinking water	citrate-stabilized Ag and Au-NPs and uncoated TiO2, CeO2, and ZnO NPs	Direct analysis	sp-ICP-MS	NR	NR	NR	Donovan et al. 2018

Table 1 contd. ...

... Table 1 contd.

Matrix	Analytes	Extraction	Determination	Recovery (%)	LOD Particle Size (nm)	LOD [] ng L^{-1}	Ref.
Water	Ag-NP	Sample on a carbon grid (TEM) Direct analysis	UV-Vis TEM-EDX ICP-MS MC-ICP-MS	NR	NR	NR	Zhang et al. 2017
Rainfall induced foliar wash-off of vine leaves	CuO-NP	Direct analysis after filtration. In some techniques, dilution.	DLS AF4-UV-MALLS (-ICP-MS) sp-ICP-MS ICP-MS (total Cu content)	67-75 % (AF4)	NR	60 ng/L (total Cu_ICP-MS)	la Calle et al. 2017
EPA MHW and MHW containing 2.5 mg L^{-1} Suwannee River fulvic acid	Au@Ag NPs	Direct analysis	sp-ICP-MS	NR	NR	0.01-50 µg L^{-1}	Merrifield et al. 2017
Environmental water and wastewater	AgNP	Centrifugation (HDC) 1/10 dilution and deposition on mesh (TEM)	TEM DLS AF4 HDC Sp-ICP-MS	NR	NR	NR	Chang et al. 2017
Ultrapurewater	Ag-NPs Au-NPs	Direct analysis	Particle number and concentration sp-ICP-MS MDG-ICP-MS	20–40 (sp) > 80 (MGD)	NR	NR	Gschwind et al. 2017
Water Wastewater	40 nm Ag-NPs	Direct analysis	FAST-sp-ICP-MS High resolution	NR	5.4	NR	Tuoriniemi et al. 2017
Constructed wetlands	Ag NP	Dilution with HNO_3 (Total Ag). Dessecation on a grid (STEM)	Paricle diameter STEM-EDX Total Ag content ICP-MS	NR	NR	20-120 ng/L (ICP-MS)	Auvinen et al. 2017
Lake water	20 nm Ag-NPs coated with CIT, PVP, LIP	Tot.Ag: HNO_3 digestion	AF4-UV-SPR sp-ICP-MS Total silver: ICP-MS	NR	NR	NR	Jiménez-Lamana and Slaveykova 2016
Water	Ag-NP	CPE with Triton 114 followed by MAD with HNO_3 (ICP-MS)	ICP-MS sp-ICP-MS	79–123 (CPE-ICP-MS)	< 2(CPE-ICP-MS)	2 ng/L (CPE-ICP-MS)	Yang et al. 2016b

Water and soil	MO-NPs	CPE HNO3 digestion: total Au	ICP-MS AF4-UV/Vis sp-ICP-MS Total Au content: ICP-MS	> 90%	30 (sp-ICP-MS)	NR	El Hadri and Hackley 2017
Water	Ag2S-NP	MSPE with Fe3O4 magnetic particles	ICP-MS	69–100	NR	68 ng/L	Zhou et al. 2016a
Lake water	Ag-NP	Direct analysis after defrozen	sp-ICP-MS (double magnetic sector MS)	NR	10	0.4 ng/L	Newman et al. 2016
Environmental water	Ag and Au NP	Dilution and homogenization in a ultrasonic bath	sp-ICP-MS	NR	20 (Ag) 19 (Au)	NR	Yang et al. 2016a
Environmental water	17–108 nm Ag-NPs CIT, MUA, PVP or MSA	SA-DLLME with Trition 114 (emulsifier) and Cl2CH2 (extractant)	ETV-ICP-MS	76	NR	2.2	Li et al. 2016
Natural water	Au-NPs	Sonicated for 15 min	sp-ICP-MS	NR	19 nm (water) 31 (0.1 μgL^{-1} Au3+ solution)	NR	Long et al. 2016
Aquatic matrices	50 and 80 nm Ag-NPs coated with CIT and PVP	Ultracentrifugal filtration (sp-ICP and DLS) and HNO_3 digestion (total content)	sp-ICP-MS DLS Total Ag content: ICP-MS	NR	> 50	NR	Telgmann et al. 2016
Water	TiO_2-NP	NR	SdFFF-MALLS-ICP-MS/MS	NR	NR	NR	Soto-Alvaredo et al. 2016
Drinking water treatment	30–50 nm ZnO and 80–200 nm CeO_2-NP	Filtration	sp-ICP-MS	NR	35–40 (ZnO) 18–20 (CeO2)	NR	Donovan et al. 2016a
Drinking water treatment	Ag, Au, TiO_2-NP	Filtration	sp-ICP-MS	NR	65–70 (TiO_2)	NR	Donovan et al. 2016b
MilliQwater, forest spring water and landfill leachate.	Fe0, FeONPs and Fe_3O_4NPs	Filtration and addition of HNO_3	Total concentration before and after treatment ICP-MS	NR	NR	NR	Peeters et al. 2016
Water	Ag, Au, Ce, Al, Ti, Zn-NPs	Filtration HNO_3 digestion (total metal)	AF4-ICP-MS Total metal content: ICP-MS	NR	NR	NR	De Klein et al. 2016
Environmental water	MO-NPs (NiO, CoO, ZnO, CuO and CeO_2)	Direct analysis (SEC) Deposition on a grid (TEM) Acid digestion (ICP-MS)	SEC-ICP-MS TEM Total metal content: ICP-MS	NR	NR	NR	Zhou et al. 2016b

Table 1 contd. ...

... Table 1 contd.

Matrix	Analytes	Extraction	Determination	Recovery (%)	LOD Particle Size (nm)	LOD [] ng L^{-1}	Ref.
Municipal wastewater (transformation processes)	80 nm Ag-NPs coated with CIT and PVP	Direct analysis Centrifugation and deposition on a grid (TEM)	sp-ICP-MS TEM TOF-SIMS XPS UV-Visible	NR	NR	NR	Azodi et al. 2016
Aquous media	40 MO-NP	Centrifugation and the supernatant was collected	sp-ICP-MS	NR	Ta, U, Ir, Rh, Th, Ce, and Hf ≤ 10 Se, Ca, and Si > 200	NR	Lee et al. 2014
Laboratory, natural, and processed waters	60–100 nm Ag-NP coated with CIT, TA and PVP	Direct analysis	sp-ICP-MS	NR	30 nm	NR	Mitrano et al. 2014
Waters	14–140 Au-NP coated with CIT, MUA, PVP, CTAB	MSPE in Al3+ immobilized Fe_3O_4@ SiO_2@IDA sequential elution of Au+ and Au-NP with $Na2S_2O_3$ and $NH_3·H_2O$	ICP-MS	73–100	0.31–0.39 ng/L	NR	Su et al. 2014
Waters	PS NP Au and AgNP coated with CIT	Direct analysis	HDC-fluorescence HDC-UV/Vis HDC-sp-ICP-MS	NR	NR	NR	Philippe and Schaumann 2014
Waters	Ag-NP	MSPE with UMPs, GMPs, DMPs Digestion with HNO_3 and HCl	ICP-MS SEM/EDS	97%	NR	NR	Mwilu et al. 2014
Natural waters (ultrapure water, lake water and seawater)	AgNPs	Sonication (CV) Cupper tape and desiccation (SEM)	Amount of Ag ions: ICP-MS Amount of particulate Ag: XPS Alternative for amount of Ag: CV Form of the particule: SEM	NR	NR	NR	Teo and Pumera 2014

Waters	PS, AgNP, AuNP (60, 40, 20, 10 nm)	Direct analysis	At high concentrations: on-line HDC-sp-ICP-MS, HDC-SDS, HDC-DLS At low concentrations: off-line HDC-sp-ICP-MS, HDC-DLS, HDC-AUC	NR	10–20 (sp-ICP-MS)	NR	Proulx and Wilkinson 2014
Litoral lake	AgNP	CPE (for AF4-sp-ICP-MS)	sp-ICP-MS AF4-sp-ICP-MS	NR	NR	NR	Furtado et al. 2014
River water	AgNP	Direct analysis	sp-ICP-MS TEM	NR	NR	NR	Telgmann et al. 2014
Roast dust leachates	Pt-NP	5 g dust and 10 mL natural stormwater runoff sonicated for 30 min at 40°C	Total Pt ICP-MS Pt-NP sp-ICP-MS	NR	7.4	0.06	Folens et al. 2018
Model dispersions with soil particles and NOM	Au-NP 30 or 100 Nm	Direct analysis	AF4-MALS-UV/Vis-ICP-MS	15–65%	NR	NR	Meisterjahn et al. 2014
Water and sediment	TiO_2-NP	MAE HNO_3+HCl +H2O_2+ HBF_4	SdFFF-ICP-MS SEM-EDX	> 95%	NR	13520	Dutschke et al. 2017
Mobility of metallic (nano) particles in leachates from landfills containing waste incineration residues	Salts and metals on NP form	EN 12457-2 to obtain the leachate (aqueous extraction). Deposition in grids (STEM) Serial filtration	Total metal analysis by ICP-MS NTA STEM-EDX	NR	NR	NR	Mitrano et al. 2017
Water and soil	Carbon	Direct injection (water) Aqueous extraction (soil)	FFF-ICP-MS (using a quadrupole) FFF-ICP-MS (using a sector field) FFF-OCD	107-122	NR	0.6 106(Q-ICP-MS) 0.3 106(SF-ICP-MS) 0.04 106(OCD)	Nischwitz et al. 2016

Table 1 contd. ...

... *Table 1 contd.*

Matrix	Analytes	Extraction	Determination	Recovery (%)	LOD Particle Size (nm)	LOD [] ng L^{-1}	Ref.
River water	MWCNT	Direct analysis	Fluorescent nanosensor based on CDs	NR	NR	370	Cayuela et al. 2015
River water	C60 and aggregates (nC60)	VALLME with 50 µL 1-octanol	LC-APCI-Q-MS	79–90	NR	80-3000	Zouboulaki and Psillakis 2016
12 wastewater	C60, C70 and NMFP	NaCl addition and LLE with toluene	LC-ESI-QTRAP-MS/MS	> 70%	NR	0.8–1.6	Zakaria et al. 2018
River water, sediments and wastewater	C60 and C70	Water is filtrated Filters/sediments: Ultrasounds with toluene Water: Add 1% NaCl and toluene	LC-APPI-HR-MS (orbitrap) NTA (nano particle number concentration and size distribution)	68–106%	NR	(2.9–17) x 10–3 34.4 × 10–3 3.2–31 × 10–3	Sanchís et al. 2017
Environmental water suspension	C60 fullerenes and their transformation products	Water is filtrated Filters: Ultrasounds with toluene Water: Add 1% NaCl and toluene	LC-APPI-HR-MS	68–106%	NR	2.9–31 × 10–3	Sanchís et al. 2018

AF4: Asymetric field flow fractionation; AUC: ultra centrifugation; CIT: citrate; CPE: cloud-point extraction; CTBA: cetyltrimethylammonium bromide; DMP: dopamine magnetic particles; ETV: electrothermal vaporization; FEG-SEM/EDX: field emission gun scanning electron microscopy in combination with energy dispersive X-ray analysis; GMP: gluthation magnetic particles; IDA: iminodiacetic acid; LIP: lipoic acid; MALLS: multi-angle laser light scattering, MHW: moderate hard water; MIH: microvawe induced heating; MSA: Mercaptosuccinic acid; MSU: 11-Mercaptoundecanoic acid; OCD: organic carbon detector; PVP: polyvinyl pyrrolidone; SdFFF: sedimentation filed flow fractionation; SEC: size exclusión chromatography; sp-ICP-MS: single particle; TA: tanninc acid; TSPP: tetrasodium pyrophosphate; UMP: unmodified magnetic particles; VALLME: vortex-assisted liquid-liquid microextraction; XPS: X-ray photoelectron spectroscopy.

is based on transmitted electrons and SEM on scattered electrons. STEM combines concepts from both. Unlike TEM, in STEM, the electron beam is focused to a fine spot (with the typical spot size 0.05–0.2 nm), which is then scanned over the sample. TEM provides details about internal composition showing many characteristics of the sample, such as morphology and crystallization, whereas SEM focus on the morphology. Energy-dispersive X-ray Spectroscopy (EDX) is commonly coupled with SEM or TEM to provide elemental analysis at nanometer scale areas. EDX attains determination of the elemental composition of individual points or conducts element mapping out of the selected area.

In electron microscopy, the sample morphology should be preserved as much as possible, which is particularly important in biological samples. Traditional steps involved dehydration, chemical or cryo-fixation and embedding in an epoxy resin. However, in the case of water, deposition on a special grid (made of carbon, cupper, etc.), is becoming the preferred procedure. Sample preparation could range from very easy to perform, just with a pipette, to more sophisticated ones requiring dedicated infrastructure. The simplest protocol involves a proper dispersion of the NMs in water, following which a small drop of the dispersion is placed on to a carbon coated copper grid and the excess of water is drawn with a tissue paper. Although this method can be applied in any laboratory, it requires high particle concentrations due to the low amount of sample that is eventually retained on the TEM grid (Auvinen et al. 2017, Chang et al. 2017, Zhang et al. 2017, Zhou et al. 2016b). An advantageous straightforward alternative involves particles' deposition directly onto TEM grids by centrifugation; therefore, particles present at low concentration are concentrated onto the grid, and ENP scan be more easily visualized than directly placing small drops (Mitrano et al. 2017). Similarly, another possibility involves concentration of NM by centrifugation and then, the concentrate was deposited onto the TEM grid (Azodi et al. 2016). In the most complicated method, the liquid samples (approximately 200 mL) were centrifuged, and the solid fraction was freeze-dried and stored at −18°C. The samples were later thawed under a gentle argon flow, packed in moist absorbing clay and ground to a fine powder with mortar and pestle. The powder was suspended in water and properly diluted and centrifuged on carbon-coated TEM grids in order to evenly distribute the NPs on the TEM grid. Due to the grinding and sonication process, no conclusions can be drawn about the association between the (transformed) Ag-NPs and other colloids/particles in the original sample, but the speciation of Ag-NPs should not change during sample preparation (Auvinen et al. 2017).

TEM has been used to imaging Au, AgO, Al, As, Ag, Mn, Fe and other metal oxides (MO), such as NiO, CuO, ZnO, CoO and Ce_2O, in artificial, surface and wastewater (Azodi et al. 2016, Baranidharan and Kumar 2018, Chang et al. 2017, Loula et al. 2018, Mitrano et al. 2017, Telgmann et al. 2014, Yang et al. 2016a, Yang et al. 2018, Zhang et al. 2017, Zhou et al. 2016b). In many of these studies, TEM is working as STEM. As an interesting example, Mitrano et al. (2017) studied the leachate from a landfill for municipal solid waste incineration residues along with the slags deposited at this site (waste incineration bottom ashes and fly ash, sewage sludge incineration bottom ash). These matrices are a source of ENP/NM/MNs to the environment. TEM images of particles observed in the natural and laboratory prepared leachates vary in terms of particle size, metal content and morphology, both within individual samples and between various leachate compositions (Fig. 2). Although the SEM resolution is lower, it has been applied in a less extent to determine silver and MO-NPs (Dutschke et al. 2017, Markus et al. 2018, Mwilu et al. 2014, Teo and Pumera 2014, Tou et al. 2017). SEM has also advantages including the ability to image a comparatively large area of the ENP/NM/MNs; the ability to image bulk materials; and the variety of analytical modes available for measuring the composition and properties of the NMs, such as field emission gun scanning electron microscopy (FEG-SEM).

Inductively Coupled Plasma Mass Spectrometry (ICP-MS)-Based Methods

Inductively Coupled Plasma Mass Spectrometry or ICP-MS is an analytical technique used for elemental determinations. Nebulized liquids or laser-ablated solids are introduced into an argon plasma, consisting of electrons and positively charged argon ions. In the plasma, elements present in the sample

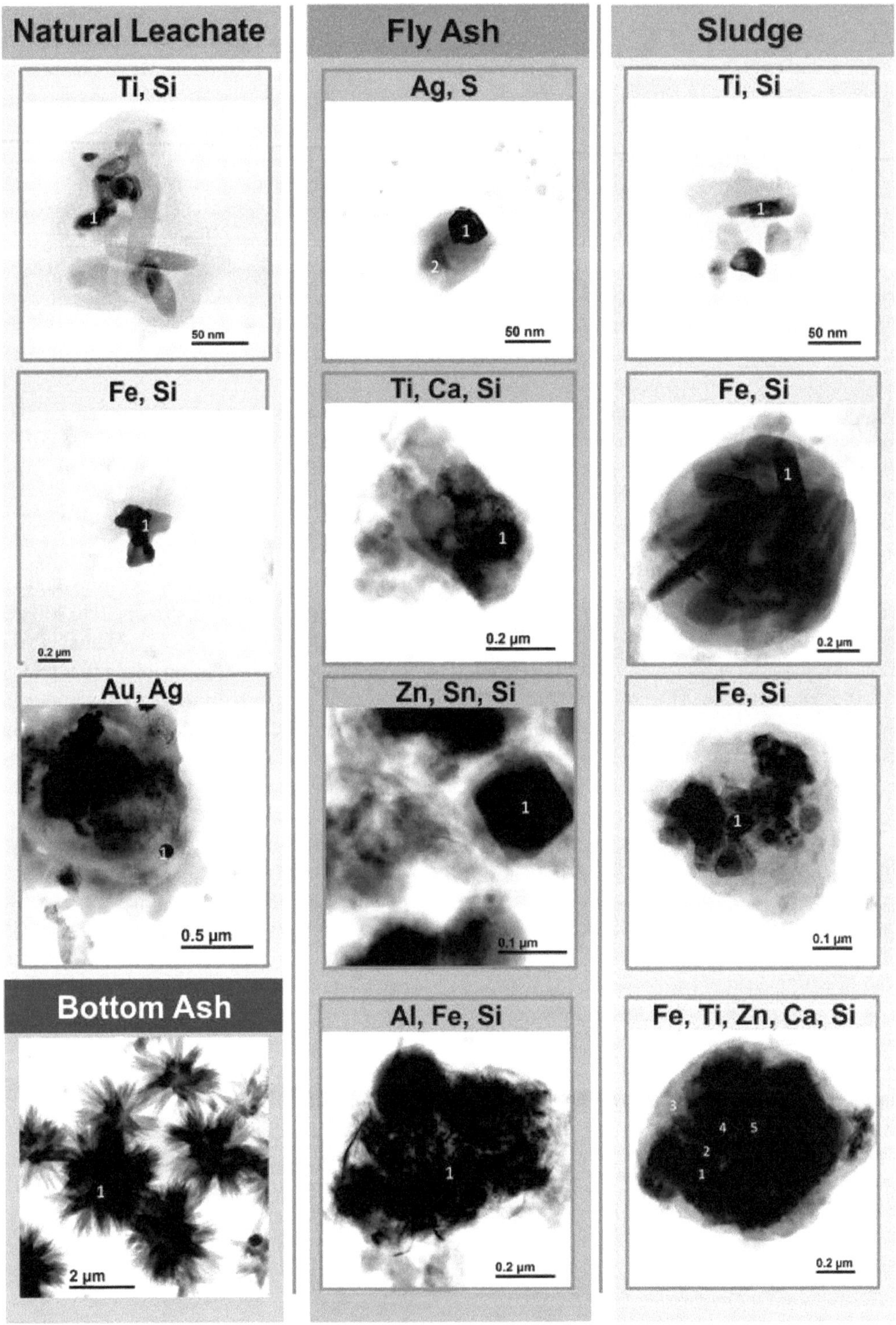

Figure 2. Selection of TEM images of particles observed in the natural and laboratory prepared leachates of MSWI bottom ash, MSWI fly ash and wastewater sludge incineration bottom ash. Particle size, metal content and morphology varied within individual samples and between various leachate compositions. A selection of images is presented here, which shows the diversity of (nano)particles observed. The metal content of each particle is given above the figure, with a number indicating where EDX analysis was performed (Mitrano et al. 2017).

are separated into individual atoms that lose electrons and become positively charged ions (anions are not detected by ICP-MS). The positive ion beam enters the mass analyzer where the ions are separated according to their mass/charge (*m/z*) ratio. ICP-MS in its most basic version has been applied to the determination of the total metal content directly in the aqueous sample or after acidic digestion. It allows simultaneous quantitative and confirmatory analysis of almost 100 metals in various matrices.

ICP-MS, because of its excellent elemental sensitivity and specificity, is an ideal detection technique for the characterization of metal-based MNs. This technique can be used in three different ways:

- after a selective extraction techniques able to separate ENP/NM/MNs of their ionic forms, such as cloud point extraction
- coupled to a separation technique, such as field-flow fractionation or liquid chromatography, to characterize nanosizes
- in single particle mode to characterize individual particles and size distribution

The three techniques have benefits and limitations, but are complementary when used together. However, it should be noted that in the single-particle analysis mode, ICP-MS may decipher more information (dissolved NPs, presence of impurities, decay with time, etc.) in a relatively short time.

ICP-MS after Selective Extraction Techniques Able to Separate ENP/NM/MNs

Due to the extremely low concentration of ENP/NM/MNs in the environment, selective extraction and preconcentration followed by detection with ICP-MS is a promising approach for its analysis. There are basically two strategies possible to address the selective extraction of NMs - liquid extraction using surfactants and magnetic solid-phase extraction.

The former group comprises cloud point extraction (CPE) and surfactant-assisted dispersive liquid–liquid micro-extraction (SA-DLLE). Both techniques use surfactants, which are organic compounds that contain hydrophobic and hydrophilic groups, soluble in organic solvents and water. They can reduce the interfacial tension between the two phases by adsorbing at the liquid–liquid interface, and serve as an emulsifier to enhance the mass-transfer rate from aqueous samples to the extraction solvent. CPE involves the addition of a surfactant at a concentration above its critical micelle concentration (CMC). If the solution temperature is increased above the cloud point temperature of the surfactant, micelles are formed via dehydration, and become less soluble in water. Two phases are obtained: viz. a surfactant-rich phase and a solvent (surfactant depleted) phase (the supernatant). Hydrophobic compounds (including NPs) are extracted with the surfactant micelles whereas ionic forms remain in the supernatant. Triton X-114 ((1,1,3,3-tetramethylbutyl) phenyl-polyethylene glycol) was the more effective surfactant to separate NPs (Yang et al. 2016b). CPE has been used to extract and concentrate MNPs (e.g., AgNPs, AuNPs, CuONPs, carbon nanotubes) in model matrices and in more complex media, such as natural water or wastewater from treatment plants, due to its relative simplicity and cost effectiveness (El Hadri and Hackley 2017, Yang et al. 2016b). The main disadvantage reported for this technique was the need, to digest with acid the surfactant prior to metal determination by ICP-MS. To overcome this disadvantage, dispersive liquid-liquid micro-extraction (DLLME) has been recently applied to extract NPs, selectively showing promising results. This procedure is based on a ternary component solvent system (an aqueous phase—the sample—an extractant—an immiscible organic solvent—and a dispersant—a water miscible organic solvent or a surfactant—to assist the formation of fine oil droplets). If the disperser is a surfactant, then the technique is named surfactant-assisted dispersive liquid–liquid microextraction (SA-DLLME). Hence, SA-DLLME was applied to the separation and preconcentration of AuNPs from environmental water samples, without acid digestion or any further dilution (Liu et al. 2016). Under optimal conditions, a detection limit of 2.2 ng L^{-1} and an enrichment factor of 152-fold was achieved for AuNPs, and their original morphology was maintained during the extraction process.

The second group of approaches are based on magnetic solid-phase extraction (MSPE), which is a new type of SPE based on the use of magnetic sorbents. MSPE form part of the so-called "third way" in analytical nanoscience and nanotechnology, that is, the application of NMs based sorbents to NMs extraction, since most of the magnetic particles are based on iron NMs. Magnetic particles are advantageous as capture media for NMs because they can be used at a high magnetic particle ratio, and then be easily separated from suspension. An interesting attempt was presented by Suet al. (2014) who prepared Al^{3+} immobilized Fe_3O_4@SiO_2@iminodiacetic acid and studied its extraction performance for AuNPs and Au ions. Both could be simultaneously retained on this adsorbent and their separation was achieved by sequential elution of Au ions and AuNPs with $Na_2S_2O_3$ and $NH_3 \cdot H_2O$, respectively. Various experimental parameters affecting MSPE of AuNPs and Au ions were investigated. AuNPs in a size range of 14–140 nm and with different coatings, including citrate, 11-mercaptoundecanoic acid (MUA), polyvinylpyrrolidone (PVP) and cetyltrimethylammonium bromide (CTAB), could be quantified. The size and shape of the AuNPs remained unchanged during the extraction process; thus, the eluent could be directly introduced into ICP-MS without digestion (Su et al. 2014). Following research on nature of magnetic particles, Mwilu et al. (2014) demonstrated that both, unmodified and surface-modified (glutathione and dopamine) magnetic particles are able to selectively capture and concentrate (up to 250-fold concentration factor) trace levels of AgNPs in water. The advantage of this method over commonly used separation techniques is its ease of separation by the use of an external magnetic field. The mixtures were subjected to magnetic separation and both, the eluate and captured particles, were extracted and analyzed for total silver by ICP-MS.

MSPE with Fe_3O_4 magnetic particles have also shown promising features for isolation and pre-concentration of Ag_2S NPs. Iron oxide magnetic particles show excellent performance and a high enrichment factor (up to 250-fold) for AgNPs. Furthermore, the Fe^{3+} constituent in the Fe_3O_4 magnetic particles makes it possible to function as the sacrificial oxidants in eluting AgNPs. Herein, these particles are capable of selectively extracting the predominant Ag-containing NPs in the environment, including AgNPs, AgCl NPs and Ag_2S NPs, in the presence of Ag^+. Then, the extracted Ag_2S NPs can be isolated from the other Ag containing NPs by sequential elution, that is, after pre-eluting AgNPs and AgCl NPs by acetic acid, Ag_2S NPs can be eluted by a mixture of thiourea/acetic acid and quantified directly by inductively coupled plasma mass spectrometry (ICP-MS) (Zhou et al. 2016a).

ICP-MS Coupled to a Separation Technique

The ICP-MS can be also coupled to a number of separation techniques that are able to separate the elements based on the size. These techniques are mostly based on asymmetric flow field fractionation (AF^4) and several types of liquid chromatography, such as size exclusion chromatography or hydrodynamic chromatography.

The separation techniques, which are coupled to ICP-MS, can also be coupled to other detectors such as UV/Vis or multi-angle light scattering (MALS). The use of these detectors is only punctual since better sensitivity, selectivity and specificity is achieved by ICP-MS. Hence, the applications of these detectors are also discussed here together with the separation technique.

Asymmetric Flow Field-Flow Fractionation (AF4)

In AF^4, NPs are separated in a trapezoidal hollow channel according to their different particle diffusion properties. Next, a perpendicular cross flow is used to allow NP samples to flow through a semipermeable membrane. The separation can then be carried out in the mobile phase with a parabolic flow and diffusion-driven transport. As a result, small NPs spend less time passing through the channel. The signal collected from the detector is used to determine the particle radius of the target NPs. Koopmans et al. (2015) used this system to quantify the concentration and size of AgNP coated with either, citrate or PVP, in combination with on-line detection by either on-line UV–vis spectroscopy

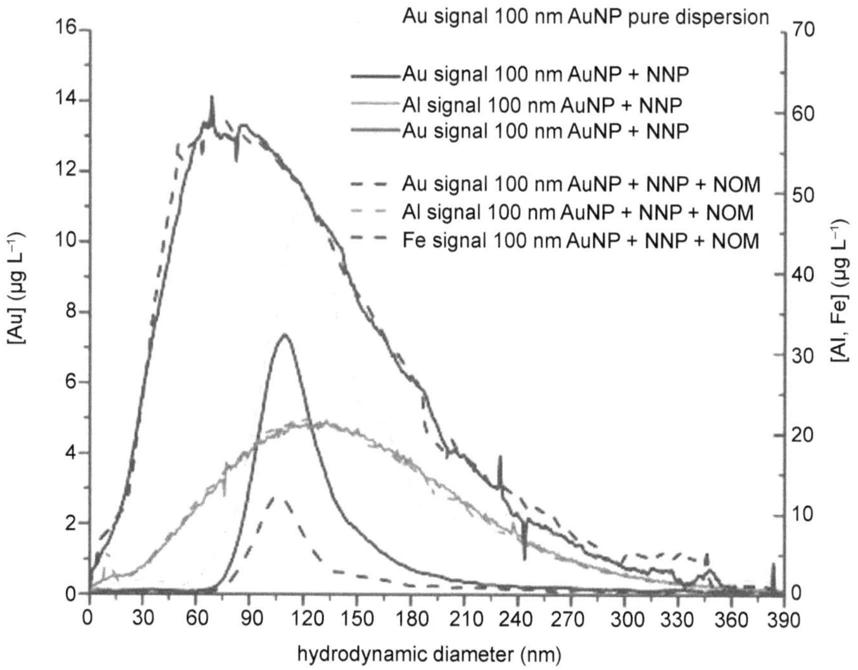

Figure 3. AF^4-ICP-MS fractograms for pure dispersions of 100 nm AuNPs and for mixed samples containing both 100 nm AuNPs and NNPs (solid lines: samples with no added NOM; dashed lines: samples with added NOM) (Meisterjahn et al. 2014).

or off-line HR-ICP-MS. The type of mobile phase was a critical factor in the fractionation of AgNP by AF^4. In synthetic systems, fractionation of a series of virgin citrate- and PVP-coated AgNPs (10–90 nm) with reasonably high recoveries could only be achieved with ultrahigh purity water as mobile phase. However, in soil-water extracts, 0.01 percent sodium dodecyl sulfate (SDS) at pH 8 was the key to a successful fractionation of the AgNP. Similarly, Meisterjahn et al. (2014) demonstrated the applicability of this technique for the detection, quantification and characterization of AuENPs in the presence of natural NPs, and different concentrations of natural organic matter (NOM) (Fig. 3). The combination of several detectors (i.e., light absorbance, MALS and ICP-MS) was useful for differentiating between heteroaggregation and homoaggregation of the NPs.

Nischwitz et al. (2016) went one step forward applying AF^4 on-line with ICP-MS for quantitative size-resolved carbon determination of ENPs and colloids in complex sample matrices, due to the key role of carbon in biological and environmental processes. This study explored the potential of online particulate carbon detection by ICP-MS to overcome limitations of UV detection or offline total organic carbon measurements. A novel organic carbon detector (OCD) was used as independent sensitive carbon detector to validate the results obtained by ICP-MS, using offline quadrupole (Q) or sector-field (SF). Limits of detection for particulate carbon are at least 10-fold higher using AF^4-ICP-MS compared to AF^4-OCD, but sufficient for monitoring of natural carbon containing NPs and colloids in real environmental samples.

Liquid Chromatography (LC)

There are different LC modes that have been combined with ICP-MS to characterize NMs. The most commonly applied methods are hydrodynamic chromatography (HDC), size exclusion chromatography (SEC) or reverse-phase-liquid phase chromatography (RP-HPLC).

SEC is a common name for liquid-chromatographic methods that separates analytes based on their effective shape and bulkiness at relatively low pressures by the differences in penetrating the different diameter pores of the stationary phases. Large particles cannot enter into many pores. Then, the larger the particles, the faster the elution. Zhou et al. (2016b) successfully applied this technique to the analysis of metal oxide NPs (MO NPs) and their released metal ions (M^{n+}) by using a 1000 Å pore size silica column, and an acetate buffer containing SDS as a mobile phase. M^{n+} and MO NPs with sizes smaller than the column pore size were baseline separated within 10 min, whereas MO NPs with sizes larger than the exclusion limit pass by the pores and elute quickly. More importantly, this mobile phase is able to avoid the dissolution of MO NPs during the SEC separation. The high recoveries (> 97%) of M^{n+} from the SEC column ensured their accurate quantification directly with the online coupled SEC-ICP-MS, while the quantification of MO NPs by subtracting the M^{n+} content from the total metal content efficiently eliminated the influence of large MO NPs retention in the column. MO NPs with different sizes and compositions, including NiO, CoO, ZnO, CuO and CeO_2 NPs, were analyzed with this method, with low detection limits (0.016–0.390 µg/L for both M^{n+} and MO NPs) and relative standard deviations (< 0.4% for retention time and < 2.2% for peak area at 50 µg/L M^{n+}). The proposed method was verified by analyzing Ni^{2+}, Ce^{4+}, NiO and CeO_2 NPs in various environmental waters, with spiked recoveries in the range of 80.8–105.1% (Zhou et al. 2016b). The method showed great potential for speciation analysis of MO NPs and their corresponding M^{n+} in environmental waters. Zhou et al. (2017) characterized size, mass quantification and composition of trace NPs in complex matrixes using a 1-step method developed through the online coupling of SEC-ICP-MS. The use of a mobile phase with a relatively high ionic strength ensured the complete elution of different-sized NPs from the column and, therefore, a size-independent response. After application of a correction for instrumental broadening by a method developed in this study, the size distribution of NPs agreed closely with that obtained from TEM analysis. Compared with TEM, this method attained a more rapid determination with a higher mass sensitivity (1 pg for gold and silver NPs) and comparable size discrimination (0.27 nm). The proposed method was also capable of identifying trace Ag_2S NPs and core-shell nanocomposite Au@Ag, as well as quantitatively tracking the dissolution and size transformation of silver NPs in serum and environmental waters (Zhou et al. 2017).

Different, HDC separates particles based on the particle's diffusion coefficients (inversely related to their hydrodynamic diameters through the Stokes–Einstein equation). It involves the application of a non-porous and non-coated silica bead-containing column that does not interact with the sample. HDC separates the NPs based on the flow rate and the velocity gradient across the particle. In the narrow conduit, small particles need more time to be transported by eluents because they transfer through the edge of these non-porous beads; on the contrary, large particles can move directly through the interspace of the conduit, allowing them to be transferred at a faster rate, resulting in a shorter retention time. Minimal sample preparation is required and, in theory, minimal sample perturbation occurs during the passage of the sample through the HDC column. Consequently, particle size separation can be achieved. The particle size can then be calculated since the retention time is highly correlated to it. Although only a few research groups have examined ENPs detection using HDC, Philippe and Schaumann (2014) evaluated HDC coupled with ICP-MS for the analysis of metal NPs in environmental samples. Using two commercially available columns, a set of well characterized calibrants and a new external time marking method, HDC, showed that flow rate and eluent composition have little influence on the size resolution and, therefore, can be adapted to the sample particularity. HDC-ICP-MS was successfully applied to samples containing a high organic and ionic background. Indeed, online combination of UV-visible, fluorescence and ICP-MS detectors allowed distinguishing between organic molecules and inorganic colloids during the analysis of Ag NPs in synthetic surface waters. The results showed that HDC-ICP-MS is a flexible, sensitive and reliable method to measure the size and the concentration of inorganic colloids in complex media, and suggest that there may be a promising future for the application of HDC in environmental science (Philippe and Schaumann 2014). As examples, the chromatograms sets for a suspension containing 5 mg L^{-1} of humic acid, the initial suspension in milli Q water and a suspension in moderate soft water containing no NOM are presented in Fig. 4a–c.

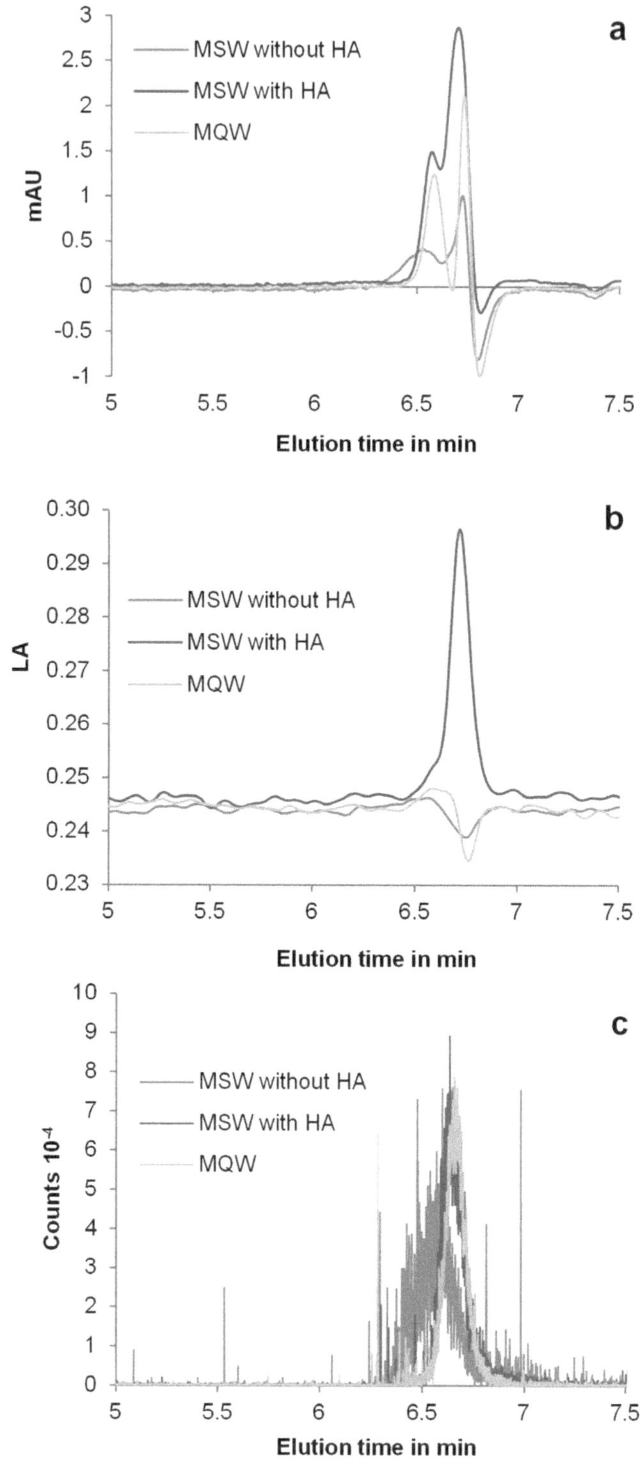

Figure 4. HDC-UVD-FLD-ICP-MS chromatograms of silver NPs in synthetic surface water. HDC chromatograms of silver NPs (core diameter: 40 nm) in milli Q water, moderate soft water and moderate soft water with 18 mg L21 of humic acid. (a) 107 Ag signal from ICP-MS detector; (b) UV-VIS signal (labs = 410 nm); (c) fluorescence signal (lext = 430 nm, lem = 500 nm). The delay between the UV-VIS detector and the ICP-MS detector was around 4 s. The dwell time of the quadrupole detector was 10 ms for all measured elements (Philippe and Schaumann 2014).

Furthermore, Proulx and Wilkinson (2014) tested the hydrodynamic separation of Au, polystyrene (PS) and Ag-ENP size standards and their mixtures. Techniques were first optimized at high concentrations with the available on-line detectors (static (SLS) and dynamic light scattering (DLS)), which provide direct information on the radii of gyration (Rg) and the hydrodynamic radii (Rh) of the ENPs. Once the experimental parameters were well defined for the elution, off-line detection of HDC fractions was carried out using DLS, analytical ultracentrifugation (AUC) and sp ICP-MS at lower ENP concentrations. Finally, the direct coupling of the HDC column to sp-ICP-MS allowed the direct detection and characterization of AgNPs spiked into a river water at the environmentally relevant concentration of 4 $\mu g\ L^{-1}$.

Reversed-phase chromatography is the most common HPLC separation technique and is used for separating compounds according to their polarity. In RP-HPLC, the stationary phase is less polar than the mobile phase and the interaction between analyte and the stationary phase has a predominantly hydrophobic (apolar) character. A robust method of HPLC-ICP-MS is established to provide the information of the characterization and quantitative determination of metallic NPs and their ionic component in complex matrices. The metallic NPs (< 40 nm) and ionic component could be effectively separated with the addition of SDS in the mobile phase and 1000 Å pore sized C18 column. The detected size of metallic NPs with this proposed method was consistent with the results of TEM, demonstrating that this accurate method is promising to characterize NPs in environmental waters. Moreover, this separation method could be used for the quantitative detection of AuNPs and Au(Ⅲ) mass concentration. The concentration of soluble Au(Ⅲ) in environmental waters could be directly detected using the proposed HPLC-ICP-MS; then the concentration of AuNPs was obtained by subtracting the Au(Ⅲ) concentration from the total Au. The concentrations of spiked AuNPs and Au(Ⅲ) in environmental water were recovered 90.2 ~ 101.2 percent and 91.6–114.6 percent, respectively, the detection limit was at sub-μg/L level. This approach provides an efficient and accurate method to study the aggregates of metallic NPs in environmental water (Yang et al. 2018).

Single Particle (sp)-ICP-MS

Single particle (sp)-ICP-MS is an emerging and powerful technique for NP analysis due to its ability to characterize NPs directly in aqueous and environmental samples, including simultaneous determination of particle size, size distribution, particle concentration and dissolved ion concentration at levels down to μgL^{-1}. The theory of (sp)-ICP-MS has been described elsewhere (Laborda et al. 2016, Montaño et al. 2016). Briefly, if a sufficiently diluted suspension of NPs is nebulized into the plasma, each ionized particle generates a discrete ion packet that can be detected as pulse (non-continuous) signals by the mass spectrometer. The area of the peak (Ip) is proportional to the particle diameter (or mass of the element within the particle) and the particle number concentration can be determined from the number of particle events (Np) during an acquisition time (t). The time-independent background signal is proportional to the dissolved element concentration.

This technique requires little sample preparation. Most of the methods for several waters, including wastewater, report direct sample analysis or analysis after a minor sample treatment as filtration, centrifugation, ultrafiltration or even dilution to ensure that ENPs concentration is sufficient to overpass instrumental sensitivity. Instead, other techniques emphasize the need to increase the stability of the ENPs suspension in water by sonication (to improve homogeneity of ENPs) or by addition of a protein, such as gelatin. Interestingly, Loula et al. (2018) studied how to enlarge short-term stability of Ag NPs in demineralized water. The study demonstrated that stability could be prolonged at least 7 h by the addition of 0.05% percent gelatin. However, the estimate of transport efficiency is distorted when concentrated dispersions of NPsare analyzed, because of the overlapping of signals of multiple NPs. This effect was observed for dispersions of concentrations greater than $1 \times 10^6\ mL^{-1}$, where an apparent decrease in transport efficiency, from an initial value 7–8 percent to 1 percent, was observed

Although (sp)-ICP-MS is a promising technique that has evolved in the last years, it still have some limitations to determine ENPs that have been remarked in the review of Laborda et al. (2016) and in some of the studies reported in Table 1. These limitations are related to:

- The velocity of data acquisition of the mass spectrometer
- The ICP pneumatic nebulization systems
- Protocols to process raw data

This system was initially developed with quadrupole mass analyzers with frequency of data acquisition dependent on applicable reading times (selected through the quadrupole dwell time per isotope) and the transmission and storage of data. The short transient signals, between 300–400 µs, generated by single particle event presents a number of challenges for conventional ICP-MS instrumentation, since dwell times were limited to the millisecond range, the 3–10 ms being commonly used, which implied recording the particle events as pulses. This can result in particle events that are missed entirely or the measurement of partial events. Multiple particle events occurring within a single measurement dwell time could result in overestimation of the particle size distribution and underestimation of the particle concentration. The ability to discriminate single particle events from the signal background is also dependent on the dwell time, as the continuous background signal increases with the dwell time, while the signals arising from single particle events do not.

Most recently, fast scanning quadrupole instruments are now commercially available with dwell times in the microsecond range, which enables the particle event to be recorded as a fully resolved transient signal, defined by multiple points, with an area proportional to the mass of the element in the particle. Double focusing instruments can also be run in fast scanning mode, by scanning the acceleration voltage instead of the magnetic field, allowing the use of 100 µs reading times. In comparison to quadrupole mass analyzers, magnetic sector instruments have improved sensitivity, which facilitates the measurement of smaller particle sizes and the ability to resolve isobaric interferences without ambiguity. An example of the capabilities of these double sector instruments is presented by Newman et al. (2016) that used a magnetic sector ICP-MS with fast, continuous data acquisition to characterize and quantify AgNPs of different sizes prepared from commercially available standard solutions, as well as AgNPs suspended in natural waters. Furthermore, Tuoriniemi et al. (2017) followed the evolution of the mass distributions and number concentrations of 40 nm Ag NPs in a mesocosm simulating a wastewater treatment plant using fast HRMS. It was thus possible to detect smaller Ag containing NPs than hitherto possible in similar studies. These small particles (ca. 5–10 nm in corresponding metallic Ag equivalent spherical diameter) were possibly dissolved $Ag+$ precipitated as Ag_2S particles. An ICP time-of-flight mass spectrometer with a temporal resolution of 30 µs and full multi-element capability was developed recently, and it is commercially available at present.

Key points are the (sp)-ICP-MS pneumatic nebulizers,which produce polydisperse aerosol that passes through a spray chamber, in which most of the liquid is lost due to collision with the chamber's wall (loss up to 99 percent). Therefore, with this setup, the quantification of the introduced NPs is only possible if the aerosol transport efficiency (TE) is known, where TE is defined as the ratio of the volume reaching the plasma to the initially nebulized one. A number of methods to measure the aerosol TE have been proposed. Liu et al. (2017) presented a comprehensive and systematic study of the accuracy, precision and robustness of (sp)-ICP-MS using the rigorously characterized reference material (RM) 8017 (Polyvinylpyrrolidone Coated Nominal 75 nm Silver NPs), recently issued by the National Institute of Standards and Technology (NIST). This study reports statistically significant differences in frequency-based and size-based measures of TE with NIST RM 8013 Gold NPs, and demonstrated that the size-based measure of TE is more robust and yields accurate results for the silver nanoparticle RM relative to TEM-based reference values. This finding could change the way to proceed because the frequency-based method to calculate TE is more widely applied than size-based measures.

In order to improve the TE efficiency, most newly developed monodisperse droplet generators (MDG) have been used for the introduction of nanoparticle suspensions. The MDG delivers discrete low-volume samples. It has been shown that MDG-ICP-MS can be used for the quantification of the mass/size of metallic and metal oxide NPs, either by internal or external calibration, without using any particle-based reference material and can be used for simultaneously accessing ionic and particulate fractions of various elements using a time-of-flight (ToF) ICP-MS. When using the MDG, the sample introduction volume can be easily increased by changing the nozzle sizes from 30 μm (droplet size ~ 40 μm) to 70 μm (droplet size ~ 90 μm), or by increasing the introduction frequency without changing the droplet TE of 100 percent (Gschwind et al. 2017).

The high elemental selectivity and sensitivity of the ICP-MS technique, combined with particle counting capability, enables both dissolved and colloidal Ag to be analyzed in a single measurement. One of the limitations of (sp)-ICP-MS analysis has been the lower size limit for detection of NPs, of around 20 nm for Ag. However, recent advances in instrumentation and manipulation of the data generated by (sp)-ICP-MS are expanding the lower range of sizes that can be detected. An interesting example of how the manipulation of data can affect results is provided in Fig. 5. Inspection of the raw data showed that the software had incorrectly assigned background signals as peaks, as shown in Fig. 5a. Using a peak threshold of 4, the contributions from the smaller particles (i.e., with diameters of 8 and 9 nm) are underestimated by (sp)-ICP-MS and this is reflected in the reported mean particle diameter (11.7 × 3.2 nm). Inspection of the raw data (Fig. 5b) shows that peaks were correctly identified. When the peak threshold was increased to 6, the software incorrectly assigned peaks to the background signal, as shown in Fig. 5c in which only the larger of the two peaks (with measured diameter of 10.49 nm) was identified as a particle event. Inspection of the raw data is required to optimize the peak search criteria and minimize contributions from false positives and false negatives to the measured particle size distribution.

There have been a handful of successful reports of (sp)-ICP-MS being used for the analysis of sizes and concentrations of NPs. For example, Peters et al. (2018) applied (sp)-ICP-MS successfully to different types of NPs in river water with average relative standard deviations of 3 percent. Vidmar et al. (2018)also used (sp)-ICP-MS for the sizing and quantification of iron zero valent ions in wastewater matrices.

Liquid Chromatography-Mass Spectrometry Approaches

The preferred method for separating aqueous fullerene aggregates is liquid chromatography (LC), and among the different detectors available for quantifying fullerenes in the aquatic environment, several mass spectrometers have been applied. The use of different mass analyzers, such as single quadrupole, triple quadrupole and high resolution mass spectrometry (HRMS) using, for example, hybrid linear trap Orbitrap (hybrid LTQ Orbitrap) have been recently reported (Sanchís et al. 2018, Sanchís et al. 2017, Zakaria et al. 2018, Zouboulaki and Psillakis 2016). Fullerenes ionize in negative ionization mode but do not provide successive fragmentations of the molecule, which explains the role that high-HRMS plays to improve the identification of compounds. Among the different ionization techniques, electrospray ionization (ESI), atmospheric pressure chemical ionization (APCI) and atmospheric pressure photoionization (APPI) were most frequently used for pristine and functionalized fullerenes determination (Li et al. 2012, Núñez et al. 2012). APPI in the negative mode, yields better sensitivity when compared to the rest as a result of the use of toluene in the mobile phase.

Liquid–liquid extraction (LLE) is one of the most commonly used methods for fullerene extraction from environmental waters. Although LLE is a multistep and time-consuming sample preparation method, requiring relatively large sample volumes, this method can easily achieve a limit of detection as low as 3 pg L^{-1} for nC60.

In 2013, solid-phase extraction (SPE) using C18 as sorbent was reported as promising approach for robust extraction of fullerenes from water (Kolkman et al. 2013). However, for reasons that are hard to understand, SPE has not been reported in more recent studies.

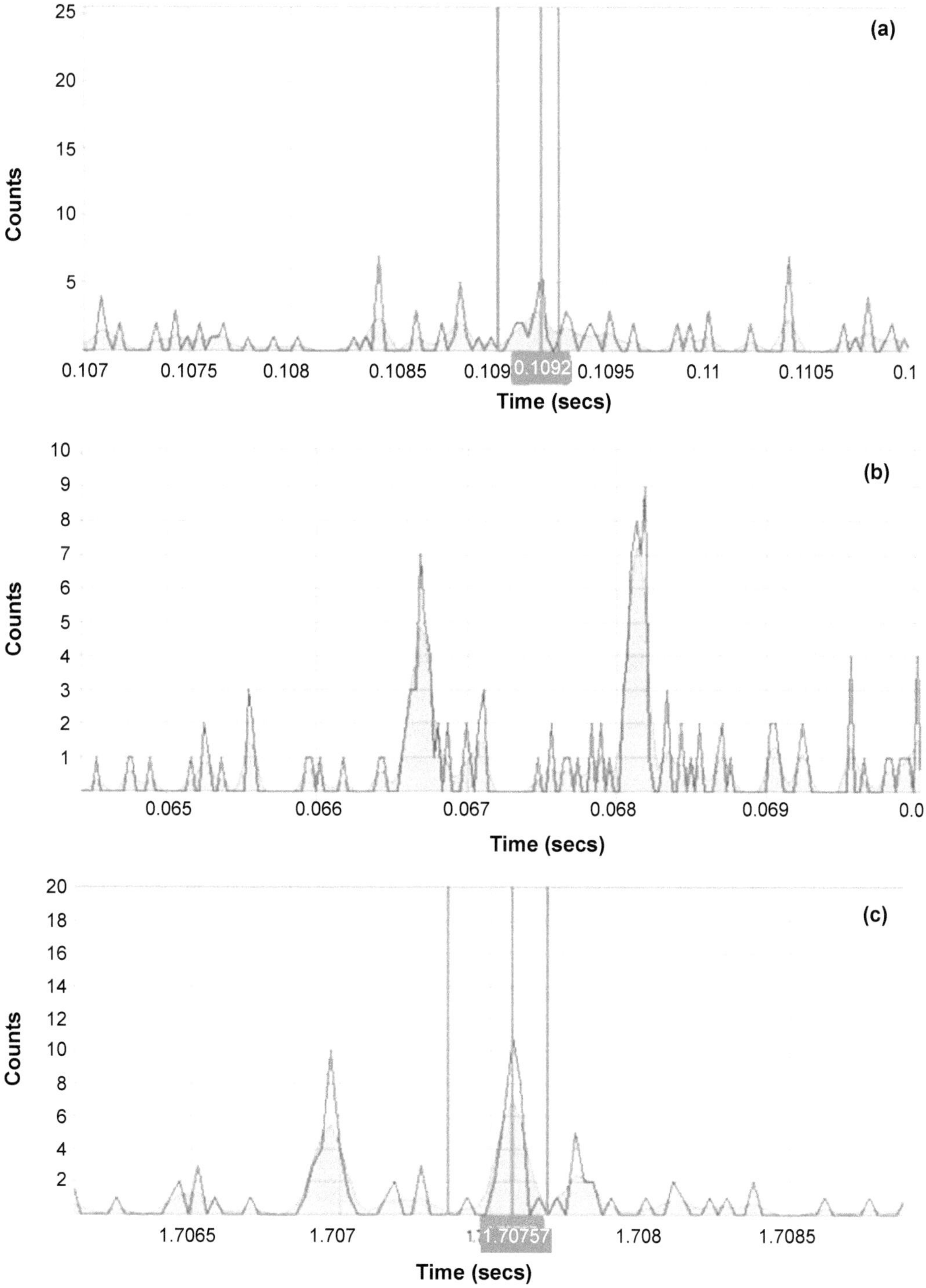

Figure 5. Raw sp-ICP-MS data for 10 nm standard (10 ng Ag L^{-1}) using a peak threshold of (a) 3, grey lines indicate peak start and end points found by the software (false positive) (b) 4, particle events correctly identified with reported particle diameters of 10.59 nm and 11.78 nm and (c) 6, only the second, larger peak is identified as a particle event (false negative); a 3 point 2 iteration rolling smooth (indicated by the shaded area) and 30 μs dwell time was used for all data sets (Newman et al. 2016).

The possibility of using a microextraction based procedure, such as vortex assisted liquid-liquid microextraction (VLLME), for the determination of fullerenes in water was recently reported by Zouboulaki and Psillakis (2016). Vortex agitation accelerated mass transfer from the aqueous sample into the octanol phase and resulted in a short extraction time. Moreover, lowering the pH of the aqueous sample and adding small amounts of salt destabilized nC60 aggregates and facilitated their mass transfer into the octanol phase. The proposed method is a low-cost and easy to use procedure that allows rapid low-level environmental determination of aqueous fullerene aggregates whilst using small sample volumes and micro-volumes of organic solvent.

A Hotchpotch of Very Specific or Recently Applied Analytical Tools

Dynamic Light Scattering (DLS)

Dynamic light scattering (DLS) is one of the most commonly employed instruments to determine the size of NPs in the liquid phase, based on the particle's Brownian motion (Chang et al. 2017). However, most of the studies that apply this detector, alone or combined with a separation technique, describe a number of interferences and consider it only useful for those samples that have high concentration of ENPs, which is not the case in most of the environmental samples (António et al. 2015, Baranidharan and Kumar 2018, Chang et al. 2017, Donovan et al. 2018, la Calle et al. 2017, Proulx and Wilkinson 2014, Telgmann et al. 2016, Yang et al. 2018).

Biosensors

Cayuela et al. (2015) described the synthesis of the first fluorescent nanosensor based on CDs for detecting water-soluble c-MWCNTs as a potential water contaminant using a simple, rapid and economic methodology. This study investigated the importance of synthesizing highly photoluminescence quantum dots (CDs) varying organic precursors by using bottom-up methodology. CDs synthesized from microcrystalline cellulose exhibit more fluorescence intensity and greater fluorescence stability against pH changes, being reported for the first time. The formation of nanohybrids between c-MWCNTs and CDs make possible the determination of such water-soluble c-MWCNTs in aqueous solutions. Thus, this type of CDs has a potential application as a photoluminescence sensor of c-MWCNTs in environmental water samples using the proposed method due to the industrial waste and diesel emissions (Cayuela et al. 2015).

Nanoparticle Tracking Analysis (NTA)

Nanoparticle tracking analysis (NTA) utilizes the properties of both light scattering and Brownian motion in order to obtain particle size distributions of samples in liquid suspension. A laser beam is passed through a prism-edged flat glass within the sample chamber. The angle of incidence and refractive index of the flat glass are designed to be such that when the laser reaches the interface between the glass and the liquid sample layer above it, the beam refracts, resulting in a compressed beam with a reduced profile and high power density. The particles in suspension in the path of this beam scatter light can be detected. NTA was performed in flow mode with temperature control, a CCD monochrome camera and an automated syringe pump.

NTA calibration was carried out with stock suspensions of polystyrene latex NPs with diameters of 100, 200 and 400 nm. A pre-scan was carried out to adjust the camera level and data threshold to the characteristics of each sample. In order to track a significant number of NPs, measurements were carried out in quintuplicate, with recordings of time periods in the range of 60–90 s. The five size population distributions were normalized and averaged to obtain the final plot. Prior to performing the NTA analysis, the samples were filtered through a 450-nm pore size filter to discard micro-sized suspended particles and then were shaken, prior to their analysis (Sanchís et al. 2017).

UV/Vis spectroscopy

Several studies are also based on monitoring ENPs optical properties (extinction spectra) using UV-vis spectroscopy (António et al. 2015, Azodi et al. 2016, Sikder et al. 2018). This technique is universal (any molecule gives signal) and then, has the problem of inter-ferences caused by other matrix components (e.g., organic matter, other salts, etc.), that could be at much higher concentrations than the ENPs.

XPS

X-ray photoelectron spectroscopy (XPS), also known as ESCA (electron spectroscopy for chemical analysis) is a surface analysis technique, which provides both elemental and chemical state information virtually without restriction on the type of material that can be analyzed. These technique has been used to get structural information in combination with other techniques (Azodi et al. 2016).

Applications

The applications of the previously reported techniques in environmental studies are mostly focus on four aspects related to ENP/NM/MNs: (i) stability in aqueous matrices, (ii) kinetics of the agglomeration process, (iii) occurrence in water and (iv) influence on the fate of organic matter.

Methods are needed to prepare stable suspensions of engineered NPs in aqueous matrices for ecotoxicity testing and ecological risk assessments. Martin et al. (2017) developed a novel method of preparing large volumes of AgNPs in suspension using a commercially available rotor–stator dispersion mill. AgNPs in powder form (PVP capped, 30–50 nm) was suspended in deionized water and natural lake water at 1 ngL^{-1}, and the addition of 0.025 percent (w/v) arabic gum (GA) increased stability over two weeks after preparation. Analysis of hydrodynamic diameters of the major peaks in suspension using DLS showed that suspensions prepared with GA were stable, and this was confirmed by single-particle ICP-MS analysis. This method for dispersing AgNPs provides an inexpensive yet reliable method for preparing suspensions for toxicity testing and ecosystem level studies of the fate and biological effects of AgNPs in aquatic ecosystems (Martin et al. 2017). Similarly, sp-ICP-MS has been used to directly measure the NP diameter and particle number concentration of suspensions containing gold–silver core–shell (Au@Ag) NPs prepared in synthetic standardized water—United Stated Environmental Protection Agency (EPA) moderately hard water (MHW), and in this water containing 2.5 mg L^{-1} of fulvic acid isolated from Suwannee River (Merrifield et al. 2017). Based on the analysis of both Au and Ag parameters (size, size distribution and particle number), concentration was shown to be a key factor in NPs behavior. At higher concentration, NPs were in an aggregation-dominated regime, while at the lower and environmentally representative concentrations, dissolution of Ag was dominant and aggregation was negligible. In addition, further formation of ionic silver as Ag NPs, in the form of AgS or AgCl, was shown to occur. Between 1 and 10 µg L^{-1}, both aggregation and dissolution were important. A simpler analysis of AgNPs dissolution, which is an important environmental transformation affecting their size, speciation and concentration in natural water systems, was carried out based on monitoring their optical properties (extinction spectra) using UV-vis spectroscopy (Sikder et al. 2018).

António et al. (2015) showed that it is possible to assess in detail the agglomeration process of silver NPs in artificial seawater by combining different characterization techniques (AF4, sp-ICP-MS, UV–Vis). In particular, the presence of alginate or humic acids differentially affects the kinetic of the agglomeration process. This study provided an experimental methodology for the in-depth analysis of the fate and behaviour of silver NPsin the aquatic environment.

Several studies confirmed the presence of any Ag, Au and TiO_2-NPs in different environmental waters using sp-ICP-MS (Gschwind et al. 2017, Loula et al. 2018, Newman et al. 2016, Venkatesan et al. 2018, Yang et al. 2016a, Yang et al. 2016b, Yang et al. 2018). These studies highlighted very frequently the influence of the coating material on NMs'behavior in water. Surface coating has a determining role on the stability of AgNPs in lake water. PVP showed the highest protective effect of the three surface coatings studied, with lipoic acid being the least protective. The enrichment of lake water with dissolved organic matter from pedogenic or aquagenic origin stabilized 20 and 50 nm citrate- or lipoic acid-coated AgNPs (lake water + extracellular polymeric substances > lake water + Suwannee River Humic Acids > lake water) (Jiménez-Lamana and Slaveykova 2016).

Other studies used selective extraction procedures such as CPE-ICP-MS (El Hadri and Hackley 2017), MSPE with Fe_3O_4 magnetic particles (Zhou et al. 2016a) and SA-DLLME (Liu et al. 2016). Few of them checked the use of separation techniques such as sdFFF (Dutschke et al. 2017, Soto-Alvaredo et al. 2016), AF4 (Furtado et al. 2014, Meisterjahn et al. 2014) or HDC (Philippe and Schaumann 2014, Proulx and Wilkinson 2014). Separation techniques incorporate the interesting feature of characterizing AgNP transformations in natural waters at toxicologically relevant concentrations. Peters et al. (2018) confirmed the presence of nano-sized Ag and CeO_2 particles and micro-sized TiO_2 particles in surface waters. NMs were detected in all samples collected in the rivers Meuse and Ijssel in the Netherlands using sp-ICP-MS as a measurement technique. A further study performed in other river in the Netherlands (the river Dommel) investigated the occurrence of seven metal-based NPs, simultaneously (Markus et al. 2018). These samples were analysed using HR-ICP-MS, ultrafiltration with a sequence of mesh sizes and SEM. There are few additional examples of determination of several metal based ENPs simultaneously (Baranidharan and Kumar 2018, Donovan et al. 2018, Zhou et al. 2016b). The most remarkable application to water samples was reported by Lee et al. (2014), who using sp-ICP-MS were able to determine 40 elements (other than a few available assessed ones).

Other applications to water samples combined the outputs of several techniques to achieve a better characterization of the NMs. In this sense, the combination of the results obtained by DLS, AF4-UV-MALLS, AF4-UV-ICP-MS (la Calle et al. 2017) and by these techniques plus TEM, HDC and sp-ICP-MS (Chang et al. 2017) demonstrated to provide a better characterization. These studies offer interesting information on the capabilities of each technique and their complementarity to identify, characterize or determine ENPs.

Zero valent iron (nZVI) is used to eliminate toxic metals in wastewaters; however, there is little information on its behavior after remediation. The influence of different iron loads (0.1, 0.25, 0.5 and 1.0 g L^{-1}) and water matrices (Milli-Q water, synthetic and effluent wastewater) on the behaviour of nZVI, their interactions with Cd^{2+} and the efficiency of Cd^{2+} removal has been studied using sp-ICP-MS. This study concluded that aggregation and sedimentation of nZVI increased with settling time and was slower in effluent wastewater than in Milli-Q water or synthetic wastewater (Vidmar et al. 2018). Model experiments performed on a laboratory scale enabled us to better understand the fate of the FeNPs after the treatment of environmental water samples. Peeters et al. (2016) showed that the Fe_3O_4NPs dispersed by mixing most effectively, removed the metals from the waters and settled faster in the investigated waters, providing interesting information on the behaviour of these ENPs in natural waters.

Dissolved organic matter (DOM) is a key factor affecting the behaviors and fate of AgNPs in the aquatic environment. However, the mechanisms for the DOM-mediated transformations of AgNPs are still not fully understood. Zhang et al. (2017) investigated the persistence of AgNPs in the aquatic environment in the presence of different concentrations of humic acids (HA) over periods of time up to 14 days. The Ag species were monitored and characterized by absorption spectrometry, TEM, ICP-MS and multicollector ICP-MS.

Up to the moment, several studies pointed out that any transformation of the AgNPs in water occurred in the suspended solids but not dissolved. Most of the larger AgNPs were removed by aggregation with large floc particles and subsequent sedimentation with the suspended particulate matter in the simulated WWTP process (Tuoriniemi et al. 2017). The retention of AgNPs were also tested in constructed wetlands and sludges (Auvinen et al. 2017). The results obtained imply that the biofilm in constructed wetlands will act as a sink for Ag-NPs, and for activated sludge. The release of (transformed) Ag-NPs during normal operation is primarily determined by the discharge of total suspended solids (Auvinen et al. 2017).

Site specific risk assessment of ENPs requires spatially resolved fate models. Validation of such models is difficult due to present limitations in detecting ENPs in the environment. An interesting study showed the progress towards validation of the spatially resolved hydrological ENP fate model NanoDUFLOW, by comparing measured and modeled concentrations of < 450 nm metal-based particles in river's waters (De Klein et al. 2016). Concentrations measured with Asymmetric Flow-Field-Flow Fractionation (AF4) coupled to ICP-MS clearly reflected the hydrodynamics of the river and showed satisfactory to good agreement with modeled concentration profiles. Together with the general applicability of the model framework, this legitimizes an optimistic view on the potential to validate such models, with important implications for the risk assessment of ENPs (De Klein et al. 2016).

Ag NPs are used in various consumer products and a significant fraction is eventually discharged with municipal wastewater. A study by Azodi et al. (2016) assessed the release of Ag from PVP- and citrate-coated 80 nm nAg in aerobic waste water effluent and mixed liquor, and the related changes in nAg size, using sp-ICP-MS. The concentration of dissolved (nonparticulate) Ag in waste water effluent was 0.89 ± 0.05 ppb at 168 h, 71 percent lower than in deionized water, in batch reactors spiked with 5×10^6 PVP-AgNPs mL^{-1} (10 μgL^{-1}), an environmentally relevant concentration. Dissolved Ag in waste water was partly reformed into ~ 22 nm nAgxSy by inorganic sulfides and organosulfur dissolved organic carbon (DOC) after 120 h, whereas the parent nAg mean diameter decreased to 65.89 ± 0.9 nm. Reformation of nAgxSy from Ag+ also occurred in cysteine solutions but not in deionized water, or humic and fulvic acid solutions. Dissolution experiments with nAg in waste water mixed liquor showed qualitatively similar dissolution trends. Time-of-flight secondary ion mass spectrometry (ToF-SIMS) and X-ray photoelectron spectroscopy (XPS) analyses indicated binding of thiol- and amine-containing DOC, as well as inorganic sulfides with Ag NP. Those wastewater components, as well as limited dissolved oxygen, decreased dissolution in wastewater (Azodi et al. 2016).

The interplay between ENP size, surface area and dissolution rate is critical in predicting ENP environmental behavior. sp-ICP-MS enables the study of ENPs at dilute (ng L^{-1}) concentrations, facilitating the measurement of ENP behavior in natural systems. Here, the utility of using sp-ICP-MS to quantitatively track the changes in particle diameter over time for 60 and 100 nm Ag ENPs (citrate, tannic acid and polyvinylpyrrolidone coated) was demonstrated. Short term (< 24 h) and intermediate term (1 week) dissolution showed that Ag+ dissolution is about 70% at the beginning but slow down by over an order of magnitude after approximately 24 h. Dissolution was measured primarily as a decrease in particle diameter over time but direct measurement of Ag+ (aq) was also completed for the experiments. The importance of water chemistry including chloride, sulfide, and DOC was demonstrated, with higher concentrations (1 mg L^{-1} Cl^-, S^{2-} and 20 mg L^{-1} DOC), resulting in negligible Ag ENP dissolution over 24 h. Slight decreases in particle diameter (< 10%) were observed with lower concentrations of these parameters (stoichiometric Cl^-, S^{2-} and 2 mg L^{-1} DOC). Capping agents showed variable effects on dissolution. ENP behavior was also investigated in natural (moderately hard water, creek water) and tap water. Water chemistry was the most significant factor affecting dissolution. Near complete dissolution was observed in chlorinated tap water within several hours. Though modeled as first-order kinetic transformations, the dissolution rates observed suggested that the dissolution kinetics might be significantly more complex. Two specific highlights of the benefits of using the sp-ICP-MS technique to measure dissolution in complex samples include: (1) the measurement of primary particle size as the metric of dissolution is more direct than attempting to measure the increase of Ag^+ in the solution, and (2) that this is possible even when known sinks for Ag^+ exist in the system (e.g., DOC, sediments, biota, sampling container) (Mitrano et al. 2014).

Up till now, all reported applications were based on the determination of metal-based NPs. However, few report the occurrence of carbon-based NMs in water, river water and wastewater (Nieschwith et al. 2016, Cayuela et al. 2015, Sanchis et al. 2017, Sanchis et al. 2018, Zakaria et al. 2018, Zouboulaki and Psillakis 2016). Qualitative and quantitative detection of particulate carbon in water and soil using AF^4-ICP-MS was successfully demonstrated (Nischwitz et al. 2016). The formation of nanohybrids between c-MWCNTs and CDs make possible the determination of such water-soluble c-MWCNTs in aqueous solutions (Cayuela et al. 2015). Fullerenes have been widely reported in different types of water by LC-MS (Sanchís et al. 2017, Zakaria et al. 2018, Zouboulaki and Psillakis 2016). In an interesting application, Sanchís et al. (2018) studied the degradation of fullerene aggregates by UHPLC-Orbitrap-MS in a series of C60 fullerene water suspensions under relevant environmental conditions, controlling the salinity, the humic substances content, the pH and the sunlight irradiation. Up to ten transformation products were tentatively identified, including epoxides and dimers with two C60 units linked via one or two adjacent furane-like rings. The knowledge is starting but still there are many gaps that needs to be filled in this area.

Conclusions

The methods for the screening of ENP/NM/MNs in environmental samples are reaching maturity. The analysis of NPs in the environment is not question of a single analytical technique, but rather a combination of multiple sophisticated procedures and instrumentation. Many of these methods rely on chromatographic separations that lack stationary phases because of the surface reactivity of particles (for example, FFF techniques). Detectors such as TOF and ICP-MS are being adapted to provide sensitive and specific detection of ENMs. Traditional electron and optical microscopy based approaches are being augmented by minimally invasive techniques, such as ESEM or STEM, which are simpler to apply and less invasive. STEM holds promise for internal imaging and 3D reconstruction with samples held under minimally destructive cryogenic conditions. The availability of routine analytical methods that address these issues is already a key to gain a better understanding of the mechanisms of NPs formation and reactivity.

Despite not being completely resolved yet, there are already protocols established for the determination of some of them. For example, in the case of metals sample dilution followed by sp-ICP-MS, or, in the case of fullerenes, extraction with toluene followed by LC-MS, are well established routine protocols. As mentioned before, existing methods are not perfect and there is still much work to be done.

Possible modifications that could determine key transformation products include biodegradation/ chemical transformation, physical attenuation such as aggregation and surface modifications. Development of analytical schemes (e.g., techniques and methodology) is needed for characterization of transformed engineering NMs in environmental matrices such as air, water, soil, sediments, biota, etc. As with conventional chemicals, to understand transport, degradation and fate, it is essential to first look at the NMs from a life cycle perspective.

Acknowledgment

This work has been supported by the Generalitat Valenciana through the project ANTROPOCEN@ (PROMETEO/2018/155) http://antropocena.com.

References

António, D.C., C. Cascio, Ž. Jakšić, D. Jurašin, D.M. Lyons, A.J.A. Nogueira et al. 2015. Assessing silver nanoparticles behaviour in artificial seawater by mean of AF4 and spICP-MS. Mar. Environ. Res. 111: 162–169.

Auvinen, H., R. Kaegi, D.P.L. Rousseau and G. Du Laing. 2017. Fate of silver nanoparticles in constructed wetlands—a microcosm study. Water Air & Soil Pollut. 228.

Azodi, M., Y. Sultan and S. Ghoshal. 2016. Dissolution behavior of silver nanoparticles and formation of secondary silver nanoparticles in municipal wastewater by single-particle ICP-MS. Environ. Sci. Technol. 50: 13318–13327.

Baranidharan, S. and A. Kumar. 2018. Preliminary evidence of nanoparticle occurrence in water from different regions of Delhi (India). Environ. Monit. Assess. 190 (art. 240).

Baun, A., P. Sayre, K.G. Steinhäuser and J. Rose. 2017. Regulatory relevant and reliable methods and data for determining the environmental fate of manufactured nanomaterials. NanoImpact 8: 1–10.
Bundschuh, M., J. Filser, S. Lüderwald, M.S. McKee, G. Metreveli, G.E. Schaumann et al. 2018. Nanoparticles in the environment: Where do we come from, where do we go to? Environ. Sci. Eur. 30 (art. 6).
Cayuela, A., M.L. Soriano and M. Valcárcel. 2015. Photoluminescent carbon dot sensor for carboxylated multiwalled carbon nanotube detection in river water. Sens. Actuators B Chem.: 596–601.
Chadha, N., S. Lal, A.K. Mishra, R. Pulicharla, M. Cledon, S.K. Brar et al. 2016. Different analytical approaches for the determination of presence of engineered nanomaterials in natural environments. J. Hazard. Toxic Radioact. Waste 20 (art. B4015002).
Chang, Y.J., Y.H. Shih, C.H. Su and H.C. Ho. 2017. Comparison of three analytical methods to measure the size of silver nanoparticles in real environmental water and wastewater samples. J. Hazard. Mat. 322: 95–104.
De Klein, J.J.M., J.T.K. Quik, P.S. Bäuerlein and A.A. Koelmans. 2016. Towards validation of the NanoDUFLOW nanoparticle fate model for the river Dommel, the Netherlands. Environ. Sci. Nano 3: 434–441.
De Marchi, L., C. Pretti, B. Gabriel, P.A.A.P. Marques, R. Freitas and V. Neto. 2018. An overview of graphene materials: Properties, applications and toxicity on aquatic environments. Sci. Total Environ. 631-632: 1440–1456.
Donovan, A.R., C.D. Adams, Y. Ma, C. Stephan, T. Eichholz and H. Shi. 2016a. Detection of zinc oxide and cerium dioxide nanoparticles during drinking water treatment by rapid single particle ICP-MS methods. Anal. Bioanal. Chem. 408: 5137–5145.
Donovan, A.R., C.D. Adams, Y. Ma, C. Stephan, T. Eichholz and H. Shi.. 2016b. Single particle ICP-MS characterization of titanium dioxide, silver, and gold nanoparticles during drinking water treatment. Chemosphere 144: 148–153.
Donovan, A.R., C.D. Adams, Y. Ma, C. Stephan, T. Eichholz and H. Shi. 2018. Fate of nanoparticles during alum and ferric coagulation monitored using single particle ICP-MS. Chemosphere 195: 531–541.
Dutschke, F., J. Irrgeher and D. Pröfrock. 2017. Optimisation of an extraction/leaching procedure for the characterisation and quantification of titanium dioxide (TiO2) nanoparticles in aquatic environments using SdFFF-ICP-MS and SEM-EDX analyses. Anal. Met. 9: 3626–3635.
El Hadri, H. and V.A. Hackley. 2017. Investigation of cloud point extraction for the analysis of metallic nanoparticles in a soil matrix. Environ. Sci. Nano 4: 105–116.
Folens, K., T. Van Acker, E. Bolea-Fernandez, G. Cornelis, F. Vanhaecke, G. Du Laing et al. 2018. Identification of platinum nanoparticles in road dust leachate by single particle inductively coupled plasma-mass spectrometry. Sci. Total Environ. 615: 849–856.
Furtado, L.M., M.E. Hoque, D.F. Mitrano, J.F. Ranville, B. Cheever, P.C. Frost et al. 2014. The persistence and transformation of silver nanoparticles in littoral lake mesocosms monitored using various analytical techniques. Environ. Chem. 11: 419–430.
Furtado, L.M., M. Bundschuh and C.D. Metcalfe. 2016. Monitoring the fate and transformation of silver nanoparticles in natural waters. Bull. Environ. Contam. Toxicol. 97: 449–455.
Giese, B., F. Klaessig, B. Park, R. Kaegi, M. Steinfeldt, H. Wigger, A. Von Gleich and F. Gottschalk. 2018. Risks, release and concentrations of engineered nanomaterial in the environment. Sci. Rep. 8 (art. 1565).
Gschwind, S., M.D.L. Aja Montes and D. Günther. 2017. Comparison of sp-ICP-MS and MDG-ICP-MS for the determination of particle number concentration. Anal. Bioanal. Chem. 407.
Hu, X., C. Ren, W. Kang, L. Mu, X. Liu, X. Li, T. Wang and Q. Zhou. 2018. Characterization and toxicity of nanoscale fragments in wastewater treatment plant effluent. Sci. Total Environ. 626: 1332–1341.
Jiménez-Lamana, J. and V.I. Slaveykova. 2016. Silver nanoparticle behaviour in lake water depends on their surface coating. Sci. Total Environ. 573: 946–953.
Kolkman, A., E. Emke, P.S. Bäuerlein, A. Carboni, D.T. Tran, T.L. Ter Laak et al. 2013. Analysis of (functionalized) fullerenes in water samples by liquid chromatography coupled to high-resolution mass spectrometry. Anal. Chem. 85: 5867–5874.
Koopmans, G.F., T. Hiemstra, I.C. Regelink, B. Molleman and R.N.J. Comans. 2015. Asymmetric flow field-flow fractionation of manufactured silver nanoparticles spiked into soil solution. J. Chromatogr. A 1392: 100–109.
la Calle ID, Pérez-Rodríguez P, Soto-Gómez D, López-Periago JE. 2017. Detection and characterization of Cu-bearing particles in throughfall samples from vine leaves by DLS, AF4-MALLS (-ICP-MS) and SP-ICP-MS. Microchem. J. 133: 293–301.
Laborda, F., E. Bolea and J. Jiménez-Lamana. 2016. Single particle inductively coupled plasma mass spectrometry for the analysis of inorganic engineered nanoparticles in environmental samples. Trends Environ. Anal. Chem. 9: 15–23.
Laux, P. et al. 2018. Challenges in characterizing the environmental fate and effects of carbon nanotubes and inorganic nanomaterials in aquatic systems. Environ. Sci. Nano 5: 48–63.
Lee, S., X. Bi, R.B. Reed, J.F. Ranville, P. Herckes and P. Westerhoff. 2014. Nanoparticle size detection limits by single particle ICP-MS for 40 elements. Environ. Sci. Technol. 48: 10291–10300.
Leopold, K., A. Philippe, K. Wörle and G.E. Schaumann. 2016. Analytical strategies to the determination of metal-containing nanoparticles in environmental waters. TrAC - Trends Anal. Chem. 84: 107–120.
Li, L., S. Huhtala, M. Sillanpää and P. Sainio. 2012. Liquid chromatography-mass spectrometry for C 60 fullerene analysis: Optimisation and comparison of three ionisation techniques. Anal. Bioanal. Chem. 403: 1931–1938.
Liu, J., K.E. Murphy, M.R. Winchester and V.A. Hackley. 2017. Overcoming challenges in single particle inductively coupled plasma mass spectrometry measurement of silver nanoparticles. Anal. Bioanal. Chem. 409: 6027–6039.
Liu, Y., M. He, B. Chen and B. Hu. 2016. Ultra-trace determination of gold nanoparticles in environmental water by surfactant assisted dispersive liquid liquid microextraction coupled with electrothermal vaporization-inductively coupled plasma-mass spectrometry. Spectrochim. Acta. Part B Atom. Spectrosc. 122: 94–102.

Long, C.L., Z.G. Yang, Y. Yang, H.P. Li and Q. Wang. 2016. Determination of gold nanoparticles in natural water using single particle-ICP-MS. J. Central South Univ. 23: 1611–1617.

Loula, M., A. Kaňa, R. Koplík, J. Hanuš, M. Vosmanská and O. Mestek. 2018. Analysis of silver nanoparticles using single-Particle inductively coupled plasma – Mass spectrometry (ICP-MS): Parameters affecting the quality of results. Anal. Let.: 1–20.

Markus, A.A., P. Krystek, P.C. Tromp, J.R. Parsons, E.W.M. Roex, P.D. Voogt and R.W.P.M. Laane. 2018. Determination of metal-based nanoparticles in the river Dommel in the Netherlands via ultrafiltration, HR-ICP-MS and SEM. Sci. Total Environ. 631-632: 485–495.

Martin, J.D., L. Telgmann and C.D. Metcalfe. 2017. A method for preparing silver nanoparticle suspensions in bulk for ecotoxicity testing and ecological risk assessment. Bull. Environ. Contam. Toxicol. 98: 589–594.

Meisterjahn, B., E. Neubauer, F. Von der Kammer, D. Hennecke and T. Hofmann. 2014. Asymmetrical flow-field-flow fractionation coupled with inductively coupled plasma mass spectrometry for the analysis of gold nanoparticles in the presence of natural nanoparticles. J. Chromatogr. A 1372: 204–211.

Merrifield, R.C., C. Stephan and J. Lead. 2017. Determining the concentration dependent transformations of Ag nanoparticles in complex media: Using SP-ICP-MS and Au@Ag core-shell nanoparticles as tracers. Environ. Sci. Technol. 51: 3206–3213.

Mitrano, D.M., J.F. Ranville, A. Bednar, K. Kazor, A.S. Hering and C.P. Higgins. 2014. Tracking dissolution of silver nanoparticles at environmentally relevant concentrations in laboratory, natural, and processed waters using single particle ICP-MS (spICP-MS). Environ. Sci. Nano 1: 248–259.

Mitrano, D.M., K. Mehrabi, Y.A.R. Dasilva and B. Nowack. 2017. Mobility of metallic (nano)particles in leachates from landfills containing waste incineration residues. Environ. Sci. Nano 4: 480–492.

Montaño, M.D., J.W. Olesik, A.G. Barber, K. Challis and J.F. Ranville. 2016. Single particle ICP-MS: Advances toward routine analysis of nanomaterials. Anal. Bioanal. Chem. 408: 5053–5074.

Mwilu, S.K., E. Siska, R.B.N. Baig, R.S. Varma, E. Heithmar and K.R. Rogers. 2014. Separation and measurement of silver nanoparticles and silver ions using magnetic particles. Sci. Total Environ. 472: 316–323.

Naasz, S., R. Altenburger and D. Kühnel. 2018. Environmental mixtures of nanomaterials and chemicals: The Trojan-horse phenomenon and its relevance for ecotoxicity. Sci. Total Environ. 635: 1170–1181.

Newman, K., C. Metcalfe, J. Martin, H. Hintelmann, P. Shaw and A. Donard. 2016. Improved single particle ICP-MS characterization of silver nanoparticles at environmentally relevant concentrations. J. Anal. Atom. Spectrom. 31: 2069–2077.

Nischwitz, V., N. Gottselig, A. Missong, T. Meyn and E. Klumpp. 2016. Field flow fractionation online with ICP-MS as novel approach for the quantification of fine particulate carbon in stream water samples and soil extracts. J. Anal. Atom. Spectrom. 31: 1858–1868.

Núñez, O., H. Gallart-Ayala, C.P.B. Martins, E. Moyano and M.T. Galceran. 2012. Atmospheric pressure photoionization mass spectrometry of fullerenes. Anal. Chem. 84: 5316–5326.

Peeters, K., G. Lespes, T. Zuliani, J. Ščančar and R. Milačič. 2016. The fate of iron nanoparticles in environmental waters treated with nanoscale zero-valent iron, FeONPs and Fe3O4NPs. Water Research 94: 315–327.

PEN. 2018. The Project on Emerging Technologies. (http://www.nanotechproject.org/).

Peters, R.J.B., G. van Bemmel, N.B.L. Milani, G.C.T. den Hertog, A.K. Undas, M. van der Lee et al. 2018. Detection of nanoparticles in Dutch surface waters. Sci. Total Environ. 621: 210–218.

Philippe, A. and G.E. Schaumann. 2014. Evaluation of hydrodynamic chromatography coupled with uv-visible, fluorescence and inductively coupled plasma mass spectrometry detectors for sizing and quantifying colloids in environmental media. PLoS ONE 9 (art. e90559).

Picó, Y., A. Alfarham and D. Barceló. 2017. Analysis of emerging contaminants and nanomaterials in plant materials following uptake from soils. TrAC—Trends Anal. Chem. 94: 173–189.

Proulx, K. and Wilkinson, K.J. 2014. Separation, detection and characterisation of engineered nanoparticles in natural waters using hydrodynamic chromatography and multi-method detection (light scattering, analytical ultracentrifugation and single particle ICP-MS). Environ. Chem. 11: 392–401.

Recommendation, C. of 18 October 2011. on the definition of nanomaterial Pages 1. Off. J. Eur. Commun.

Sanchís, J., C. Bosch-Orea, M. Farré and D. Barceló. 2017. Nanoparticle tracking analysis characterisation and parts-per-quadrillion determination of fullerenes in river samples from Barcelona catchment area. Anal. Bioanal. Chem. 407: 4261–4275.

Sanchís, J., Y. Aminot, E. Abad, A.N. Jha, J.W. Readman and M. Farré. 2018. Transformation of C60 fullerene aggregates suspended and weathered under realistic environmental conditions. Carbon 128: 54–62.

Scott-Fordsmand, J.J. et al. 2017. Environmental risk assessment strategy for nanomaterials. Int. J. Environ. Res. Public Health 14 (art. 1251).

Schaumann, G.E. et al. 2015. Understanding the fate and biological effects of Ag- and TiO<inf>2</inf>-nanoparticles in the environment: The quest for advanced analytics and interdisciplinary concepts. Sci. Total Environ. 535: 3–19.

Séby, F., J. Dumont, C. Gleyzes, M. Menta, V. Vacchina and M. Bueno. 2015. Determination of chemical species and nanoparticles in water samples: Analytical methods, preconcentration and validation. Rev. Sci. L'Eau 28: 27–32.

Shevlin, D., N. O'Brien and E. Cummins. 2018. Silver engineered nanoparticles in freshwater systems – Likely fate and behaviour through natural attenuation processes. Sci. Total Environ. 621: 1033–1046.

Sikder, M., J.R. Lead, G.T. Chandler and M. Baalousha. 2018. A rapid approach for measuring silver nanoparticle concentration and dissolution in seawater by UV-Vis. Sci. Total Environ. 618: 597–607.

Soto-Alvaredo, J., F. Dutschke, J. Bettmer, M. Montes-Bayón, D. Pröfrock and A. Prange. 2016. Initial results on the coupling of sedimentation field-flow fractionation (SdFFF) to inductively coupled plasma-tandem mass spectrometry (ICP-MS/MS) for the detection and characterization of TiO2 nanoparticles. J. Anal. Atom. Spectrom. 31: 1549–1555.

Su, S., B. Chen, M. He, Z. Xiao and B. Hu. 2014. A novel strategy for sequential analysis of gold nanoparticles and gold ions in water samples by combining magnetic solid phase extraction with inductively coupled plasma mass spectrometry. J. Anal. Atom. Spectrom. 29: 444–453.

Telgmann, L. and C.D. Metcalfe and H. Hintelmann. 2014. Rapid size characterization of silver nanoparticles by single particle ICP-MS and isotope dilution. J. Anal. Atom. Spectrom. 29: 1265–1272.

Telgmann, L., M.T.K. Nguyen, L. Shen, V. Yargeau, H. Hintelmann and C.D. Metcalfe. 2016. Single particle ICP-MS as a tool for determining the stability of silver nanoparticles in aquatic matrixes under various environmental conditions, including treatment by ozonation. Anal. Bioanal. Chem. 408: 5169–5177.

Teo, W.Z. and M. Pumera. 2014. Fate of silver nanoparticles in natural waters; Integrative use of conventional and electrochemical analytical techniques. RSC Ad. 4: 5006–5011.

Tou, F., Y. Yang, J. Feng, Z. Niu, H. Pan, Y. Qin et al. 2017. Environmental Risk Implications of Metals in Sludges from Waste Water Treatment Plants: The Discovery of Vast Stores of Metal-Containing Nanoparticles. Environ. Sci. Technol. 51: 4831–4840.

Tuoriniemi, J., M.D. Jürgens, M. Hassellöv and G. Cornelis. 2017. Size dependence of silver nanoparticle removal in a wastewater treatment plant mesocosm measured by FAST single particle ICP-MS. Environ. Sci. Nano 4: 1189–1197.

Uclés, A., M.J. Martínez Bueno, M.M. Ulaszewska, M.D. Hernando, C. Ferrer and A.R. Fernández-Alba. 2013a. Quantitative determination of poly(amidoamine) dendrimers in urine by liquid chromatography/electrospray ionization hybrid quadrupole linear ion trap mass spectrometry. Rapid Commun. Mass Spectrom. 27: 2519–2529.

Uclés, A., M.M. Ulaszewska, M.D. Hernando, M.J. Ramos, S. Herrera, E. García et al. 2013b. Qualitative and quantitative analysis of poly(amidoamine) dendrimers in an aqueous matrix by liquid chromatography-electrospray ionization-hybrid quadrupole/time-of-flight mass spectrometry (LC-ESI-QTOF-MS). Anal. Bioanal. Chem. 405: 5901–5914.

Ulaszewska, M.M., M.D. Hernando, A.U. Moreno, A.V. García, E.G. Calvo and A.R. Fernández-Alba. 2013. Identification and quantification of poly(amidoamine) PAMAM dendrimers of generations 0 to 3 by liquid chromatography/hybrid quadrupole time-of-flight mass spectrometry in aqueous medium. Rapid Commun. Mass Spectrom. 27: 747–762.

Ulaszewska, M.M., M.D. Hernando, A. Uclés, R. Rosal, A. Rodríguez, E. Garcia-Calvo and A.R. Fernández-Alba. 2012. Chemical and ecotoxicological assessment of dendrimers in the aquatic environment. pp. 197–233. Compr. Anal. Chem..

Venkatesan, A.K., R.B. Reed, S. Lee, X. Bi, D. Hanigan, Y. Yang et al. 2018. Detection and sizing of Ti-containing particles in recreational waters using single particle ICP-MS. Bull. Environ. Contam. Toxicol. 100: 120–126.

Vidmar, J., P. Oprčkal, R. Milačič, A. Mladenovič and J. Ščančar. 2018. Investigation of the behaviour of zero-valent iron nanoparticles and their interactions with Cd^{2+} in wastewater by single particle ICP-MS. Sci. Total Environ. 634: 1259–1268.

Yang, Y., C.L. Long, H.P. Li, Q. Wang and Z.G. Yang. 2016a. Analysis of silver and gold nanoparticles in environmental water using single particle-inductively coupled plasma-mass spectrometry. Sci. Total Environ. 563-564: 996–1007.

Yang, Y., L. Luo, H.P. Li, Q. Wang, Z.G. Yang and C.L. Long. 2016b. Separation and determination of silver nanoparticle in environmental water and the UV-induced photochemical transformations study of AgNPs by cloud point extraction combined ICP-MS. Talanta 161: 342–349.

Yang, Y., L. Luo, H.P. Li, Q. Wang, Z.G. Yang, Z.P. Qu and R. Ding. 2018. Analysis of metallic nanoparticles and their ionic counterparts in complex matrix by reversed-phase liquid chromatography coupled to ICP-MS. Talanta 182: 156–163.

Zakaria, S., E. Fröhlich, G. Fauler, A. Gries, S. Weiß and S. Scharf. 2018. First determination of fullerenes in the Austrian market and environment: Quantitative analysis and assessment. Environ. Sci. Pollut. Res. 25: 562–571.

Zhang, T., D. Lu, L. Zeng, Y. Yin, Y. He, Q. Liu et al. 2017. Role of secondary particle formation in the persistence of silver nanoparticles in humic acid containing water under light irradiation. Environ. Sci. Technol. 51: 14164–14172.

Zhou, X., J. Liu, C. Yuan and Y. Chen. 2016a. Speciation analysis of silver sulfide nanoparticles in environmental waters by magnetic solid-phase extraction coupled with ICP-MS. J. Anal. Atom. Spectrom. 31: 2285–2292.

Zhou, X.X., J.F. Liu and F.L. Geng. 2016b. Determination of metal oxide nanoparticles and their ionic counterparts in environmental waters by size exclusion chromatography coupled to ICP-MS. NanoImpact 1: 13–20.

Zhou, X.X., J.F. Liu and G.B. Jiang. 2017. Elemental mass size distribution for characterization, quantification and identification of trace nanoparticles in serum and environmental waters. Environ. Sci. Technol. 51: 3892–3901.

Zouboulaki, R. and E. Psillakis. 2016. Fast determination of aqueous fullerene C60 aggregates by vortex-assisted liquid-liquid microextraction and liquid chromatography-mass spectrometry. Anal. Met. 8: 4821–4827.

Index

Color Plate Section

Chapter 3

Table 4. Heat map relating doses (mg/L) with effects (colours) of NPs, additives and corresponding ionic and bulk forms on haemocytes and gill cells based on cell viability assays (NR and MTT assays). - : not tested; o: no effect.

	Haemocytes		Gill Cells	
	NR	MTT	NR	MTT
Ionic Ag	1	0.1	1	0.1
Bulk Ag	10	10	25	10
Mal-Ag20 NPs	10	10	0.1	1
Mal-Ag40 NPs	10	10	10	10
Mal-Ag100 NPs	10	10	10	10
Maltose	o	0.01	o	o
Ag20 NPs	25	25	25	10
Ag80 NPs	25	25	25	25
Ionic Cu	1	0.5	0.5	0.5
Bulk CuO	10	10	10	10
CuO NPs	1	1	0.5	1
Ionic Cd	1	0.01	1	1
Bulk CdS	10	10	10	10
CdS QDs	1	10	10	10
Ionic Zn	10	10	10	10
Bulk ZnO	25	50	25	50
ZnO<130-EcodisP90 NPs	10	10	10	10
ZnO<280-EcodisP90 NPs	10	10	10	10
Ecodis P90	25	10	10	10
TiO_2 BRU	10	10	50	50
TiO_2 RUAN	10	10	1	1
WtC10 TiO_2 NPs	10	1	10	1
WtC40 TiO_2 NPs	10	1	0.1	10
WtC60 TiO_2 NPs	50	0.1	1	50
Pl100 TiO_2 NPs	10	0.1	50	10
Mi60 TiO_2 NPs	10	1	10	10
DSLS	50	1	50	10
P25 TiO_2 NPs	10	1	10	10
Ionic Au	-	25	-	-
Bulk Au	-	o	-	-
Au5-Cit NPs	-	50	-	-
Au15-Cit NPs	-	50	-	-
Au40-Cit NPs	-	50	-	-
Na-citrate	-	50	-	-
Ionic Si	100	25	o	25
Bulk Si	o	100	o	o
$SiO_2$15-Larg NPs	50	50	100	25
$SiO_2$30-Larg NPs	100	50	100	50
$SiO_2$70-Larg NPs	100	100	o	50
L-arginine	o	o	o	o

>100	100->10	10->1	1->0.1	≤0.1	mg/L
not toxic	low toxicity	moderate	toxic	very toxic	

Table 6. Heat map relating doses (mg/L) with effects (colours) of NPs and corresponding ionic and bulk forms on different cellular processes of haemocytes and gill cells. o: no effect. ROS: Reactive oxygen species production assay; CAT: Catalase activity assay; DNA: DNA damage by Comet assay; AcP: Acid phosphatase activity assay; MXR: Multixenobiotic resistance transport activity assay; PHAGO: Phagocytic activity assay; Na-K-ATPase: Na-K-ATPase activity assay.

	Haemocytes						Gill cells					
	ROS	**CAT**	**DNA**	**AcP**	**MXR**	**PHAGO**	**ROS**	**CAT**	**DNA**	**AcP**	**MXR**	**Na-K-ATPase**
Ionic Ag	0.03	0.03	0.06	0.06	0.25	o	0.03	0.06	0.12	0.5	0.06	0.06
Bulk Ag	o	5	10	10	o	o	o	10	o	o	0.62	5
Mal-Ag20 NPs	0.62	1.25	1.25	0.15	o	1.25	1.25	1.25	2.5	o	0.31	1.25
Ionic Cu	0.2	0.2	0.2	0.12	2	o	0.2	1	1.5	0.25	0.02	0.25
Bulk CuO	7.5	o	10	o	o	o	10	o	o	o	10	o
CuO NPs	0.05	2.75	2.75	2.75	2.75	2.75	0.5	5	2.75	5	2.75	1.25
Ionic Cd	0.1	0.1	1	0.5	2	o	0.25	2	1	2	2	0.25
Bulk CdS	1.25	1.25	10	10	5	2.5	2.5	5	10	o	5	o
CdS QDs	2.5	5	2.5	1.25	0.31	1.25	1.25	o	5	2.5	2.5	o

> 100	100->10	10->1	1->0.1	≤ 0.1	mg/L
not toxic	**low toxicity**	**moderate**	**toxic**	**very toxic**	

Chapter 5

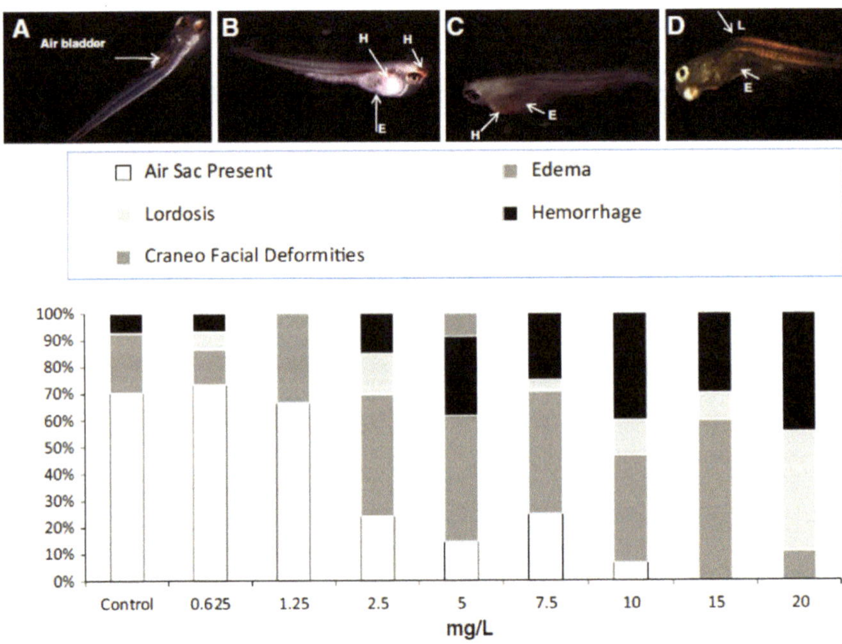

Figure 6. Captured images of fathead minnow larvae exposed to silver nanoparticles (Ag NPs) (a control; b and c exposed to 10 mg/L of NanoAmor and Sigma Ag NPs, respectively; and d exposed to 2.5 mg/L of NanoAmor Ag NPs). Note the absence of air bladder in treatment larvae. Bar graph summarizes the types and frequency of deformities per treatment group with N = 30 observed compared to controls (no Ag NP added).

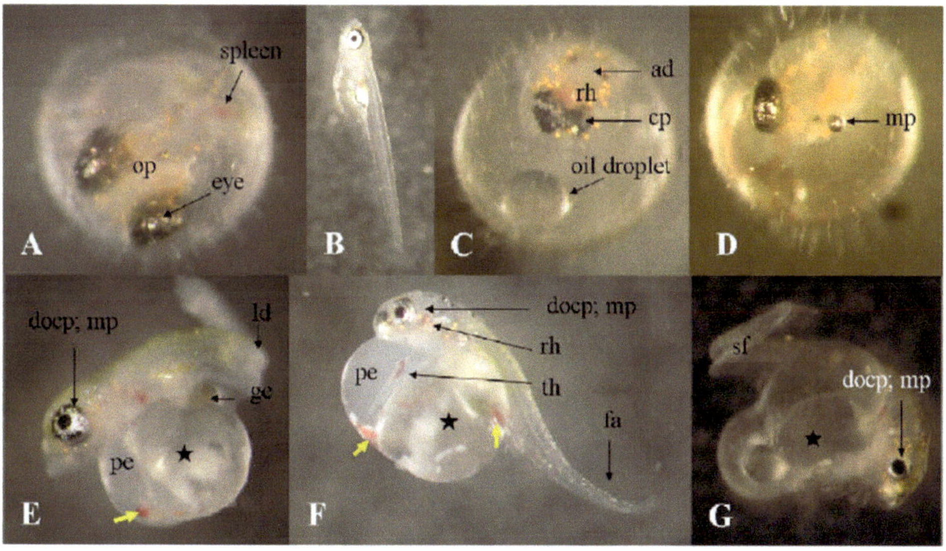

Figure 7. Representative morphological abnormalities of medaka larvae exposed to 62.5–1000 mg/L AgNPs during the embryonic stage. Embryos normally developed in the control (A and B). However, various abnormalities were observed in the AgNP-treated groups (C–G). ad, arrested development; cp, cyclopia; docp, decreased optic cup pigmentation; fa, finfold abnormality; ge, gallbladder oedema; Id, lordosis; mp, microphthalmia; op, optic tectum; pe, pericardial oedema; rh, reduced head; sf, skeletal flexture; and th, tubular heart. Stars mark the opaque and edematous yolk sac, whereas yellow arrowheads indicate hemostasis.

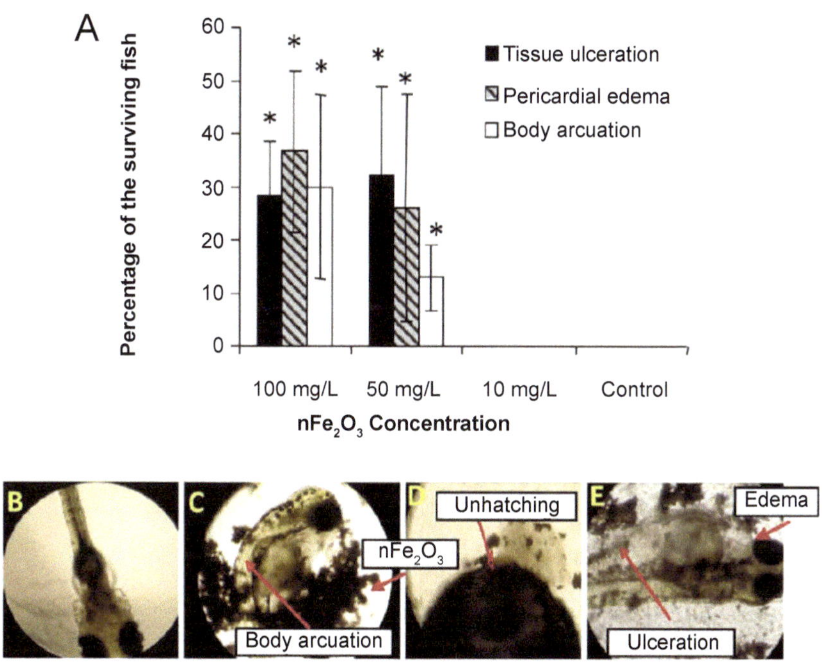

Figure 16. Malformations (e.g., pericardial edema, tissue ulceration, and body arcuation) induced by nFe_2O_3 at 168 hpf. (A) Malformation percentage in the surviving fish; (B) control fish; (C) hatching fish with body arcuation, treated with 50 mg/L of nFe_2O_3 aggregates; (D) unhatching embryos, treated with 50 mg/L of n Fe_2O_3 aggregate, dead at 168 hpf (E) hatching fish with pericardial edema, treated with 100 mg/L of nFe_2O_3 aggregates. Error bars represent 6 one standard deviation from the mean of three replicates. Significance indicated by: *p, 0.05.

Chapter 6

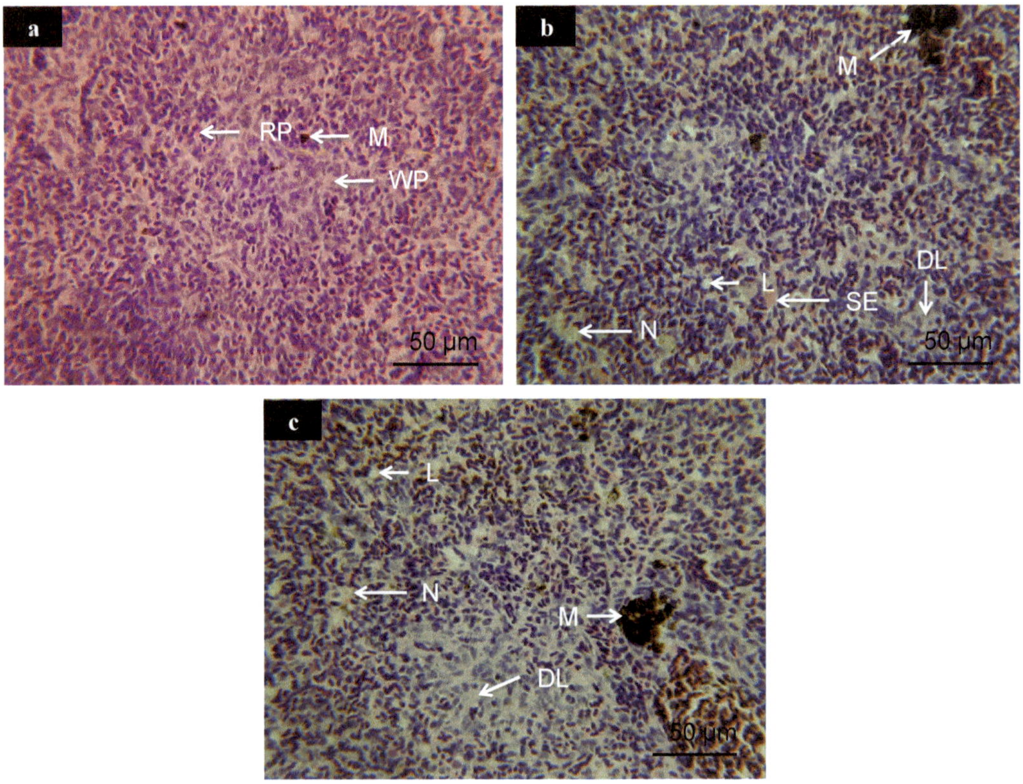

Figure 2. Spleen morphology in rainbow trout following exposure to $CuSO_4$ or Cu ENMs for 4 d. Panels include (a) control, (b) 100 µg L^{-1} $CuSO_4$, (c) 100 µg L^{-1} Cu ENMs. Spleen from control fish showed normal histology, with defined red (RP) and white (WP) pulp. All treatments showed similar types of lesions. These lesions include necrosis (N), depletion of lymphoid tissues (DL), melanomacrophage deposits (M), Lipidosis (L), and swollen erythrocyte (SE). Scale bar indicates magnification; sections were 7 µm thick and stained with H&E. Note the termination of the 100 µg L^{-1} $CuSO_4$ treatment after 4 d for ethical reasons (Al-Bairuty and Handy, unpublished).

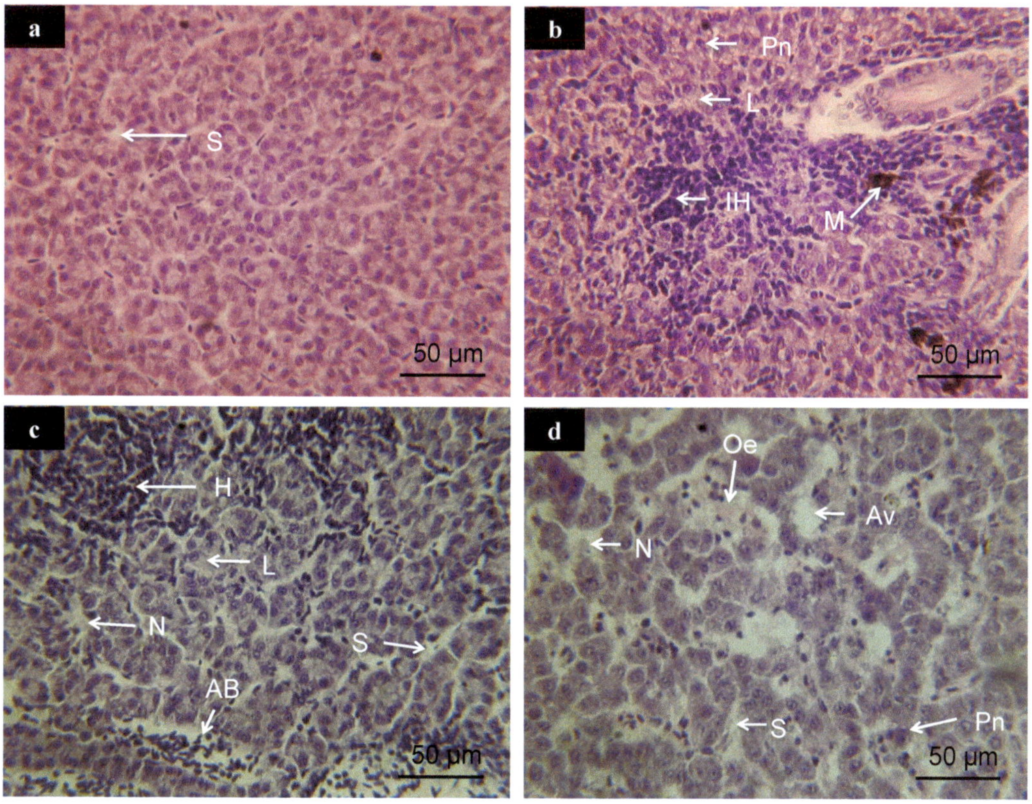

Figure 3. Liver morphology in rainbow trout following waterborne exposure to $CuSO_4$ or Cu ENMs for 10 d. Panels include (a) control, (b) 20 µg L^{-1} $CuSO_4$, (c) 20 µg L^{-1} Cu ENMs and (d) 100 µg L^{-1} Cu ENMs. The livers of control fish showed normal histology with sinusoid space (S). Both materials caused similar types of lesions, although these were severe in the equivalent Cu ENM treatment. These lesions include cells with pyknotic nuclei (Pn), foci of hepatitis-like injury (H), foci of melanomacrophages (M), lipidosis (L), vacuole formation (V), necrosis (N), oedema in the tissue (Oe), and aggregation of blood cell (AB). Scale bar indicates magnification; sections were 7 µm thick and stained with haematoxylin and eosin. Note the termination of the 100 µg L^{-1} $CuSO_4$ treatment after 4 d for ethical reasons (modified from Al-Bairuty et al. 2013).

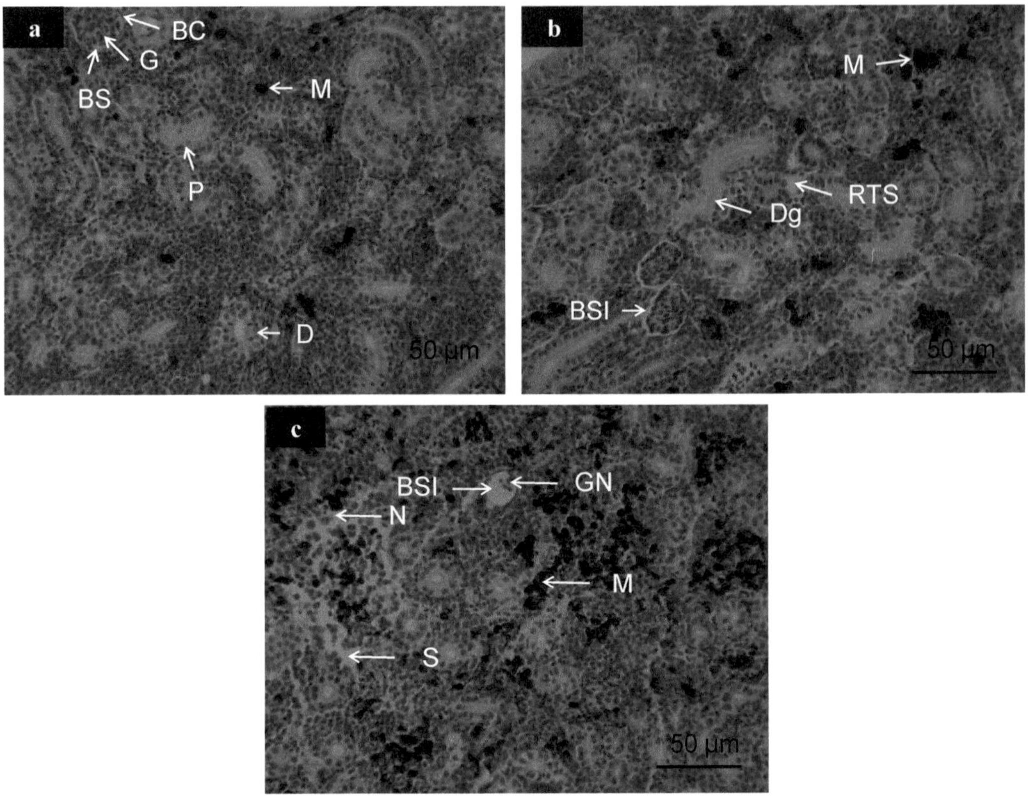

Figure 4. Kidney morphology in rainbow trout following waterborne exposure to (a) control (b) 20 µg L^{-1} of Cu as $CuSO_4$, and (c) 20 µg L^{-1} of Cu as Cu ENMs for 10 d. Kidney of control fish showed normal histology structure with parietal epithelium of Bowman's capsule (BC), glomerulus (G), Bowman's space (BS), proximal tubules (P), distal tubules (D) and melanomacrophages (M). The types of pathologies were similar for $CuSO_4$ and Cu ENMs, but with more deposit of melanomacrophage in the latter. These lesions include degeneration of renal tubule (Dg), increased Bowman's space (BSI), melanomacrophage aggregate (M), sinusoids were enlarged (S), renal tubular separation (RTS), necrosis of haematopoietic tissue (N) and glomerular necrosis (GN). Scale bar indicates magnification, section were 7 µm thick and stained with haematoxylin and eosin (modified from Al-Bairuty et al. 2013).

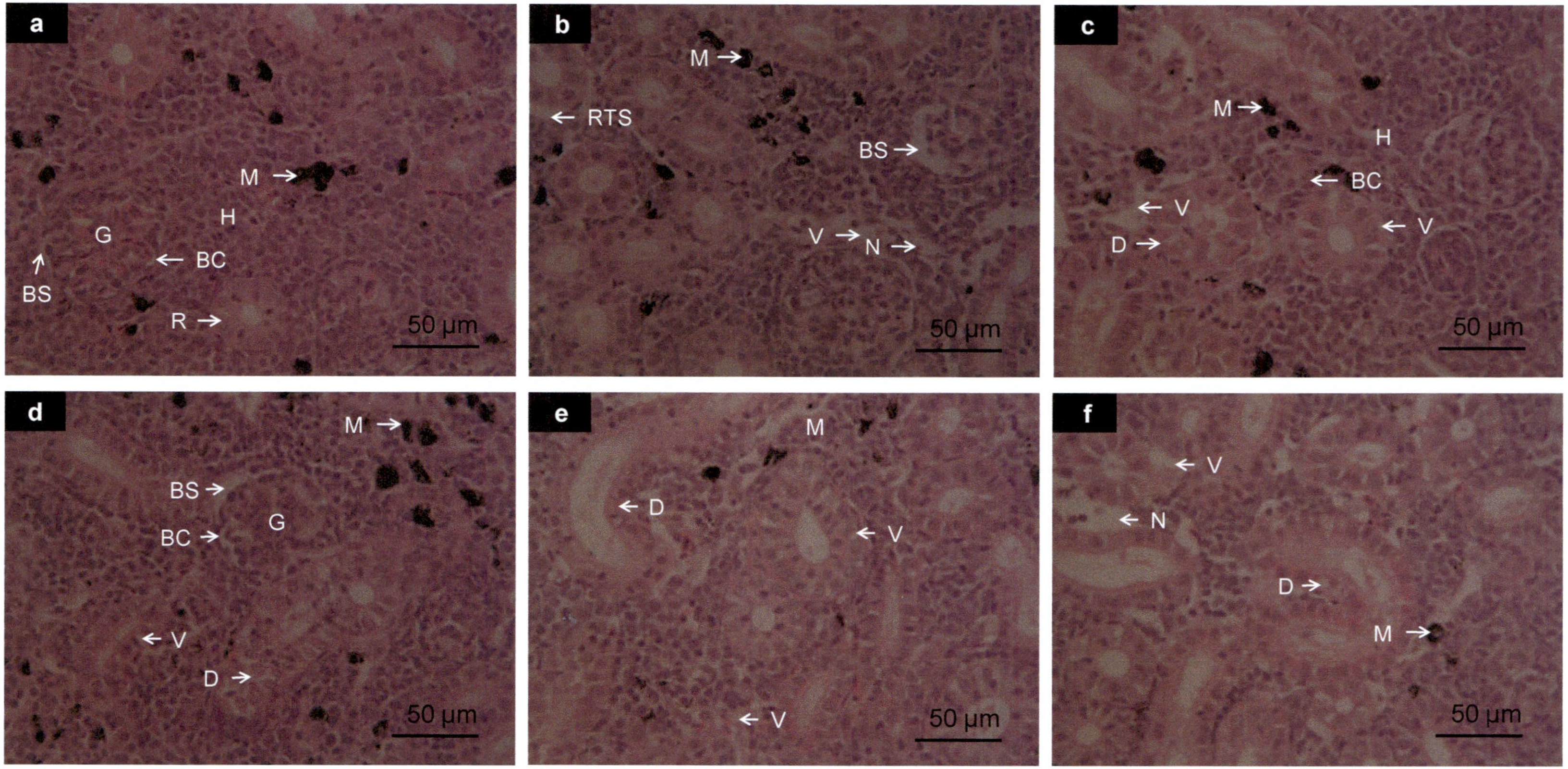

Figure 5. Kidney morphology in rainbow trout after 7 d of waterborne exposure to (a) control (b) $CuSO_4$, (c) Cu ENMs at pH 7 and (d) control (e) $CuSO_4$ (f) Cu ENMs at pH 5. Kidney of control fish showed normal histology with parietal epithelium of Bowman's capsule (BC), glomerulus (G), Bowman's space (BS), renal tubules (R), haematopoietic tissues (H) and melanomacrophages deposits (M). Similar pathologies were observed with $CuSO_4$ and Cu ENMs at pH 7 and 5, with worse effects at pH 5 than 7. These pathologies include necrosis (N), vacuolisation (V), and degeneration of renal tubule (D). Scale bar indicates magnification, section were 7 μm thick and stained with haematoxylin and eosin (Al-Bairuty and Handy, unpublished observations from the exposure reported in Al-Bairuty et al. 2016).

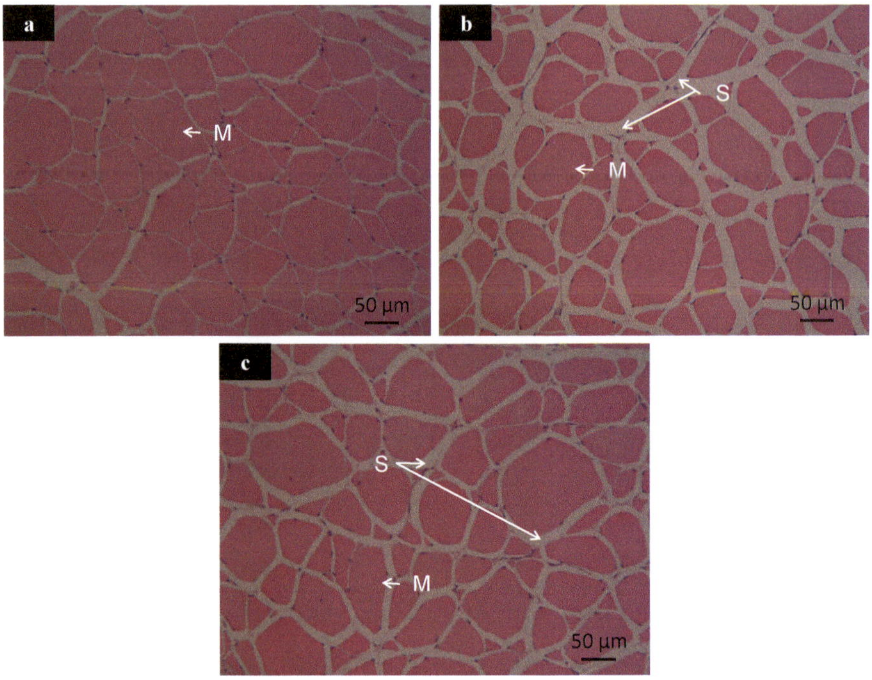

Figure 6. Muscle morphology in trout after 14 d of exposure to (a) control (b) 1 mg L^{-1} bulk TiO_2 (c) 1 mg L^{-1} TiO_2 ENMs. Muscles of control groups showed normal structure of muscle fibres (M). Both treatments showed increased space among muscles fibres (S). Scale bar indicates magnification. Sections were 7 µm thick and stained with haematoxylin and eosin. (Al-Bairuty, Boyle and Handy, unpublished observations).

Chapter 7

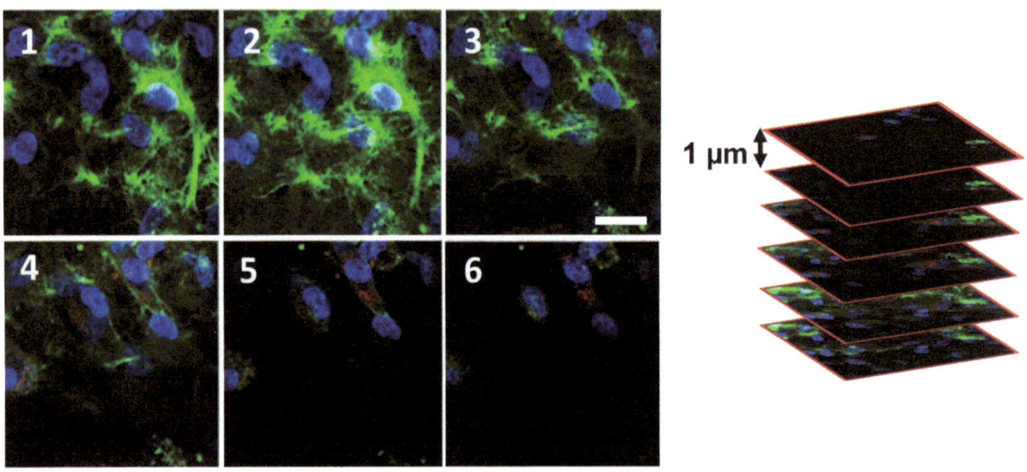

Figure 1. CLSM 2D image of osteogen differentiated human stem cells under chronic exposure to polystyrene NPs over 21 days (1 µg/ml). Nano- and microplastics are a present topic in the field of ecotoxicology, because the number of microplastics in the environment, especially in the sea, increases dramatically due to plastic waste. Image 1–6 show the individual stacks of a 1 µm-z-stack as overlay of the fluorescence images. Green: Hydroxylapatite, Blue: DAPI staining of nucleus, Red: Polystyrene NPs labeled with red fluorecence. Scale bar 25 µm. Image from Patrizia Komo (Fraunhofer Institute for Biomedical Engineering, Sulzbach, Germany), "Effect of nanoparticles on the differentiation of mesenchymal stem cells in 2D and 3D *in vitro* model".

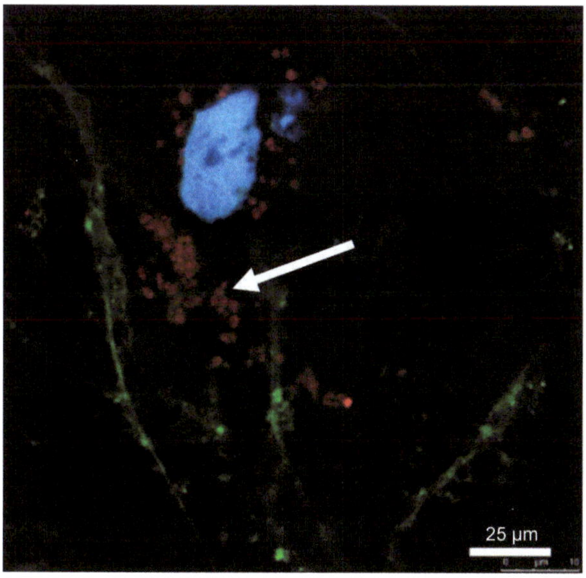

Figure 2. CLSM 2D image of adipogenc differentiated human stem cells under chronic exposure to polystyrene NPs over 21 days (1 μg/ml). Image shows the overlay of the fluorescence images. Blue: DAPI staining of nucleus, Green: Concanavalin A 488 staining of the cell structure. Red: Polystyrene NPs labeled with red fluorescence. Scale bar 25 μm. Image from Patrizia Komo (Fraunhofer Institute for Biomedical Engineering, Sulzbach, Germany), "Effect of nanoparticles on the differentiation of mesenchymal stem cells in 2D and 3D *in vitro* model".

Chapter 8

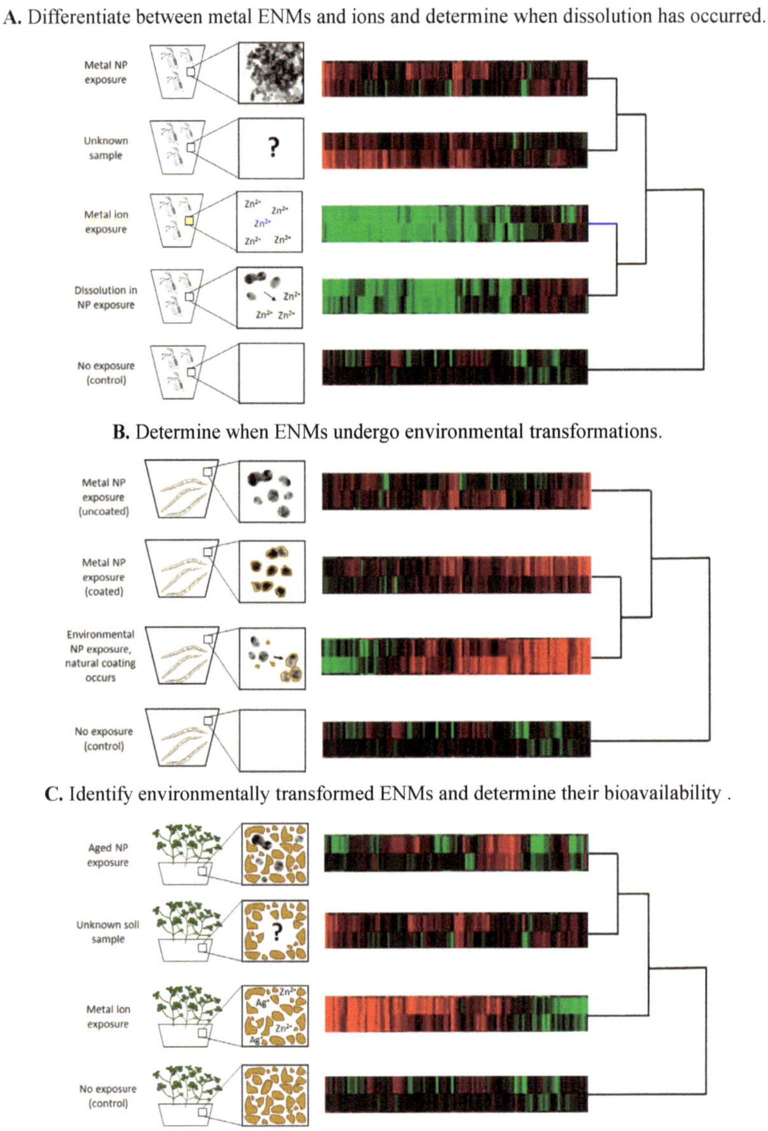

Figure 2. Potential applications of 'omic technologies for environmental monitoring of ENMs. (A) 'Omic expression profiles are able to distinguish between metal NP and metal ion exposures (Poynton et al. 2011). When dissolution of NPs occurs, gene expression profiles resemble ionic metal exposures (Adam et al. 2015). These results imply that 'omic expression profiles will differentiate between ionic metal and metal NP exposures in unknown environmental samples (although in cases where NPs act through a "Trojan-horse" like mechanism, this may not be possible, e.g., Poynton et al. 2013). In this example, the unknown environmental sample produces an 'omic profile similar to the metal NP exposure and is classified as a NP exposure. (B) 'Omic expression profiles are able to distinguish between pristine NPs and coated NPs (Park et al. 2015). This example suggests that if NPs undergo environmental transformations that result in organic coating of the NPs, 'omic profiles would differentiate the sample from pristine particles. In this example, the environmental sample contains NPs that were coated by natural organic matter and the 'omic profile is most similar to the coated NP exposure. (C) 'Omic profiles represent the bioavailability of ENMs in aged soils (Chen et al. 2015). In environmental samples, 'omic profiles should be able to distinguish between metal ion exposures and metal NP exposure even following environmental transformations. In this example, the unknown environmental samples produces an 'omic profile most similar to the aged ENMs exposure. For all panels, the different exposures are shown on the left in each panel, and heatmaps representative of different 'omic profiles are shown in the middle. The relationships between 'omic profiles are shown here using heatmaps and hierarchical clustering for simple visualization (Gehlenborg et al. 2010); however, more sophisticated classification models are usually employed for mTIE applications. For a discussion of different methods, see Kostich (2017).